Isolation and Structure Elucidation of Bioactive Compounds (*Dedicated to the memory of the late Professor Charles D. Hufford*)

Isolation and Structure Elucidation of Bioactive Compounds (*Dedicated to the memory of the late Professor Charles D. Hufford*)

Special Issue Editors

Muhammad Ilias
Charles L. Cantrell

MDPI • Basel • Beijing • Wuhan • Barcelona • Belgrade

MDPI

Special Issue Editors
Muhammad Ilias   Charles L. Cantrell
University of Mississippi   USDA-ARS
USA   USA

Editorial Office
MDPI
St. Alban-Anlage 66
4052 Basel, Switzerland

This is a reprint of articles from the Special Issue published online in the open access journal *Molecules* (ISSN 1420-3049) from 2018 to 2019 (available at: https://www.mdpi.com/journal/molecules/special_issues/Isolation_Elucidation)

For citation purposes, cite each article independently as indicated on the article page online and as indicated below:

LastName, A.A.; LastName, B.B.; LastName, C.C. Article Title. *Journal Name* **Year**, *Article Number,* Page Range.

ISBN 978-3-03897-780-3 (Pbk)
ISBN 978-3-03897-781-0 (PDF)

Contents

About the Special Issue Editors

Muhammad Ilias (Research Professor, University of Mississippi) obtained Bachelors and Masters degrees in Pharmacy (1976 and 1978, respectively) from the University of Dhaka, Bangladesh. He obtained a Ph. D. from the Department of Pharmaceutical Chemistry (Phytochemistry) from the University of Strathclyde, UK, in 1985, and post-doctoral ressearch training in Nattural Products (NP) from the Department of Pharmacognosy, University of Mississippi from 1987-89. He worked as a teaching faculty at the University of Dhaka (1979–81 & 1985–87) and a Researcher, King Saud University, Saudi Arabia from 1989 to 1998). Since 1998, Muhammad has been working as a Research Faculty at the National Center for Natural Products Research, School of Pharmacy, University of Mississippi, where currently as a Research Professor since 2012. His primary research area is natural products drug discovery, focused on anticancer, anti-infective and neuroprotective agents. He has received various grants in these areas for last 20 years, funded by NIH, DoD, MMV, and USDA-ARS., where he served as a PI/ CO-PI/Collaborator. He is collaborating with Children Brain Tumor Center of Excellence (CBTCE), University of Cambridge, U.K., and Institute of Pharmaceutical Sciences, King's College, London, on anticancer and anti-microbial NP drug discovery projects, and also pursuing to develop a library of natural by-products for neurodegenerative diseases. He has invented/ developed new environment-friendly centrifugal chromatographic technology and devices, named "Chromatorotor" (UM #8270) and "Spin Chromatography System" (UM #8390), for the application of Centrifugal Planar Chromatography using various types of binder- free sorbents, especially reversed- phase (C18) silica gel. Muhammad has examined bioactive compounds from plants belong to verious biodiverse regions of the world, including Asia, Africa and Americas, which has enriched his NP respository. He has contributed to many of scientific meetings on bioactive NP's among researchers in multiple countries.

Charles L. Cantrell obtained a Ph.D. in Organic Chemistry from LSU in 1998 under the direction of Dr. Nikolaus H. Fischer. From 1998 to 2000 he conducted postdoctoral research in the Laboratory of Drug Discovery Research and Development, National Cancer Institute, in Frederick, Maryland under the direction of Dr. Kirk Gustafson. His research consisted of the isolation and structure elucidation of cytotoxic constituents isolated from marine plants and animals. This research led to the discovery of the chondropsins, a new class of cytotoxic macrolides from the marine sponge Chondropsis sp. Following this postdoctoral position, he spent 2 years as a Research Chemist with the United States Department of Agriculture, Agricultural Research Service (USDA-ARS) in Peoria, Illinois. Dr. Cantrell left ARS in 2001 to pursue a career in the private sector as Associate Director of Research and Development for Tanical Therapeutics, Inc. Research at Tanical focused on the identification of small-molecule drug candidates from plants with a history of traditional usage against diseases such as cancer, rheumatoid arthritis, and benign prostate hyperplasia. Dr. Cantrell has also worked for both Hauser, Inc. and Sandoz Pharmaceuticals. Dr. Cantrell served as Treasurer of the Phytochemical Society of North America (PSNA) from 2002 to 2005, Editor of The Cornucopia from 2000 to 2004, Associate Editor of Pest Management Science from 2012 to present, and President of the PSNA in 2010. Dr. Cantrell is currently a Research Chemist for the USDA-ARS in Oxford, MS. As part of the Natural Products Utilization Research Unit, he is responsible for the discovery of natural product-based pesticides. Dr. Cantrell has co-authored over 130 peer-reviewed publications, 7 book chapters, 3 US patents, and 2 international patents.

Preface to "Isolation and Structure Elucidation of Bioactive Compounds (*Dedicated to the memory of the late Professor Charles D. Hufford*)"

Throughout the history of mankind, plants have played an indispensable role for the benefit of human health. From ancient folk medicine to modern drugs, dietary supplements, and even agrochemicals, the chemistry of plant constituents has become more important for us in understanding life-threatening human diseases and chemoprevention practices. The absence of curative drugs in cognitive health has led the world's aging population toward traditional medicine-based supplements. Isolation and structure determination of these diverse bioactive molecules that only nature can produce has become a key area for generating new scaffolds for various biological explorations in medicinal chemistry and agrochemical discovery. This is an area currently under threat due to the disappearance of natural resources. For these reasons, we think it is appropriate to document some recent research in this field.

We wish to dedicate this Special Issue Book Version to the memory of the late Professor Charles D. Hufford, a true research pioneer in the fields of natural products and pharmacognosy, who made immense contributions to the field. Making advances in the isolation and structure elucidation of bioactive compounds using NMR technology, especially secondary plant metabolites, was a passion of Professor Hufford, and therefore, this Special Issue of the journal Molecules has been dedicated to his memory.

Muhammad Ilias, Charles L. Cantrell
Special Issue Editors

molecules

MDPI

Editorial

Tribute to the Late Dr. Charles D. Hufford

Alice M. Clark

Department of BioMolecular Sciences, Division of Pharmacognosy; School of Pharmacy, The University of Mississippi, Oxford, MS 38677, USA; amclark@olemiss.edu

Published: 7 February 2019

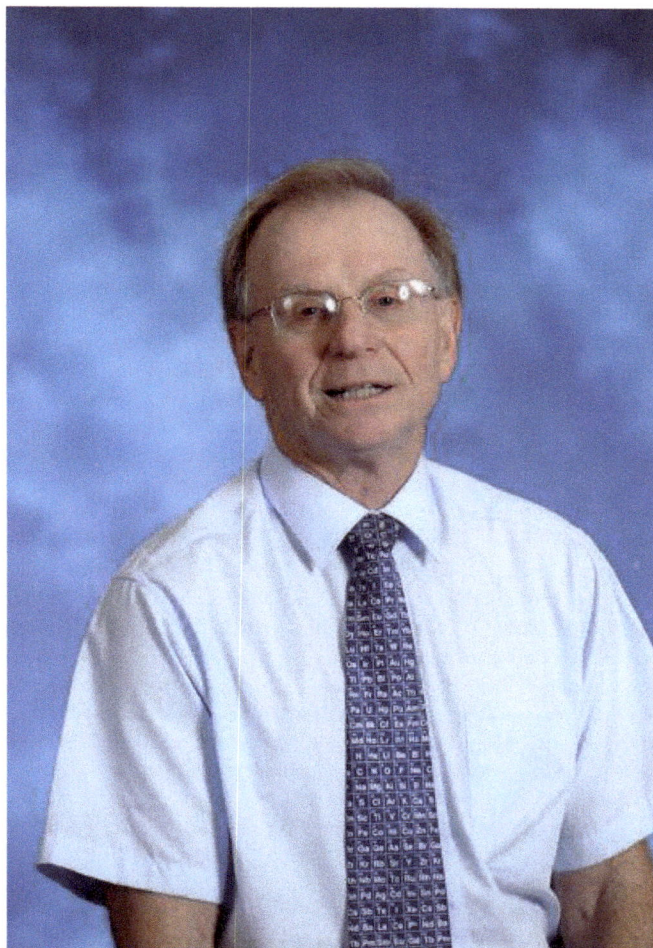

This Special Issue is dedicated to the late Dr. Charles (Charlie) D. Hufford, former Professor of Pharmacognosy and Associate Dean for Research and Graduate Studies at the University of Mississippi. Dr. Hufford passed away, May 15, 2017 at the age of 72.

Charlie was born in the small community of Sycamore, Ohio in 1944 to Charles and Magdalena Hufford. His father, Charley, was an avid and skilled amateur botanist, teaching Charlie about the flora of the northern U.S. and Canada as they hunted and fished the areas. Charlie's love of botany, developed as a child, eventually led him to a lifetime of work devoted to understanding the chemistry and biology of plants. After graduating from Mohawk High School in 1962, he enrolled at The Ohio State University (OSU), where he found both his career and his lifelong devotion to Buckeye sports! He graduated with a pharmacy degree from The Ohio State University in 1967.

While at OSU, he had the good fortune to meet Jack Beal, who recognized his talent for pharmacognosy and encouraged him to pursue graduate studies. Charlie obtained his Ph.D. in pharmacognosy in 1972 under the direction of Ray Doskotch. While pursuing his graduate studies, he also served as a pharmacist in the Air Force Reserve. It was during his graduate studies that Charlie developed his passion for NMR spectroscopy, eventually becoming a leader in the application and interpretation of NMR to the structure elucidation of novel natural products.

He joined the faculty of the University of Mississippi (UM) School of Pharmacy as an assistant professor in the summer of 1972. Throughout his career he served the school in several roles, including two terms of service as department chair before becoming the school's first associate dean for research and graduate studies in 1995. For 42 years he was a tireless advocate for the School of Pharmacy graduate students, faculty, and staff. He originated important research programs that continue today, was one of the principal investigators for the long-running NIH-funded antifungal research program, and was the singular driving force behind building the school's capacity in NMR spectroscopy. He retired from the university in January 2015 to devote his time to his two highest priorities: grandsons and bowling!

Charlie was a quiet leader who worked out of the spotlight to support the careers of countless students and colleagues. Throughout his own career, he collaborated with many other scientists to identify hundreds of natural products representing a wide array of chemotypes and an impressive range of bioactivities. He was also a pioneer in the use of microorganisms to predict mammalian metabolism of drugs, and was instrumental in advocating that the University of Mississippi School of Pharmacy incorporate the understanding of dietary supplements into its research and educational initiatives several years before the Dietary Supplement Health and Education Act in 1994.

Charlie's greatest professional achievement was his role in developing scores of scientists, educators, pharmacists, and leaders. He was a patient and gifted mentor and teacher who set many graduate students on their way to fulfilling and influential lives in academia, industry, government, and nonprofit organizations throughout the world. His devotion to graduate education in pharmacognosy continues through the Charles D. Hufford Graduate Student Fellowship Endowment at The University of Mississippi, which he established when he retired in order to support in perpetuity graduate education in pharmacognosy.

Charlie received many honors and awards for both his professional and personal accomplishments, including the 1994 UM School of Pharmacy Outstanding Researcher of the Year and the 1995 OSU College of Pharmacy Jack Beal Award. He was an elected Fellow of the American Association of Pharmaceutical Scientists, and served in every leadership position of the American Society of Pharmacognosy.

He also had a passion and skill for bowling—he bowled more than 30 perfect 300 games over his bowling career, the latest on December 14, 2016. In that sphere as well, he was widely known and admired as a mentor, coach, teammate, and friend to all.

I had the unique privilege of being partner to one of the greatest minds and most generous hearts in science—Charlie always gave more than he got. Like so many others, I was the beneficiary of his skill and expertise, his patient and methodical approach to our work, and his unwavering commitment to the highest standards.

It is fitting that this Special Issue comprises papers from former students, collaborators, and colleagues throughout the world who have utilized an impressive array of techniques to probe every possible natural source to identify a variety of chemotypes with a wide spectrum of biological activities—with some new microbial transformations and insights into dietary supplements as well. It is an appropriate reflection of Charlie's career that spanned 40+ years, during which he contributed to both our scientific knowledge and the development of scientists throughout the world.

molecules

MDPI

Article

Microbial Oxidation of the Fusidic Acid Side Chain by *Cunninghamella echinulata*

Abdel-Rahim S. Ibrahim [1], Khaled M. Elokely [2,3], Daneel Ferreira [4] and Amany E. Ragab [1,*

1   Department of Pharmacognosy, Faculty of Pharmacy, Tanta University, Tanta 31527, Egypt;
     arsib16@hotmail.com
2   Department of Pharmaceutical Chemistry, Faculty of Pharmacy, Tanta University, Tanta 31527, Egypt;
     kelokely@pharm.tanta.edu.eg
3   Institute for Computational Molecular Science and Department of Chemistry, Temple University,
     Philadelphia, PA 19122, USA
4   Department of BioMolecular Sciences, Division of Pharmacognosy, School of Pharmacy,
     The University of Mississippi, University, MS 38677-1848, USA; dferreir@olemiss.edu
*   Correspondence: amany.ragab@pharm.tanta.edu.eg; Tel.: + 20-403-336-007 (ext. 266)

Academic Editor: Charles L. Cantrell
Received: 10 April 2018; Accepted: 20 April 2018; Published: 21 April 2018

Abstract: Biotransformation of fusidic acid (**1**) was accomplished using a battery of microorganisms including *Cunninghamella echinulata* NRRL 1382, which converted fusidic acid (**1**) into three new metabolites **2–4** and the known metabolite **5**. These metabolites were identified using 1D and 2D NMR and HRESI-FTMS data. Structural assignment of the compounds was supported via computation of ^{1}H- and ^{13}C-NMR chemical shifts. Compounds **2** and **3** were assigned as the 27-hydroxy and 26-hydroxy derivatives of fusidic acid, respectively. Subsequent oxidation of **3** afforded aldehyde **4** and the dicarboxylic acid **5**. Compounds **2**, **4** and **5** were screened for antimicrobial activity against different Gram positive and negative bacteria, *Mycobacterium smegmatis*, *M. intercellulare* and *Candida albicans*. The compounds showed lower activity compared to fusidic acid against the tested strains. Molecular docking studies were carried out to assist the structural assignments and predict the binding modes of the metabolites.

Keywords: fusidic acid; *Cunninghamella echinulata*; C-26-oxidation; C-27-oxidation

1. Introduction

Fusidic acid (**1**) is a natural antibacterial steroid-like compound without any steroidal activity [1,2]. It was first isolated and identified from the fungus *Fusidium coccineum* [1,2] and introduced into the market in the 1960s as the corresponding sodium salt for clinical use. It has activity against Gram positive bacteria, particularly methicillin resistant *Staphylococcus aureus* (MRSA), and modest activity against anaerobic Gram-negative bacteria [3,4]. Fusidic acid (**1**) acts through inhibition of protein synthesis by binding to the elongation factor EF-G [5]. The specific and narrow spectrum of activity of fusidic acid makes it an ideal target for investigating possible biotransformation pathways and the effects of the metabolites on the activity spectrum and/or efficacy. Here, we explored the metabolic fate of fusidic acid using several organisms among which the fungus *Cunninghamella echinulata* was the most proficient in the biotransformation of this antibiotic.

Fusidic acid (Figure 1) is metabolized into a dicarboxylic acid derivative in mammals. Other detected metabolites include 3-didehydrofusidic acid and fusidic acid 21-*O*-glucuronide conjugate [6,7]. Fusidic acid was reported to undergo oxygenation and oxidation by microbial transformation to yield 6-hydroxy, 7-hydroxy, 3-didehydro and 6-oxofusidic acid [8–10] or deacetylation to produce the 16β-hydroxy derivative, which spontaneously converts into the biologically inactive lactone analog [11].

16-De-*O*-acetyl-7β-hydroxyfusidic acid was isolated from the fungus *Acremonium crotocinigenum* [12]. Biotransformation of the side chain functionalities of fusidic acid is rare. Several microbial strains were harnessed for studying the metabolism of drugs as a mimic of the phase-1 mammalian metabolism stage [13]. *Cunninghamella echinulata* is one of the established microbial models for investigating bioconversions of xenobiotics [13]. This study describes the formation of new metabolites emanating from microbial transformation of the side chain functionalities of fusidic acid using *C. echinulata*.

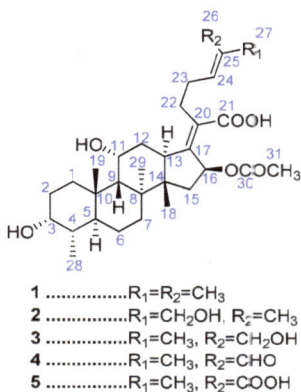

Figure 1. Structures of fusidic acid (**1**) and the isolated metabolites.

2. Results and Discussion

Compound **2** (Figure 1) showed a potassium adduct ion at *m/z* 571.3030 using high resolution electrospray ionization Fourier transform mass spectrometry (HRESI-FTMS) which, in conjunction with the ^{1}H- and ^{13}C-NMR data (Tables 1 and 2), corresponds to a molecular formula of $[C_{31}H_{48}O_7 + K]^+$ (calculated 571.3032). The molecular formula of fusidic acid is $C_{31}H_{48}O_6$ and the observed molecular formula of **2** thus indicates the mono-oxygenation of fusidic acid. By comparison of the ^{13}C-NMR data of fusidic acid and compound **2**, C-27 was deshielded from δ_C 25.7 to 68.6 suggesting its conversion from CH_3 to CH_2-O-, and thus resulted in deshielding of C-25 and shielding of C-26 (Table 2).The DEPT 135 experiment showed nine methylene carbons compared to the eight of fusidic acid. The ^{1}H-NMR spectrum of compound **2** in CDCl$_3$ (Table 1) showed the presence of a singlet at δ_H 3.9 integrating for two protons with the absence of the singlet at δ_H 1.67 for Me-27 in the spectrum of fusidic acid. This shift is consistent with a methylene group carrying an electronegative atom, thus, indicating the structure of compound **2** as 27-hydroxyfusidic acid. Other proton and carbon signals were highly similar to those of fusidic acid (Tables 1 and 2). The 2D HSQC NMR spectrum showed correlation of the proton singlet of CH$_2$-27 (δ_H 3.9) and C-27 at δ_C 68.6 which confirmed the site of oxygenation in compound **2** at C-27 (Figure 1). The ^{1}H-^{1}H COSY spectrum of fusidic acid (Supplementary Materials) indicated the correlation of the protons at C-24 and C-27 which disappeared in the ^{1}H-^{1}H COSY spectrum of compound **2** suggesting that compound **2** is a (24*E*) isomer. The experimental and computed chemical shifts of compound **2** were compared to assign the degree of fitness (Supplementary Materials), using the mean absolute error (MAE) and regression analysis (R^2) for that purpose. The absolute error for the computed ^{13}C-NMR data of **2** was calculated as 68.14 and the MAE as 2.198 supporting the assignment of **2** as the (24*E*) geometrical isomer of the new 27-hydroxyfusidic acid.

Table 1. ^{1}H-NMR data of fusidic acid and the isolated metabolites. δ_H ppm (J = Hz).

Position	Compound				
	1 *	2 (300 MHz)	3 (500 MHz)	4 (300 MHz)	5 (500 MHz)
1	1.51(m)/2.17 (m)	1.49 (m)/2.07 (m)	1.50 (m)/2.09 (m)	1.50 (m)/2.09 (m)	1.51 (m)/2.17 (m)
2	1.75 (m)/1.86 (m)	1.71 (m)/1.82 (m)	1.63 (m)	1.70 (m)/1.83 (m)	1.62 (m)/1.88 (m)
3	3.76 (s)	3.72 (s)	3.73 (s)	3.75 (d, 1.54)	3.68 (s)
4	1.58 (m)	1.46 (brs)	1.51 (m)	1.46 (m)	1.55 (m)
5	2.11 (m)	2.17 (m)	2.17 (m)	2.10 (m)	2.17 (m)
6	1.13 (m)/1.59 (m)	1.09 (m)/1.60 (m)	1.12 (m)/1.60 (m)	1.11 (m)/1.70 (m)	1.16 (m)/1.73 (m)
7	1.12 (m)/1.74 (m)	1.68 (m)/1.11(m)	1.68 (m)/1.12(m)	1.24 (m)/1.83 (m)	1.16 (m)/1.8 (m)
8	-	-	-	-	-
9	1.57 (s)	1.55 (s)	1.59 (s)	1.56 (s)	1.62 (s)
10	-	-	-	-	-
11	4.35 (brs)	4.36(brs)	4.36 (brs)	4.34 (brs)	4.34 (brs)
12	1.85 (m)/2.33 (m)	1.82 (m)/2.44 (m)	2.44 (m)	1.87 (m)/2.43 (m)	1.89 (m)/2.32 (m)
13	3.06 (d, 10.91)	2.95 (d, 13.0)	3.06 (d, 10.66)	3.08 (d, 11.10)	3.11 (d, 11.20)
14	-	-	-	-	-
15	1.30 (d, 14.20)/2.19 (m)	1.27 (d, 14.0)/2.17 (m)	1.27 (m)/2.19 (m)	2.10 (m)/1.40 (m)	1.27(d, 4.20)/2.18 (m)
16	5.88(d, 8.32)	5.88 (d, 8.2)	5.86 (d, 7.02)	5.90 (d, 8.3)	5.85 (d, 8.17)
17	-	-	-	-	-
18	0.89 (s)	0.88 (s)	0.91 (s)	0.91 (s)	0.96 (s)
19	0.96 (s)	0.96 (s)	0.98 (s)	0.96 (s)	1.02 (s)
20	-	-	-	-	-
21	-	-	-	-	-
22	2.46 (m)	2.55 (m)	2.55 (m)	2.43 (m)/2.61 (m)	2.57(m)/2.65 (m)
23	2.07 (m)/2.17(m)	2.05 (m)/2.20 (m)	2.30 (m)	2.61 (m)	2.33 (m)
24	5.10 (t, 6.97)	4.49 (t, 7.2)	5.26 (t, 7.2)	6.49 (t, 7.9)	6.80 (t, 8.0)
25	-	-	-	-	-
26	1.60 (s)	1.62 (s)	4.03 (d, 11.75), 4.13 (d, 11.75)	9.36 (s)	
27	1.67 (s)	3.9 (s)	1.77 (s)	1.73 (s)	1.84 (s)
28	0.90 (d, 5.8)	0.89 (d, 7.8)	0.90 (d,7.28)	0.90 (d, 6.31)	0.92 (d, 6.43)
29	1.38 (s)	1.34 (s)	1.38 (s)	1.36 (s)	1.41 (s)
31	1.96 (s)	1.97 (s)	1.99 (s)	1.96 (s)	1.98 (s)

* Data of fusidic acid (1) taken from reference [14].

The HRESI-FTMS data of compound 3 showed a potassium adduct ion at m/z 571.3030 which, in conjunction with ^{13}C-NMR data, corresponds to a molecular formula of $[C_{31}H_{48}O_7 + K]^+$ (calculated 571.3032) suggesting the oxygenation of fusidic acid. Comparing the ^{13}C-NMR data of fusidic acid and compound 3, C-26 was deshielded from δ_C 17.8 to 61.2 which resulted in deshielding of C-24, C-25 and shielding of C-27 (Table 2). The DEPT 135 spectrum showed nine methylene carbons with the chemical shift of the carbon at δ_C 61.2 suggesting oxygenation at C-26. The ^{1}H-NMR data of compound 3 in CDCl$_3$ showed two doublets at δ_H 4.03 and 4.13 (^3J = 11.75 Hz), characteristic for geminal coupling, replacing the singlet (δ_H 1.60) for Me-26 in the spectrum of fusidic acid. This shift is reminiscent of a methylene group attached to an electronegative atom suggesting the structure of compound 3 as 26-hydroxyfusidic acid. Other proton and carbon signals were similar to those of fusidic acid (Tables 1 and 2). The gradient HMQC data showed correlation of the proton doublets at δ_H 4.03 and 4.13 and C-26 (δ_C 61.2) which confirmed the site of oxygenation in compound 3 at C-26 (Figure 1). The ^{1}H-^{1}H COSY spectrum showed the correlation of the protons at C-24 and C-27, indicating that compound 3 is a (24Z) isomer. The computed ^{13}C-NMR spectrum of 3 showed an absolute error of 83.157 with an MAE of 2.682 matching the assignment of the structure of compound 3 as the new (24Z)-26-hydroxyfusidic acid.

Table 2. ^{13}C-NMR data of fusidic acid and the isolated metabolites. δ_C ppm.

Position	Compound				
	1 *	2 (75 MHz)	3 (125 MHz)	4 (75 MHz)	5 (125 MHz)
1	30.2	30.1	30.2	30.5	31.4
2	29.8	30.4	30.1	30.2	31.4
3	71.5	71.9	72.0	71.9	72.9
4	36.4	36.1	35.7	36.7	38.6
5	36.0	37.1	37.2	36.3	37.2
6	20.9	21.4	23.0	21.2	22.8
7	32.1	32.1	32.0	31.9	33.3
8	39.5	49.1	39.9	39.3	41.1
9	49.3	50.0	49.8	49.7	51.1
10	36.9	39.8	36.9	37.3	38.2
11	68.2	69.0	68.5	68.5	69.0
12	35.6	35.5	35.7	36.0	37.8
13	44.3	44.7	44.5	44.9	45.6
14	48.7	49.1	49.2	49.2	50.4
15	38.9	39.2	39.3	39.3	40.4
16	74.5	74.9	74.9	74.8	76.1
17	150.7	150.2	150.4	152.8	150.6
18	17.8	18.0	18.0	18.3	18.5
19	23.0	23.7	23.6	23.4	24.2
20	129.6	129.9	130.2	128.8	131.6
21	174.4	173.9	173.2	173.2	174.0
22	28.8	27.2	28.1	29.6	29.0
23	28.5	27.9	28.8	27.7	30.3
24	123.1	124.0	127.1	152.2	142.5
25	132.6	136.2	135.6	140.3	130.2
26	17.8	14.1	61.2	195.8	172.9
27	25.7	68.6	21.8	9.6	13.0
28	15.9	16.3	16.3	16.3	16.9
29	23.9	24.1	24.0	24.3	24.0
30	170.7	171.4	171.6	171.2	172.0
31	20.6	21.0	21.0	21.0	21.1

* Data of fusidic acid (1) taken from reference [14].

We next investigated the phenomenon of the 27-hydroxymethylene protons in **2** resonating as a singlet and the 26-hydroxymethylene protons in **3** as two one-proton doublets. The lowest energy conformers were analyzed to investigate the relative chemical environment of these protons in each case (Figure 2). Owing to strong hydrogen bonding between the C-11 and C-27 hydroxy groups, two major orientations of the C-27 protons of compound **2** were observed (Figure 2, panels A and B). This creates similar average chemical environments and results in a singlet resonance for the geminal hydrogen atoms. In **3**, the hydrogen bonding between the C-11 and C-26 hydroxy groups anchored the C-26 methylene (Figure 2, panel C) group to such an extent as to create diastereotopic-like protons culminating in two one-proton doublets in the ^{1}H-NMR spectrum.

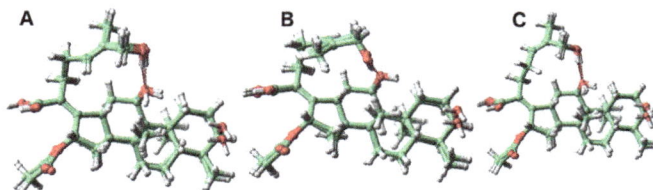

Figure 2. The most abundant conformers of **2** (**A** and **B**) and **3** (**C**).

The HRESI-FTMS data of compound **4** revealed a sodium adduct ion at *m/z* 553.3131 which, in conjunction with the ^{13}C-NMR data, accounts for a molecular formula of $[C_{31}H_{46}O_7 + Na]^+$

(calculated 553.3135), again indicative of the presence of an oxidation product of fusidic acid. By comparison of the ^{13}C-NMR data of fusidic acid and compound **4**, C-26 was deshielded from δ_C 17.8 to 195.8 indicating the oxidation of Me-26 into a formyl group which resulted in deshielding of C-24, C-25 and shielding of C-27 (Table 2). The DEPT 90 spectrum of compound **4** showed nine methine carbons in contrast to the eight of fusidic acid. The ^{1}H-NMR data of compound **4** in CDCl$_3$ showed the presence of a one-proton singlet at δ_H 9.36 and the absence of the Me-26 singlet (δ_H 1.60) in the spectrum of fusidic acid. This shift is reminiscent of formyl group formation suggesting the structure of compound **4** as 26-formylfusidic acid. The HSQC spectrum showed correlation of the proton singlet (δ_H 9.36) and C-26 (δ_C 195.8), and, thus, confirmed the structure of the new compound **4** (Figure 1). The ^{1}H-^{1}H COSY spectrum showed the correlation of the protons at C-24 and C-27, indicating that compound **4** is a (24Z) isomer. The calculated absolute error, MAE and (R^2) supported the structural assignment of the new compound **4** as (24Z)-26-formylfusidic acid (Supplementary Materials).

The molecular formula of compound **5** was determined as C$_{31}$H$_{46}$O$_8$ via its ^{13}C-NMR and HRESI-FTMS data which showed a sodium adduct ion at m/z 569.3072 for [C$_{31}$H$_{46}$O$_8$ + Na]$^+$ (calculated 569.3084). The molecular formula of compound **5** has one extra oxygen atom compared to compound **4** which is reminiscent of an oxidation product of fusidic acid. By comparison of the ^{13}C-NMR data of fusidic acid and compound **5**, the C-26 resonance was deshielded from δ_C 17.8 to 172.9 which strongly suggests oxidation at C-26, thus resulted in deshielding of C-24, C-25 and shielding of C-27 (Table 2). The DEPT 90 and 135 spectra of compound **5** evidenced one fewer methyl group compared to fusidic acid which implied the presence of a hydroxycarbonyl functional group. The ^{1}H-NMR data of compound **5** in methanol-d_4 showed the disappearance of the Me-26 singlet (δ_H 1.60) present in the spectrum of fusidic acid. This is consistent with the presence of a carboxylic group, and hence the structure of compound **5** as 26-carboxyfusidic acid (Figure 1) which matched the literature data [7]. 2D NMR data of compound **5** supported the deduced structure.

Compounds **3** and **4** may be considered as intermediates towards the formation of compound **5** and this is the first report of their formation and structural elucidation. Von Daehne et al. reported as "unpublished observations" that compound **2** was chemically synthesized by Godtfredsen and Vangedal via oxidation of fusidic acid with selenium oxide in *t*-butanol [14], followed by reduction with sodium borohydride to yield compound **2**. The oxygenation step of the 26-Me and 27-Me diastereotopic ligands in the side chain of fusidic acid using *C. echinulata* does not exhibit regioselectivity, whereas subsequent oxidation of the mixture of **2** and **3** into the formyl and hydroxycarbonyl fusidic acid derivatives **4** and **5** proceeded regiospecifically at C-26.

The antimicrobial activity testing of compounds **2**, **4** and **5** revealed that oxidation of fusidic acid at C-26 to the formyl derivative **4** diminishes the activity, whilst further oxidation to the carboxylic acid **5** abolishes the activity completely. The oxygenation at C-27 decreased the antimicrobial activity of fusidic acid (Table 3). These results showed that the methyl groups in the side chain of fusidic acid are crucial for maximum activity.

Table 3. Antimicrobial activity testing of fusidic acid and the isolated metabolites.

Microorganism	Compound, MIC (µg/mL)			
	Fusidic acid 1	2	4	5
Streptomyces faecalis	1.50	50	50	-ve *
Streptomyces durans	6.00	25	25	-ve
Staphyllococcus aureus	0.38	2.5	2.5	-ve
Bacillus subtlis	0.38	100	50	-ve
Escherichia coli	-ve	-ve	-ve	-ve
Pseudomonas aeruginosa	-ve	-ve	-ve	-ve
Mycobacterium smegmatis	12.5	100	-ve	-ve
Mycobacterium intercellulare	12	-ve	-ve	-ve
Candida albicans	1.25	-ve	-ve	-ve

* -ve (no antimicrobial activity) at the highest tested concentration (100 µg/mL).

A docking simulation was carried out using the crystal structure of *Thermus thermophilus* EF-G (PDB accession code: 4V5F). Fusidic acid showed the best docking score of −4 kcal/mol, while compounds **2, 3, 4** and **5** exhibited docking scores of −2.5, −2.6, −2.8 and −0.36 kcal/mol, respectively. The simulated binding poses of compounds **2, 3** and **4** were studied and compared with that of fusidic acid (Figures 3–6). Compounds **1, 2, 3** and **4** exhibited non-covalent interactions with the amino acid residues of the ligand binding pocket, mostly in the form of electrostatic and Van der Waals contacts.

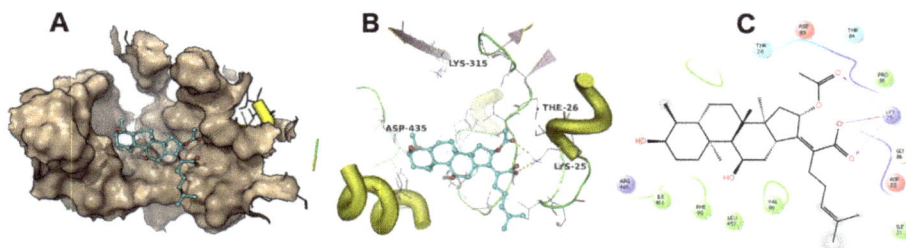

Figure 3. The binding mode of fusidic acid (**1**). The ligand binding pocket is shown as surface (**A**). The amino acid residues involved in ligand interaction are shown as lines (**B**). A 2D ligand interaction profile is demonstrated in (**C**).

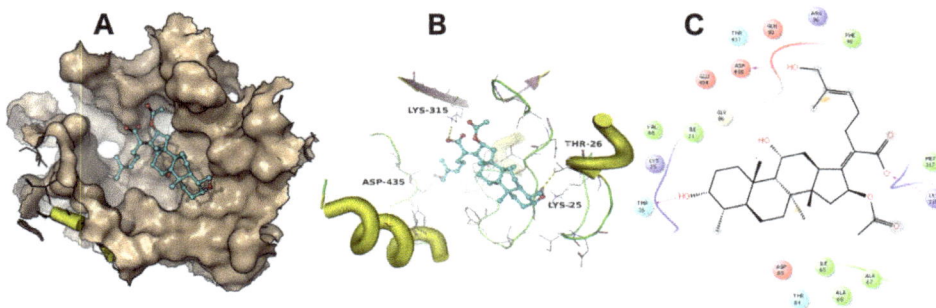

Figure 4. The binding mode of **2**. The ligand binding pocket is shown as surface (**A**). The amino acid residues involved in ligand interaction are shown as lines (**B**). A 2D ligand interaction profile is demonstrated in (**C**).

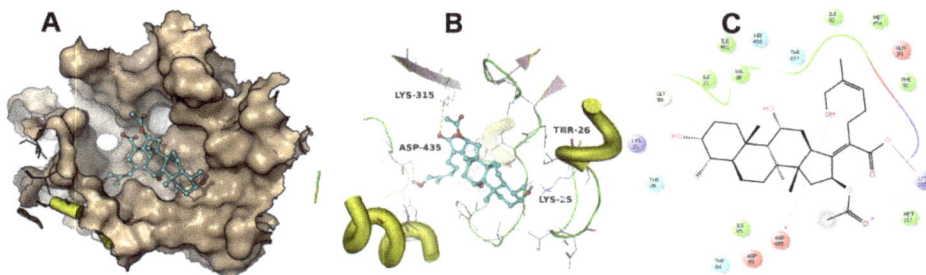

Figure 5. The binding mode of **3**. The ligand binding pocket is shown as surface (**A**). The amino acid residues involved in ligand interaction are shown as lines (**B**). A 2D ligand interaction profile is demonstrated in (**C**).

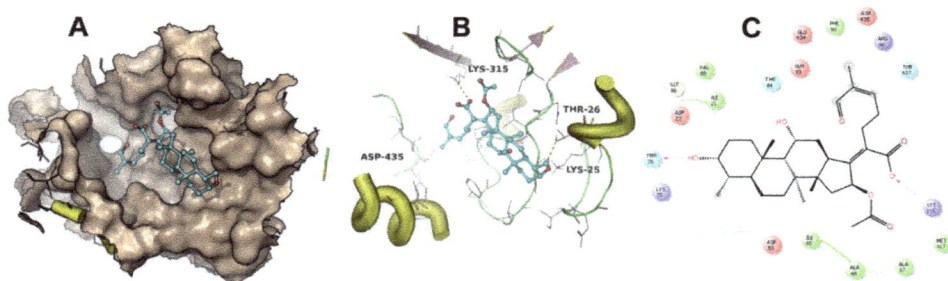

Figure 6. The binding mode of **4**. The ligand binding pocket is shown as surface (**A**). The amino acid residues involved in ligand interaction are shown as lines (**B**). A 2D ligand interaction profile is demonstrated in (**C**).

The amino acid residues involved in ligand interaction include Thr26, Lys25, Ile21, Val88, Arg96, Asp435, Glu434, Met317, Lys315, Ala68, Ile65, Ala67, Asp83, Thr84, Thr437 and Phe90. Lys315 and Thr26 form conserved hydrogen bonds, while Asp435 forms a hydrogen bond only with **2**. Lys25 showed a strong ionic interaction with **1**. This simulation indicated that fusidic acid fits best in the binding pocket with non-covalent and ionic interactions, while compounds **2–4** showed less binding affinity which may account for their decreased activity. The docking score of compound **5** (−0.36 kcal/mol) implies weak or no binding which explains the complete loss of activity.

3. Materials and Methods

3.1. General Experimental Procedures

Sodium fusidate was purchased from Leo Pharmaceutical Company (Ballerup, Denmark). IR spectra were recorded on a Perkin Elmer IR spectrophotometer (PerkinElmer Inc., Waltham, MA, USA). UV data were acquired using a 60/PC ultraviolet spectrophotometer (Shimadzu, Kyoto, Japan). NMR spectra were recorded using Varian XL300 (Varian Inc., Palo Alto, CA, USA) and Bruker Avance 500 spectrophotometers (Bruker, Billerica, MA, USA) using CDCl$_3$ and methanol-d_4 as solvents and tetramethyl silane (TMS) as internal standard. ^{1}H-NMR spectra were recorded at 300 or 500 MHz, and ^{13}C-NMR spectra at 75 or 125 MHz. DEPT, COSY and HETCOR analyses were obtained using Varian Pulse Sequences at 300 or 500 MHz. HR-ESIFTMS data were acquired using a Bruker Bioapex FT-mass spectrometer (Bruker, Billerica, MA, USA) in ESI mode. Thin layer chromatography (TLC) was carried out using precoated silica gel 60 F$_{254}$ plates (0.25 mm layer, E. Merck, Darmstadt, Germany) and visualization was by spraying with *p*-anisaldehyde reagent followed by heating at 110 °C.

3.2. Preparation of Fusidic Acid

Sodium fusidate was dissolved in water (50 mg/mL) and acidified with acetic acid. The precipitated fusidic acid was filtered, washed acid-free with distilled water, and dried to constant weight in a vacuum desiccator. The NMR and MS data were identical to reported data [14,15].

3.3. Microorganisms and Culture Conditions

Microbial transformation experiments were conducted according to published procedures [16]. For the initial screening experiments, 25 microbial cultures belonging to the genera *Aspergillus*, *Candida*, *Cunninghamella*, *Saccharomyces*, *Rhizopus*, *Penicillium*, *Streptomyces*, *Gymnascella*, *Lindera*, and *Rhodotorula* were used. The tested strains were obtained from either The American Type Culture Collection (ATCC, Manassas, VA, USA) or the National Center for Agricultural Utilization Research

(NCAUR, Peoria, IL, USA). The strains were maintained at 4 °C on Sabouraud dextrose agar slants and subcultured quarterly.

3.4. Culture Media

In all fermentations, the medium consists of 10 mL/L glycerol, 10 g/L glucose, 5 g/L peptone, 5 g/L yeast extract, 5 g/L NaCl, and 5 g/L K_2HPO_4 in distilled water. The pH was adjusted to 6.0 before autoclaving at 121 °C for 15 min.

3.5. Initial Biotransformation Screening Experiments

Cells of the tested microorganisms were transformed from two-week old slants into sterile liquid medium (50 mL/250 mL flask) and kept on a gyratory shaker at 28 °C and 200 rpm for 72 h to give stage I culture. Stage I culture (5 mL) was used as an inoculum for stage II culture (50 mL/250 mL flask). After 24 h of incubation of stage II culture, sodium fusidate (10 mg) was added as a solution in absolute ethanol (250 µL) to each flask. Samples were taken after 3 and 6 days of incubation, acidified with a few drops of 10% HCl, filtered and the filtrate was extracted with an equal volume of chloroform. After evaporation of the chloroform, the residues were chromatographed on precoated silica gel plates using chloroform-methanol (5:1) or benzene-ethyl acetate- formic acid (3 mL:7 mL:1 drop) as mobile phase and detection was carried out by UV light visualization and *p*-anisaldehyde spray reagent. Both substrate and organism-free controls were also prepared and processed in the same way. The results of preliminary screening using fusidic acid were identical to those of using sodium fusidate. Amongst the tested strains, *C. echinulata* NRRL 1382 and *C. elegans* 1392 displayed the best transformations. This paper discusses the metabolites obtained from transformation using *C. echinulata*.

3.6. Large Scale Fermentation

Stage I cultures were prepared by inoculating culture media with two weeks old Sabouraud dextrose agar slants of *C. echinulata* and incubated at 28 °C, and 200 rpm for 72 h. Stage II cultures were initiated by inoculating stage I culture (5 mL) into new culture media (50 mL in 250 mL flasks) and incubated at 28 °C, and 200 rpm for 24 h. Sodium fusidate, dissolved in absolute ethanol (2.7 g/67 mL), was added to 270 stage II cultures to give a 0.02% *w/v* final concentration, and incubation continued for six days. Substrate and organism free control cultures were prepared. The cultures were pooled, acidified with 10% HCl (1 mL/30 mL culture), filtered and the filtrate was extracted twice with an equal volume of chloroform. The chloroform extract was dried over anhydrous sodium sulfate and evaporated under vacuum to give an amber-colored residue (3.4 g). TLC was carried out using chloroform-methanol (5:1) or benzene-ethyl acetate-formic acid (3 mL:7 mL:1 drop) as mobile phases and detection was carried out by UV light visualization and *p*-anisaldehyde spray reagent.

3.7. Isolation of Metabolites

The residue obtained from the chloroform extract after evaporation (3.4 g) was loaded onto a silica gel column (300 g) and eluted with a gradient of ethyl acetate in benzene (0–60%) containing 0.2% formic acid and fractions of 100 mL were collected. The percentage of formic acid was increased to 0.4% starting from fraction no. 107 and similar fractions were pooled to give three groups of fractions.

3.7.1. Fractions 80–106

The residue obtained upon pooling and evaporation of these fractions (360 mg) was rechromatographed on a silica gel column (40 g) using a gradient of methanol/chloroform (0–10%), and 50 mL fractions were collected. Fractions 47–64 afforded compound **4** (110 mg) and fractions 73–136 gave compound **5** (72 mg).

3.7.2. Fractions 122–142

The residue of these fractions (300 mg) was rechromatographed on a silica gel column (40 g) using a gradient of methanol/chloroform (0–10%) and 50 mL fractions were collected. Fractions 46–84 afforded compound **2** (115 mg).

3.7.3. Fractions 143–190

The residue of these fractions (320 mg) was partially purified using Sephadex LH-20 column (200 mL bed volume) chromatography followed by silica gel column chromatography (40 g) using a gradient of methanol/chloroform (0–4%) and collecting 50 mL fractions. Fractions 79–114 yielded compound **3** which was recrystallized from n-hexane/chloroform mixture to provide 21 mg of pure **3**.

3.7.4. 27-Hydroxyfusidic Acid (**2**)

White powder; UV (MeOH) λ_{max} 223 nm; IR ν_{max} (KBr disc) cm^{-1}: 3440, 2880, 1725, 1395, 1275; ^{1}H and ^{13}C-NMR (CDCl$_3$): see Tables 1 and 2; HRESI-FTMS (*m/z*): 571.3030 [M + K]$^+$ (calc. for C$_{31}$H$_{48}$O$_7$K, 571.3032).

3.7.5. 26-Hydroxyfusidic Acid (**3**)

White powder; UV (MeOH) λ_{max} 223 nm; IR ν_{max} (KBr disc) cm^{-1}: 3432, 2936, 1717, 1638, 1443, 1379, 1260; ^{1}H and ^{13}C-NMR (CDCl$_3$): see Tables 1 and 2; HRESI-FTMS (*m/z*): 571.3030 [M + K]$^+$ (calc. for C$_{31}$H$_{48}$O$_7$K, 571.3032).

3.7.6. 26-Formylfusidic Acid (**4**)

White powder; UV (MeOH) λ_{max} 218 nm; IR ν_{max} (KBr disc) cm^{-1}: 3500, 2970, 2910, 1720, 1690, 1465, 1385, 1265; ^{1}H and ^{13}C-NMR (CDCl$_3$): see Tables 1 and 2; HRESI-FTMS (*m/z*): 553.3131 [M + Na]$^+$ (calc. for C$_{31}$H$_{46}$O$_7$Na, 553.3135).

3.7.7. 26-Carboxyfusidic Acid (**5**)

White powder; UV (MeOH) λ_{max} 223 nm; IR ν_{max} (KBr disc) cm^{-1}: 3435, 3169, 2939, 1700, 1641, 1381, 1260; ^{1}H and ^{13}C-NMR (CDCl$_3$): see Tables 1 and 2; HRESI-FTMS (*m/z*): 569.3072 [M + Na]$^+$ (calc. for C$_{31}$H$_{46}$O$_8$Na, 569.3084).

3.8. Antimicrobial Activity

Samples were tested according to the National Committee of Clinical Laboratory Standard (NCCLS, 1994) using ATCC strains.

3.9. Assignment of Relative Configuration

To assign the relative configuration of the compounds, all possible chemical structures were sketched and energy minimized in Maestro. MacroModel with the OPLS3 forcefied was used to generate the conformers of the proposed structures. We used the stochastic conformational search approach of MacroModel and the Monte Carlo multiple minimum method to allow for better torsional sampling. The energy window for selecting the conformers was defined at 10.04 kcal mol^{-1}. Geometry optimization and frequencies were calculated for all optimized conformers, based on Boltzmann analysis, using Gaussian 09 at the M06-2X/6-31+G(d,p) level. Gaussian 09 at the B3LYP/6-311+G(2d,p) level was used to compute the NMR shielding tensors using the gauge-independent (or including) atomic orbitals (GIAO) method. In all DFT calculations we used the integrated equation formalism polarized continuum model (IEFPCM) was used.

3.10. Protein Preparation

The protein crystal structure of *T. thermophilus* EF-G (PDB accession code: 4V5F) was obtained from the protein databank (www.rcsb.org). The protein structure was prepared for docking by PrepWizard of the Schrödinger suite. Missing hydrogen atoms, amino acid side chains and loops were added. To account for correction of hydrogen bond networks, the orientations of amide groups (Asn and Gln), hydroxy groups (Tyr, Thr and Ser), and protonation states of imidazole moiety (His) were adjusted. No energy minimization was conducted.

3.11. Ligand Preparation

The compounds were sketched and converted into 3D structures in Maestro. The molecules were then prepared to address all possible protonation and tautomerization states using LigPrep with the OPLS3 forcefield. Only the lowest energy conformer for each ligand was kept.

3.12. Receptor Grid Preparation

The make receptor module of OpenEye scientific software (www.eyesopen.com) was used to construct the receptor grid. The native ligand was used to define the centroid of the docking box. The volume and dimensions of the grid box were defined as 7374 Å^3 (17.27 Å × 19.14 Å × 22.31 Å). The dimensions of the outer contour of the docking region was 3140 Å^3.

3.13. Docking Simulation

The multi-conformers' compound database was docked using FRED of the OpenEye scientific software with standard docking precision was used. One best pose was saved for each compound.

4. Conclusions

Among the screened strains, *C. echinulata* was the only organism that metabolized fusidic acid (**1**) in a regioselective fashion targeting the allylic Me-26 and the Me-27 groups of the hydrophobic side chain. The microorganism seems to detoxify the antibiotic fusidic acid (**1**) by regioselective oxidation of the methyl groups of the hydrophobic side chain into hydroxymethyl, formyl and hydroxycarbonyl functionalities in order to minimize the antimicrobial activity. The dicarboxylic acid may eventually undergo decarboxylation to norfusidic acid, which, however is yet to be isolated and assessed for antimicrobial activity. The intermediate oxidation products **2–4** may be exploited to develop antibiotic ligands with better activity and lower toxicity. These data indicate the presence of an interesting oxidation system in *C. echinulata* which targeted the side chain of fusidic acid in contrast to *C. elegans* which targeted ring B in our previous work [17].

Supplementary Materials: The following are available online. Figures S1–S26 are NMR and mass spectra of the isolated compounds. Excel sheets contain the NMR (proton and carbon) calculation results.

Acknowledgments: We are grateful for OpenEye Scientific Software for supporting the academic license. We also thank Ahmed Galal for his help in HR-ESIFTMS analysis of the isolated metabolites at the School of Pharmacy, University of Mississippi, USA.

Author Contributions: A.-R.S.I., K.M.E. and A.E.R. designed and performed research, A.-R.S.I., K.M.E., A.E.R. and D.F. analyzed data, A.-R.S.I., K.M.E., A.E.R. and D.F. wrote and revised the paper.

Conflicts of Interest: The authors declare no conflict of interest.

References

1. Godtfredsen, W.O.; Jahnssen, S.; Lorck, H.; Roholt, K.; Tybring, L. Fusidic acid: A new antibiotic. *Nature* **1962**, *10*, 193–897. [CrossRef]
2. Godtfredsen, W.O.; Vangedal, S. The structure of fusidic acid. *Tetrahedron* **1962**, *18*, 1029–1048. [CrossRef]
3. Turnridge, J. Fusidic acid pharmacology, pharmacokinetics, and pharmacodynamics. *Int. J. Antimicrob. Agents* **1999**, *12*, S23–S34. [CrossRef]

4. Collignon, P.; Turnidge, J. Fusidic acid in vitro activity. *Int. J. Antimicrob. Agents* **1999**, *12*, S45–S58. [CrossRef]

5. Berchtold, H.; Reshetinova, L.; Reiser, O.A.; Schirmer, N.K.; Sprinzl, M.; Hilgenfeld, R. Crystal structure of the active elongation factor Tu reveals major domain rearrangements. *Nature* **1993**, *365*, 126–132. [CrossRef] [PubMed]

6. Reeves, D.S. The pharmacokinetics of fusidic acid. *J. Antimicrob. Chemother.* **1987**, *20*, 467–476. [CrossRef] [PubMed]

7. Godtfredsen, W.O.; Vangedal, S. On the metabolism of fusidic acid in man. *Acta Chem. Scand.* **1966**, *20*, 1599–1607. [CrossRef] [PubMed]

8. Hadara, K.; Tomita, K.; Fujii, K.; Sato, N.; Uchida, H.; Yazawa, K.; Mikami, Y. Inactivation of fusidic acid by pathogenic Nocardia. *J. Antibiot.* **1999**, *52*, 335–339. [CrossRef]

9. Von Daehne, W.; Lorch, H.; Godtfredsen, W.O. Microbial transformation of fusidane-type antibiotics, A correlation between fusidic acid and and helvolic acid. *Tetrahedron Lett.* **1968**, *47*, 4843–4846. [CrossRef]

10. Dvnoch, W.; Greenspan, G.; Alburn, H.E. Microbiological oxidation of fusidic acid. *Experientia* **1966**, *22*, 517. [CrossRef]

11. Von der Harr, B.; Schrefmp, H. Purification and characterization of a novel extracellular *Streptomyces lividans* 66 enzyme inactivating fusidic acid. *J. Bacteriol.* **1995**, *177*, 152–155. [CrossRef]

12. Evans, L.; Hedger, J.N.; Brayford, D.; Stavri, M.; Smith, E.; O'Donnell, G.; Gray, A.I.; Griffith, G.W.; Simmon, G. Antibacterial hydroxy fusidic acid analogue from *Acremonium crotocinigenum*. *Phytochemistry* **2006**, *67*, 2110–2114. [CrossRef] [PubMed]

13. Yang, W.; Jiang, T.; Acosta, D.; Davis, P.J. Microbial models of mammalian metabolism: Involvement of cytochrome P450 in the *N*-demethylation of *N*-methylcarbazole by *Cunninghamella echinulata*. *Xenobiotica* **1993**, *23*, 973–982. [CrossRef] [PubMed]

14. Rastrup-Andersen, N.; Duvold, T. Reassignment of the ^{1}H-NMR spectrum of fusidic acid and total assignment of ^{1}H and ^{13}C-NMR spectra of some selected fusidane derivatives. *Mag. Res. Chem.* **2002**, *40*, 471–473. [CrossRef]

15. Von Daehne, W.; Godtfredsen, W.O.; Rasmussen, P.R. Structure-activity relationships in fusidic acid–type antibiotics. *Adv. Appl. Microbiol.* **1979**, *25*, 95–145. [PubMed]

16. Galal, A.M.; Ibrahim, A.S.; Mossa, J.S.; El-Feraly, F.S. Microbial transformation of parthenolide. *Phytochemistry* **1999**, *51*, 761–765. [CrossRef]

17. Ibrahim, A.S.; Ragab, A.E. Fusidic acid ring B hydroxylation by *Cunninghamella elegans*. *Phytochem. Lett.* **2018**, *25*, 86–89. [CrossRef]

Sample Availability: Samples of the compounds are not available from the authors.

molecules

MDPI

Article

A Chemical Investigation of the Leaves of *Morus alba* L.

Xiao-yan Chen [1], Ting Zhang [2], Xin Wang [3], Mark T. Hamann [1], Jie Kang [4], De-quan Yu [4] and Ruo-yun Chen [4,*]

[1] Department of Drug Discovery and Biomedical Sciences, College of Pharmacy, Medical University of South Carolina, Charleston, SC 29425, USA; chenxiaoyan8615@gmail.com or chenxi@musc.edu (X.-y.C.); hamannm@musc.edu (M.T.H.)

[2] Institute of Medical Information & Library, Chinese Academy of Medical Sciences and Peking Union Medical College, Beijing 100020, China; brendatingting@126.com

[3] Beijing Key Laboratory of Bioactive Substances and Function Foods, Beijing Union University, Beijing 100191, China; shtwangxin@buu.edu.cn

[4] State Key Laboratory of Bioactive Substance and Function of Natural Medicines, Institute of Materia Medica, Chinese Academy of Medical Sciences and Peking Union Medical College, Beijing 100050, China; jiekang@imm.ac.cn (J.K.); dqyu@imm.ac.cn (D.-q.Y.)

* Correspondence: rych@imm.ac.cn; Tel.: +86-10-8316-1622

Received: 31 March 2018; Accepted: 24 April 2018; Published: 26 April 2018

Abstract: The leaves of *Morus alba* L. are an important herbal medicine in Asia. The systematic isolation of the metabolites of the leaves of *Morus alba* L. was achieved using a combination of liquid chromatography techniques. The structures were elucidated by spectroscopic data analysis and the absolute configuration was determined based on electronic circular dichroism (ECD) spectroscopic data and hydrolysis experiments. Their biological activity was evaluated using different biological assays, such as the assessment of their capacity to inhibit the aldose reductase enzyme; the determination of their cytotoxic activity and the evaluation of their neuroprotective effects against the deprivation of serum or against the presence of nicouline. Chemical investigation of the leaves of *Morus alba* L. resulted in four new structures **1–4** and a known molecule **5**. Compounds **2** and **5** inhibited aldose reductase with IC_{50} values of 4.33 µM and 6.0 µM compared with the potent AR inhibitor epalrestat (IC_{50} 1.88×10^{-3} µM). Pretreatment with compound **3** decreased PC12 cell apoptosis subsequent serum deprivation condition and pretreatment with compound **5** decreased nicouline-induced PC12 cell apoptosis as compared with control cells ($p < 0.001$).

Keywords: *Morus alba* L.; aldose reductase inhibitor; neuroprotective agent; natural products

1. Introduction

The species *Morus alba* L., known as white mulberry, belongs to the genus *Morus* of the family Moraceae, is native to China and now is cultivated throughout the world [1,2]. All parts of this plant have been used medicinally in Traditional Chinese Medicine including the leaves, root bark, stem and fruits [3,4]. In East Asia the leaves have been an important herbal medicine for treatment of cold, fever, headache, cough and rheumatic diseases for thousands of years [3,5]. Extracts or constituents of *M. alba* L. leaves were reported to possess anti-inflammatory, antioxidant, antiobesity, antidiabetic, and hypolipidemic properties [5,6]. Phytochemical investigations of the leaves of *M. alba* reported the presence of flavonoids, lignans, pyrrole alkaloids, polyphenols, fatty acids, and anthocyanin [6,7]. Previous phytochemical studies and the pharmacological potentials of constituents of the genus *Morus* have been reviewed by Yang et al. [8].

Aldose reductase is the first and rate-controlling enzyme in the polyol pathway that reduces glucose into sorbitol and then fructose. Intracellular excess sorbitol is thought to lead to diabetic complications, including neuropathy, nephropathy, retinopathy, and cataract [9]. Fructose can lead to the formation of 3-deoxyglucosone, a key intermediate known to accelerate the formation of Advanced Glycation End products (AGEs) [10]. The presence and accumulation of AGEs contribute to the development of atherosclerosis and promotes renal damage, diabetic nephropathy, and a series of cancers [11–14]. Besides reducing glucose, aldose reductase is also involved in the reduction of oxidative stress-generated lipid aldehydes and their conjugate with GSH, which can alter cellular signals by mediating transcription factors such as NF-Kb and AP1 [15]. Aldose reductase inhibitors have been shown to be an effective multi-disease target to prevent diabetic complications, cancers, cardiovascular diseases, and inflammatory complications [16]. This study investigated the acetone and chloroform fractions of the ethanol extract of the leaves of *M. alba* L., which showed effects against diabetes and human cancer cell lines in our previous study, leading to the identification of four new structures, namely a sesquiterpenoid glucoside **1**, an aromatic glucoside **2**, a farnesylacetone derivative **3**, a flavan **4**, and a known compound, (9R)-hydroxyl-(10E,12Z,15Z)-octadecatrienoic acid (**5**). In addition, we report the results of aldose reductase inhibitory and neuroprotective activity evaluations.

2. Results and Discussion

2.1. Characterization

The crude extract of the leaves of *Morus alba*. L was divided into four fractions by flash silica gel column chromatography. The generated acetone and chloroform fractions were further isolated by the combination of resin column chromatography, silica gel column chromatography, medium pressure liquid chromatography (MPLC), and high performance liquid chromatography (HPLC), generating four new compounds and a known one (Figure 1).

Moralsin (**1**) was obtained as a white powder, $[\alpha]_D^{20}$ −72 (*c* 0.19, MeOH). The IR spectrum of **1** showed the presence of hydroxy (2957 cm^{-1}), alkyl (3402 cm^{-1}) and ester carbonyl (1760 cm^{-1}) functional groups. The molecular formula, $C_{21}H_{30}O_8$, was determined from its sodium adduct ion in the HRESIMS (433.1826 [M + Na]$^+$, calcd. 433.1833), corresponding to seven indices of hydrogen deficiency. The ^{1}H-NMR and ^{13}C-NMR spectra (Tables 1 and 2) revealed the presence of two trisubstituted double bonds [δ_H 6.82 (t, *J* = 3.0 Hz, H-4), 5.66 (d, *J* = 9.3 Hz, H-8); δ_C 132.4 (C-4), 130.7 (C-5), 129.8 (C-8), 138.8 (C-9)], one terminal double bond [δ_H 6.44 (dd, *J* = 17.5, 10.5 Hz, H-10), 5.37 (d, *J* = 17.5 Hz, H-11a), 5.20 (d, *J* = 10.5 Hz, H-11b)], one ester carbonyl (δ_C 168.8), two oxygenated methines [δ_H 4.41 (m, H-3), 5.12 (t, *J* = 9.3 Hz, H-7); δ_C 70.4 (C-3), 76.8 (C-7)], three methyls [δ_H 0.90 (3H, s), 0.87 (3H, s), 1.88 (3H, s)], one methylene [δ_H 1.86 (m, H-2α), 1.60 (dd, *J* = 15.0, 6.5 Hz, H-2β)], two methines [δ_H 2.57 (m, H-6), 5.12 (t, *J* = 9.3 Hz, H-7)], one tertiary carbon group [δ_C 29.5 (C-1)] and a glucopyranosyl unit [δ_H 4.35 (d, *J* = 8.0 Hz, H-1′), 2.9-3.7 (6H, H-2′-6′)]. In combination with analysis of ^{1}H-^{1}H COSY spectrum, the NMR date displayed there were two spin systems C2-C3-C4 and C6-C7-C8 in **1**. In the HMBC spectrum, the correlations of H-12, H-13/C-1, C-2, C-6; H-4/C-5, C-14; H-6/C-4; H-11/C-9; and H-15/C-8, C-10 determined the monocyclofarnesane carbon skeleton containing a 14,7-olide ring. In addition, the correlations of H-1′/C-3 indicated the glucopyranosyl unit was connected to C-3 of the monocyclofarnesane-type sesquiterpenoid aglycone.

A β-anomeric configuration for the glucosyl unit was assigned via its large $^3J_{1,2}$ coupling constant (8.0 Hz). The D-configuration of the glucose was determined by GC analysis of the trimethylsilyl L-cysteine derivatives after acid hydrolysis of **1**. The E-configuration of the 8,9-double bond was demonstrated by the nuclear Overhauser effect (NOE) effect of H-7/H-15 and H-8/H-10 in the ROESY 1D experiment. NOE effect of H-3/H-2β, H-7/H-12, and H-6/H-8, H-2β, and H-13 showed that H-2β, H-3, H-6, C-13, and C-8 were cofacial, assigned as the β-orientation, and H-7 and C-12 were α-oriented. The absolute configuration of aglycone moiety was assigned by analysis of the electronic circular dichroism (ECD) spectroscopy using excitation chirality method [17]. The ECD spectrum showed positive Cotton effect at 263 nm and negative Cotton effect at 224 nm arising from coupling between conjugated diene and α,β-unsaturated ester chromophores (Supplementary Materials). Such a pattern was in agreement of a negative chirality of **1** as depicted in Figure 2. Thus the absolute configuration of **1** was unequivocally assigned as (3*S*, 6*S*, 7*R*) and the structure of moralsin was determined as **1**.

Table 1. ^{1}H-NMR Spectroscopic Data of Compounds **1–4** [a].

	1		2		3		4
Position	δ_H (*J* in Hz)	Position	δ_H (*J* in Hz)	Position	δ_H (*J* in Hz)	Position	δ_H (*J* in Hz)
2α	1.86, (overlapped)	1	2.25, s	2	5.27, dd, (9.6, 1.8)	3	6.65, d, (8.4)
2β	1.60, dd, (15.0, 6.5)	3	6.09, d, (16.5)	3a	2.20, m	4	7.18, dd, (8.4)
3	4.41, m	4	7.45, dd, (16.5, 11.5)	3b	1.85, m	5	6.55, d, (8.4)
4	6.82, t, (3.0)	5	6.23, d, (11.5)	4a	2.86, m	2′, 6′	7.47, m
6	2.57, m	7	3.93, t, (6.5)	4b	2.63, m	3′, 5′	7.36, m
7	5.12, t, (9.3)	8	1.47, m	5	6.79, d, (8.4)	4′	7.36, m
8	5.66, d, (9.3)	9	1.97, t, (7.5)	6	6.30, d, (8.4)	7′a	5.33, d, (12.6)
10	6.44, dd, (17.5, 9.0)	11	5.03, t, (7.0)	3′	6.40, d, (2.4)	7′b	5.22, d, (12.6)
11a	5.37, d, (17.5)	12	2.12, dd, (7.5, 7.0)	5′	6.42, dd, (8.4, 2.4)	1″	4.84, d, (7.2)
11b	5.20, d, (10.5)	13	2.42, t, (7.5)	6′	7.25, d, (8.4)	2″	3.21, m
12	0.90, s	15	1.06, s	3″	3.73, dd, (7.2, 6.0)	3″	3.49, m
13	0.87, 3H, s	16	1.85 (3H, s)	4″a	2.92, dd, (17.4, 7.8)	4″	3.09, t, (8.7)
15	1.88, s	17	1.62 (3H, s)	4″b	2.54, dd, (17.4, 5.4)	5″	3.25, m
1′	4.35, d, (8.0)			5″	1.22, s [b]	6″a	3.86, d, (8.1)
2′	2.91, t, (8.0)			6″	1.31, s [b]	6″b	3.44, m
3′	3.15, m			-OMe	3.75, s	1‴	4.83, d, (3.0)
4′	3.03, t, (8.0)					2‴	3.75, d, (3.0)
5′	3.15, m					4‴a	3.89 (1H, d, 9.0)
6′a	3.68, dd, (10.5, 1.0)					4‴a	3.59 (1H. d, 9.0)
6′b	3.44, dd, (10.5, 6.0)					5‴	3.35 (1H, m)

[a] ^{1}H–NMR data (δ) were measured at 500 MHz in DMSO-d_6 for **1**, **2**, **4**, and at 600 MHz in methanol-d_4 for **3**. The assignments were based on HSQC and HMBC experiments; [b] Interchangeable.

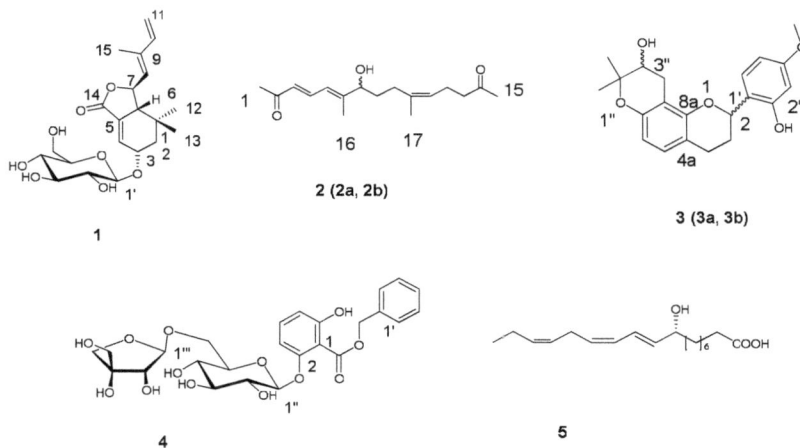

Figure 1. Structures of Compounds **1–5**.

Table 2. ^{13}C-NMR Spectroscopic Data of Compounds **1–4** [a].

	1		2		3		4
Position	δ_C, Type	Position	δ_C, Type	Position	δ_C, Type	Position	δ_C, Type
1	29.5, C	1	27.1, CH$_3$	2	74.2, CH	1	120.0, C
2	43.2, CH$_2$	2	198.3, CH	3	29.9, CH$_2$	2	155.3, C
3	70.4, CH	3	129.9, CH	4	25.8, CH$_2$	3	105.5 [b], CH

Table 2. *Cont.*

	1		2		3		4
Position	δ_c, Type	Position	δ_c, Type	Position	δ_c, Type	Position	δ_c, Type
4	132.4, CH	4	139.4, CH	5	128.5, CH	4	131.0, CH
5	130.7, C	5	122.3, CH	6	109.7, CH	5	109.4 [b], CH
6	52.3, CH	6	153.1, C	7	153.0, C	6	155.4, C
7	76.8, CH	7	74.7, CH	8	109.3, C	7	165.8, C
8	129.8, CH	8	33.3, CH$_2$	4a	114.4, C	1'	136.2, C
9	138.8, C	9	27.5, CH$_2$	8a	154.5, C	2', 6'	127.8, CH
10	140.0, CH	10	135.3, C	1'	122.4, C	3', 5'	128.3, CH
11	115.6, CH$_2$	11	124.0, CH	2'	156.1, C	4'	127.8, CH
12	21.0, CH$_3$	12	21.8, CH$_2$	3'	102.1, CH	7'	66.0, CH$_2$
13	28.9, CH$_3$	13	43.1, CH$_2$	4'	161.4, C	1''	100.4, CH
14	168.8, C	14	208.1, C	5'	105.7, CH	2''	73.27, CH
15	12.3, CH$_3$	15	29.8, CH$_3$	6'	128.0, CH	3''	75.6, CH
1'	102.7, CH	16	13.4, CH$_3$	2''	77.4, C	4''	70.0, CH
2'	73.5, CH	17	23.2, CH$_3$	3''	70.5, CH	5''	76.8, CH
3'	76.8, CH			4''	27.3, CH$_2$	6''	67.8, CH2
4'	70.1, CH			5''	20.8	1'''	109.4, CH
5'	76.9, CH			6''	25.8	2'''	75.9, CH
6'	61.2, CH$_2$ b 4.29, dd, (11.5, 6)			-OMe	55.7	3'''	78.7, C
						4'''	73.30, CH$_2$
						5'''	63.2, CH$_2$

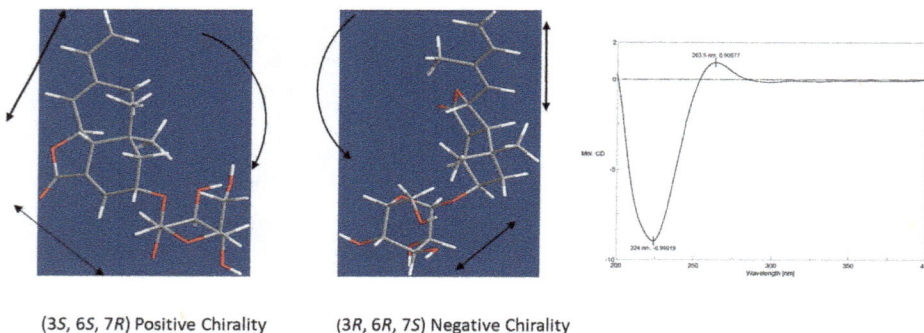

(3*S*, 6*S*, 7*R*) Positive Chirality   (3*R*, 6*R*, 7*S*) Negative Chirality

Figure 2. ECD Spectrum of compound **1**. The positive Cotton effect of ECD spectrum is in agreement with the negative chirality of (3*S*, 6*S*, 7*R*) diastereoisomer of compound **1**.

Compound **2** was obtained as a yellow oil. The IR spectrum of **2** showed the presence of hydroxy (3419 cm^{-1}), alkyl (2931 cm^{-1}), and carbonyl (1712 cm^{-1}) functional groups. The molecular formula, $C_{17}H_{26}O_3$, was determined from its sodium adduct ion in the HRESIMS (301.1785 [M + Na]$^+$, calcd. 301.1774), corresponding to five indices of hydrogen deficiency. The ^{1}H-NMR (Table 1), ^{13}C-NMR (Table 2), and DEPT (Supplementary Information) spectra revealed the presence of three double bonds at δ_H 6.09 (d, *J* = 16.5 Hz, H-3), 7.45 (dd, *J* = 16.5, 11.5 Hz, H-4), 6.23 (d, *J* = 11.5 Hz, H-5), 5.03 (t, *J* = 7.0 Hz, H-11) and δ_C 129.9 (C-3), 139.4 (C-4), 122.3 (C-5), 153.1 (C-6), 135.3 (C-10), 124.0 (C-11), a oxymethine group at δ_H 3.93 (t, *J* = 6.5 Hz, H-7), four methylene groups at δ_H 1.47 (m, H-8), 1.97 (t, *J* = 7.5 Hz, H-9), 2.12 (m, H-12), 2.42 (t, *J* = 7.5 Hz, H-13), four terminal methyl groups at δ_H 2.25 (s, H-1), 2.06 (s, H-15), 1.85 (s, H-16), 1.62 (s, H-17), and two carbonyl groups at δ_C 198.3 (C-2), 208.1 (C-14). The coupling patterns in ^{1}H-NMR spectrum and the correlations in ^{1}H-^{1}H COSY spectrum showed the presence of three spin systems of C-3-C-4-C-5, C-7-C-8-C-9, and C-11-C-12-C-13. In the HMBC spectrum the correlations of H-1, H-3, H-4/C-1, H-4, H-8/C-6, H-5/C-7, H-8, H-12/C-10, H-11/C-9, and H-12, H-14, H-15/C-14 clarified the connections of the two terminal methyl groups (C-1, 15), the two carbonyl groups and the three spin systems, suggesting a linear structure of pentadecatrien-2,14-dione. The HMBC correlations of H-5, H-7/C-16, H-16/C-5, C-7, H-9, H-11/C-17, and H-17/C-9, C-10, C-11 allowed the attachments of C-16 to C-6 and C-17 to C-10, which also further supported the presence of a pentadecatrien-2,14-dione moiety.

A 3*E* geometry was assigned via its large $^3J_{3,4}$ coupling constant (16.5 Hz). The NOESY correlations of H-5/H-7 and H-11/H-17 revealed that the geometries of the C-5 and C-10 olefins in **2** were 5*E* and 10*Z*. Compound **2** was characterized as (3*E*,5*E*,10*Z*)-7-hydroxy-6, 10-dimethyl-pentadecatrien-2,14-dione. The optical rotation of **2** that is close to 0 suggested **2** is a pair of enantiomers, which was proved by its separation on HPLC using a chiral chromatography column.

Compound **3** was obtained as a yellow oil. The molecular formula, $C_{21}H_{24}O_5$, was established from its proton adduct ion in the HRESIMS (357.16873 [M + H]$^+$, calcd. 357.1702), which was also supported by NMR data. Analysis of the ^{1}H-NMR spectrum (Table 1) of **3** revealed the presence of a set of ABX system aromatic protons at δ_H 6.40 (d, *J* = 2.4 Hz, H-3'), 6.42 (dd, *J* = 8.4, 2.4 Hz, H-5'), and 7.25 (d, *J* = 8.4 Hz, H-6'), two *ortho*-coupled doublet aromatic protons at δ_H 6.79 (d, *J* = 8.4 Hz, H-5) and 6.30 (d, *J* = 8.4 Hz, H-6), and a set of aliphatic proton signals at δ_H 5.27 (dd, *J* = 9.6, 1.8 Hz, H-2), 2.20 (m, H-3a), 1.85 (m, H-3b), 2.86 (m, H-4a), 2.63 (m,H-4b), suggesting a flavan skeleton for **3**, which was consistent with the ^{13}C-NMR data. The proton signals at δ_H 3.73 (dd, *J* = 7.2, 6.0 Hz, H-3''), 2.92 (dd, *J* = 17.4, 7.8 Hz, H-4''a), 2.54 (dd, *J* = 17.4, 5.4 Hz, H-4''b), 1.31 (s, 3H) and 1.22 (s, 3H), in combination with ^{13}C-NMR signals at δ_C at 77.4 (C-2'''), 70.5 (C-3'''), 27.3 (C-4'''), and 25.8, 20.8 (-Me) showed the presence of a 3''-hydroxyprenyl residue forming a furan ring with a hydroxyl group. The correlations of H-3'', H-4''a, H-4''b/C-8 confirmed the isoprenyl substituent was located at C-8, cyclizing onto 7-hydroxyl group. Accordingly, the structure of **3** was assigned as shown in Figure 1.

The optical rotation of **3** that was close to 0, suggesting **3** is a pair of enantiomers, which was proved by the separation on HPLC using a chiral chromatography column.

Compound **4** was obtained as a yellow oil. The IR spectrum of **4** showed the presence of hydroxy (3347 cm^{-1}), carbonyl (1726 cm^{-1}) and aromatic (1606 and 1466 cm^{-1}) functional groups. The molecular formula, $C_{25}H_{30}O_{13}$, was determined from its sodium adduct ion in the HRESIMS (561.1583 [M + Na]$^+$, calcd. 561.1579) and also supported by the NMR spectroscopic data. The ^{1}H-NMR (Table 1) spectrum showed signals attributable to a monosubstituted aromatic ring at δ_H 7.47 (2H, m, H-2', 6') and 7.36 (3H, m, H-3', 4', 5'), and an oxymethylene group at δ_H 5.33 (d, *J* = 12.6 Hz, H-7'a) and 5.22 (d, *J* = 12.6 Hz, H-7'b), revealing the presence of a benzyl group in combination with the correlations of H-7'/C-2',6' in HMBC spectrum. An ABC spin system attributed to anomeric protons at δ_H 6.65 (d, *J* = 8.4 Hz, H-3), 7.18 (dd, *J* = 8.4 Hz, H-4), and 6.55 (d, *J* = 8.4 Hz, H-5). The correlations of H-3, H-5/C-7 (δ_C 165.8) in HMBC spectrum showed the presence of a 2,6-bisubstituted benzoyl moiety. Together with the coupling patterns of oxymethylene and oxymethine protons resonating between δH 3.09 and 4.84 indicated the presence of a glucopyranosyl and a apiofuranosyl units. The correlations of H-7'/C-7, H-1''/C-2, H-1'''/C-6'' in HMBC spectrum determined a moiety of apiofuranosyl(1→6)-glucopyranose was attached to C-2 of benzoyl moiety and the benzyl group was connected to C-7of the benzoyl moiety. A β-anomeric configuration for the glucosyl unit was assigned via its large $^3J_{1'',2''}$ coupling constant (7.2 Hz). The β-configuration for apiofuranosyl unit was assigned via its $^3J_{1''',2'''}$ coupling constant (3.0 Hz) and the chemical shift of anomeric carbon (δ_C 109.4) [6,18]. The D-configurations of glucopyranosyl and a apiofuranosyl units were determined by gas chromatography (GC) analysis of the trimethylsilyl L-cysteine derivatives after acid hydrolysis of **4**. On the basis of the above data, **4** was characterized as benzyl 2-*O*-[β-D-apiofuranosyl(1→6)-β-D-glucopyranosyl]-2,6-dihydroxy-benzoate.

The known compound was identified as (9*R*)-hydroxyl-(10*E*,12*Z*,15*Z*)-octadecatrienoic acid (**5**) by NMR analysis and comparison with literature data [19].

2.2. Aldose Reductase Inhibitory Effects of **2** *and* **5** *and Neuroprotective Effects of Compounds* **1–5**

The aldose reductase inhibitory and neuroprotective bioactivities of compounds **1–5** were assessed. Compounds **2** and **5** possessed inhibition activities against aldose reductase, with IC_{50} values of 4.33 μM and 6.0 μM compared with the potent AR inhibitor epalrestat (IC_{50} 1.88×10^{-3} μM) [20]. Compound **3** exhibited neuroprotective activity against PC12 cell damage induced by serum deprivation and **5** appeared to protect against PC12 cell damage caused by nicouline (Table 3), an assay extensively used in screening active agents for Parkinson's disease [21–24]. Compounds **1–5** were evaluated for their cytotoxic activities against eight human cancer cell lines (human colon carcinoma cell line HCT-8, hepatocellular carcinoma cell line Bel-7402, human renal cell carcinoma cell line KETR3, Human cervical carcinoma cell line HELA, human gastric cancer cell line BGC-823, human ovarian carcinoma cell line A2780, human breast cancer cell line MCF-7, and human lung carcinoma cell line A549) by means of the MTT assay [25], using paclitaxel and 5-fluouracil as positive controls. Nevertheless, all the isolated compounds resulted to be inactive.

Table 3. Neuroprotective Effects of Compounds **1–5** at concentration of 10^{-5} M (means ± SD, $n = 6$).

Sample	Serum Deprivation (%)	Nicouline 4 μM (%)
control	100.0 ± 3.7	100.0 ± 1.4
model	41.4 ± 3.8 ###	74.3 ± 1.4 ###
1	50.2 ± 12.7	78.6 ± 2.9
2	66.2 ± 12.6	78.4 ± 2.0 *
3	59.8 ± 2.7***	75.7 ± 3.0
4	50.9 ± 7.8	76.9 ± 3.6
5	70.2 ± 16.1	86.2 ± 7.6 ***

$p < 0.001$ vs. control,* $p < 0.05$,** $p < 0.001$, *** $p < 0.0001$ vs. model.

2.3. Discussion

We found it was helpful to subject the chloroform fraction of the ethanol extract to flash silica gel column chromatography repeatedly before isolation to remove pigments. The 80% MeOH-H_2O mobile phase of Sephadex LH-20 column chromatography worked well for all kinds of structures in our study. We obtained two pairs of enantiomers, (**2a,2b**) and (**3a,3b**). Yang et al. reported an interesting phenomenon whereby the *R* and *S* configurations of C-2 of a similar flavan were interconvertible, which (**3a,3b**) may be subject to [5]. In addition, considering the wide range of examples that the isomers of enantiomers showed differences in pharmacological processes, further separation and research of the two pairs of enantiomers is needed.

Polyphenols from *Morus* plants have indicated extensive antioxidative activities, especially the kind of Diels-Alder type adducts [8,26]. Oxidative stress played a key role in neurodegenerative diseases, which implied the potential of polyphenols from *Morus* plants against neurodegenerative disorders [27]. Unfortunately, most of the previous studies on *Morus* polyphenols had been focused on their anti-oxidant properties. More efforts are suggested to explore the neuroprotective action of constituents of *Morus* plants.

3. Materials and Methods

3.1. Plant Material

The leaves of *Morus alba* L. were collected in the Anding Mulberry Garden (Beijing, China), in July 2011, and identified by Prof. Lin Ma (Institute of Materia Medica, Chinese Academy of Medical Sciences & Peking Union Medical College, Beijing, China). A voucher specimen (No. ID-S-2543) has been deposited at the Herbarium of Institude of Materia Medica, Chinese Academy of Medical Sciences & Peking Union Medical College.

3.2. General Experimental Procedures

Optical rotations were measured with a P-2000 polarimeter (Jasco, Tokyo, Japan) and UV spectra with a Jasco V-650 spectrophotometer. ECD spectra were measured on a Jasco J-815 spectrometer. IR spectra were recorded on a model 5700 spectrometer (Nicolet, Madison, SD, USA) by an FT-IR microscope transmission method. NMR measurements were performed using VNS-600 (Varian Medical Systems, Inc., Palo Alto, CA, USA), Mercury-300 (Varian Medical Systems, Inc., Palo Alto, CA, USA), Bruker-AV-III-500 (Bruker Corporation, Karlsruhe, Germany), and Inova-500 (Varian Medical Systems, Inc., Palo Alto, CA, USA) spectrometers. ESIMS was performed on Agilent 1100 Series LC/MSD Trap SL mass spectrometer and HRESIMS data were obtained using an Agilent 6520 Accurate-Mass Q-TOF LC/MS (Agilent Technologies, Ltd., Santa Clara, CA, USA). Gas chromatography (GC) was operated on Agilent 7890A system. HPLC was performed on a Lumtech instrument (Lumiere Tech Ltd. Beijing, China) equipped with a 500 ELSD detector (Alltech, Deerfield, IL, USA) and a YMC-Pack ODS-A column (250 × 20 mm, 5 μm, YMC, Tokyo, Japan). Silica gel (200−300 mesh, Qingdao Marine Chemical Factory, Qingdao, China), Sephadex LH-20 (GE), and ODS (50 μm, YMC) were used for column chromatography. TLC was carried out with GF254 plates (Qingdao Marine Chemical Factory). Spots were visualized by spraying with 10% H_2SO_4 in EtOH followed by heating.

3.3. Cell Lines, Chemicals and Biochemicals

PC12 cells (adrenal gland; pheochromocytoma) were purchased from the American Type Culture Collection (Manassas, VA, USA). Dimethyl sulphoxide (DMSO), nicouline, more commonly known as rotenone, and 3-(3,4-dimehylthiazol-2-yl)-2,5-diphenyl-tetrazolium bromide (MTT) were obtained from Sigma (St. Louis, MO, USA). Dulbecco's Modified Eagle's Medium (DMEM), fetal bovine serum (FBS), and horse serum were purchased from Gibco BRL (New York, NY, USA). Epelrestat was purchased from Dayin Marine Bio-Pharmaceutical Co., Ltd. (Rongcheng, Shandong, China). NADPH-Na4, paclitaxel and 5-fluorouracil were purchased from Sigma-Aldrich (Beijing, China). All other chemicals were of analytical grade and were commercially available.

3.4. Extraction and Isolation

Air-dried leaves of *Morus alba* L. (30 kg) were exhaustively extracted with 95% aqueous EtOH (3 × 100 L, 2 h) at reflux. The combined extracts were concentrated under reduced pressure to dryness. The residue (2.9 kg) was subjected to column chromatography on silica gel and eluted with petroleum ether, chloroform, acetone and methanol. The acetone residue (423 g) was subjected to D101 macroporous resins column chromatography by a gradient elution with EtOH/H_2O (0:100, 30:70, 60:40, 95:5) to yield four fractions (fractions A–D). The separation of fraction B (120 g) was carried out on silica gel column chromatography eluted with $CHCl_3$/MeOH (10:1–3:1) to provide three subfractions B1–B3. Subfraction B2 (80 g) was further purified by MPLC (ODS, 50 μm, YMC) and eluted with 15, 35, 55, 75 and 100% MeOH−H_2O, to afford 40 subfractions. Fraction B2-18 (42 mg) was purified by preparative HPLC using 20% MeCN−H_2O (8 mL/min) as the mobile phase to yield compound **1** (4 mg) and fraction B2-20 (20 mg) was purified by preparative HPLC using 25% MeCN−H_2O (8 mL/min) to yield compound **4** (18 mg). The chloroform residue (355 g) was subjected to silica gel column chromatography by a gradient elution with petroleum ether/acetone (100:0, 95:5, 90:10, 80:20, 70:30, 60:40) to yield six fractions (fractions E–M). The separation of fraction M (38 g) was carried out by MPLC (ODS, 50 μm, YMC) and eluted with 5, 15, 35, 55, 75, and 100% MeOH−H_2O, to afford 25 subfractions. Fraction M-7 (100 mg) was purified by preparative HPLC using 20% MeOH−H_2O (8 mL/min) as the mobile phase to yield compound **5** (45 mg). The separation of fraction N (20 g) was carried out by MPLC (ODS, 50 μm, YMC) and eluted with 10, 30, 55, 75, and 100% MeOH−H_2O, to afford 28 subfractions. Fraction N-6 (50 mg) was purified by preparative HPLC using 30% MeOH−H_2O (8 mL/min) as the mobile phase to yield compound **2** (10 mg). The separation of fraction g (139 g) was carried out by MPLC (ODS, 50 μm, YMC) and eluted with 15, 35, 55, 75, 85, and 100% MeOH-H2O, to afford 38 subfractions. Fraction G-10 (385 mg) was subjected to fractionation using Sephadex LH-20 column chromatography (80% MeOH-H2O) to provide 40 subfractions. Fraction G-10-18 (42 mg) was purified by the preparative HPLC using 35% MeCN−H2O (8 mL/min) as the mobile phase to yield compound **3** (15 mg).

3.5. Characterization

Moralsin (**1**): white powder; $[\alpha]_D^{20}$ −72 (*c* 0.19, MeOH); UV(MeOH): λ_{max} (log ε) 228 (2.10) nm; IR ν_{max} 3402, 2957, 1760, 1606, 1080 cm^{-1}; ^{1}H-NMR (DMSO-d_6, 500 MHz) and ^{13}C-NMR (DMSO-d_6, 125 MHz) see Table 2; positive-ion HRESIMS *m/z* 433.1826 [M + Na]$^+$ (calcd. 433.1833).

(3E,5E,10Z)-7-Hydroxy-6,10-dimethyl-pentadecatrien-2,14-dione (**2**): yellow oil; UV(MeOH): λ_{max} (log ε) 207 (4.31) nm; 247 (0.11) nm; IR ν_{max} 3347, 2926, 1726, 1606, 1466, 1051, 1025 cm^{-1}; ^{1}H-NMR (DMSO-d_6, 500 MHz) and ^{13}C-NMR (DMSO-d_6, 125 MHz) see Tables 1 and 2; positive-ion HRESIMS *m/z* 561.1583 [M + Na]$^+$ (calcd. 561.1579).

2'-Hydroxy-4'-methoxyl-2H-(2''', 2'''-dimethyl-3'''-hydroxy)-pyran-(5''',6''':8,7)-flavane (**3**): yellow oil; UV(MeOH): λ_{max} (logε) 206 (3.90) nm, 280 (2.80) nm; IR ν_{max} 3372, 1718, 1602 cm^{-1}; ^{1}H-NMR (MeOH-d_4, 600 MHz) Table 1; ^{13}C-NMR (MeOH-d_4, 150 MHz) Table 2; positive-ion HRESIMS *m/z* 357.16873 [M + H]$^+$ (calcd. for $C_{21}H_{25}O_5$, 357.1702).

Benzyl 2-O-[β-D-apiofuranosyl(1→6)-β-D-glucopyranosyl]-2,6-dihydroxybenzoate (**4**): yellow oil; UV(MeOH): λ_{max} (log ε) 207 (4.32) nm, 247 (0.1) nm; IR ν_{max} 3347, 2926, 1726, 1606, 1466, 1051, 1025 cm^{-1}; ^{1}H-NMR (DMSO-d_6, 300 MHz) and ^{13}C-NMR (DMSO-d_6, 125 MHz) see Tables 1 and 2; positive-ion HRESIMS *m/z* 561.1583 [M + Na]$^+$ (calcd. 561.1579).

(9R)-Hydroxy-(10E,12Z,15Z)-octadecatrienoic acid (**5**): Yellow powder, $[\alpha]_D^{20}$ −3.04° (0.58 CHCl3); ESIMS *m/z* 317.2 [M + Na]$^+$, ^{1}H-NMR (DMSO-d_6, 500 MHz) δ: 2.18 (2H, t, 9.6), 1.47 (2H, m, H-3), 1.38 (2H, m, H-8), 2.04 (2H, m, H-17), 0.92 (3H, t, *J* = 9.6 Hz, H-18), 3.97 (1H, m, H-9), 5.66 (1H, dd, *J* = 15.5, 6.0 Hz, H-10), 6.44 (1H, dd, *J* = 15.5, 11.5 Hz, H-11), 5.26-5.40 (3H, m, H-12, 15, 16), 5.96 (1H, t, *J* = 11.0 Hz, H-13), 2.88 (2H, dd, *J* = 11.0, 7.5, Hz, H-14), 1.24 (6H, m, H-4-6). ^{13}C-NMR (DMSO-d_6, 125MHz) δ: 174.5 (C-1), 138.6 (C-10), 131.7 (C-16), 128.7 (C-13), 128.3 (C-12), 126.7 (C-15), 123. 5 (C-11), 70.44 (C-9), 37.2 (C-8), 33.7 (C-2), 28.5, 28.8, 28.9 (C-4-6), 25.5 (C-14), 24.9 (C-7), 24.5 (C- 3), 20.0 (C-17), 14.1 (C-18).

3.6. Acid Hydrolysis of the Saponins and Determination of the Absolute Configuration of the Monosaccharides

Compound **1** (2 mg) was hydrolyzed in 2 M HCl/H2O at 80 °C for 2 h. The residue was reacted sequentially with L-cysteine methyl ester hydrochloride and N-trimethylsilylimidazole. The resulting monosaccharide N-trimethylsilylimidazole derivatives were analyzed by GC. D-Glucose was confirmed by comparison of the retention time of the derivatives with those of authentic sugars derivatized in a similar way, which showed retention times of 27.93 min. The constituent sugars of compounds **4** were identified by the same method as **1**. Retention times of authentic samples were detected at 17.87 min (D-apiofuranose) and at 27.93 (D-glucose). The reaction and GC conditions were as described in the literature [28].

3.7. Aldose Reductase Assay

The assay was operated in 96 well culture plate. A 100 μL mixture that contained 10 mM DL-glyceraldehyde, 0.16 mM NADPH-Na4 and aldose reductase in 0.1 M sodium phosphate buffer (pH 6.2), with or without test compounds was prepared at 0 °C. Appropriate blank were employed for corrections. The assay mixture was incubated at 25 °C. After 10 min of incubation, the plate was immediately cooled at −20 °C for 5 min to stop the reaction. The change in the absorbance at 340 nm due to NADPH oxidation was measured in a plate reader [29].

Molecules **2018**, *23*, 1018

3.8. Neuroprotection Bioassays

The PC12 cells were cultured in DMEM medium supplemented with 5% horse serum and 5% fetal bovine serum. Then, 100 µL of cells with an initial density of 5×10^4 cells/mL was seeded in each well of a poly-L-lysine-coated, 96-well culture plate and precultured for 24 h. The medium was then replaced by different fresh medium including the control (complete medium), the model (complete medium with 4 µM rotenone or serum free medium), and the sample (the test compounds with different drug concentrations, 10, 1, and 0.1 µM, were added to the aforementioned model medium), and the cells were cultured for 48 h. Then, 10 µL of MTT (5 mg/mL) was added to each well. After incubation for 4 h, the medium was removed, and 150 µL of DMSO was added to dissolve the formazan crystals. The optical density (OD) of the PC12 cells was measured on a microplate reader at 570 nm [30].

4. Conclusions

The plants of genus *Morus* have been extensively investigated for their medicinal constituents and a series of unique structures were characterized [31–37]. This studied focused on the fractions that had been rarely researched before and generated four new structures with aldose reductase inhibitory or neuroprotective activities. The known molecule **5** was isolated for the first time from the genus *Morus* and its neuroprotective activity reported for the first time. The compounds reported here provide new potential aldose reductase inhibitory or neuroprotective agents for further research.

Supplementary Materials: The following are available online.

Author Contributions: D.-q.Y. and R.-y.C. conceived and designed the experiments; X.-y.C. performed the experiments and wrote the paper; T.Z. and X.W. contributed data analysis, M.T.H. and J.K. contributed writing.

Acknowledgments: The author is grateful to Li Li who helped measure the CD spectra, Zhu-Fang Shen for the aldose reductase assay and Nai-Hong Chen for the neuroprotection bioassay. This work was supported by CAMS Innovation Fund for Medical Sciences (CIFMS) (2016-I2M-1-010).

Conflicts of Interest: The authors declare no conflict of interest.

References

1. Wu, Z.Y.; Raven, P.H.; Hong, D.Y. *Flora of China*, 2003 ed.; Science Press: Beijing, China, 2003; Missouri Botanical Garden Press: St. Louis, MI, USA, 2003; Volume 5, pp. 22–26.
2. Gao, L.; Li, Y.D.; Zhu, B.K.; Li, Z.Y.; Huang, L.B.; Li, X.Y.; Wang, F.; Ren, F.C.; Liao, T.G. Two new prenylflavonoids from *Morus alba*. *J. Asian Nat. Prod. Res.* **2017**, *23*, 117–121. [CrossRef] [PubMed]
3. Chinese Pharmacopoeia Commission. *Pharmacopoeia of the People's Republic of China*; China Medical Science Press: Beijing, China, 2010; Volume 1, pp. 279–281.
4. Pel, P.; Chae, H.S.; Nhoek, P.; Kim, Y.M.; Chin, Y.W. Chemical Constituents with Proprotein Convertase Subtilisin/Kexin Type 9 mRNA Expression Inhibitory Activity from Dried Immature *Morus alba* Fruits. *J. Agric. Food Chem.* **2017**, *65*, 5316–5321. [CrossRef] [PubMed]
5. Yang, Z.Z.; Wang, Y.C.; Wang, Y.; Zhang, Y. Bioassay-guided screening and isolation of α-glucosidase and tyrosinase inhibitors from leaves of *Morus alba*. *F. Food Chem.* **2012**, *131*, 617–625. [CrossRef]
6. Zhao, G.J.; Xi, Z.X.; Chen, W.X.; Li, X.; Sun, L.; Sun, L.N. Chemical constituents from Tithonia diversifolia and their chemotaxonomic significance. *Biochem. Syst. Ecol.* **2012**, *44*, 250–254. [CrossRef]
7. Li, H.X.; Jo, E.; Myung, C.S.; Kim, Y.H.; Yang, S.Y. Lipolytic effect of compounds isolated from leaves of mulberry (*Morus alba* L.) in 3T3-L1 adipocytes. *Nat. Prod. Res.* **2017**. [CrossRef] [PubMed]
8. Yang, Y.; Tan, Y.X.; Chen, R.Y.; Kang, J. The latest review on the polyphenols and their bioactivities of Chinese *Morus* plants. *J. Asian Nat. Prod. Res.* **2014**, *16*, 690–702. [CrossRef] [PubMed]
9. Sangshetti, J.; Chouthe, R.; Sakle, N.; Gonjari, I.; Shinde, D. Aldose Reductase: A Multi-disease Target. *Curr. Enzym. Inhib.* **2014**, *10*, 2–12. [CrossRef]
10. Shen, B.; Vetri, F.; Mao, L.; Xu, H.L.; Paisansathan, C.; Pelligrino, DA. Aldose reductase inhibition ameliorates the detrimental effect of estrogen replacement therapy on neuropathology in diabetic rats subjected to transient forebrain ischemia. *Brain Res.* **2010**, *1342*, 118–126. [CrossRef] [PubMed]

11. Goldin, A.; Beckman, J.A.; Schmidt, AM.; Creager, MA. Advanced Glycation End Products Sparking the Development of Diabetic Vascular Injury. *Circulation* **2006**, *114*, 597–605. [CrossRef] [PubMed]

12. Thomas, M.C. Advanced glycation end products. *Contrib. Nephrol.* **2011**, *170*, 66–74. [CrossRef] [PubMed]

13. Ishiguro, H.; Nakaigawa, N.; Miyoshi, Y.; Fujinami, K.; Kubota, Y.; Uemura, H. Receptor for advanced glycation end products (RAGE) and its ligand, amphoterin are overexpressed and associated with prostate cancer development. *Prostate* **2005**, *64*, 92–100. [CrossRef] [PubMed]

14. Kuniyasu, H.; Oue, N.; Wakikawa, A.; Shigeishi, H.; Matsutani, N.; Kuraoka, K.; Ito, R.; Yokozaki, H.; Yasui, W. Expression of receptors for advanced glycation end-products (RAGE) is closely associated with the invasive and metastatic activity of gastric cancer. *J. Pathol.* **2002**, *196*, 163–170. [CrossRef] [PubMed]

15. Ramana, K.V. ALDOSE REDUCTASE: New Insights for an Old Enzyme. *Biomol. Concepts* **2011**, *2*, 103–114. [CrossRef] [PubMed]

16. Srivastava, K.S.; Yadav, U.; Reddy, A.; Saxena, A.; Tammali, R.; Mohammad, S.; Ansari, H.N.; Bhatnagar, A.; Petrash, J.M.; Srivastava, S.; et al. Aldose reductase inhibition suppresses oxidative stress-induced inflammatory disorders. *Chem. Biol. Interact.* **2011**, *191*, 330–338. [CrossRef] [PubMed]

17. Harada, N.; Nakanishi, K. Exciton chirality method and its application to configurational and conformational studies of natural products. *Acc. Chem. Res.* **1972**, *5*, 257–263. [CrossRef]

18. Kitagawa, I.; Hori, K.; Sakagami, M. Saponin and Sapogenol. XLIX. On the Constitutents of the Roots of Glycyrrhiza inflata BATALIN from Xinjiang, China. Characterization of Two Sweet Oleanane-Type Triterpene Oligoglycosides, Apioglycyrrhizin and Araboglycyrrhizin. *Chem. Pharm. Bull.* **1993**, *41*, 1350–1357. [CrossRef] [PubMed]

19. McLean, S.F.; Reynolds, W.; Tinto, W.F.; Chan, W.R.; Shepherd, V. Complete ^{13}C and ^{1}H Spectral Assignments of Prenylated Flavonoids and a Hydroxy Fatty Acid from the Leaves of Caribbean *Artocarpus communis*. *Magn. Reson. Chem.* **1996**, *34*, 719–722. [CrossRef]

20. Ramirez, M.A.; Borja, N.L. Epalrestat: An aldose reductase inhibitor for the treatment of diabetic neuropathy. *Pharmacotherapy* **2008**, *28*, 646–655. [CrossRef] [PubMed]

21. Yoon, I.S.; Au, Q.; Barber, J.R.; Ng, S.C.; Zhang, B. Development of a high-throughput screening assay for cytoprotective agents in rotenone-induced cell death. *Anal. Biochem.* **2010**, *407*, 205–210. [CrossRef] [PubMed]

22. Bansol, P.K.; Deshmukh, R. *Animal Models of Neurological Disorders*; Springer Nature: Singapore, 2017; p. 30, ISBN 978-981-10-5980-3.

23. Javed, H.; Azimullah, S.; Abul Khair, S.B.; Ojha, S.; Haque, M.E. Neuroprotective effect of nerolidol against neuroinflammation and oxidative stress induced by rotenone. *BMC Neurosci.* **2016**, *17*, 58. [CrossRef] [PubMed]

24. Ramkumar, M.; Rajasankar, S.; Gobi, V.V.; Dhanalakshmi, C.; Manivasagam, T.; Justin Thenmozhi, A.; Essa, M.M.; Kalandar, A.; Chidambaram, R. Neuroprotective effect of Demethoxycurcumin, a natural derivative of Curcumin on rotenone induced neurotoxicity in SH-SY 5Y Neuroblastoma cells. *BMC Complement. Altern. Med.* **2017**, *17*, 217. [CrossRef] [PubMed]

25. Ni, G.; Zhang, Q.J.; Zheng, Z.F.; Chen, R.Y.; Yu, D.Q. 2-Arylbenzofuran Derivatives from *Morus cathayana*. *J. Nat. Prod.* **2009**, *72*, 966–968. [CrossRef] [PubMed]

26. Dai, S.J.; Wu, Y.; Wang, Y.H.; He, W.Y.; Chen, R.Y.; Yu, D.Q. New Diels-Alder type adducts from *Morus macroura* and their anti-oxidant activities. *Chem. Pharm. Bull.* **2004**, *52*, 1190–1193. [CrossRef] [PubMed]

27. Kim, G.H.; Kim, J.E.; Rhie, S.J.; Yoon, S. The Role of Oxidative Stress in Neurodegenerative Diseases. *Exp. Neurobiol.* **2015**, *24*, 325–340. [CrossRef] [PubMed]

28. Liang, D.; Hao, Z.Y.; Zhang, G.J.; Zhang, Q.J.; Chen, R.Y.; Yu, D.Q. Cytotoxic Triterpenoid Saponins from Lysimachia clethroides. *J. Nat. Prod.* **2011**, *74*, 2128–2136. [CrossRef] [PubMed]

29. Kang, J.; Tang, Y.; Liu, Q.; Guo, N.; Zhang, J.; Xiao, Z.; Chen, R.; Shen, Z. Isolation, modification, and aldose reductase inhibitory activity of rosmarinic acid derivatives from the roots of Salvia grandifolia. *Fitoterapia* **2016**, *112*, 197–204. [CrossRef] [PubMed]

30. Zhang, F.; Yang, Y.N.; Song, X.Y.; Shao, S.Y.; Feng, Z.M.; Jiang, J.S.; Li, L.; Chen, N.H.; Zhang, P.C. Forsythoneosides A-D, Neuroprotective Phenethanoid and Flavone Glycoside Heterodimers from the Fruits of Forsythia suspensa. *J. Nat. Prod.* **2015**, *78*, 2390–2397. [CrossRef] [PubMed]

31. Ercisli, S.; Orhan, E. Chemical composition of white (*Morus alba*), red (*Morus rubra*) and black (*Morus nigra*) mulberry fruits. *Food Chem.* **2007**, *103*, 1380–1384. [CrossRef]
32. Butta, M.S.; NazirbM, A.; Sultana, M.T.; Schroën, K. *Morus alba* L. nature's functional tonic. *Trends Food Sci. Technol.* **2008**, *19*, 505–512. [CrossRef]
33. Arabshahi-Delouee, S.; Urooj, A. Antioxidant properties of various solvent extracts of mulberry (*Morus indica* L.) leaves. *Food Chem.* **2007**, *102*, 1233–1240. [CrossRef]
34. Chang, L.W.; Juang, L.J.; Wang, B.S.; Wang, M.Y.; Tai, H.M.; Hung, W.J.; Chen, Y.J.; Huang, M.H. Antioxidant and antityrosinase activity of mulberry (*Morus alba* L.) twigs and root bark. *Food Chem. Toxicol.* **2011**, *49*, 785–790. [CrossRef] [PubMed]
35. Singab, A.N.; El-Beshbishy, H.A.; Yonekawa, M.; Nomura, T.; Fukai, T. Hypoglycemic effect of Egyptian *Morus alba* root bark extract: Effect on diabetes and lipid peroxidation of streptozotocin-induced diabetic rats. *J. Ethnopharmacol.* **2005**, *100*, 333–338. [CrossRef] [PubMed]
36. Doi, K.; Kojima, T.; Makino, M.; Kimura, Y.; Fujimoto, Y. Studies on the Constituents of the Leaves of *Morus alba* L. *Chem. Pharm. Bull.* **2001**, *49*, 151–153. [CrossRef] [PubMed]
37. Oku, T.; Yamada, M.; Nakamura, M.; Sadamori, N.; Nakamura, S. Br Inhibitory effects of extractives from leaves of *Morus alba* on human and rat small intestinal disaccharidase activity. *J. Nutr.* **2006**, *95*, 933–938. [CrossRef]

Sample Availability: Sample of the compound **5** is available from the authors.

molecules

MDPI

Article

MS/MS-Guided Isolation of Clarinoside, a New Anti-Inflammatory Pentalogin Derivative

Coralie Audoin [1], Adam Zampalégré [1], Natacha Blanchet [1], Alexandre Giuliani [2,3], Emmanuel Roulland [4], Olivier Laprévote [4,5] and Grégory Genta-Jouve [4,*]

[1] Laboratoires Clarins, 5 rue Ampère, 95300 Pontoise, France; coralie.audoin@clarins.com (C.A.); adam.zampalegre@clarins.com (A.Z.); natacha.blanchet@clarins.com (N.B.)
[2] DISCO Beamline, Synchrotron SOLEIL, 91192 Gif-sur-Yvette, France; alexandre.giuliani@synchrotron-soleil.fr
[3] UAR1008, CEPIA, INRA, 44316 Nantes, France
[4] C-TAC, UMR 8638 CNRS, Faculté de Pharmacie de Paris, Université Paris Descartes, Sorbonne Paris Cité, 4 Avenue de l'Observatoire, 75006 Paris, France; emmanuel.roulland@parisdescartes.fr (E.R.); olivier.laprevote@parisdescartes.fr (O.L.)
[5] Department of Biochemistry, Hôpital Européen Georges Pompidou, AH-HP, 75015 Paris, France
* Correspondence: gregory.genta-jouve@parisdescartes.fr; Tel.: +33-173-531-585

Received: 30 March 2018; Accepted: 19 May 2018; Published: 22 May 2018

Abstract: Re-investigation of the chemical composition of the annual plant *Mitracarpus scaber* Zucc. led to the identification of clarinoside, a new pentalogin derivative containing a rare quinovose moiety, and the known compound harounoside. While the planar structure was fully determined using tandem mass spectrometry (MS) and quantum mechanics (QM) calculations, the tridimensional structure was unravelled after isolation and NMR analysis. The absolute configuration was assigned by comparison of experimental and theoretical synchrotron radiation circular dichroism spectra. Both compounds were tested for anti-inflammatory activity, and compound **1** showed the ability to inhibit the production of interleukin-8 (Il-8) with an IC_{50} value of 9.17 μM.

Keywords: *Mitracarpus scaber* Zucc.; pentalogin; anti-inflammatory; MS/MS; Il-8

1. Introduction

Mass spectrometry (MS) has become a very convenient technique for the targeted search of new bioactive metabolites [1,2], and the recent introduction of the Global Natural Product Social Molecular Networking (GNPS) Web platform (http://gnps.ucsd.edu) has enabled the quick and automatic spectral mining of MS/MS spectra [3]. In our ongoing research for bioactive compounds, we decided to re-investigate the chemical composition of *Mitracarpus scaber* Zucc. using a MS/MS-guided approach. *M. scaber* is an annual plant used in African traditional medicine endowed with antifungal, antimicrobial and anti-inflammatory properties [4,5]. Indeed, in West Africa, the leaves of *M. scaber* are widely used for headache, toothache, amenorrhea, dyspepsia, hepatic diseases, venereal diseases, leprosy, and for the treatment of skin diseases such as scabies, infectious dermatitis, and eczema. It is well known to contain phenols [5], flavonoid glycosides [5], furanocoumarines [5], terpenes [6], alkaloids [7], and pentalongin derivatives [8,9]. Herein, we report the identification of clarinoside (**1**), a new pentalongin derivative exhibiting the rare quinovose moiety along with the known harounoside (**2**) (Figure 1). Both compounds were tested for anti-inflammatory activity by evaluating their ability to inhibit the production of interleukin-8 (Il-8).

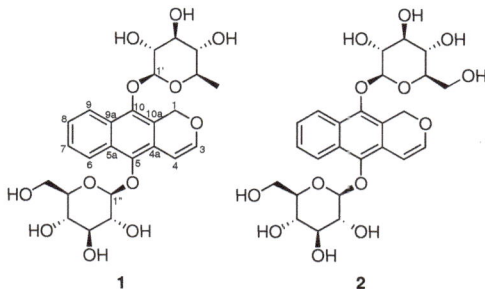

Figure 1. Structures of clarinoside (**1**) and harounoside (**2**).

2. Results

The analysis started with the creation of a molecular network of the ethanolic extract of *M. scaber.* The data-dependent analysis (DDA) LC-MS/MS data were uploaded to the GNPS platform, and a network was generated using the parameters listed in the Materials and Method section below (Figure 2). As an anchor (reference) compound, harounoside (**2**) was used. Its node was quickly "illuminated" using the high-resolution MS data and fragmentation pattern. The *m/z* value at 561.159 corresponding to the [M + Na]$^+$ adduct of **2** was identified, and two diagnostic MS/MS fragments were present on the spectrum (see Supplementary Materials): the first resulting from the cleavage of the glycosidic bond between the aglycone and a glucose moiety at *m/z* 399.1080 [M-Glc + Na]$^+$, and the second at *m/z* 236.0452 resulting from the cleavage of the second O–C between the aglycone and the other glucose [M-2Glc + Na]$^{\bullet+}$. In order to find the structurally related compounds, the cluster was further studied by annotating the edges with *m/z* differences corresponding to known (bio)chemical modifications implemented in the MetaNetter 2 package [10]. Out of the 307 nodes of the network, 1 node directly connected to harounoside (**2**), with a *m/z* difference of −15.995 attracting our attention. According to the biotransformation list available in the MetaNetter package, this difference corresponded to a dehydroxylation.

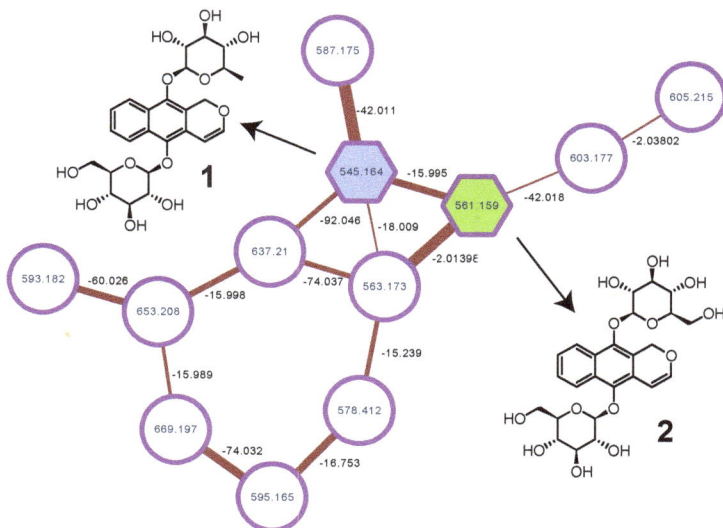

Figure 2. Selected cluster containing clarinoside (**1**) and harounoside (**2**).

Considering the structure of **2**, dehydroxylation could occur at several positions of each of the two sugars; in order to know on which side of the compound **2** the dehydroxylation site was located, the fragmentation of both glucose moieties was studied using quantum mechanics (QM). After examination of the MS/MS spectrum, the sequential losses of two glucoses were observed. The nature of the sugar being the same at C-5 and C-10, the energy level of the two O–C bonds was only related to the position on the aglycone. In order to confirm this hypothesis, the energy profile of the homolytic dissociation was predicted using the B3LYP method at the STO-3G level (see Supplementary Materials). The calculations predicted a difference of ca. +2.5 eV in favor of the O–C1″, indicating that the first fragment observed at *m/z* 399.1080 was related to the loss of one glucose at C-10. This energy difference was very supportive; on the basis of these theoretical results, an energy-resolved mass spectrometry (ERMS) study [11] was undertaken in order to determine the stability of the two O–C bonds (O–C1″ and O–C1′) of compound **2**. After selection of the parent ion at *m/z* 561.16, the intensity of the ion at *m/z* 399.11 was recorded using an increasing value of collision energy (). After the complete extinction of the parent ion (Figure 3A), the daughter ion at *m/z* 399.11 was then fragmented into one major ion at *m/z* 185.04 using the same approach (Figure 3B). As shown in Figure 3, the O–C1′ bond linking the aglycone to the glucose moiety was weaker than the O–C1″ bond, as it required a lower collision energy value for a 50% dissociation (ca. 15 and 17 for O–C1′ and O–C1″, respectively).

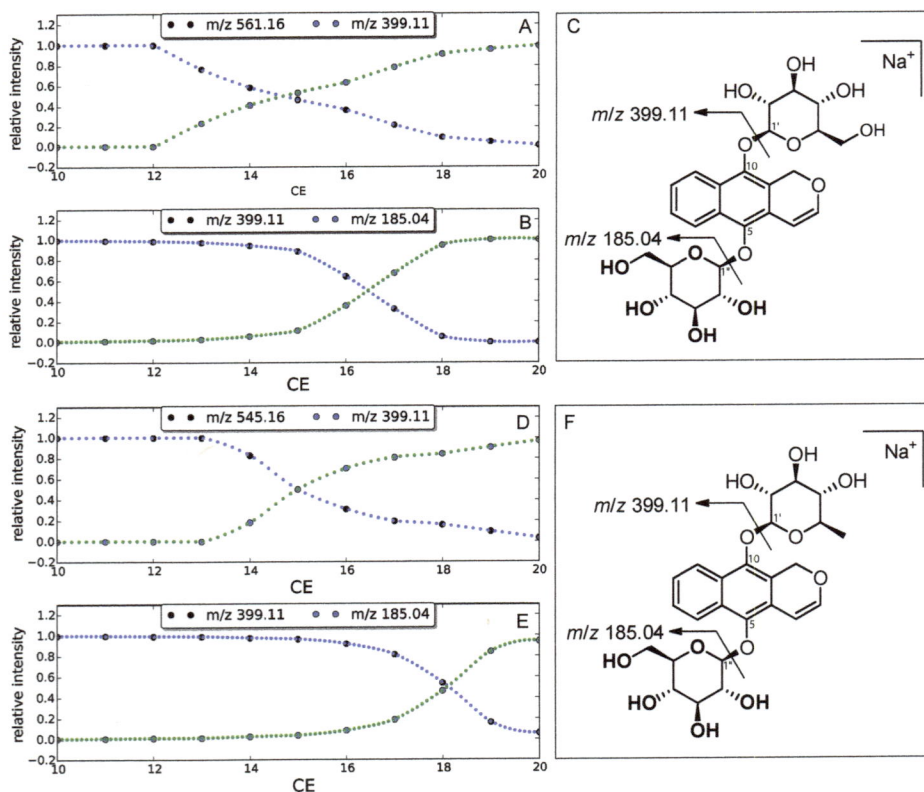

Figure 3. Plot of relative ion current vs collision energy corresponding to *m/z* 561.16 vs. 399.10 (**A**) and *m/z* 399.10 vs. 185.04 (**B**). (**C**) MS/MS fragments of **2**. Plot of relative ion current vs. collision energy corresponding to *m/z* 545.16 vs. 399.10 (**D**) and *m/z* 399.10 vs. 185.04 (**E**). (**F**) MS/MS fragments of **1**.

The experimental data confirmed the QM predicted values, and the same ERMS approach was used for compound **1**. As observed in Figure 3D,E, an increase in the collision energy value was required in order to produce the ion at *m/z* 185.05 (ca. 15 and 18 for O–C1′ and O–C1″, respectively). These results clearly confirmed the position of the deoxyhexose moiety at C-10. The planar structure was further confirmed by the identification of the neutral loss of 146.0605 Da resulting from the difference between the [M + Na]⁺ ion at *m/z* 545.1685 and the fragment at *m/z* 399.1080. In parallel with the loss of 162.05 Da corresponding to a glucose moiety (observed for **2**), the loss of 146.0605 was consistent with a deoxyhexose such as fucose, rhamnose, or quinovose. Unfortunately, and despite the use of a recent methodology to distinguish the mono-saccharides using MS/MS [12], it was not possible to determine the relative stereochemistry of the sugar moieties using MS analysis only; thus the isolation of compound **1** was undertaken.

After a reverse-phase high-performance liquid chromatography (HPLC) purification, 84 mg of compound **1** was obtained, and a full set of NMR experiments was performed. The structure of the aglycone was confirmed by comparison of the ¹H and ¹³C NMR chemical shifts (see Table 1). The nature of the sugar was determined by taking advantage of the newly published methodology by Giner et al. [13], which is based on the acid-promoted hydrolysis of the studied compound performed directly in the deuterated NMR solvent. Looking at the ¹H NMR spectrum, the doublets at δ 4.66 (J = 7.7 Hz) and 4.79 (J = 7.8 Hz) ppm clearly confirmed a glucose and a quinovose moiety (see Supplementary Materials). According to the ERMS data, the quinovose was located at C-10, and this was confirmed by the two ³J coupling between H-1′/C-10 and H-1″/C-5 on the HMBC spectrum. The detailed NMR data are given in the Table 1.

Table 1. ¹H and ¹³C NMR data for **1** at 600 MHz in CD₃OD (δ_H in ppm).

	Clarinoside (1)						Harounoside (2) [8]	
No.	δ_H (Multiplicity, J)	δ_C	No.	δ_H (Multiplicity, J)	δ_C	No.	δ_H	δ_C
1	5.29 (dd, 40.7, 13.8 Hz)	65.2	1′	4.66 (d, 7.8)	106.2	1	5.39; 5.30	65.4
3	6.67 (dd, 14.4, 5.9 Hz)	147.8	2′	3.61 (dd, 9.0, 7.8)	75.9	3	6.68	147.8
4	6.66 (dd 14.4, 5.9 Hz)	102.1	3′	3.38 (t, 9.0)	77.7	4	6.64	102.2
4a		121.6	4′	3.11 (t, 9.0)	73.5	4a		121.7
5		143.3	5′	3.11 (m)	77.8	5		143.4
5a		131.0	6′	1.21 (d, 5.4)	18.1	5a		131.0
6	8.43 (d, 8.2 Hz)	124.7	1″	4.79 (d, 7.8)	106.9	6	8.42	124.7
7	7.44 (dt, 8.2, 1 Hz)	127.0	2″	3.65 (m)	75.8	7	7.43	127.0
8	7.40 (dt, 8.2, 1 Hz)	126.2	3″	3.46 (m)	71.5	8	7.39	126.3
9	8.41 (d, 8.2 Hz)	123.7	4″	3.14 (m)	76.9	9	8.43	123.7
9a		129.1	5″	3.47 (t, 9.0)	78.0	9a		129.1
10		144.9	6″	3.67 (m)	62.7	10		145.0
10a		122.6				10a		122.7

The absolute configuration of compound **1** was determined by comparison of a synchrotron radiation circular dichroism (SRCD) spectrum with a time-dependent density functional theory (TD DFT) theoretical electronic circular dichroism (ECD) spectrum (Figure 4). Unexpectedly, the ECD spectrum was quite complex with four Cotton effects of alternative signs. The calculations were run on the four diasteroisomers, that is, D-Glc/D-Qui, D-Glc/L-Qui, L-Glc/D-Qui, and L-Glc/L-Qui. While the absolute D configuration of the glucose moiety was expected, as it is well known that higher plants produce only this enantiomer [14], the absolute configuration of the quinovose moiety was not obvious because the quinovose can originate from either D-glucose [15] or L-fucose [16].

A very good agreement was observed between the D-Glc/D-Qui theoretical and the experimental spectra (Figure 4). Compound **1** could be named as 5,10-dihydroxy-2*H*-naphtho[2,3-b] -pyran-5-β-D-glucopyranosyl-10-β-D-quinovopyranoside.

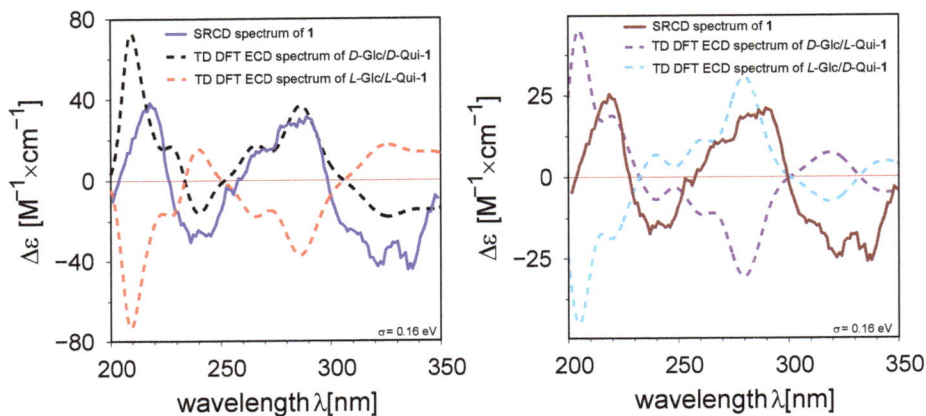

Figure 4. Overlay of synchrotron radiation circular dichroism (SRCD) and TD DFT spectra of **1**.

Both compounds **1** and **2** were tested for anti-inflammatory activity by measuring their ability to inhibit the production of Il-8, one of the key mediators associated with inflammation [17,18]. After exposure to the tumor necrosis factor alpha (TNF-α) at 0.5 ng·mL^{-1} for 24 h, the production of Il-8 was measured and compared to the known anti-inflammatory standard epigallocatechin gallate (EGCG). As shown in Figure 5, the TNF-α induced the production of Il-8 of 398.37 ± 24.09 pg/mg of total protein, while the addition of EGCG at 21.8 µM allowed a return to the basal threshold of 33.01 ± 2.12 pg/mg of total protein.

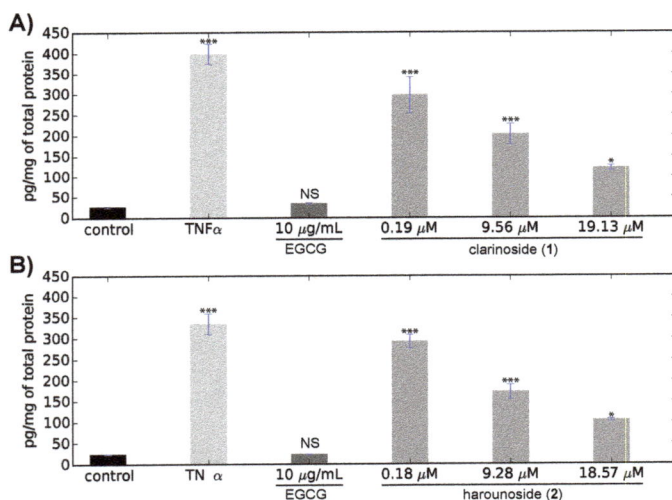

Figure 5. Inhibition of the interleukin-8 (Il-8) production: (**A**) clarinoside (**1**); (**B**) harounoside (**2**). Means $\pm$ SD are shown. NS: not significant; *$p < 0.05$; ***$p < 0.001$.

Although the two compounds were tested during two independent tests, the differences in the measured concentrations (i.e., the production of Il-8 and its inhibition) were observed in both experiments. A good correlation ($R^2 = 0.996$) was observed between the inhibition of the production of Il-8 and the concentration of clarinoside (**1**). An IC$_{50}$ value of 9.17 µM was measured, and a total inhibition of the Il-8 production at 36 µM could be extrapolated. The IC$_{50}$ value of **1** was of the same order of magnitude as that of EGCG (10.9 µM [19]), although it was slightly lower. Interestingly, the IC$_{50}$ value of **2** (9.21 µM) was very similar to that measured for compound **1**, indicating that the structural modification had no impact on its biological activity.

To conclude, this study enabled the rapid identification of one new compound from *M. scaber*. The biological activity evaluation highlighted the ability of compounds **1** and **2** to inhibit the production of Il-8, confirming the importance of *M. scaber* metabolites and their possible uses in cosmetics and personal care products.

3. Materials and Methods

3.1. General Procedure

The preparative HPLC was performed on a VWR LaPrep P110 system using a C-630 Büchi UV detector. NMR spectra acquisition was realized using a 600 MHz Bruker Avance spectrometer equipped with Z-gradients and a triple resonance TXI probe. The signals were referenced in ppm to the residual solvent signals (CD$_3$OD, at δ_H 3.31 and δ_C 49.0). The infrared spectrum was acquired on a Nicolet IS50 FT-IR spectrophotometer. The specific rotation was measured using an Anton Paar MCP150 polarimeter.

3.2. Plant Material

The flowered aerial parts of *M. scaber* were collected in Burkina Faso in the town of Poun and then dried in the same area.

3.3. Extraction and Purification

An ethanolic extraction was performed on a 300 g sample of the dried plant with a plant/solvent ratio of 1/7 yielding 16.5 g of crude extract, which was then directly processed by reverse-phase HPLC with an XBridge Prep C18, 5 µm (OBD 30 × 250 mm) preparative HPLC column. A gradient H$_2$O/MeOH (from 90/10 to 70/30 in 30 min at 100 mL·min^{-1}) was used to afford compounds **1** (84 mg) and **2** (113 mg).

3.4. LC-MS Data Acquisition and Processing

An XEVO-G2 XS QTOF (Waters) equipped with an electrospray ionization (ESI) source was used for the qualitative analysis of the extract. A first screening analysis was performed using the MS^E technology (Waters) on a mass range from 50 to 1500 Da. The optimal ionization-source working parameters were as follows: capillary voltage of 3.0 kV; sampling cone of 40 V; extraction cone of 6.0 V; source temperature of 150 °C; desolvation temperature of 600 °C; cone gas flow of 50 L/h; desolvation gas flow of 1000 L/h. MS/MS data were obtained using a DDA with the same ionization parameters as above using three different collision energies: 10, 20, and 40 V.

3.5. Construction of the Molecular Network

A molecular network was created using the online workflow at GNPS [3]. The data were filtered by removing all MS/MS peaks within ±17 Da of the precursor *m/z*. MS/MS spectra were window filtered by choosing only the top six peaks in the ±50 Da window throughout the spectrum. The data were then clustered with MS Cluster with a parent mass tolerance of 2.0 Da and a MS/MS fragment ion tolerance of 0.5 Da to create consensus spectra. Furthermore, consensus spectra that contained fewer than two spectra were discarded. A network was then created in which edges were filtered to have a cosine score of above 0.7 and more than three matched peaks. Furthermore, edges between two nodes were kept in the network if and only if each of the nodes appeared in each other's respective top 10 most similar nodes. The spectra in the network were then searched against GNPS's spectral libraries. The library spectra were filtered in the same manner as the input data. All matches kept between network spectra and library spectra were required to have a score of above 0.7 and at least six matched peaks.

3.6. Energy-Resolved Mass Spectrometry

The LTQ-Orbitrap XL mass spectrometer (Thermo Scientific (Bremen), Bremen, Germany) was used for the ERMS study. The analysis was performed in positive-ion mode with a mass range of m/z 100–1100. The optimized ESI parameters were set as follows: capillary temperature of 250 °C; sheath gas (nitrogen) flow of 30 arb.; auxiliary gas (nitrogen) flow of 10 arb.; source voltage of 4.25 kV; capillary voltage of 25 V; tube lens voltage of 110 V. The resolution of the Orbitrap mass analyzer was set at 30,000. The isolation width was 2 amu, and the normalized collision energy (CE) was set from 10 to 20. Collision-induced dissociation (CID) was conducted in LTQ with an activation q value of 0.25 and activation time of 30 ms. All instruments were controlled by the Xcalibur data system, and the data acquisition was carried out by analyst software Xcalibur (version 2.1) (Waltham, MA, USA) from Thermo Electron Corp.

3.7. Synchrotron Radiation Circular Dichroism

The SRCD experiments were carried out on the SRCD station [20] DISCO beamline [21] at the SOLEIL synchrotron (Gif-sur-Yvette, France). The samples were placed in calcium fluoride cells of 100 micron optical path lengths and measured at 0.2 mol/L in methanol. (+)-Camphor-10-sulfonic acid (CSA) solution was used to calibrate the SRCD signal. For each sample, three spectra were collected in the 350–200 nm range with a 1 nm step and 1200 ms integration time. The molar circular dichroism $\Delta\epsilon$ is expressed in M^{-1} cm^{-1}.

3.8. Computational Details

All QM calculations were carried out using Gaussian 16 [22]. The energy scan of the C–O bonds was performed using the Hartree–Fock method at the STO-3G level and a 0.1 Å bond-length step. The GMMX package was used for the conformational analysis (force field: MMFF94). The TD DFT calculations were performed using the B3LYP method at the 6-31G(d) level for 20 excited states. SpecDis 1.71 software was used to plot the ECD spectrum [23].

3.9. Cell Culture

HaCaT keratinocyte cells were cultured under standard conditions in DMEM supplemented with 10% fetal calf serum. The medium was changed every second day. Confluent cultures were removed by trypsin incubation, and then the cells were counted. They were seeded into 96-well culture microplates at a density of 30,000 cells per well (200 μL) and kept at 37 °C for 24 h.

3.10. Interleukine Release Measurement

The release of Il-8 in cell supernatants was determined by ELISA. After TNF-α incubation (0.5 ng/mL), cell supernatants were harvested and stored at −20 °C until use for measurements. The quantity of released Il-8 was measured according to the manufacturer's instructions (Kit ELISA Human CXCL8 / IL8 R&D Systems). The decrease in Il-8 production by EGCG (10 μg/mL) validated the method.

3.11. Statistical Analyses

All statistical analyses were performed using R 3.5.0 [24]. Cell samples were analyzed by repeated measures (n = 4) one-way analysis of variance (ANOVA) followed by a Tukey's range test. Significant differences for both clarinoside (**1**) and harounoside (**2**) were relative to control as indicated (NS: not significant; * $p < 0.05$; *** $p < 0.001$).

Molecules **2018**, *23*, 1237

3.12. Compound Characterization

1: White, amorphous solid; $[\alpha]_D^{20}$ +12.8 (c 0.1, CH$_3$OH); UV (DAD) λ_{max} 223, 245, 284, 346 nm; ^{1}H NMR and ^{13}C NMR data: see Table 1; HRESIMS (+) m/z 545.1685 [M + Na]$^+$ (545,16295 calcd. for C$_{25}$H$_{30}$O12Na, Δ −1.8 ppm).

Supplementary Materials: Supplementary materials, including HRMS, 1D and 2D NMR spectra for compound **1**, and computational details for **2**, are available online.

Author Contributions: G.G.-J. and C.A. conceived and designed the experiments; C.A., A.Z., N.B., E.R., and G.G.-J. performed the experiments; C.A., O.L., and G.G.-J. analyzed the data; A.G and E.R. contributed reagents, materials, and analysis tools; all the authors wrote the paper.

Acknowledgments: SOLEIL support is acknowledged under Proposal No. 20170521. We also thank the technical staff of SOLEIL for smooth and efficient running of the facility. The authors are grateful to Pascale Leproux and Karim Hammad for MS and NMR spectra recording. The Région Île de France is also acknowledged for its support through the funding of the DIM Analytics Project PRIMEVEGE (ANA2014-ML-001).

Conflicts of Interest: The authors declare no conflict of interest.

References and Notes

1. Olivon, F.; Allard, P.M.; Koval, A.; Righi, D.; Genta-Jouve, G.; Neyts, J.; Apel, C.; Pannecouque, C.; Nothias, L.F.; Cachet, X.; et al. Bioactive Natural Products Prioritization Using Massive Multi-informational Molecular Networks. *ACS Chem. Biol.* **2017**, *12*, 2644–2651. [CrossRef] [PubMed]

2. Nothias, L.F.; Nothias-Esposito, M.; da Silva, R.; Wang, M.; Protsyuk, I.; Zhang, Z.; Sarvepalli, A.; Leyssen, P.; Touboul, D.; Costa, J.; et al. Bioactivity-Based Molecular Networking for the Discovery of Drug Leads in Natural Product Bioassay-Guided Fractionation. *J. Nat. Prod.* **2018**, *81*, 758–767. [CrossRef] [PubMed]

3. Wang, M.; Carver, J.J.; Phelan, V.V.; Sanchez, L.M.; Garg, N.; Peng, Y.; Nguyen, D.D.; Watrous, J.; Kapono, C.A.; Luzzatto-Knaan, T.; et al. Sharing and community curation of mass spectrometry data with Global Natural Products Social Molecular Networking. *Nat. Biotechnol.* **2016**, *34*, 828–837. [CrossRef] [PubMed]

4. Ekpendu, T.O.; Akah, P.A.; Adesomoju, A.A.; Okogun, J.I. Antiinflammatory and Antimicrobial Activities of *Mitracarpus scaber* Extracts. *Int. J. Pharmacogn.* **1994**, *32*, 191–196. [CrossRef]

5. Bisignano, G.; Sanogo, R.; Marino, A.; Aquino, R.; D'angelo, V.; Germano, M.P.; De Pasquale, R.; Pizza, C. Antimicrobial activity of *Mitracarpus scaber* extract and isolated constituents. *Lett. Appl. Microbiol.* **2000**, *30*, 105–108. [CrossRef] [PubMed]

6. Gbaguidi, F.; Accrombessi, G.; Moudachirou, M.; Quetin-Leclercq, J. HPLC quantification of two isomeric triterpenic acids isolated from Mitracarpus scaber and antimicrobial activity on Dermatophilus congolensis. *J. Pharm. Biomed. Anal.* **2005**, *39*, 990–995. [CrossRef] [PubMed]

7. Okunade, A.L.; Clark, A.M.; Hufford, C.D.; Oguntimein, B.O. Azaanthraquinone: An Antimicrobial Alkaloid from *Mitracarpus scaber. Planta Med.* **1999**, *65*, 447–448. [CrossRef] [PubMed]

8. Harouna, H.; Faure, R.; Elias, R.; Debrauwer, L.; Saadou, M.; Balansard, G.; Boudon, G. Harounoside a pentalongin hydroquinone diglycoside from Mitracarpus scaber. *Phytochemistry* **1995**, *39*, 1483–1484. [CrossRef]

9. Pialat, J.P.; Hoffmann, P.; Moulis, C.; Fouraste, I.; Labidalle, S. Synthesis and Extraction of Pentalongin, A Naphthoquinoid From Mitracarpus Scaber. *Nat. Prod. Lett.* **1998**, *12*, 23–30. [CrossRef]

10. Burgess, K.; Borutzki, Y.; Rankin, N.; Daly, R.; Jourdan, F. MetaNetter 2: A Cytoscape plugin for ab initio network analysis and metabolite feature classification. *J. Chromatogr. B* **2017**, *1071*, 68–74. [CrossRef] [PubMed]

11. Menachery, S.P.M.; Laprévote, O.; Nguyen, T.P.; Aravind, U.K.; Gopinathan, P.; Aravindakumar, C.T. Identification of position isomers by energy-resolved mass spectrometry. *J. Mass Spectrom.* **2015**, *50*, 944–950. [CrossRef] [PubMed]

12. Xia, B.; Zhou, Y.; Liu, X.; Xiao, J.; Liu, Q.; Gu, Y.; Ding, L. Use of electrospray ionization ion-trap tandem mass spectrometry and principal component analysis to directly distinguish monosaccharides. *Rapid Commun. Mass Spectrom.* **2012**, *26*, 1259–1264. [CrossRef] [PubMed]

13. Giner, J.L.; Feng, J.; Kiemle, D.J. NMR Tube Degradation Method for Sugar Analysis of Glycosides. *J. Nat. Prod.* **2016**, *79*, 2413–2417. [CrossRef] [PubMed]

14. Genta-Jouve, G.; Weinberg, L.; Cocandeau, V.; Maestro, Y.; Thomas, O.P.; Holderith, S. Revising the Absolute Configurations of Coatlines via Density Functional Theory Calculations of Electronic Circular Dichroism Spectra. *Chirality* **2013**, *25*, 180–184. [CrossRef] [PubMed]

15. Han, A.R.; Park, S.R.; Park, J.W.; Lee, E.Y.; Kim, D.M.; Kim, B.G.; Yoon, Y.J. Biosynthesis of Glycosylated Derivatives of Tylosin in Streptomyces venezuelae. *J. Microbiol. Biotechnol.* **2011**, *21*, 613–616. [CrossRef] [PubMed]

16. De Castro, C.; Kenyon, J.J.; Cunneen, M.M.; Molinaro, A.; Holst, O.; Skurnik, M.; Reeves, P.R. The O-specific polysaccharide structure and gene cluster of serotype O:12 of the *Yersinia pseudotuberculosis* complex, and the identification of a novel l-quinovose biosynthesis gene. *Glycobiology* **2013**, *23*, 346–353. [CrossRef] [PubMed]

17. Baggiolini, M.; Clark-Lewis, I. Interleukin-8, a chemotactic and inflammatory cytokine. *FEBS Lett.* **1992**, *307*, 97–101. [CrossRef]

18. Pease, J.E.; Sabroe, I. The Role of Interleukin-8 and its Receptors in Inflammatory Lung Disease. *Am. J. Respir. Med.* **2002**, *1*, 19–25. [CrossRef] [PubMed]

19. This concentration of EGCG has been measured on a previous experiment.

20. Réfrégiers, M.; Wien, F.; Ta, H.P.; Premvardhan, L.; Bac, S.; Jamme, F.; Rouam, V.; Lagarde, B.; Polack, F.; Giorgetta, J.L.; et al. DISCO synchrotron-radiation circular-dichroism endstation at SOLEIL. *J. Synchrotron Radiat.* **2012**, *19*, 831–835. [CrossRef] [PubMed]

21. Giuliani, A.; Jamme, F.; Rouam, V.; Wien, F.; Giorgetta, J.L.; Lagarde, B.; Chubar, O.; Bac, S.; Yao, I.; Rey, S.; et al. DISCO: A low-energy multipurpose beamline at synchrotron SOLEIL. *J. Synchrotron Radiat.* **2009**, *16*, 835–841. [CrossRef] [PubMed]

22. Frisch, M.J.; Trucks, G.W.; Schlegel, H.B.; Scuseria, G.E.; Robb, M.A.; Cheeseman, J.R.; Scalmani, G.; Barone, V.; Petersson, G.A.; Nakatsuji, H.; et al. *Gaussian 16 Revision B.01*; Gaussian Inc.: Wallingford, CT, USA, 2016.

23. Bruhn, T.; Schaumlöffel, A.; Hemberger, Y.; Bringmann, G. SpecDis: Quantifying the Comparison of Calculated and Experimental Electronic Circular Dichroism Spectra. *Chirality* **2013**, *25*, 243–249. [CrossRef] [PubMed]

24. R Core Team. *R: A Language and Environment for Statistical Computing*; R Foundation for Statistical Computing: Vienna, Austria, 2018.

Sample Availability: Samples of the compounds **1** and **2** are available from the authors.

![molecules](molecules logo)

molecules

MDPI

Article

N-oxide alkaloids from *Crinum amabile* (Amaryllidaceae)

Luciana R. Tallini [1], Laura Torras-Claveria [1], Warley de Souza Borges [2], Marcel Kaiser [3,4], Francesc Viladomat [1], José Angelo S. Zuanazzi [5] and Jaume Bastida [1,*]

[1] Group of Natural Products, Faculty of Pharmacy, University of Barcelona, Av. Joan XXIII, 27-31, 08028-Barcelona, Spain; lucianatallini@gmail.com (L.R.T.); lauratorras@hotmail.com (L.T.-C.); fviladomat@ub.edu (F.V.)
[2] Department of Chemistry, Federal University of Espírito Santo, Av. Fernando Ferrari 514, 29075-915 Vitoria ES, Brazil; warley000@yahoo.com.br
[3] Medicinal Parasitology and Infection Biology, Swiss Tropical Institure, Socinstrasse 57, 4051 Basel, Switzerland; marcel.kaiser@unibas.ch
[4] University of Basel, Petersplatz 1, 4001 Basel, Switzerland
[5] Faculty of Pharmacy, Federal University of Rio Grande do Sul, Av. Ipiranga 2752, 90610-000 Porto Alegre RS, Brazil; zuanazzi@ufrgs.br
[*] Correspondence: jaumebastida@ub.edu; Tel.: +34-934-020-268

Academic Editors: Muhammad Ilias and Charles L. Cantrell
Received: 9 April 2018; Accepted: 24 May 2018; Published: 26 May 2018

Abstract: Natural products play an important role in the development of new drugs. In this context, the Amaryllidaceae alkaloids have attracted considerable attention in view of their unique structural features and various biological activities. In this study, twenty-three alkaloids were identified from *Crinum amabile* by GC-MS and two new structures (augustine *N*-oxide and buphanisine *N*-oxide) were structurally elucidated by NMR. Anti-parasitic and cholinesterase (AChE and BuChE) inhibitory activities of six alkaloids isolated from this species, including the two new compounds, are described herein. None of the alkaloids isolated from *C. amabile* gave better results than the reference drugs, so it was possible to conclude that the *N*-oxide group does not increase their therapeutic potential.

Keywords: *Crinum amabile*; augustine *N*-oxide; buphanisine *N*-oxide; biological activities

1. Introduction

Natural products play an important role in the development of new drugs [1]. For example, between 1940 and 2014, 49% of the small molecules approved for the treatment of cancer were developed or directly derived from natural products [1]. The isoquinoline-type alkaloids found in the Amaryllidaceae plant family represent an interesting source of new drugs due to their diverse biological activities [2]. The most important Amaryllidaceae alkaloid is galanthamine, which was approved by the Food and Drug Administration (FDA) for the clinical treatment of mild to moderate Alzheimer's disease (AD) in 2001, due to its potential acetylcholinesterase inhibitory activity [3]. According to the most recent botanical classification, the Amaryllidaceae are now a subfamily known as the Amaryllidoideae, which together with the Agapanthoideae and Allioideae belong to the Amaryllidaceae family [4]. Amaryllidoideae includes 59 genera and about 850 species, with centers of diversity in South Africa, South America, particularly in the Andean region, and in the Mediterranean [5].

Within the Amaryllidoideae, the pantropical *Crinum* genus is of commercial, economical and medicinal importance [6,7]. This genus contains approximately 65 species, which are widely distributed in diverse habitats, including coastal areas, pans (seasonally flooded depressions), sandy and aquatic areas, and swamps [8]. *Crinum* seeds are highly buoyant, with corky, water-repellent surfaces, allowing them to be dispersed by water [8,9]. Extracts from *Crinum* species have been used in folk medicine to treat fever, pain, swelling, sores, wounds, cancer and malaria [10]. The biological activities of *Crinum* species, including antitumor, immunostimulating, analgesic, antiviral, antibacterial, and antifungal, are attributed to their alkaloid content [7,9].

Known as a decorative plant, *Crinum amabile* has also long been used in Vietnamese folk medicine as an emetic and a remedy for rheumatism and earache [11]. Fifteen alkaloids have been previously identified in *C. amabile*: amabiline, ambelline, augustine, buphanisine, crinamabine, crinamine, crinidine, 4a-dehydroxycrinamabine, flexinine, galanthamine, galanthine, hippeastrine, lycorine, narvedine and tazettine [11–13]. Among these, amabiline, augustine, buphanisine, crinamine and lycorine have been isolated from this species and assessed for their antimalarial and cytotoxic potential, with augustine being the most active [13].

Tropical diseases such as malaria, leishmaniasis, Chagas disease and African trypanosomiasis affect more than one billion people and cost developing economies billions of dollars every year [14]. As these diseases prevail in areas where poverty limits access to prevention and treatment interventions, the pharmaceutical industry has little interest in investing in tackling them by drug development [15]. On the other hand, dementia affects around 50 million citizens worldwide, 60–70% of whom suffer from Alzheimer's disease, for which the current clinical treatment offers only palliative effects [16,17]. Thus, all these diseases require more research on effective treatment, in which the Amaryllidaceae alkaloids may potentially play an important role.

The aim of this work was to perform a detailed study of the alkaloid constituents of *C. amabile*, utilizing spectroscopic and chromatographic methods, including GC-MS and NMR. Two new alkaloids were isolated and chemically characterized by spectroscopic methods and twenty-three known alkaloids were identified by GC-MS. Due to the potential of Amaryllidaceae alkaloids in the clinical treatment of Alzheimer's disease [3], as well as the activity of augustine against malaria [13], we decided to check the cholinesterase-acetylcholinesterase (AChE) and butyrylcholinesterase (BuChE)-inhibitory activities and the antiprotozoal capacity of six alkaloids isolated from *C. amabile*, including the two new alkaloids. The role of *N*-oxide compounds in these biological activities was explored.

2. Results and Discussion

2.1. Alkaloids Identified by GC-MS

Twenty-three known alkaloids from *Crinum amabile* were identified by GC–MS (Table 1 and Figure 1) by comparison of the Rt, fragmentation patterns and spectral data using our home database. This database was built from single alkaloids isolated and identified by spectroscopic and spectrometric methods (NMR, UV, CD, IR, MS) in the Natural Products Laboratory, University of Barcelona, Spain. Also used were the NIST 05 Database and literature data [18–22].

2.2. Structural Elucidation

Two new alkaloids were identified in *C. amabile*: augustine *N*-oxide (**1**) and buphanisine *N*-oxide (**2**), both *N*-oxides of the structures augustine (**13**) and buphanisine (**8**), respectively. *N*-oxides occur as natural products and are not artefacts formed during the isolation procedures [23,24]. Ungiminorine *N*-oxide, homolycorine *N*-oxide, *O*-methyllycorenine *N*-oxide, galanthamine *N*-oxide, sanguinine *N*-oxide, lycoramine *N*-oxide and undulatine *N*-oxide are examples of Amaryllidaceae alkaloid *N*-oxides also reported as natural products [25–28].

Table 1. Alkaloids identified in *Crinum ariabile* by GC-MS.

Alkaloid	RI	M+	MS
Lycorene (**3**)	2102.2	255 (52)	254 (100), 227 (17), 226(20), 211 (15), 183(14), 181(10)
Ismine (**4**)	2124.3	257 (28)	239(16), 238 (100), 196 (10), 168 (10)
Demethylismine (**5**)	2128.8	243 (22)	225(21), 224 (100), 167 (10), 166 (15), 154 (11), 77 (12)
Demethylmesembrenol (**6**)	2177.0	275 (5)	206 (9), 205 (76), 115 (6), 70 (100)
Galanthamine (**7**)	2262.8	287 (85)	286 (100), 244(29), 216 (45), 174(39), 165(16), 141 (14), 128 (21), 115 (31)
Buphanisine (**8**)	2283.7	285 (95)	273 (54), 272 (43), 254 (40), 215 (100), 157 (39), 129 (35), 128 (55), 115 (64)
Sanguinine (**9**)	2288.3	273 (100)	272 (85), 202 (40), 165(20), 160 (50), 131 (20), 128 (19),115 (28), 77(20),
Crinine (10)	2326.7	271 (100)	228 (24), 200 (30), 199(81), 187 (76), 173 (28), 129 (34), 128(44), 115 (47), 56 (32)
8-*O*-Demethylmaritidine (**11**)	2373.8	273 (100)	230 (25), 202 (29), 201 (80), 189 (65), 175 (29), 129 (24), 128 (30), 115 (32), 56 (30)
3-*O*-Acetylsanguinine (**12**)	2387.1	315 (37)	256 (100), 255 (42), 254 (37), 212(26), 165 (33), 152 (23), 115 (30), 96 (67)
Augustine (**13**)	2411.6	301 (93)	228 (36), 187 (30), 175 (300), 173 (24), 159 (38), 143 (57), 128 (259, 115 (75)
Buphanisine *N*-oxide (**2**)	2429.8	301 (nv)	285 (100), 270 (33), 254 (35), 216 (21), 215 (82), 201 (24), 157 (20), 128 (22)
Haemanthamine (**14**)	2436.9	301 (55)	257 (54), 227 (80), 225 (98), 224(80), 181 (100), 153 (46), 152 (46), 115 (64)
11,12-Dehydroanhydrolycorine (**15**)	2448.5	249 (55)	248 (100), 191 (13), 190 (31), 189 (11), 95 (14)
Crinamine (**16**)	2497.6	273 (17)	272 (100), 242 (12), 214 (11), 186 (12), 128 (15)
Hamayne (**17**)	2551.7	259 (14)	258 (100), 242 (11), 214 (10), 211 (12), 181 (14), 128 (19)
1-*O*-Acetyllycorine (**18**)	2563.1	329 (20)	299(15), 268 (28), 250 (17), 244 (20), 227 (56), 226 (100), 240 (11)
Augustine *N*-oxide (**1**)	2571.8	317 (nv)	301 (100), 228 (34), 187 (22), 175 (77), 173 (17), 159 (27), 143 (37), 115 (37)
Lycorine (**19**)	2592.2	287 (19)	286 (13), 268 (18), 250 (10), 227 (60), 226 (100), 147 (11)
Undulatine (**20**)	2594.4	331 (100)	258 (37), 219 (22), 217 (36), 205 (71), 203 (37), 189 (43), 173 (39), 115 (35)
Ambelline (**21**)	2621.1	331 (69)	287 (100), 260 (81), 257 (62), 255 (74), 254 (52), 241 (51), 239 (61), 211 (69)
Augustamine (**22**)	2628.7	301 (76)	300 (100), 245(16), 244(84), 215(33), 201 (32), 188 (14), 115 (22), 70 (21)
6-Hydroxybuphanidrine (**23**)	2631.3	331 (35)	277 (16), 276 (100), 261 (30), 218 (17), 217 (23), 216 (24), 115 (18), 56 (25)
N-Formylnorgalanthamine (**24**)	2649.1	301 (100)	225 (26), 211 (29), 181 (19), 165 (14), 129 (18), 128 (22), 115 (30), 77 (15)
2-*O*-Acetyllycorine (**25**)	2676.2	329 (21)	328 (17), 270 (40), 269 (72), 268 (100), 252 (43), 250 (73), 227 (27), 226 (67)

* not visible.

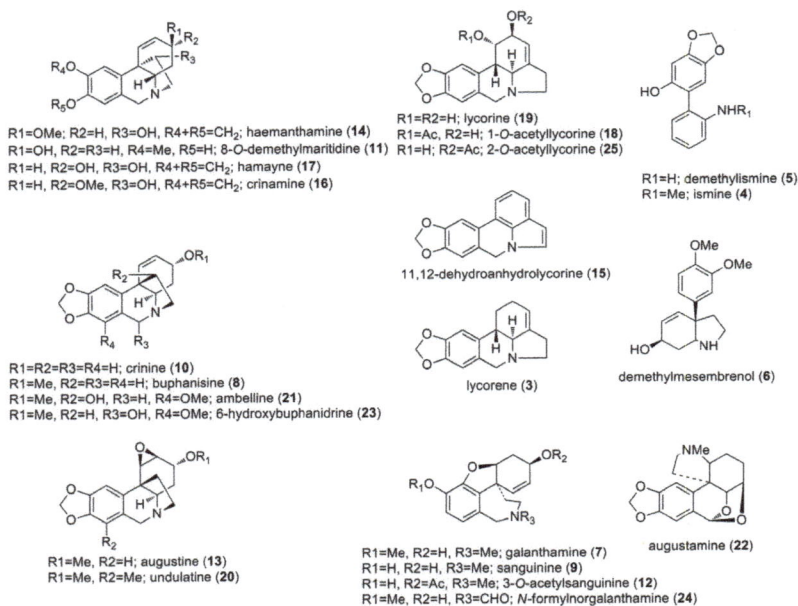

Figure 1. Alkaloids identified in *C. amabile* by GC-MS.

2.2.1. Augustine *N*-oxide (**1**)

The ^{1}H-NMR signals of compound **1** (Table 2) were consistent with the structure of augustine [29]. However, H-4α and H-4a and H-6α, H-6β, and H-12*endo* and H-12*exo* were deshielded between 1.21 and 0.36 ppm. These deshielding effects are congruent with the salt or *N*-oxide form of augustine. HR-ESI-MS analysis was carried out to confirm an additional oxygen atom in the structure. Compound **1** exhibited a parent [M + H]$^+$ ion at *m/z* 318.1335 in its HR-ESI-MS spectrum, suggesting the molecular formula $C_{17}H_{20}NO_5$ (calcd. 318.1336) and confirming compound **1** as augustine *N*-oxide.

The absolute configuration of this structure was determined by CD. The curve and shape were qualitatively similar to those of known crinine-type alkaloids, with the 5,10b-ethano bridge in a β-orientation, having a maximum of around 245 nm and a minimum of 295 nm. The COSY spectrum showed a benzylic coupling between H-7 and H-6, which allowed us to determine the H-7 proton location in the ^{1}H-NMR spectrum. The two C-6 protons were differentiated as an AB system with a geminal coupling of around 15.7 Hz. H-4a showed a NOESY correlation with H-6α, which turned out to be crucial for the assignment of its orientation. We were able to determine the H-4β orientation from the large coupling constant between H-4a and H-4β (around 13.8 Hz), and the α-orientation of the methoxy group at C-3 from the small constant between H-4β and H-3 (around 2.7 Hz). The β-orientation of the epoxy group was assignable based on the low values of the H-1 and H-2 constants (3.5 and 3.2 Hz, respectively). The NOESY contour plot between H-4β and H-11*exo* and H-12*exo* allowed us to determine the H-11*exo* and H-12*exo* locations in the ^{1}H-NMR spectrum. The quaternary carbons C-6a, C-10a, C-8 and C-9 were ascribed by means of their respective three-bond HMBC correlations with H-10, H-7, H-10 and H-7. Finally, the singlet resonance signal at δ = 43.95 ppm in the ^{13}C spectrum was assigned to C-10b, taking into account the three-bond connectivities to H-10, H-4α and H-4β (Figure 2) in the HMBC experiment.

Table 2. NMR data for compounds **1** and **2** (400 MHz for ^{1}H and 100 Hz for ^{13}C, CDCl$_3$).

No.	δ_C, type	δ_H (*J* in Hz)	δ_C, type	δ_H (*J* in Hz)
		1		**2**
1	52.06	3.68, d (3.5)	129.59	6.39, d (10.0)
2	55.06	3.42, ddd (3.2, 2.4, 0.7)	127.18	6.08, ddd (10.0, 5.3, 1.0)
3	73.43	4.12, dd (2.7, 2.5)	71.38	3.95, ddd (5.6, 3.6, 2.0)
4α	19.72	2.91, dt (14.1, 3.1)	23.46	3.13 ddt (13.6, 4.2, 2.4)
4β	19.72	1.54, ddd (13.8, 13.8, 2.9)	23.46	1.72 ddd (13.7, 13.7, 4.0)
4a	72.59	3.51, dd (13.4, 3.6)	74.14	3.74, dd (13.2, 4.3)
6α	67.56	4.83, dd (15.7, 1.8)	76.55	4.84, d (15.6)
6β	67.56	4.68, d (15.7)	76.55	4.72, d (15.6)
6a	122.33	-	121.71	-
7	106.40	6.57, s	106.44	6.54, s
8	147.13	-	147.20	-
9	147.79	-	147.79	-
10	102.49	6.90, s	102.98	6.81, s
10a	133.92	-	134.75	-
10b	43.95	-	46.60	-
11endo	35.64	1.99, ddd (12.4, 9.4, 5.1)	40.02	2.11, ddd (12.5, 8.0, 6.0)
11exo	35.64	2.79, ddd (12.4, 12.4, 6.9)	40.02	2.26, ddd (12.2, 10.8, 8.6)
12endo	67.56	3.81, ddd (12.8, 9.4, 7.0)	68.97	3.88, m
12exo	67.56	3.73, dddd (12.5,12.5, 5.0, 2.2)	68.97	3.85, m
OCH$_2$O	101.58	5.99, d (1.3), 5.98 d (1.3)	101.51	5.95, d (1.3), 5.93 d (1.3)
OMe	57.92	3.47, s	57.23	3.39, s

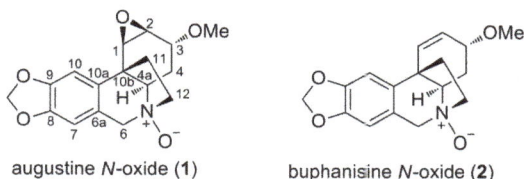

augustine *N*-oxide (**1**)   buphanisine *N*-oxide (**2**)

Figure 2. Structures of the two new alkaloids elucidated from *C. amabile*.

2.2.2. Buphanisine *N*-oxide (**2**)

The ^{1}H-NMR spectrum of compound **2** (Table 2) was similar to that of buphanisine [30]. However, the H-4α, H-4a, H-6α, H-6β, H-12*endo* and H-12*exo* protons were assigned as 0.80, 0.24, 0.32, 0.82, 0.86 and 0.29 ppm more deshielded, respectively, than their homologs in buphanisine. HR-ESI-MS analysis allowed us to confirm the presence of *N*-oxide in this structure. Compound **2** exhibited a parent [M + H]$^+$ ion at *m/z* 302.1385 in its HR-ESI-MS spectrum, suggesting the molecular formula C$_{17}$H$_{20}$NO$_4$ (calcd. 302.1387) and confirming compound **2** as buphanisine *N*-oxide.

The absolute configuration of this structure was determined by CD. The spectrum curve had a maximum of around 245 nm and a minimum of around 292 nm, confirming a crinine-type alkaloid with the 5,10b-ethano bridge in a β-orientation. The assignment of the aromatic protons was based on the benzylic coupling between H-6 and H-7 observed by a 2D COSY experiment. The ^{1}H-NMR spectrum showed two doublets at δ 4.84 and 4.72 ppm with a coupling constant of around 15.6 Hz. The first one was assigned as H-6α due to the NOESY contour plot of the proton H-4a. The large coupling constant between H-4a and H-4β (around 13.7 Hz) allowed the H-4β proton location to be determined in the ^{1}H-NMR spectrum. The NOESY contour plot between H-4β and H-11*exo* and H-12*exo* enabled us to determine their location in the ^{1}H-NMR spectrum. In the HMBC spectrum, the three-bond correlations observed for H-7 to C-9, H-10 to C-3, H-7 to C-10a and H-10 to C-6a allowed us to identify the location of the quaternary carbons C-8, C-9, C-6a and C-10a in the ^{13}C spectrum.

The three-bond coupling between C-10b and H-2, H-10 and H-4 permitted its location to be assigned at δ = 46.60 ppm in the ^{13}C spectrum (Figure 2).

2.3. Biological Activities

The alkaloid augustine presents significant activity against chloroquine-sensitive and chloroquine-resistant strains of *Plasmodium falciparum* (IC$_{50}$ = 0.46 and 0.60 μM, respectively) [13]. We consequently decided to verify *in vitro* the activity of six alkaloids isolated from *C. amabile*, augustamine (**22**), augustine (**13**), augustine *N*-oxide (**1**), buphanisine (**8**), buphanisine *N*-oxide (**2**) and crinine (**10**), against four different protozoa, *Trypanosoma brucei rhodesiense*, *Trypanosoma cruzi*, *Leishmania donovani* and *Plasmodium falciparum*, which are related to sleeping sickness, Chagas disease, visceral leishmaniasis and malaria, respectively. These alkaloids are structurally very similar, with the exception of augustamine, which is a unique kind of Amaryllidaceae alkaloid previously isolated from other *Crinum* species, including *C. augustum*, *C. kirkii* and *C. latifolium* [28,31,32], and completely elucidated in 2000 [33]. The rareness of this structure motivated us to isolate it and check its biological activity. In addition, due to the potential effectiveness of Amaryllidaceae alkaloids in the clinical treatment of Alzheimer's disease, the alkaloids were also tested *in vitro* against acetyl- and butyrylcholinesterase.

2.3.1. AChE and BuChE Inhibitory Activities

All the results for cholinesterase inhibitory activities are shown in Table 3. No tested alkaloid presented BuChE inhibitory activity. AChE inhibitory activity was moderate in augustine (**13**), and low in buphanisine (**8**). The structures of these two alkaloids are very similar: between C-1 and C-2 augustine (**13**) presents an epoxy group and buphanisine (**8**) an olefin group (Figure 1). Interestingly, the augustine (**13**) epoxy group seems to increase the AChE inhibitory activity, more than the olefin in buphanisine (**8**). The AChE inhibitory activity is also slightly improved by the presence of a hydroxyl group at C-3, as occurs in the crinine (**10**) alkaloid, but not by the methoxy group in the same substituent, as occurs in buphanisine (**8**). Unfortunately, the *N*-oxide group did not increase the AChE inhitory activities of augustine *N*-oxide (**1**) and buphanisine *N*-oxide (**2**). Furthermore, augustamine (**22**) did not show any cholinesterase inhibitory activities.

Table 3. Results of AChE and BuChE inhibitory activities of the alkaloids isolated from *C. amabile*.

Alkaloid	AChE *	BuChE *
Augustine *N*-oxide (**1**)	79.64 ± 5.26	>200
Buphanisine *N*-oxide (**2**)	>200	>200
Agustamine (**22**)	>200	>200
Augustine (**13**)	45.26 ± 2.11	>200
Buphanisine (**8**)	183.31 ± 36.64	>200
Crinine (**10**)	163.89 ± 15.69	>200
Galanthamine (**7**)	0.45 ± 0.03	3.88 ± 0.19

* all results are in μg mL^{-1}.

2.3.2. Antiprotozoal Activity

All the alkaloids isolated from *C. amabile* showed low activity against all the protozoa tested (Table 4). Buphanisine (**8**) showed significant inhibitory activity against the NF54 strain of *P. falciparum* (with a 50% inhibitory concentration (IC$_{50}$) of 4.28 ± 0.18 μg mL^{-1}. The presence of an *N*-oxide group in augustine *N*-oxide (**1**) and buphanisine *N*-oxide (**2**) appears to decrease their activity against *T. brucei* and *P. falciparum* compared to augustine (**13**) and buphanisine (**8**), respectively. In this experiment, the epoxy group at C-1 and C-2 probably decreases the activity of augustine (**13**) against *P. falciparum* compared to buphanisine (**8**), which has a double bond between C-1 and C-2. Furthermore, the presence of a methoxy group at C-3 seems to increase the activity of buphanisine (**8**) against *P. falciparum* compared to crinine (**10**), which has a hydroxyl group in the same position.

Table 4. *In vitro* antiprotozoal and cytotoxic activities of **1** and **2**. Values expressed in IC$_{50}$ (μg mL^{-1}).

Parasite	T. brucei rhodesiense	T. cruzi	L. doncvani	P. falciparum	Cytotoxicity
Reference drug	0.003 ± 0.001 [a]	0.865 ± 0.08 [b]	0.515 ± 0.06 [c]	0.004 ± 0.0007 [d]	0.004 ± 0.0007 [e]
Augustine (**13**)	15.05 ± 1.06	56.00 ± 0.71	>100	14.20 ± 0.14	>100
Augustine *N*-oxide (**1**)	58.85 ± 11.53	66.25 ± 11.81	>100	36.65 ± 4.74	>100
Buphanisine (**8**)	16.5 ± 0.57	55.55 ± 4.60	>100	4.28 ± 0.18	72.85 ± 5.02
Buphanisine *N*-oxide (**2**)	55.25 ± 4.31	64.05 ± 1.34	>100	32.55 ± 0.07	>100
Crinine (**10**)	18.95 ± 0.78	57.45 ± 6.86	>100	30.95 ± 2.19	>100
Augustamine (**22**)	19.20 ± 2.97	54.00 ± 4.53	>100	20.35 ± 0.21	81.55 ± 0.64

[a] melarsoprol; [b] benznidazole; [c] miltefosine; [d] chloroquine; [e] podophyllotoxin.

3. Materials and Methods

3.1. Plant Material

Bulbs of *Crinum amabile* Donn. were collected in Vitoria (Espírito Santo, Brazil) in September 2016. The sample was authenticated by Dr. Alan Meerow at the Subtropical Horticulture Research Station (Miami, FL, USA). A specimen voucher (VIES 39506) has been deposited in the Herbarium of the Universidade Federal do Espirito Santo (UFES; Vitoria, Brazil).

3.2. Equipment

About 2 mg of each alkaloid extract was dissolved in 1000 μL of methanol (MeOH) and/or chloroform (CHCl$_3$) and injected directly into the GC-MS apparatus (Agilent Technologies, Santa Clara CA, USA) operating in the EI mode at 70 eV. A Sapiens-X5 MS column (30 m × 0.25 mm i.d., film thickness 0.25 μm) was used. The temperature gradient performed was the following: 2 min at 100 °C, 100–180 °C at 15 °C min^{-1}, 180–300 °C at 5 °C min^{-1} and 10 min hold at 300 °C. The injector and detector temperatures were 250 °C and 280 °C, respectively, and the flow-rate of carrier gas (He) was 1 mL min^{-1}. A split ratio of 1:10 was applied and the injection volume was 1 μL. The alkaloids were identified by GC-MS and the mass spectra were deconvoluted using the software AMDIS 2.64. Kovats retention indices (RI) were recorded with a standard calibration *n*-hydrocarbon mixture (C9–C36) using AMDIS 2.64 software.

^{1}H-NMR, ^{13}C-NMR, COSY, NOESY, HSQC, and HMEC spectra were recorded on a Bruker 400 MHz Avance III equipped with CryoProbe Prodigy (Bruker, Billerica, MA, USA), using CDCl$_3$ as the solvent and tetramethylsilane (TMS) as the internal standard. Chemical shifts are reported in units of δ (ppm) and coupling constants (J) are expressed in Hz. CD, UV and IR spectra were recorded on Jasco-J-810 (Jasco, Easton, MD, USA), Dinko UV2310 (Dinko Instruments, Barcelona, USA) and Thermo Scientific Nicolet iN10 MX spectrophotometers (Thermo Fisher Scientific, Waltham, MA, USA), respectively. HR-ESI-MS spectra were obtained on an LC/MSD-TOF (2006) mass spectrometer (Agilent Technologies) operating in the positive mode, applying 4 kV in the capillary, 175 V in the fragmentor, a gas temperature of 325 °C, and N$_2$ as the nebulizing gas (15 psi) and drying gas (flow = 7.0 L min^{-1}). Silica gel SDS chromagel 60 A CC (6–35 μm) was used for VLC, and silica gel 60 F$_{254}$ (Merck, Darmstadt, Germany) for analytics and prep. Spots on chromatograms were detected under UV light (254 nm) and by Dragendorff's reagent stain.

3.3. Extraction

Fresh bulbs (2.2 kg) and leaves (1.3 kg) of *C. amabile* were collected and macerated with MeOH (3 × 1.0 L) at room temperature for 4 days. The combined macerate was filtered and the solvent evaporated to dryness under reduced pressure. The bulb and leaf crude extracts (485 and 390 g, respectively) were then acidified to pH 3 with diluted sulfuric acid, H_2SO_4 (2%, *v/v*). The neutral material was removed with Et_2O (3 × 200 mL) and extracted with ethyl acetate (EtOAc) (3 × 200 mL) to obtain the acid EtOAc extracts (4.58 and 2.8 g, respectively). The aqueous solutions were basified up to pH 9–10 with ammonium hydroxide, NH_4OH (25%, *v/v*) and extracted with *n*-hexane, *n*-Hex (3 × 150 mL) to give the *n*-Hex extracts (1.16 and 0.40 g, respectively, and finally extracted with EtOAc (2 × 200 mL) to obtain the EtOAc extracts (5.11 and 1.15 g, respectively).

The extracts were subjected to a combination of chromatographic techniques, including vacuum liquid chromatography (VLC) [34], Sephadex, thin layer chromatography (TLC) and semi-preparative TLC. The VLC is an effective methodology to rapidly and inexpensively separate large or small quantities of compounds from extracts [35]. A silica gel 60 A (6–35 μm) column was used with a height of 4 cm and a variable diameter according to the amount of sample (2.5 cm for 400–1000 mg; 1.5 cm for 150–400 mg). Alkaloids were eluted with n-Hex containing increasing EtOAc concentrations, followed by neat EtOAc, which was gradually enriched with MeOH (reaching a maximum concentration of 20%, *v/v*). Fractions of 10–15 mL were collected, monitored by TLC (UV 254 nm, Dragendorff's reagent), and combined according to their profiles. For semi-preparative TLC, silica gel 60F^{254} was used (20 cm × 20 cm × 0.25 mm) together with different solvent mixtures depending on each particular sample (EtOAc:MeOH, 9:1, *v/v*; EtOAc:MeOH, 8:2, *v/v*; or EtOAc:CHCl$_3$:MeOH, 6:4:2, *v/v/v*), always in an environment saturated with ammonia. The alkaloids were each identified by GC-MS and the two new alkaloids had their structure elucidated by NMR.

Exclusion chromatography was carried out using a Sephadex LH-20 column (2.5 cm × 40 cm) to clean and separate the alkaloids in the n-Hex bulb extract (1.16 g). It was eluted with 100% MeOH, producing 52 fractions, each one containing about 2 mL, which were monitored by TLC and grouped in four fractions. Fraction 3 (1.00 g) was subjected to a VLC column (2.5 cm × 4.0 cm), starting the elution with 100% n-Hex, and gradually increasing the polarity by adding concentrations of EtOAc up to 100%. The MeOH percentage in the mixture was then increased up to a ratio of EtOAc:MeOH (80:20, *v/v*) and finally, keeping the MeOH percentage stable, the EtOAc percentage was decreased and the CHCl$_3$ percentage increased to a ratio of EtOAc:MeOH:CHCl$_3$ (60:20:40, *v/v/v*). 48 fractions (100 mL each) were collected, analyzed by TLC and grouped in twelve fractions.

Fraction B (28.5 mg) - eluted with n-Hex:EtOAc (eluted with 70:30 until 60:40, *v/v*), fraction D (26.1 mg)-eluted with n-Hex: EtOAc (40:60, *v/v*), fraction F (40.0 mg) - eluted with EtOAc:MeOH (eluted with 96:4 until 92:8, *v/v*), fraction H (10.5 mg) - eluted with EtOAc:MeOH (eluted with 88:12 until 83:17, *v/v*) and fraction J (25.0 mg) - eluted with EtOAc:MeOH:CHCl$_3$ (eluted with 80:20:0 until 71:20:9, *v/v/v*) were subject to different semi-preparative TLC using a mobile phase consisting of EtOAc:MeOH:CHCl$_3$ (60:20:40, *v/v/v*) in an environment saturated with ammonia. 12.0 mg of augustamine (22) was isolated from fraction B, 9.1 mg of augustine (13) from fraction D, 6.0 mg of buphanisine (8) from fraction F, 2.9 mg of crinine (10) from fraction H, and 4.0 mg of augustine *N*-oxide (1) and 12.0 mg of buphanisine *N*-oxide (2) from fraction J.

3.4. Characterization of Compounds

Augustine N-oxide (**1**): Amorphous solid; $[\alpha]_D^{22}$ −24.0 (*c* 0.001, CHCl$_3$); UV (MeOH) λmax (log ε): 292.0 (3.69), 240.5 (5.57) nm; CD (MeOH, 20 °C) $\Delta\varepsilon_{245}$ + 5739, $\Delta\varepsilon_{295}$ −5169; IR v_{max} 3363, 2986, 2851, 1740, 1504, 1489, 1464, 1391, 1374, 1241, 1147, 1079, 1034, 926, 846 and 814 cm^{-1}; ^{1}H-NMR (CDCl$_3$, 400 MHz) and ^{13}C-NMR (CDCl$_3$, 100 MHz) see Table 2; ESI-MS data shown in Table 1; HR-ESI-MS of [M + H]$^+$ *m/z* 318.1335 (calcd. for $C_{17}H_{20}NO_5$, 318.1336).

Buphanisine N-oxide (**2**): Amorphous solid; $[\alpha]_D^{22}$ −1.0 (*c* 0.001, CHCl$_3$); UV (MeOH) λmax (log ε): 292.0 (3.57), 240.5 (3.45) nm; CD (MeOH, 20 °C) $\Delta\varepsilon_{245}$ + 7271, $\Delta\varepsilon_{292}$ −7223; IR v_{max} 3350, 3037, 2977, 2824, 1652, 1488, 1402, 1377, 1255, 1241, 1097, 1071, 1034, 966, 931 and 854 cm^{-1}; ^{1}H-NMR (CDCl$_3$, 400 MHz) and ^{13}C-NMR (CDCl$_3$, 100 MHz) see Table 2; ESI-MS data shown in Table 1; HR-ESI-MS of [M + H]$^+$ *m/z* 302.1385 (calcd. for $C_{17}H_{20}NO_4$, 302.1387).

3.5. Biological Activities

3.5.1. Antiprotozoal Activities

In vitro tests for the biological activity of the alkaloids isolated from *C. amabile* against *Trypanosoma brucei rhodesiense* (trypomastigotes forms, STIB 900 strain), *Trypanosoma cruzi* (axenic grown amastigotes forms, Tulahuen C4 strain), *Leishmania donovani* (amastigotes forms, MHOM-ET-67/L82 strain), and *Plasmodium falciparum* (intraerythrocytic forms, IEF, NF54 strain) and a cytotoxicity test against the mammalian L6 cell line from rat skeletal myoblasts were carried out at the Swiss Tropical and Public Health Institute (Swiss TPH, Basel, Switzerland) according to established protocols as described by Orhan and co-workers [36]. The reference drugs used in these assays were melarsoprol, benznidazole, miltefosine, chloroquine and podophyllotoxin, respectively.

3.5.2. Acetylcholinesterase and Butyrylcholinesterase Inhibitory Activities

Cholinesterase inhibitory activities were analyzed as by Ellman and co-workers [37] with some modifications as by López and co-workers [38]. Fifty microliters of AChE or BuChE phosphate buffer (8m M K_2HPO_4, 2.3 mM NaH_2PO_4, 0.15 M NaCl, pH 7.5) and 50 μL of the sample dissolved in the same buffer were added to the wells. The plates were incubated for 30 min at room temperature before 100 μL of the substrate solution (0.1 M Na_2HPO_4, 0.5 M DTNB, and 0.6 mM acetylthiocholine iodide, ATCI, or 0.24 mM butyrylthiocholine iodide, BTCI, in Millipore water, pH 7.5) was added. The absorbance was read in a Labsystem microplate reader (Helsinki, Finland) at 405 nm after 10 min. Galanthamine served as positive control. In a first step, samples were assessed at 10, 100 and 200 μg mL^{-1} towards both enzymes. Samples with an IC$_{50}$ > 200 μg mL^{-1} were considered inactive. Samples with an IC$_{50}$ < 200 μg mL^{-1} were further analyzed to determine the IC$_{50}$ values. Enzyme activity was calculated as a percentage compared to an assay using a buffer without any inhibitor. The cholinesterase inhibitory data were analyzed with the software Microsoft Office Excel 2010.

4. Conclusions

Twenty-five alkaloids were identified in *C. amabile*, including two new alkaloids, augustine *N*-oxide and buphanisine *N*-oxide. This is the first time that augustamine and *N*-oxide structures have been described in this species. These alkaloids, together with augustine, buphanisine and crinine, were isolated, but none showed remarkable biological activity.

Author Contributions: Jaume Bastida designed the experiments, analyzed the data and wrote the paper. Luciana R. Tallini and Laura Torras-Claveria performed the experiments, analyzed the data and wrote the paper. Warley de Souza Borges collected the plant, performed the extraction and wrote the paper. Marcel Kaiser performed the antiprotozoal experiments and wrote the paper. Francesc Viladomat and José Angelo S. Zuanazzi wrote the paper.

Funding: The authors (Research Group 2017-SGR-604) thank CCiTUB and Programa CYTED (416RT0511) for technical and financial support, respectively. WB is thankful to CAPES (Processo CSF-PVE-S 88887115334/2015-00) and FAPES (Processo Universal 80708382/18) for technical and financial support.

Acknowledgments: We thank M. Cal, S. Keller-Märki and R. Rocchetti for assistance with parasitic assays. JASZ acknowledges CNPq (Brazil) for a research fellowship. LRT is thankful to CAPES (Coordenação de Pessoal de Nível Superior – Bolsista CAPES, Processo n° 13553135) for a doctoral fellowship. Plant identification by Professor Alan Meerow is highly appreciated.

Conflicts of Interest: The authors declare no conflict of interest.

References

1. Newman, D.J.; Cragg, G.M. Natural products as sources of new drugs from 1981 to 2014. *J. Nat. Prod.* **2016**, *79*, 629–661. [CrossRef] [PubMed]
2. Bastida, J.; Lavilla, R.; Viladomat, F. Chemical and biological aspects of *Narcissus* Alkaloids. In *The Alkaloids: Chemistry and Biology*; Cordell, G.A., Ed.; Elsevier: Amsterdam, Netherlands, 2006; Volume 63, pp. 87–179.

3. Maelicke, A.; Samochocki, M.; Jostock, R.; Fehrenbacher, A.; Ludwig, J.; Albuquerque, E.X.; Zerlin, M. Allosteric sensitization of nicotinic receptors by galantamine, a new treatment strategy for Alzheimer's disease. *Biol. Psychiat.* **2001**, *49*, 279–288. [CrossRef]

4. APG III. An update of the angiosperm phylogeny group classification for the orders and families of flowering plants: APG III. *Bot. J. Linn. Soc.* **2009**, *161*, 105–121.

5. Rønsted, N.; Symonds, M.R.E.; Birkholm, T.; Christensen, S.B.; Meerow, A.W.; Molander, M.; Mølgaard, P.; Petersen, G.; Rasmussen, N.; van Staden, J.; et al. Can phylogeny predict chemical diversity and potential medicinal activity of plants? A case study of Amaryllidaceae. *BMC Evol. Biol.* **2012**, *12*, 182–194. [CrossRef] [PubMed]

6. Fennell, C.W.; van Staden, J. *Crinum* species in traditional and modern medicine. *J. Ethnopharmacol.* **2001**, *78*, 15–26. [CrossRef]

7. Tram, N.T.N.; Titorenkova, T.V.; Bankova, V.S.; Handjieva, N.V.; Popov, S.S. *Crinum* L. (Amaryllidaceae). *Fitoterapia* **2002**, *73*, 183–208. [CrossRef]

8. Kwembeya, E.G.; Bjora, C.S.; Stedje, B.; Nordal, I. Phylogenetic relationships in the genus *Crinum* (Amaryllidaceae) with emphasis on tropical African species: Evidence from *trnL-F* and nuclear ITS DNA sequence data. *Taxon* **2007**, *56*, 801–810. [CrossRef]

9. Maroyi, A. A review of ethnobotany, therapeutic value, phytochemistry and pharmacology of *Crinum macowanii* Baker: A highly traded bulbous plant in Southern Africa. *J. Ethnopharmacol.* **2016**, *194*, 595–608. [CrossRef] [PubMed]

10. Presley, C.C.; Krai, P.; Dalal, S.; Su, Q.; Cassera, M.; Goetz, M.; Kingston, D.G.I. New potently bioactive alkaloids from *Crinum erubescens*. *Bioorgan. Med. Chem.* **2016**, *24*, 5418–5422. [CrossRef] [PubMed]

11. Pham, L.H.; Döpke, W.; Wagner, J.; Mügge, C. Alkaloids from *Crinum amabile*. *Phytochemistry* **1998**, *48*, 371–376. [CrossRef]

12. Murav'eva, D.A.; Popova, O.I. Alkaloid composition of the bulbs of *Crinum amabile*. *Khim. Prir. Soedin.* **1982**, *2*, 263–264.

13. Likhitwitayawuid, K.; Angerhofer, C.K.; Chai, H.; Pezzuto, J.M.; Cordell, G.A. Cytotoxic and antimalarial alkaloids from the bulbs of *Crinum amabile*. *J. Nat. Prod.* **1993**, *56*, 1331–1338. [CrossRef]

14. WHOa. World Health Organization Neglected Tropical Diseases. Available online: http://www.who.int/ neglected_diseases/diseases/en/ (accessed on 4 May 2018).

15. Klug, D.M.; Gelb, M.H.; Pollastri, M.P. Repursposing strategies for tropical disease drug discovery. *Bioorg. Med. Chem. Lett.* **2016**, *26*, 2569–2576. [CrossRef] [PubMed]

16. WHOb. World Health Organization Dementia. Available online: http://www.who.int/en/news-room/fact-sheets/detail/dementia (accessed on 4 May 2018).

17. Wu, W.Y.; Dai, Y.C.; Li, N.G.; Dong, Z.X.; Gu, T.; Shi, Z.H.; Xue, X.; Tang, Y.P.; Duan, J.A. Novel multitarget-directed tacrine derivatives as potential candidates for the treatment of Alzheimer's disease. *J. Enzym. Inhib. Med. Ch.* **2017**, *32*, 572–587. [CrossRef] [PubMed]

18. de Andrade, J.P.; Pigni, N.B.; Torras-Claveria, L.; Berkov, S.; Codina, C.; Viladomat, F.; Bastida, J. Bioactive alkaloid extracts from *Narcissus broussonetii*: Mass spectral studies. *J. Pharmaceut. Biomed.* **2012**, *70*, 13–25. [CrossRef] [PubMed]

19. de Andrade, J.P.; Guo, Y.; Font-Bardia, M.; Calvet, T.; Dutilh, J.; Viladomat, F.; Codina, C.; Nair, J.J.; Zuanazzi, J.A.S.; Bastida, J. Crinine-type alkaloids from *Hippeastrum aulicum* and *H. calyptratum*. *Phytochemistry* **2014**, *103*, 188–195. [CrossRef] [PubMed]

20. Torras-Claveria, L.; Berkov, S.; Codina, C.; Viladomat, F.; Bastida, J. Metabolomic analysis of bioactive Amaryllidaceae alkaloids of ornamental varieties of *Narcissus* by GC-MS combined with k-means cluster analysis. *Ind. Crop. Prod.* **2014**, *56*, 211–222. [CrossRef]

21. Tallini, L.R.; de Andrade, J.P.; Kaiser, M.; Viladomat, F.; Nair, J.J.; Zuanazzi, J.A.S.; Bastida, J. Alkaloid constituents of the Amaryllidaceae plant *Amaryllis belladonna* L. *Molecules* **2017**, *22*, 1437. [CrossRef] [PubMed]

22. Tallini, L.R.; Osorio, E.H.; dos Santos, V.D.; Borges, W.D.S.; Kaiser, M.; Viladomat, F.; Zuanazzi, J.A.S.; Bastida, J. *Hippeastrum reticulatum* (Amaryllidaceae): Alkaloids profiling, biological activities and molecular docking. *Molecules* **2017**, *22*, 2191.

23. Phillipson, J.D.; Handa, S.S.; El-Dabbas, S.W. *N*-Oxides of morphine, codeine and thebaine and their occurrence in *Papaver* species. *Phytochemistry* **1976**, *15*, 1297–1301. [CrossRef]

24. Dembitsky, V.M.; Gloriozova, T.A.; Poroikov, V.V. Naturally occurring plant isoquinoline *N*-oxide alkaloids: Their pharmacological and SAR activities. *Phytomedicine* **2015**, *22*, 183–202. [CrossRef] [PubMed]

25. Suau, R.; Gómez, A.I.; Rico, R.; Tato, M.P.V.; Castedo, L.; Riguera, R. Alkaloids *N*-oxides of Amaryllidaceae. *Phytochemistry* **1988**, *27*, 3285–3287. [CrossRef]

26. Kobayashi, S.; Satoh, K.; Numata, A.; Shingu, T.; Kihara, M. Alkaloid *N*-oxides from *Lycoris sanguinea*. *Phytochemistry* **1991**, 675–677. [CrossRef]

27. Bessa, C.D.P.B.; de Andrade, J.P.; de Oliveira, R.S.; Domingos, E.; Heloa, S.; Romão, W.; Bastida, J.; Borges, W.S. Identification of Alkaloids from *Hippeastrum aulicum* (Ker Gawl.) Herb. (Amaryllidaceae) Using CGC-MS and Ambient Ionization Mass Spectrometry (PS-MS and LS-MS). *J. Braz. Chem. Soc.* **2017**, *28*, 819–830. [CrossRef]

28. Hanh, T.T.H.; Anh, D.H.; Huong, P.T.T.; Thanh, N.V.; Trung, N.Q.; Cuong, T.V.; Mai, N.T.; Cuong, N.T.; Cuong, N.X.; Nam, N.H.; Minh, C.V. Crinane, augustamine, and β-carboline alkaloids from *Crinum latifolium*. *Phytochem. Lett.* **2018**, *24*, 27–30. [CrossRef]

29. Frahm, A.W.; Ali, A.A.; Kating, H. Relative configuration of the alkaloid augustine. *Phytochemistry* **1981**, *20*, 1735–1738. [CrossRef]

30. Viladomat, F.; Codina, C.; Bastida, J.; Mathee, S.; Campbell, W.E. Further alkaloids from *Brunsvigia josephinae*. *Phytochemistry* **1995**, *40*, 961–965. [CrossRef]

31. Ali, A.A.; Hambloch, H.; Frahm, A.W. Relative configuration of the alkaloid augustamine. *Phytochemistry* **1983**, *22*, 283–287. [CrossRef]

32. Machocho, A.K.; Bastida, J.; Codina, C.; Viladomat, F.; Brun, R.; Chhabra, S.C. Augustamine type alkaloids from *Crinum kirkii*. *Phytochemistry* **2004**, *65*, 3143–3149. [CrossRef] [PubMed]

33. Joshi, B.S.; Pelletier, S.W.; Ali, A.A.; Holt, E.M.; Bowen, J.P.; Ehlers, T. Crystal and molecular structure of augustamine. *J. Chem. Crystallogr.* **2000**, *30*, 135–138. [CrossRef]

34. Coll, J.C.; Bowden, B.F. The application of vacuum liquid chromatography to the separation of terpene mixtures. *J. Nat. Prod.* **1986**, *49*, 934–936. [CrossRef]

35. Targett, N.M.; Kilcoyne, J.P.; Green, B. Vacuum liquid chromatography: An alternative to common chromatographic methods. *J. Org. Chem.* **1979**, *44*, 4962–4964. [CrossRef]

36. Orhan, I.; Şener, B.; Kaiser, M.; Brun, R.; Tasdemir, D. Inhibitory activity of marine sponge-derived natural products against parasitic protozoa. *Mar. Drugs.* **2010**, *8*, 47–58. [CrossRef] [PubMed]

37. Ellman, G.L.; Courtney, K.D.; Andres, V., Jr.; Featherstone, R.M. A new and rapid colorimetric determination of acetylcholinesterase activity. *Biochem. Pharmacol.* **1961**, *7*, 88–95. [CrossRef]

38. López, S.; Bastida, J.; Viladomat, F.; Codina, C. Acetylcholinesterase inhibitory activity of some Amaryllidaceae alkaloids and *Narcissus* extracts. *Life Sci.* **2002**, *71*, 2521–2529. [CrossRef]

Sample Availability: Not available.

molecules

MDPI

Article

Efficacy of Prosopilosidine from *Prosopis glandulosa* var. *glandulosa* against *Cryptococcus neoformans* Infection in a Murine Model

Mohammad K. Ashfaq [1,*], Mohamed Sadek Abdel-Bakky [1,2], Mir Tahir Maqbool [1], Volodymyr Samoylenko [1,3], Aziz Abdur Rahman [1,4] and Ilias Muhammad [1,*]

1   National Center for Natural Product Research, Research Institute of Pharmaceutical Sciences, School of Pharmacy, University of Mississippi, University, MS 38677, USA; abdelbakkym@yahoo.com (M.S.A.-B.); tmmir@olemiss.edu (M.T.M.); voxa@outlook.com (V.S.); aziz2002@ru.ac.bd (A.A.R.)
2   Faculty of Pharmacy, Al-Azhar University, Cairo 11651, Egypt
3   Department of Arts and Sciences, Keiser University, 2085 Vista Pkwy, West Palm Beach, FL 33411, USA
4   Department of Pharmacy, University of Rajshahi, Rajshahi 6205, Bangladesh
*   Correspondence: mkashfaq@olemiss.edu (M.K.A.); milias@olemiss.edu (I.M.); Tel.: +1-(662)915-1577 (M.K.A.); Fax: +1-(662)915-7062 (M.K.A.)

Received: 1 June 2018; Accepted: 6 July 2018; Published: 10 July 2018

Abstract: In this study, 2,3-dihydro-1*H*-indolizinium alkaloid-prosopilosidine (PPD), that was isolated from *Prosopis glandulosa*, was evaluated against *C. neoformans* in a murine model of cryptococcosis. In vitro and in vivo toxicity of indolizidines were also evaluated. Mice were infected via the tail vein with live *C. neoformans*. Twenty-four hours post-infection, the mice were treated with PPD once a day (i.p.) or twice a day (*bid*) orally, or with amphotericin B (Amp B) intraperitoneally (IP), or with fluconazole (Flu) orally for 5 days. The brains of all of the animals were aseptically removed and the numbers of live *C. neoformans* were recovered. In vitro toxicity of indolizidine alkaloids was determined in HepG2 cells. PPD showed to be potent in vivo activity against *C. neoformans* at a dose of 0.0625 mg/kg by eliminating ~76% of the organisms compared to ~83% with Amp B (1.5 mg/kg). In addition, PPD was found to be equally efficacious, but less toxic, at either 0.125 or 0.0625 mg/kg compared to Amp B (1.5 mg/kg) when it was administered *bid* (twice a day) by an i.p. route. When tested by an oral route, PPD (10 mg/kg) showed potent activity in this murine model of cryptococcosis with ~82% of organisms eliminated from the brain tissue, whereas Flu (15 mg/kg) reduced ~90% of the infection. In vitro results suggest that quaternary indolizidines were less toxic as compared to those of tertiary bases. PPD (20 mg/kg) did not cause any alteration in the plasma chemistry profiles. These results indicated that PPD was active in eliminating cryptococcal infection by oral and i.p. routes at lower doses compared to Amp B. or Flu.

Keywords: *Cryptococcus neoformans*; cryptococcosis; HepG2; *Prosopis glandulosa*; prosopilosidine; amphotericin B; fluconazole

1. Introduction

Cryptococcus neoformans is a dimorphic fungus that causes serious infection leading to pneumonia and life-threatening diseases of the central nervous system (CNS). The virulence of the organism has been linked to the presence of a thick carbohydrate capsule and its pigment melanin [1,2]. Both immunocompetent and immunocompromised individuals are affected; immunocompetent patients generally develop pulmonary cryptococcosis as a sole manifestation of the disease [3]. However, in immune compromised and elderly patients, both lung and brain infections cause high morbidity and mortality [4,5]. In advanced cases of HIV infections, the incidence of cryptococcal infection ranged from 10–15% in developed countries and was even higher in the underdeveloped countries [6,7]. With the advent of antiretroviral therapy (ART), the immune system of patients with HIV/AIDS became less vulnerable to fungal infections or other infections, and this reduced the incidence of *C. neoformans* infections in people with advanced HIV/AIDS, especially those from developed countries [8]. Thus, fungal infections in HIV/AIDS patients declined significantly in the U.S. [9,10]. However, it still remains a major problem in developing countries due to limited healthcare facilities [11]. It is estimated that globally, cryptococcal meningitis is responsible for 15% of AIDS-related deaths, with an estimate of 220,000 new cases of cryptococcal meningitis each year, resulting in 181,000 deaths [11]. Hence, cryptococcosis still remains a concern for people living with HIV/AIDS.

Only a limited number of antifungals are available for the treatment of Cryptococcal infections. Amphotericin B (Amp B) and fluconazole (Flu) are among the commonly used antifungals for patients with cryptococcal infections of the CNS. However, the emergence of Flu-resistant fungal pathogens is a concern [12]. In vitro investigations on various strains of *C. neoformans* indicate that mutation to Flu resistance is a dynamic and heterogeneous process that involves multiple mechanisms [13,14]. Recently, concerns have been raised about the efficacy of initial Flu treatments of cryptococcal meningitis in HIV patients, as resistance to Flu leads to a relapse of cryptococcal meningitis symptoms [15,16]. In such cases, a combination of Amp B and Flu or voriconazole has been suggested [17,18].

According to World Health Organization (WHO) guidelines, the treatment for cryptococcal meningitis should be comprised of a combination of therapies, starting with Amp B and flucytosine for 2-weeks, followed by Flu [17,18]. All of these anti-fungal agents have different mechanisms of action against Cryptococcal infections. Amp B makes the yeast membranes porous by binding with ergosterol [19], flucytosine prevents protein synthesis by intercalating into fungal RNA [20,21], and Flu acts by binding and inhibiting 14-α demethylase- enzyme which is important for ergosterol synthesis in fungal cells [22]. Combinational therapy of these drugs with different mechanisms of action makes it difficult for fungal cells to develop resistance against these drugs during the course of treatment [23]. However, flucytosine is associated with bone marrow suppression [24]. Also, flucytosine is excessively expensive (>$500/day) and is not licensed in many countries [25]. Therefore, WHO recommended the use of Flu in place of flucytosine [17]. The emergence of resistance to Flu and the toxicity of Amp B [26,27] underscore the need to search for new compounds that have anti-cryptococcal activity.

Natural products have played a significant role in building the armory of anti-infectives. However, natural products of plant origin with anti-infective property have rarely been introduced as drug(s). *Prosopis glandulosa*, a medium-sized tree, is one of the two varieties of honey mesquites that is available in North America [28,29]. The genus *Prosopis* contains indolizidine alkaloids substituted with alkyl piperidine unit(s), of which juliprosopine exhibited strong in vitro antimicrobial, anti-dermatophytic, and amebicidal activities [30,31]. In continuation to our earlier work on indolizidine alkaloids [32], herein we report the in vivo anticryptococcal activity of PPD, which was isolated from *Prosopis glandulosa*, in a murine model. In addition, we also report the in vitro toxicity on HepG2 cells of indolizidine alkaloids (Prosopilosidine (**1**), isoprosopilosidine (**2**), juliprosine (**3**), prosopilosine (**4**), isoprosopilosine (**5**), and juliprosopine (**6**) (Figure 1) which were isolated from *P. glandulosa* that was collected from Nevada and Texas, USA.

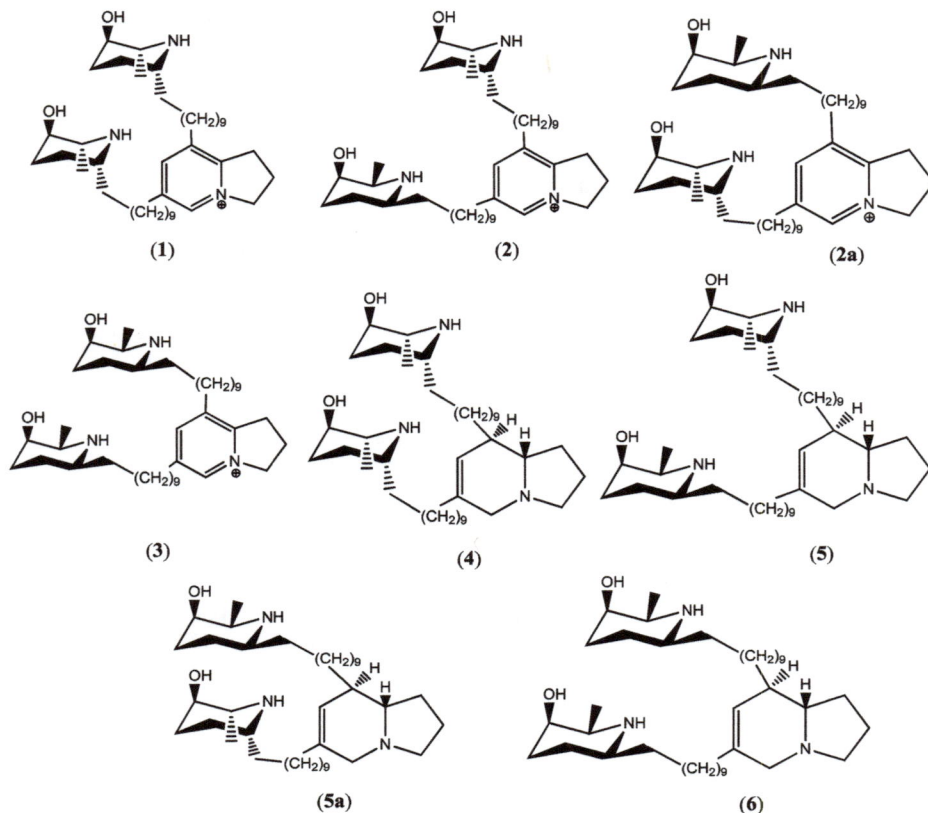

Figure 1. Chemical structures of indolizidine alkaloids isolated from *P. glandulosa* var. *gladulosa*: prosopilosidine (**1**), isoprosopilosidine (**2/2a**), juliprosine (**3**), prosopilosine (**4**), isoprosopilosine (**5/5a**), and juliprosopine (**6**).

2. Results and Discussion

2.1. *In Vitro Toxicity of* **1–6** *against HepG2 Cells*

In the MTS assay (Abcam, Cambridge, MA, USA), the treatment of HepG2 with tertiary indolizidine alkaloids, prosopilosine (**4**), isoprosopilosine (**5**), and juliprosopine (**6**) were found to be toxic at a concentration of >1 μg/mL. On the other hand, the two quaternary indolizidines, PPD and juliprosine (**3**), both with symmetrical piperidenyl side chains, did not show any toxicity up to 50 μg/mL, while their non-symmetrical diasterioisomer isoprosopilosidine (**2**) was found to be toxic at 50 μg/mL (Figure 2). Collectively, the evaluation of the cytotoxicity of indolizidine alkaloids against HepG2 cells revealed that the quaternary PPD and juliprosine (**3**) were found to be less toxic to those of tertiary indolizidine alkaloids (**4–6**), a phenomenon that is consistent with their toxicities against VERO cells (Samoylenko et al., 2009) [29].

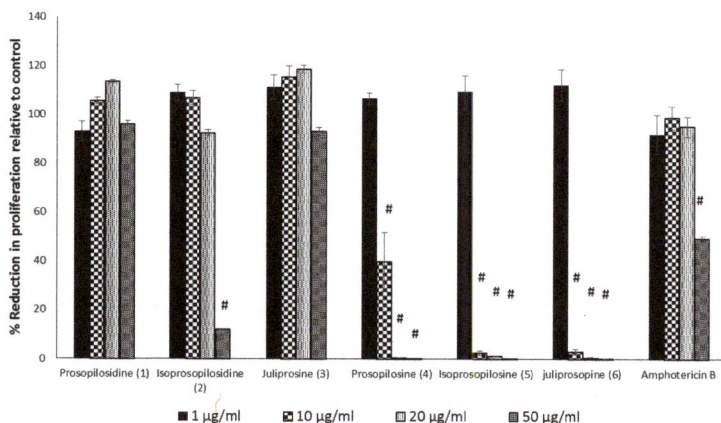

Figure 2. The effect of different compounds, prosopilosidine (**1**), isoprosopilosidine (**2**), juliprosine (**3**), prosopilosine (**4**), isoprosopilosine (**5**), juliprosopine (**6**) and Amphotericin B (Amp B) on cell viability. HepG2 cells were seeded on 96-well plates and were treated with 1, 10, 20, and 50 µg/mL of the test compounds or with vehicle only for 24 h. The MTS assay for cell viability and proliferation was performed with the CellTiter 96®AQueous Non-Radioactive Cell Proliferation Assay Kit (Promega). Data are shown as mean ± SEM and were analyzed with ANOVA. [#] $p < 0.001$ is considered statistically significant compared to the untreated cells (100%).

2.2. In Vivo Anti-Cryptococcal Activity of PPD

Two 2,3-dihydro-1*H*-indolizinium alkaloids, PPD and isoprosopilosidine (**2**), which were isolated from *P. glandulosa*, have demonstrated potent in vitro antifungal activity against *C. neoformans* and antibacterial activity against methicillin-resistant *S. aureus* and *M. intracellulare* [32]. Due to its potent in vitro activity against *C. neoformans* and its lower toxicity in mammalian VERO cells [32], anti-cryptococcal activity was conducted on PPD in vivo. The maximum tolerated i.p. dose in mice was determined to be 2.5 mg/kg. At this dose, no deaths were observed by the intraperitoneal route. At 5.0 and 10.0 mg/kg, deaths were observed (Table 1).

Table 1. Mean body weights of the mice injected i.p. with different doses of prosopilosidine (PPD). All values are expressed as mean ± SD (*n* = 5).

Dose of 1	Mean Body Weight of Mice on Day					
	1	2	3	4	5	6
Vehicle	24.40 ± 1.36	24.58 ± 1.4	24.37 ± 1.37	24.47 ± 1.41	24.55 ± 1.39	25.28 ± 0.88
1.0 mg/kg	24.83 ± 1.79	24.79 ± 1.81	23.66 ± 1.02	23.81 ± 0.95	24.28 ± 0.96	24.39 ± 0.93
2.5 mg/kg	24.8 ± 0.58	25.01 ± 0.68	25.1 ± 0.62	24.89 ± 0.37	24.85 ± 0.29	24.94 ± 0.28
5.0 mg/kg	23.55 ± 0.77	-	-	-	-	-
10 mg/kg	25.12 ± 0.66	-	-	-	-	-

Based on the MTD results, an exploratory in vivo experiment to determine the anti-cryptococcal activity of PPD by i.p. route was performed in a rodent model. In this exploratory experiment, treatment with PPD showed a dose dependent effect at low doses. In order to determine the maximum anti-cryptococcal effect of PPD, additional experiments were conducted with the addition of Amp B as a reference standard (1.5 mg/kg; i.p.), with doses ranging between 0.125 and 1.0 mg/kg/day for 5 days. PPD showed potent activity against *C. neoformans* at 0.125 mg/kg/day, with ~75% of organisms eliminated from the brain tissue, whereas Amp B at 1.5 mg/kg/day reduced ~83% of infection. Thus, the dose-dependent antifungal effect at lower doses, as was observed in our preliminary study, was reconfirmed in this experiment. Finally, the lowest dose for the maximum inhibitory effect, together with the dose-response relationships, were determined by another in vivo anti-cryptococcal experiment with 4 doses of 0.03125, 0.0625, 0.125, and 0.25 mg/kg/day. The maximum activity was observed at 0.0625 mg/kg, followed by 0.125 mg/kg, 0.25 mg/kg, 0.03125 mg/kg, where 80, 76, 71, and 68% of the organisms were eliminated from the brain tissue, respectively, vs. 83% by Amp B (Figure 3). However, increasing the dose further to 0.125 mg/kg/day or decreasing it to 0.03125 mg/kg/day did not change the antifungal activity significantly.

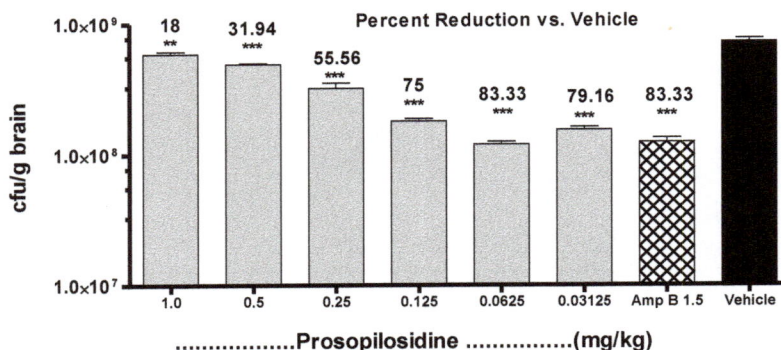

Figure 3. Treatment of *Cryptococcus neoformans* infection with different doses of prosopilosidine administered for 5 days in a murine model. Graph showing live colony forming units (CFU) of *C. neoformans* per gram of brain tissue with percent reduction in infection by prosopilosidine treatment given at doses 0.03125, 0.06215, 0.125, 0.25, 0.5, and 1.0 mg/kg body weight as compared to the vehicle control. Amp B: Amphotericin B (1.5 mg/kg body weight). Data are shown as mean + SEM ($n = 5$) and was analyzed with ANOVA. ** $p < 0.01$, *** $p < 0.001$ is considered statistically significant compared to the vehicle control.

The i.p. administration of PPD showed increasing activity against *C. neoformans* with decreasing concentration. This phenomenon appears to be similar to the "hormesis hypothesis" which was reported earlier [33,34], where high doses of a compound showed no observable effect, while at low doses, the effect was observed. It is unclear from the present study what could be the cause for this phenomenon. However, explanations can be speculated. It has been reported that certain compounds at different dose levels can induce different CYP enzymes in mice; at low doses (<0.25 mg/kg), perfluorodecanoic acid (PFDA) increases Cyp4A14 expression. Whereas, at higher doses (>10 mg/kg), in addition to CYP4A14, PFDA elevates the expression of Cyp2B10 [35]. It is possible that in our study, the i.p. administration of PPD at higher doses induced a set of CYP enzymes that metabolized it and rendered it ineffective against *C. neoformans*. At a low dose, such enzymes may not be induced and thus, a beneficial effect was observed at low doses. Another possibility lies in the transport of the compound inside the fungal cells. The channels by which this compound reaches the cryptococcal cells may be blocked at higher doses. It is known that many compounds up or down regulate different genes at different dose levels in yeast organisms [35]. One or more of the above hypothetical explanations may be responsible for the hormesis phenomenon that was observed in this study.

2.3. Bid Treatment with PPD

In an attempt to see if two doses per day would increase the antifungal effect of PPD, we considered *bid* (twice a day) in vivo treatment. Mice were infected via the i.v. route on day 0 and were then treated with PPD (24 h post infection) via the intraperitoneal route twice a day for 5 days. Figure 4 shows the comparison of PPD and Amp B in the mice. The number of *C. neoformans* organisms was substantially reduced in the brain by a *bid* dose of 0.0625 or 0.125 mg/kg. At these doses, the infection was reduced by 79.21% and 71.87%, respectively, with Amp B showing a 81.16% reduction in the number of *Cryptococcus* organisms in the brain tissue. It is important to mention that none of the animals that were treated with PPD showed any signs of distress during this dose regimen. Amp B treated animals showed rough hair coat throughout the course of the treatment, which is indicative of stress. The vehicle control group did not show any other sign of distress.

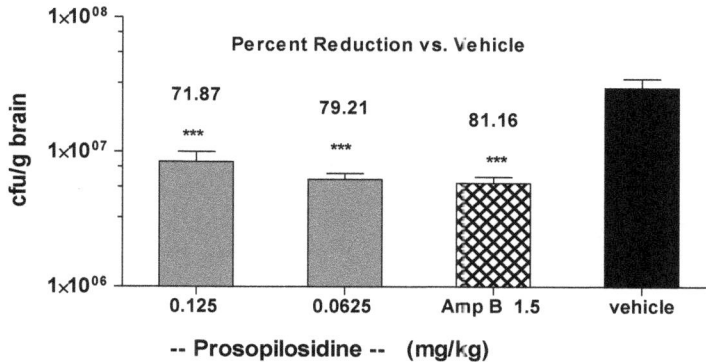

Figure 4. *Cryptococcus neoformans* infection in the mice treated with prosopilosidine administered *bid*. Graph showing live colony forming units (CFU) of *C. neoformans* per gram of brain tissue and percent reduction in infection by prosopilosidine treatment given at doses 0.06215 and 0.125 mg/kg body weight and amphotericin B (1.5 mg/kg body weight), as compared to the vehicle control. Amp B: Amphotericin B (1.5 mg/kg body weight). Data are shown as mean + SEM (n = 5) and was analyzed with ANOVA. *** $p < 0.001$ was considered statistically significant compared to the vehicle control.

2.4. Oral Administration of PPD

In a separate set of experiments, we attempted oral administration of PPD. These mice did not show any signs of distress for up to 4 h after PPD administration via the oral route. Then, the animals were sacrificed by CO_2 asphyxiation. At necropsy, all visceral organs appeared to be normal. No remarkable lesion was seen in the GI tract on gross examination. The blood clinical chemistry showed no shift in any of the parameters listed in Table 2.

Table 2. Blood chemistry profile of mice administered Prosopilosidine at 20 mg/kg via the oral route. All values are expressed as mean $\pm$ SEM (n = 3).

Parameters	Vehicle	Prosopilosidine (20 mg/kg)
Albumin (g/dL)	3.5 $\pm$ 0	3.1 $\pm$ 0
Alanine transaminase (ALT) U/L	35.0 $\pm$ 0	54.5 $\pm$ 26.5
Total bilirubin (mg/dL)	0.3 $\pm$ 0	0.2 $\pm$ 0
Blood urea nitrogen (BUN) mg/dL	21.0 $\pm$ 1	17.5 $\pm$ 0.5
Creatinine (CRE) mg/dL	0.245 $\pm$ 0.05	0.3 $\pm$ 0
BUN/CRE	89.3 $\pm$ 15.97	58.33 $\pm$ 1.67
Calcium (mg/dL)	10.65 $\pm$ 0.05	10.0 $\pm$ 0.3
Phosphate (mg/dL)	7.3 $\pm$ 0.1	9.5 $\pm$ 0
Glucose (mg/dL)	70.0 $\pm$ 9	145.5 $\pm$ 29.5
Na^+ (mmol/L)	156.5 $\pm$ 2.5	150.5 $\pm$ 0.5
K^+ (mmol/L)	8.4 $\pm$ 0.2	5.45 $\pm$ 0.25
Total protein (g/dL)	5.4 $\pm$ 0	5.0 $\pm$ 0
Globulin (g/dL)	1.95 $\pm$ 0.05	1.9 $\pm$ 0

This data confirmed that PPD was not toxic if given orally up to 20 mg/kg. We then conducted an efficacy study with the oral administration of PPD in the infected mice. The mice were infected by *Cryptococcus neoformans* inoculum via the i.v. route on day 0 and were then treated with PPD (24 h post infection) via the oral route once a day for 5 days. Oral administration of PPD showed a dose dependent activity against *C. neoformans*. Figure 5 shows the number of live organisms (CFU) that were recovered from the brains of the mice who were treated with PPD (using doses of 2.5, 5.0, and 10 mg/kg/day), Flu (15 mg/kg/day), or the vehicle. PPD, Flu, and the vehicle were administered orally in a volume of 100 μL per mouse via oral gavage needle. The vehicle treated group showed an average of 3.5×10^8 CFUs in their brain tissues. In comparison, orally administered Flu at 15 mg/kg eliminated the organism in the brain, showing over 93% reduction in infection compared to the vehicle control. At doses 2.5, 5.0, and 10 mg/kg, a direct dose-effect relationship was observed with a maximum 82% reduction in infection which was observed at 10 mg/kg. At doses 2.5 and 5.0 mg/kg, the reduction in infection was 48% and 65%, respectively. It is evident from these results that PPD reached the target organ after absorption and was able to reduce the *C. neoformans* load in the brain tissue. It is also clear from these results that PPD was well tolerated, since no deaths or apparent signs of toxicity were observed.

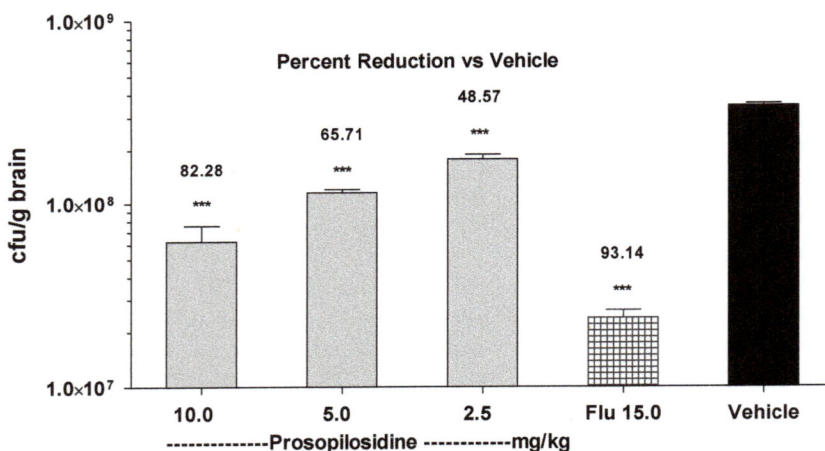

Figure 5. *Cryptococcus neoformans* infection in mice treated with prosopilosidine given via the oral route. Graph showing live colony forming units (CFU) of *C. neoformans* per gram of brain tissue and the percent reduction in infection by prosopilosidine treatment given at doses 2.5, 5.0, and 10.0 mg/kg body weight and Fluconazole (Flu) (15 mg/kg body weight) as compared to the vehicle control. Data are shown as mean ± SEM (*n* = 5) and was analyzed with ANOVA. *** $p < 0.001$ was considered significant as compared to the vehicle control.

3. Materials and Methods

3.1. Compounds and Chemicals

Prosopilosidine (**1**), isoprosopilosidine (**2**), prosopilosine (**4**), isoprosopilosine (**5**), and juliprosopine (**6**) were isolated from *P. glandulosa* (Fam. Leguminosae), which was collected from Nevada, USA, as described previously by Samoylenko et al. (2009) [29]. In addition, compounds (**3**) and (**6**) were isolated from the *P. glandulosa* sample that was collected in Texas. Amp B and Flu were purchased from Sigma-Aldrich (St. Louis, MO, USA). Amphotericin B and PPD were separately dissolved as a stock solution (1 mg/mL containing 13.3 μL DMSO, 13.3 μL Tween-20, and 973.4 μL distilled water) as colored and clear solutions, respectively. They were protected from light and were stored in the refrigerator. Different doses were prepared from this stock and were used daily. CellTiter 96® AQueous Non-Radioactive Cell Proliferation Assay kit (MTS) was purchased from Promega (Madison, WI, USA).

3.2. Animals

Female mice (CD-1) weighing 20–25g were obtained from Envigo (Indianapolis, IN, USA). They were quarantined on arrival for at least 3 days at the University of Mississippi vivarium. All of the animals were housed in plastic cages with fiberglass filter tops and were provided with food and water *ad-libitum*. They were maintained according to the Institutional Animal Care and Use Committee (IACUC) guidelines of the University. All experiments reported here were approved by the IACUC protocol number 07-011.

3.3. Inoculum

C. neoformans ATCC 90113 was grown on Saburaud Dextrose Agar (SDA) at 30 °C for 48 h to check for purity. A single colony was inoculated to 20 mL Saburaud Dextrose Broth (SDB) in a 100 mL flask and was kept at 30 °C in a shaker incubator overnight. The broth culture was centrifuged in a 50 mL tube and was washed three times with PBS. The pellet was suspended in 10 mL PBS and was adjusted to approximately 1–5×10^6 fungal cells/mL. This inoculum was kept on ice until all of the animal inoculations were completed. Serial dilutions from the inoculum were grown on SDA at 30 °C to confirm the inoculum size by determining the live colony forming units (CFU).

3.4. In Vitro Cytotoxicity Assay

Human hepatoma (HepG2) cells were maintained at 37 °C in equilibration with 5% CO_2-95% air in 75-cm^2 flasks containing maintenance medium plus 10% fetal bovine serum (FBS). The maintenance media (DMEM) consisted of 10% fetal bovine albumin, 1% nonessential amino acids, 1% L-glutamine and 100 U/mL penicillin, and 10 mg/mL streptomycin. Subcultures of the cells for use in the experiments were obtained from a 1:4 split of the confluent monolayers. The cells were seeded on a flat bottom 96 well plate (2×10^4 cells per well). The cells were exposed to different doses of compounds **1–6** and tryptamine for 24 h, and cytotoxicity was determined using MTS assay kit according to the manufacturer's instructions (Promega, Madison, WI, USA). Tryptamine was evaluated as a control, which was found to be non-toxic up to 50 µg/mL.

3.5. Maximum Tolerated Dose

The maximum tolerated dose (MTD) in the mice was determined by giving intraperitoneal (i.p.) injections of PPD, with doses ranging from 1 to 10 mg/kg, daily for 5 days. The body weights of the mice and their deaths were recorded. Necropsy was performed on all mice and gross changes were recorded. In a separate experiment, the mice were administered PPD orally at 20 mg/kg using oral gavage needles. Blood was collected 4 h later and was subjected to diagnostic profile Vetscan2 (Abaxis, Union City, CA, USA) to observe any change in the clinical chemistry parameters.

3.6. Experimental Design

Mice were inoculated with 100 µL of the inoculum intravenously (i.v.) via the tail vein and were distributed in various groups ($n = 5$/group). Twenty-four hours post inoculation, the mice were administered their respective treatments (different doses of PPD, Amp B, and vehicle) i.p. or orally using oral gavage needles for 5 days. The animals were dosed either once a day or twice a day at 12 h intervals (*bid*). On day 6 post inoculation, all mice were sacrificed by CO_2 asphyxiation, as approved by the IACUC. Their brains were aseptically removed, weighed, and homogenized in 5 mL of PBS. Serial dilutions of these homogenates were made in PBS and were cultured in duplicates on SDA. After 48 h of incubation at 30 °C, CFU from each homogenate was enumerated. The number of CFU of

C. neoformans per gram of brain tissue was determined for each mouse. Percent reduction in CFU's per gram of brain tissue was calculated by formula:

$$\% \text{ reduction} = \frac{\text{Vehicle} - \text{Sample}}{\text{Vehicle}} \times 100$$

3.7. In Vivo Clinical Chemistry of PPD

In order to assess the toxic potential of PPD by the oral route, a dose of 20 mg/kg (twice the maximum oral anti-cryptococcal dose used in this study) was given to mice ($n = 5$). Blood was sampled 4 h after the oral dose and liver profiles were determined by using VetScan2 (Abaxis, Union City, CA, USA). The major parameters included albumin, alanine transaminase (ALT), Blood urea nitrogen (BUN), glucose, total bilirubin, total protein, globulin, and electrolytes. Blood from three mice from each group was analyzed for clinical chemistry.

3.8. Statistical Analysis

Differences between the groups were analyzed using ANOVA followed by Dunnett's multiple comparisons test using Graph Pad Prism 5.0, and the minimum criterion for statistical significance was set at $p < 0.05$ for all of the comparisons.

4. Conclusions

This is the first report of in vivo anti-cryptococcal activity of PPD isolated from *P. glandulosa*. The results indicate that i.p. administration of PPD against *C. neoformans* infection showed better efficacy at lower doses. In addition, no signs of discomfort were observed in the mice who were treated with PPD, which was administered either once or twice a day. Higher doses of PPD were equally effective when they were given orally.

Author Contributions: M.K.A. and I.M. formulated the research idea; M.S.A.-B. and V.S. carried out the experimental work; M.K.A., M.T.M., and A.A.R. interpreted the data and prepared the manuscript. All of the authors have read and approved the final manuscript.

Funding: This work was supported by the Overhead Funds of NCNPR and in part by the USDA-ARS Specific Cooperative Agreement No. 58-6408-1-603.

Acknowledgments: The authors sincerely thank Alice M. Clark, Vice-Chancellor for Research and sponsored programs, UM, for her valuable advice and suggestions on the antifungal activity of PG compounds during new drugs for OI infection meetings. The authors acknowledge Jon F. Parcher for editing the manuscript for the English language. The authors also thank Mohamed Ali Ibrahim for drawing the structurers in Figure 1 of the manuscript. Last but not least, the authors also thank Penny Bolton for her assistance in the maintenance of the animals.

Conflicts of Interest: The authors declare no conflicts of interest.

References

1. O'Meara, T.R.; Alspaugh, J.A. The *Cryptococcus neoformans* Capsule: A Sword and a Shield. *Clin. Microbiol. Rev.* **2012**, *25*, 387–408. [CrossRef] [PubMed]

2. Alspaugh, J.A. Virulence Mechanisms and *Cryptococcus neoformans* pathogenesis. *Fungal Genet. Biol.* **2015**, *78*, 55–58. [CrossRef] [PubMed]

3. Fisher, J.F.; Valencia-Rey, P.A.; Davis, W.B. Pulmonary Cryptococcosis in the Immunocompetent Patient-Many Questions, Some Answers. *Open Forum Infect. Dis.* **2016**, *3*, ofw167. [CrossRef] [PubMed]

4. Lui, G.; Lee, N.; Ip, M.; Choi, K.W.; Tso, Y.K.; Lam, E.; Chau, S.; Lai, R.; Cockram, C.S. Cryptococcosis in apparently immunocompetent patients. *QJM* **2006**, *99*, 143–151. [CrossRef] [PubMed]

5. Sabiiti, W.; May, R.C. Mechanisms of infection by the human fungal pathogen *Cryptococcus neoformans*. *Future Microbiol.* **2012**, *7*, 1297–1313. [CrossRef] [PubMed]

6. Currie, B.P.; Casadevall, A. Estimation of the prevalence of cryptococcal infection among HIV infected individuals in New York city. *Clin. Infect. Dis.* **1994**, *19*, 1029–1033. [CrossRef] [PubMed]

7. Chayakulkeeree, M.; Perfect, J.R. Cryptococcosis. *Infect. Dis. Clin. N. Am.* **2006**, *20*, 507–544. [CrossRef] [PubMed]
8. McKenney, J.; Smith, R.M.; Chiller, T.M.; Detels, R.; French, A.; Margolick, J.; Klausner, J.D. Prevalence and correlates of cryptococcal antigen positivity among AIDS patients—United States, 1986–2012. *Morb. Mortal. Wkly. Rep.* **2014**, *63*, 585–587. [CrossRef] [PubMed]
9. Mirza, S.A.; Phelan, M.; Rimland, D.; Graviss, E.; Hamill, R.; Brandt, M.E.; Gardner, T.; Sattah, M.; de Leon, G.P.; Baughman, W.; et al. The changing epidemiology of cryptococcosis: An update from population-based active surveillance in 2 large metropolitan areas, 1992–2000. *Clin. Infect. Dis.* **2003**, *36*, 789–794. [CrossRef] [PubMed]
10. Kaplan, J.E.; Hanson, D.; Dworkin, M.S.; Frederick, T.; Bertolli, J.; Lindegren, M.L.; Holmberg, S.; Jones, J.L. Epidemiology of human immunodeficiency virus-associated opportunistic infections in the United States in the era of highly active antiretroviral therapy. *Clin. Infect. Dis.* **2000**, *30*, S5–S14. [CrossRef] [PubMed]
11. Rajasingham, R.; Smith, R.M.; Park, B.J.; Jarvis, J.N.; Govender, N.P.; Chiller, T.M.; Denning, D.W.; Loyse, A.; Boulware, D.R. Global Burden of Disease of HIV-Associated Cryptococcal Meningitis: An Updated Analysis. *Lancet Infect. Dis.* **2017**, *17*, 873–881. [CrossRef]
12. Sanglard, D. Emerging Threats in Antifungal-Resistant Fungal Pathogens. *Front. Med. Lausanne* **2016**, *3*, 11. [CrossRef] [PubMed]
13. Xu, J.; Onyewu, C.; Yoell, H.J.; Ali, R.Y.; Vilgalys, R.J.; Mitchell, T.G. Dynamic and Heterogeneous Mutations to fluconazole resistance in *Cryptococcus neoformans*. *Antimicrob. Agent. Chemother.* **2001**, *45*, 420–427. [CrossRef] [PubMed]
14. Yamazumi, T.; Pfaller, M.A.; Messer, S.A.; Houston, A.K.; Boyken, L.; Hollis, R.J.; Furuta, I.; Jones, R.N. Characterization of heteroresistance to fluconazole among clinical Isolates of *Cryptococcus neoformans*. *J. Clin. Microbiol.* **2003**, *41*, 267–272. [CrossRef] [PubMed]
15. Bicanic, T.; Harrison, T.; Niepieklo, A.; Dyakopu, N.; Meintjes, G. Symptomatic relapse of HIV-associated cryptococcal meningitis after initial fluconazole monotherapy: The role of fluconazole resistance and immune reconstitution. *Clin. Infect. Dis.* **2006**, *43*, 1069–1073. [CrossRef] [PubMed]
16. Seilmaier, M.; Hecht, A.; Guggemos, W.; Rüdisser, K. Cryptococcal Meningoencephalitis Related to HIV Infection with Resistance to Fluconazole, Relapse, and IRIS. *Med. Klin. Munich* **2009**, *104*, 58–62. [CrossRef] [PubMed]
17. World Health Organization. *Rapid Advice: Diagnosis, Prevention and Management of Cryptococcal Disease in HIV-Infected Adults, Adolescents and Children*; World Health Organization: Geneva, Switzerland, 2011. Available online: http://www.who.int/hiv/pub/cryptococcal_disease2011/en/ (accessed on 25 January 2018).
18. Day, J.N.; Chau, T.T.; Wolbers, M.; Mai, P.P.; Dung, N.T.; Mai, N.H.; Phu, N.H.; Nghia, H.D.; Phong, N.D.; Thai, C.Q.; et al. Combination antifungal therapy for cryptococcal meningitis. *N. Engl. J. Med.* **2013**, *368*, 1291–1302. [CrossRef] [PubMed]
19. Gray, K.C.; Palacios, D.S.; Dailey, I.; Endo, M.M.; Uno, B.E.; Wilcock, B.C.; Burke, M.D. Amphotericin primarily kills yeast by simply binding ergosterol. *Proc. Natl. Acad. Sci. USA* **2012**, *109*, 2234–2239. [CrossRef] [PubMed]
20. Polak, A.; Scholer, H.J. Mode of action of 5-fluorocytosine and mechanisms of resistance. *Chemotherapy* **1975**, *21*, 113–130. [CrossRef] [PubMed]
21. Diasio, R.B.; Bennett, J.E.; Myers, C.E. Mode of action of 5-fluorocytosine. *Biochem. Pharmacol.* **1978**, *27*, 703–707. [CrossRef]
22. Odds, F.C.; Brown, A.J.; Gow, N.A. Antifungal agents: Mechanisms of action. *Trends Microbiol.* **2003**, *11*, 272–279. [CrossRef]
23. Bicanic, T.; Meintjes, G.; Wood, R.; Hayes, M.; Rebe, K.; Bekker, L.G.; Harrison, T. Fungal burden, early fungicidal activity, and outcome in cryptococcal meningitis in antiretroviral-naive or antiretroviral-experienced patients treated with amphotericin B or fluconazole. *Clin. Infect. Dis.* **2007**, *45*, 76–80. [CrossRef] [PubMed]
24. Rajasingham, R.; Rolfes, M.A.; Birkenkamp, K.E.; Meya, D.B.; Boulware, D.R. Cryptococcal meningitis treatment strategies in resource-limited settings: A cost-effectiveness analysis. *PLoS Med.* **2012**, *9*, e1001316. [CrossRef] [PubMed]
25. Kauffman, C.A.; Frame, P.T. Bone marrow toxicity associated with 5-fluorocytosine therapy. *Antimicrob. Agents Chemother.* **1977**, *11*, 244–247. [CrossRef] [PubMed]

26. Bicanic, T.; Bottomley, C.; Loyse, A.; Brouwer, A.E.; Muzoora, C.; Taseera, K.; Jackson, A.; Phulusa, J.; Hosseinipour, M.C.; van der Horst, C.; et al. Toxicity of Amphotericin B Deoxycholate-Based Induction Therapy in Patients with HIV-Associated Cryptococcal Meningitis. *Antimicrob. Agents Chemother.* **2015**, *59*, 7224–7231. [CrossRef] [PubMed]

27. Falci, D.R.; da Rosa, F.B.; Pasqualotto, A.C. Hematological toxicities associated with amphotericin B formulations. *Leuk. Lymphoma* **2015**, *56*, 2889–2894. [CrossRef] [PubMed]

28. Bukhart, A.A. Monograph of the genus *Prosopis. J. Arnold Arbor.* **1976**, *57*, 450–525.

29. Hilu, Y.W.; Boyd, S.; Felker, P. Morphological diversity and toaxonomy of *California mesquites* (*Prosopis*, Lepminosae). *Madrono* **1982**, *29*, 237–254.

30. Snider, B.B.; Neubert, B.J. Syntheses of Ficuseptine, Juliprosine, and Juliprosopine by Biomimetic Intramolecular Chichibabin Pyridine Syntheses. *Org. Lett.* **2005**, *7*, 2715–2718. [CrossRef] [PubMed]

31. Ahmad, A.; Ahmad, V.; Khalid, M.S.; Ansari, F.A.; Khan, K.A. Study on the Antifungal efficacy of Juliflorine and a Benzene-insoluble alkaloidal fraction of *Prosopis juliflora. Philipp. J. Sci.* **1997**, *126*, 175–182.

32. Samoylenko, V.; Ashfaq, M.K.; Jacob, M.R.; Tekwani, B.L.; Khan, S.I.; Manly, S.P.; Joshi, V.C.; Walker, L.A.; Muhammad, I. Indolidine, Antiinfective and Antiparasitic Compounds from *Prosopis glandulosa* var. *glandulosa. J. Nat. Prod.* **2009**, *72*, 92–98. [CrossRef] [PubMed]

33. Calabrese, E.J. Primer on BELLE. In *Biological Effects of Low Level Exposures: Dose-Response Relationships*; Calabrese, E.J., Ed.; CRC/Lewis Publishers: Boca Raton, FL, USA, 1994; pp. 27–42.

34. Calabrese, E.J.; Baldwin, L.A. Hormesis as a biological hypothesis. *Environ. Health Perspect.* **1998**, *106*, 357–362. [CrossRef] [PubMed]

35. Cheng, X.; Klaassen, C.D. Perfluorocarboxylic Acids Induce Cytochrome P450 Enzymes in Mouse Liver through Activation of PPAR-α and CAR Transcription Factors. *Toxicol. Sci.* **2008**, *106*, 29–36. [CrossRef] [PubMed]

Sample Availability: Samples of compounds **1–6** mentioned in this manuscript are available from the authors.

![molecules logo]

![MDPI logo]

Article

Preparation and Characterization of Resveratrol Loaded Pectin/Alginate Blend Gastro-Resistant Microparticles

Oihane Gartziandia [1,2], **Arrate Lasa** [3,4,*], **Jose Luis Pedraz** [1,2], **Jonatan Miranda** [3,4], **Maria Puy Portillo** [3,4], **Manoli Igartua** [1,2] and **Rosa Maria Hernández** [1,2,*]

1   NanoBioCel Group, Laboratory of Pharmaceutics, School of Pharmacy, University of the Basque Country (UPV/EHU), Vitoria-Gasteiz 01006, Spain; oihane.gartziandia@ehu.eus (O.G.); joseluis.pedraz@ehu.eus (J.L.P.); manoli.igartua@ehu.eus (M.I.)

2   Biomedical Research Networking Centre in Bioengineering, Biomaterials and Nanomedicine (CIBER-BBN), Vitoria-Gasteiz 01006, Spain

3   Nutrition and Obesity Group, Department of Nutrition and Food Science, Faculty of Pharmacy and Lucio Lascaray Research Institute, University of País Vasco (UPV/EHU), Vitoria-Gasteiz 01006, Spain; jonatan.miranda@ehu.eus (J.M.); mariapuy.portillo@ehu.es (M.P.P.)

4   CIBEROBN Physiopathology of Obesity and Nutrition, Institute of Health Carlos III (ISCIII), Vitoria-Gasteiz 01006, Spain

*   Correspondence: arrate.lasa@ehu.eus (A.L.); rosa.hernandez@ehu.eus (R.M.H.); Tel.: +34-9450-14070 (A.L.); +34-9450-13095 (R.M.H.); Fax: +34-9450-13040 (R.M.H.)

Academic Editors: Muhammad Ilias and Charles L. Cantrell
Received: 5 July 2018; Accepted: 27 July 2018; Published: 28 July 2018

Abstract: Background: The use of resveratrol as a dietary supplement is limited because it is easily oxidized and, after oral ingestion, it is metabolized into enterocytes and hepatocytes. Thus, new formulations are needed in order to improve its oral bioavailability. Objective: The objective of this study was to develop and characterize a gastro-resistant formulation of resveratrol for oral administration as a dietary supplement. Method: Resveratrol was encapsulated in Eudragit-coated pectin-alginate microparticles. Results: The microparticle size was about 1450 μm, with an encapsulation efficiency of 41.72% ± 1.92%. The dissolution assay conducted, as specified in the European Pharmacopoeia for delayed-release dosage forms, revealed that our microparticles were gastro-resistant, because the resveratrol percentage released from microparticles in acid medium was less than 10%. In addition, the high-performance liquid chromatographic (HPLC) method developed for resveratrol content quantification in the microparticles was validated according to International Council for Harmonisation (ICH) Q2 (R1) guidelines. Finally, the biological activity of resveratrol was investigated in 3T3-L1 mature adipocytes, concluding that the encapsulation process does not affect the activity of resveratrol. Conclusion: In summary, the gastro-resistant microparticles developed could represent a suitable method of including resveratrol in dietary supplements and in functional foods used in obesity therapy.

Keywords: resveratrol; dietary supplement; gastro-resistant; microparticles; obesity; HPLC

1. Introduction

Resveratrol (3,5,4′-trihydroxystilbene), a polyphenolic phytoalexin, is naturally produced in the fruits of several plant species such as grapes, mulberries, and peanuts in response to exogenous stress factors such as injuries, fungal infections, or UV irradiation [1]. It was first found in 1940 in the roots of white hellebore (*Veratrum grandiflorum*). Later, in 1963, it was isolated from the roots of *Polygonum cuspidatum*, a plant used in traditional Chinese medicine. Nowadays, it is mainly extracted from red grapes, whose fresh skin is estimated to contain about 50–100 µg of resveratrol per gram.

This polyphenol is a stilbene derivate, which exists in *cis*- and in *trans*- stereoisomeric forms (Figure 1). The most common form in nature, the *trans* form, is relatively stable, but it can undergo isomerisation to the *cis*- form when exposed to ultraviolet irradiation [2]. This isomerisation is not desirable, because the *trans*- form is the steric form, which is responsible for the beneficial effects of this compound. Therefore, it is important to maintain its stability in order to retain its biological and pharmacological activities [1].

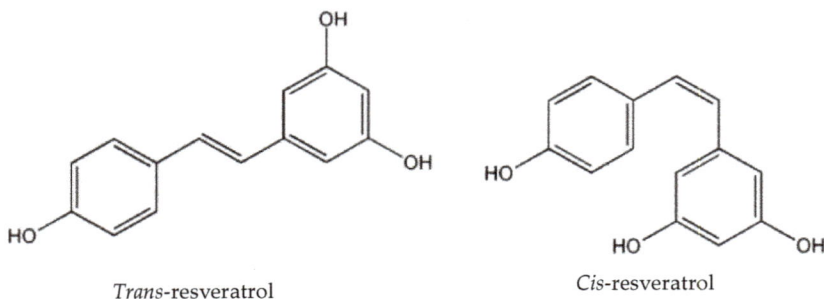

Trans-resveratrol   *Cis*-resveratrol

Figure 1. Chemical structure of *trans*- and *cis*-resveratrol.

The pharmacological activity of resveratrol takes the form of an antioxidant and anti-inflammatory effect on a wide range of disorders associated with inflammation, such as diabetes mellitus, obesity, cardiovascular disease, neurodegenerative diseases, or cancer [2]. Currently, resveratrol is mainly used as an antioxidant dietary supplement to protect against cardiovascular problems and some alterations associated with aging, and as a supplement in the treatment of cancer, obesity, diabetes, or hypercholesterolemia [3]. The doses of resveratrol used in these supplements range from 50 to 500 mg per capsule.

Nevertheless, the use of resveratrol is limited by its being easily oxidable and extremely photosensitive. In addition, its low solubility in water and its rapid metabolism in enterocytes and hepatocytes (where sulphate and glucuronic conjugates are produced), as well as its rapid elimination, make the oral resveratrol bioavailability very low [4,5]. As a result, several strategies are being carried out to increase its bioavailability, such as gastro-resistant microparticles that delay release until it reaches the distal portions of the intestinal tract [6].

Microencapsulation consists of coating drugs with different materials to obtain particles of micrometric size. It is a commonly used technique in pharmaceutical, cosmetic, and food industries, because it increases the stability of the active component and the release of the drug is controlled [1].

In this context, pectin (a natural polysaccharide) may be considered as a suitable substance for the development of gastro-resistant microparticles, as it is specifically degraded by pectinolytic enzymes produced by colonic microbiota, while resisting enzymes present in the stomach and intestine. This natural polysaccharide is composed of partially methoxylated poly (1-4-α-D-galacturonic acids) with some 1–2 linked L-rhamnose groups [7]. It is present in the cell wall of most edible plants, and so it is approved for its use as an excipient in oral formulations [8]. Moreover, pectins with low levels of methoxylation can be cross-linked with calcium ions (Ca^{2+}, divalent cations) to produce Ca-pectinate networks, therefore, these have been used as vehicles for drug delivery [9].

In addition to pectin, alginate has also been used for the development of microparticles because of its bio/mucoadhesive properties. Thus, the preparation of ca-pectinate/alginate microparticles is an interesting approach, because pectin can protect from enzymatic digestion and alginate can prolong the residence time of particles in the administration site, and so achieve controlled drug release [10].

However, the use of pectin and alginate as the sole components of microparticles is not enough to completely avoid resveratrol release until it reaches the distal intestinal tract. For this reason, in order to obtain gastro-resistant microparticles, they were coated with Eudragit® FS-30D, a methacrylic acid copolymer that dissolves at pH 7. This polymer avoids the dissolution of the microparticles in the first portions of the gastrointestinal tract, and the consequent release of the drug in the portions of the intestine with a pH $\geq$ 7 [11].

With these considerations in mind, the objective of this study was to develop and characterize resveratrol loaded pectin/alginate blend gastro-resistant microparticles to be orally administered as a dietary supplement.

2. Results

2.1. Validation of the Quantification Technique of Resveratrol by High-Performance Liquid Chromatographic (HPLC)

Following ICH Q2(R1) guidelines, we have demonstrated that our method meets the linearity criteria described in Table 1. The results show an excellent correlation between the areas of the chromatographic peak and the concentration of the analyte in the concentration range of 5–60 µg/mL.

Table 1. Acceptance criteria.

Parameter		Especification	
	Selectivity	Identification	-
		Resolution	Rs > 1.5
		Absence of interference	no interferences
	Linearity	Correlation coefficient	r $\geq$ 0.999
		C.V. response factors	C.V. $\leq$ 2%
		Relative error percentage	$\leq$2%
		Slope linearity test	$t_{exp} > t_{table}$
		Slope confidence intervals	No include 0
		Test of proportionality	$t_{exp} < t_{table}$
		Intercept confidence intervals	Include 0
	Repeatability of Instrumental System	C.V.	C.V. < 1.37%
	Repeatability of the Method	C.V.	C.V. < 1.94%
	Intermediate Precision	C.V. individuals	C.V. < 1.94%
		C.V. intermediate	C.V. < 3.88%
	Recovery percentage	Recovery	98.0–102.0%
		Relative error percentage	$\leq$2%
		Test for equality of variances	$G_{exp} < G_{table}$
Robustness	Influence of analyst	C.V.	C.V. < 3.88%
	Influence of analysis temperature	C.V.	C.V. < 1.94%
Stability	Standard	Concentration in relation to time	98.0–102.0%
	Sample	0	

C.V.: Coefficient of variation; G_{exp}: Experimental G value; G_{table}: G value in tables.

We also demonstrated that our method was selective, precise, accurate, and specific because all parameters met the established acceptance criteria (Table 1).

In addition, we proved that although the resveratrol standard solution was stable for 96 h at room temperature, the sample test solution was only stable for 24 h, taking into account the established acceptance criteria (Table 1).

2.2. Particle Size and Morphology

After measuring the diameter of randomly selected fifty dry microparticles, as expected, the mean particle size was 1443.45 µm $\pm$ 126.11 µm, as we expected, as a result of using the ionotropic gelation method with a 25G needle to prepare the microparticles [9].

In terms of morphology, the newly prepared microparticles showed a spherical and uniform shape. However, with dehydration the morphology became more irregular, and particle size was reduced by about 500 μm, from 2000 to 1500 μm (Figure 2).

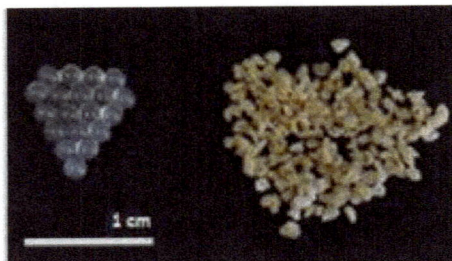

Figure 2. Comparison of morphology and particle size between newly prepared (left) and dried (right) microparticles.

2.3. Encapsulation Efficiency (EE)

EE (%) determined by HPLC was 41.72% ± 1.92%. The low EE was most probably linked to the lipophility of resveratrol, which is why during gelation resveratrol could undergo diffusion from pectin-alginate mixture to gelation solution ($CaCl_2$). Similar EE results were obtained by other authors using the same method for encapsulation of Cwp84 antigen [12].

2.4. Drug Release Study: Resveratrol Release from Microparticles

As we can see in the Figure 3, in the first 2 h of the study, when particles were in contact with acidic medium, the percentage of resveratrol released from microparticles was less than 10%, meeting the criteria established in the European Pharmacopoeia monograph for gastro-resistant formulations [13]. After 2 h, when the medium pH was changed from acidic to basic, the microspheres showed a faster drug release rate, until reaching 70% of resveratrol total content in 24 h.

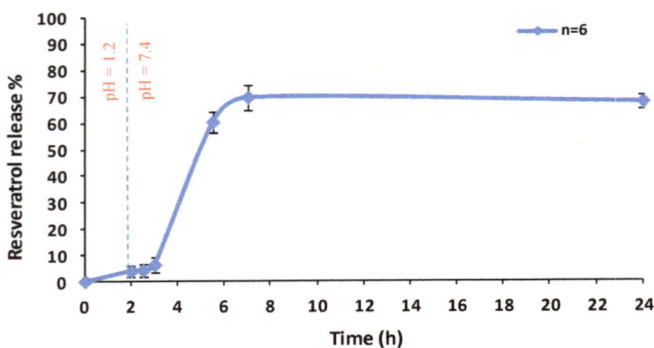

Figure 3. Percentage of resveratrol release from microparticles. The microparticles were dispersed in 750 mL of an acid medium (pH 1.2) for 2 h, followed by a basic medium (pH 7.4). Values represent the means ($n = 6$ for each test) ± the standard error.

2.5. In Vitro Bioactivity Study: Measurement of Triacylglycerol Content in Adipocytes

After incubation of adipocytes with resveratrol for 24 h, a reduction in triacylglycerol content was observed for both concentrations used (1 and 10 μM). However, a dose-dependent effect was not observed. When comparing the effect of resveratrol released from the microparticles with that of free resveratrol, no differences were found between both compounds (Figure 4).

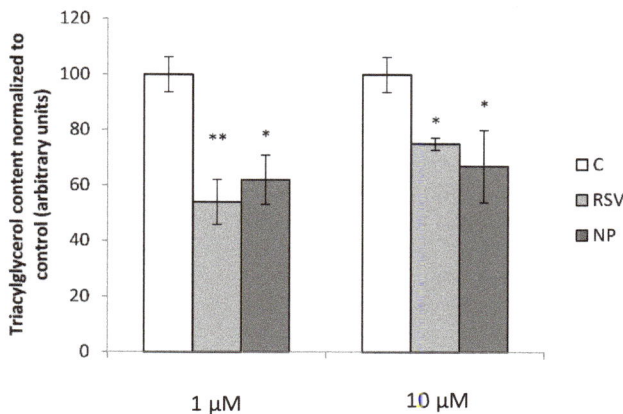

Figure 4. Triacylglycerol content in 3T3-L1 mature adipocytes after treatment with 1 and 10 μM free resveratrol (RSV) and RSV released from microparticles (NP). Values are means ± SEM. Comparison between each treatment with the control was analyzed by Student's *t*-test. The asterisks represent differences versus the control (* $p < 0.05$; ** $p < 0.01$).

3. Discussion

In the present study resveratrol loaded pectin/alginate blend gastro-resistant microparticles were prepared as a new formulation to improve the oral bioavailability of resveratrol. Pectin resists the enzymes present in the stomach and intestine, so it could protect resveratrol from metabolism until it reaches the more distal parts of the intestine. Furthermore, alginate has properties which can prolong the residence time of the microparticles in the administration site, and achieve a controlled release, thanks to its bio/mucoadhesive properties [10]. The combination of these compounds is a good strategy for the development of microparticles because they can be cross-linked with calcium ions (Ca^{2+}, divalent cations) obtaining Ca-pectinate-Ca-alginate networks. Therefore, the ionotropic gelation method has been used for the preparation of pectin/alginate blend microparticles [9]. The resulting microparticles were subsequently coated with the enteric polymer Eudragit®FS-30D to avoid completely the release of resveratrol until it reaches the distal intestinal tract, since this methacrylic acid copolymer dissolves to pH 7 [11].

In order to demonstrate that we succeeded in producing gastro-resistant microparticles, we carried out the dissolution assay described in the European Pharmacopoeia for delayed-release dosage forms [14]. As the release study revealed, when particles were in contact with acid medium, the percentage of resveratrol released from microparticles was less than 10%, meeting the criteria established in the European Pharmacopoeia monograph for gastro-resistant formulations. While when the pH of the medium was changed from acid to basic (pH 7.4), a faster dissolution rate which reached 70% of the total content in 24 h was observed. These results demonstrated that Eudragit FS-30D enteric coating protected resveratrol loaded microparticles from acid pH, and therefore, that resveratrol will be released into the distal portions of the intestinal tract after oral ingestion of miroparticles [15].

The advantages of microencapsulation to protect resveratrol and to increase its bioavailability have been described in this manuscript, but when this strategy is used, it is necessary to check whether it affects the biological activities of this compound. For his purpose, in the present study the effects of encapsulated resveratrol and free resveratrol (used as a control) on triacylglycerol content in 3T3-L1 mature adipocytes was analyzed because, according to the results obtained in in vitro and in vivo studies [16–18], resveratrol has been proposed as a potential anti-obesity compound. It seems to mimic the effects of energy restriction, thus leading to reduced body fat and improved insulin sensitivity [19]. Resveratrol supplements usually provide mg of this compound [20–22]. After absorption, intestinal and hepatic metabolism, the amounts of resveratrol found in plasma and tissues are in the range of 1–2 µM [23,24]. Taking this is mind, the amounts of resveratrol used in the present study for adipocyte culture were 1 and 10 µM.

Both forms of resveratrol, free and encapsulated, reduced triacylglycerol content at the two doses used, and no differences were observed between them, meaning that in fact resveratrol was not negatively affected by microencapsulation. In both cases the effect of 1 µM was greater than that of 10 µM comparing to control. Although this seems surprising, the relationship between the dose and the effectiveness is quite complex in polyphenol studies. In the case of resveratrol, this situation (greater effects of lower doses than those of higher doses) has been already found in in vivo studies by other authors. Thus, Cho et al., observed greater anti-obesity effect of resveratrol when mice were treated with this phenolic compound at a dose of 0.005% in the diet than when they were treated at a dose of 0.02% [25].

Taken as a whole, the present results show that pectin/alginate gastro-resistant microparticles protect resveratrol from acid pH, allowing the phenolic compound to be released into the distal portions of the intestinal tract after oral ingestion of miroparticles. In addition, resveratrol remains active after the encapsulation process. Consequently, this type of microencapsulation could be a useful strategy in the development of dietary supplements or functional foods that may be beneficial for the prevention or treatment of obesity.

4. Materials and Methods

4.1. Materials

Pectin (from apple) (CAS Number: 9000-69-5), alginate (from Brown algae) (CAS Number: 9005-38-3), calcium chloride (CAS Number: 10043-52-4), and pectinase (from Aspergillus aculeatus) (MDL number: MFCD00131809) were purchased from Sigma-Aldrich Química S.A (Madrid, Spain). *Trans*-resveratrol (Resveratrol 95% (HPLC) from Polygonum cuspidatum) was supplied by Monteloeder (Alicante, Spain). Eudragit FS-30D was obtained from Evonik (Essen, Alemania). Triethyl citrate (TEC) (CAS Number: 77-93-0) was donated by Morflex. Methanol HPLC grade (CAS Number: 67-56-1), Formic acid (98–100%, reagent grade) (CAS Number: 64-18-6), HCl 35% *w/w* (CAS Number: 7647-01-0), NaOH (CAS Number: 1310-73-2) and tri-Sodium phosphate monohydrate were purchased from Scharlau (Barcelona, Spain). Dulbecco´s modified Eagle´s medium (CAS Number: 103130-21) was supplied by GIBCO (BRL Life Technologies, Grand Island, NY, USA). Triacylglcyerols (TG) were determined by Infinity Triglycerides reagent (Thermo Electron Corporation, Rockford, IL, USA) and protein concentrations of cell extracts were measured with BCA reagent (Thermo Scientific, Rockford, IL, USA).

4.2. Preparation of Microparticles

Microparticles were prepared using the ionotropic gelation method [9]. Firstly, pectin (2%, (*w/v*); 2 g in 100 mL) and alginate (1%, (*w/v*); 1 g in 100 mL) were dissolved in deionized water and resveratrol (10%, *w/w*; 0.3 g) was dispersed in it. This mixture was then added dropwise through a needle of 25G on a gently agitated $CaCl_2$ dissolution (5%, (*w/v*); 10 g in 200 mL), using a peristaltic pump (ecoline, ISMATEC, 0.4 mL/min). Subsequently, microparticles were obtained by the gelation of the pectin and the alginate and held in magnetic stirring for 3 h. The microparticles were then separated; washed with distilled water; and dried on a fluidized bed, Mini Glatt 4® (for 10 min at 70 °C).

The dried microparticles were coated with a Eudragit® FS-30D (20%, (*w/w*)) and TEC (1.5%, (*v/w*)) solution, on the fluidized bed, Glatt® (70 °C, 0.6–0.7 bar). The coating solution was atomized into microparticles using a peristaltic pump (0.3 mL/min) until the desired coating thicknesses were achieved, with an increase in weight of 20%.

Empty microparticles were prepared by the same procedure, but without adding resveratrol to the pectin and alginate mixture.

4.3. Development and Validation of an HPLC Method for the Quantification of Resveratrol

Quantitative high-performance liquid chromatographic (HPLC) analysis was performed on a Waters HPLC system (Waters Corporation, Milford, MA, USA) equipped with a binary HPLC pump (Waters 1525), a dual λ Absorbance UV-visible detector (Waters 2487), a column oven (Waters Column Heater Module), and an auto sampler (Waters 717 plus Autosampler) controlled by software (Empower 3 Software), which was used for data analysis and processing.

An HPLC method was developed for the quantification of resveratrol in the microparticles. It was carried out on a XBridge column (4.6 × 75 mm, C18 2.5 μm) with column oven temperature of 35 °C using a gradient elution system consisting of MeOH and Formic acid 0.1%. The proportion of Formic acid/MeOH in the gradient was 50:50, 50:50, 10:90, 10:90, 50:50, and 50:50, at 0, 0.5, 2, 4.5, 4.51, and 8.5 minutes, respectively. The retention time of *trans*- and *cis*-resveratrol were 2.42 min and 3.84 min, respectively, with a flow rate of 0.7 mL/min. The analysis was carried out at 305 nm and 285 nm wavelength for *trans*- and *cis*-resveratrol, respectively, with a run time of 8.5 min.

The method was validated according to the ICH Q2(R1) guideline [13] in terms of linearity in the concentration range of 5–60 μg/mL, selectivity, precision, accuracy, specificity and stability. With the results obtained in each of the trials, it was seen whether the different parameters met the established acceptance criteria, described in Table 1.

4.4. Characterization of Microparticles

4.4.1. Particle Size

The size of the dry microparticles was measured by an optical microscope (Nikon ECLIPSE TE2000-S), set at 4X objective. Fifty microparticles were randomly selected, and the diameter of each one was measured with the help of the scale of the Eclipse Net software, after capturing the images through a digital camera (Nikon Digital Sight DS-U1) connected with the microscope. The average size of the microparticles was expressed as the mean diameter (μm) ± standard deviation (SD).

4.4.2. Encapsulation Efficiency (EE%)

Resveratrol EE was determined by a direct method. Briefly, 10 mg of microparticles was added to a topaz flask of 10 mL with a mix of MeOH–H_2O–Formic acid (45:55:0.1) and pectinase (1%, *w/v*). The mixture was left in magnetic stirring until the dissolution of the particles, in order to completely extract the resveratrol from microparticles. The concentration of resveratrol in the flask was determined by HPLC. The experiment was performed in triplicate. EE was expressed as the actual resveratrol mass percentage, compared with the total mass of resveratrol added initially.

4.4.3. Dissolution Assay

The assay was conducted as described in the European Pharmacopoeia for delayed-release dosage forms under Apparatus 1 (Sotax dissolution tester), using the specified media at 37 °C and 50 rpm [14].

Firstly, the test was performed in an acid medium (HCl, 0.1 N). For this, seven vessels of the apparatus were filled with 750 mL of 0.1 N hydrochloric acid, and microparticles were placed in the baskets (empty microparticles in the first vessel, and resveratrol loaded in the rest). After 2 h of operation, an aliquot of each vessel was withdrawn and replaced with the same volume of medium.

Immediately, 250 mL of tri-Sodium phosphate monohydrate was added to the vessels and the pH was adjusted to 7.4 with NaOH 2 N, to continue with the test in a basic medium. At different times (1 h, 2.5 h, 5 h, and 24 h) aliquots were withdrawn and replaced with the same volume of medium.

All aliquots were analyzed by HPLC to obtain the concentration of resveratrol, and results were expressed as cumulative percentage of resveratrol released from the microparticles at a given time, against the initial resveratrol loading in the microparticle sample.

4.5. In Vitro Bioactivity Assay

4.5.1. Experimental Design

The 3T3-L1 pre-adipocytes, supplied by American Type Culture Collection (Manassas, VA, USA), were cultured in DMEM containing 10% foetal calf serum (FCS). Two days after confluence (day 0), the cells were stimulated to differentiation with DMEM containing 10% *v/v* FCS, 10 μg/mL insulin, 0.5 mM isobutylmethylxanthine (IBMX), and 1 μM dexamethasone for two days. On day 4, the differentiation medium was replaced by FBS/DMEM medium (10%) containing 0.2 μg/mL insulin. This medium was changed every two days until cells were harvested. All media contained 1% *v/v* Penicillin/Streptomycin (10,000 U/mL), and the media for differentiation and maturation contained 1% (*v/v*) of Biotin and Panthothenic Acid. Cells were maintained at 37 °C in a humidified 5% CO_2 atmosphere.

4.5.2. Cell Treatment

For the treatment of mature adipocytes, cells grown in six-well plates were incubated with free resveratrol and resveratrol released from microparticles, at 1 and 10 μM (diluted in 95% ethanol, final ethanol concentration in the medium 0.1%) on day 12 after differentiation because on that day, >90% of cells had matured, with visible lipid droplets. After 24 h, supernatant was removed and cells were used for triglyceride (TG) determination. This experiment was repeated three times.

4.5.3. Measurement of Triacylglycerol Content in Adipocytes

Mature adipocytes were washed extensively with phosphate-buffered saline (PBS) and incubated three times with 800 μL of hexane/isopropanol (2:1). The total volume was then evaporated by nitrogen gas and the pellet was resuspended in 200 μL Tritón X-100 in 1% distilled water. Afterwards, TG were disrupted by sonication and the content was measured by a commercial kit. For protein determinations, cells were lysed in 0.3 N NaOH, 0.1% SDS. Protein measurements were performed using the BCA reagent. TG content results were obtained as mmol glycerol/mg protein and were converted to arbitrary units.

4.5.4. Statistical Analysis

Results are presented as mean + standard error of the mean. Statistical analysis was performed using SPSS 19.0 (SPSS Inc., Chicago, IL, USA). Statistical analysis was determined by unpaired Student's unpaired *t*-test (two-tailed). Statistical significance was set-up at the $p < 0.05$ level.

5. Conclusions

Taking into account the results obtained, it may be concluded that the developed gastro-resistant microparticles can represent an appropriate strategy for the inclusion of resveratrol in dietary supplements and functional foods with potential beneficial effects in the prevention or treatment of obesity.

Author Contributions: O.G. carried out the preparation and characterization of microparticles. A.L. performed in vitro experiments. J.L.P. and J.M. analyzed all data. M.P.P., M.I., and R.M.H. wrote the manuscript.

Funding: This research was partially funded by the Basque Government (Consolidated Groups, IT-407-07; IT-572-13; SAIOTEK SA-2011-00118) and the Spanish Government (INNPACTO, IPT-2012-0602-300000, 2012). The authors gratefully acknowledge the support provided by the University of the Basque Country UPV/EHU (UFI11/32).

Acknowledgments: Oihane Gartziandia thanks the University of the Basque Country (UPV/EHU) for the fellowship grant.

Conflicts of Interest: There are no conflicts of interest to declare.

References

1. Shi, G.; Rao, L.; Yu, H.; Xiang, H.; Yang, H.; Ji, R. Stabilization and encapsulation of photosensitive resveratrol within yeast cell. *Int. J. Pharm.* **2008**, *349*, 83–93. [CrossRef] [PubMed]
2. Wu, C.; Yang, J.; Wang, F.; Wang, X. Resveratrol: Botanical origin, pharmacological activity and applications. *Chin. J. Nat. Med.* **2013**, *11*, 1–15. [CrossRef]
3. Frémont, L. Biological effects of resveratrol. *Life Sci.* **2000**, *66*, 663–673. [CrossRef]
4. Andres-Lacueva, C.; Macarulla, M.T.; Rotches-Ribalta, M.; Boto-Ordóñez, M.; Urpi-Sarda, M.; Rodríguez, V.M.; Portillo, M.P. Distribution of resveratrol metabolites in liver, adipose tissue, and 2 skeletal muscle in rats fed different doses of this polyphenol. *J. Agric. Food Chem.* **2012**, *60*, 4833–4840. [CrossRef] [PubMed]
5. Rotches-Ribalta, M.; Andres-Lacueva, C.; Estruch, R.; Escribano, E.; Urpi-Sarda, M. Pharmacokinetics of resveratrol metabolic profile in healthy humans after moderate consumption of red wine and grape extract tablets. *Pharmacol. Res.* **2012**, *66*, 375–382. [CrossRef] [PubMed]
6. Amri, A.; Chaumeil, J.C.; Sfar, S.; Charrueau, C. Administration of resveratrol: What formulation solutions to bioavailability limitations? *J. Controlled Release* **2012**, *158*, 182–193. [CrossRef] [PubMed]
7. Liu, L.; Fishman, M.L.; Kost, J.; Hicks, K.B. Pectin-based systems for colon-specific drug delivery via oral route. *Biomaterials* **2003**, *24*, 3333–3343. [CrossRef]
8. Cantalapiedra, J.; Dirección General de Farmacia y Productos Sanitarios; Servicio de Información de Medicamentos; Servicio de Gestión de Banco de Datos. *Diccionario de Excipientes de las Especialidades Farmacéuticas Españolas*; Ministerio de Sanidad y Consumo: Madrid, Spain, 1989; pp. 9–109.
9. Das, S.; Ng, K. Colon-specific delivery of resveratrol: Optimization of multi-particulate calcium-pectinate carrier. *Int. J. Pharm.* **2010**, *385*, 20–28. [CrossRef] [PubMed]
10. Simonoska Crcarevska, M.; Glavas Dodov, M.; Goracinova, K. Chitosan coated Ca–alginate microparticles loaded with budesonide for delivery to the inflamed colonic mucosa. *Eur. J. Pharm. Biopharm.* **2008**, *68*, 565–578. [CrossRef] [PubMed]
11. Cerchiara, T.; Bigucci, F.; Corace, G.; Zecchi, V.; Luppi, B. Eudragit-coated albumin nanospheres carrying inclusion complexes for oral administration of indomethacin. *J. Incl. Phenom. Macrocycl. Chem.* **2011**, *71*, 129–136. [CrossRef]
12. Sandolo, C.; Péchiné, S.; Le Monnier, A.; Hoys, S.; Janoir, C.; Coviello, T.; Alhaique, F.; Collignon, A.; Fattal, E.; Tsapis, N. Encapsulation of Cwp84 into pectin beads for oral vaccination against clostridium difficile. *Eur. J. Pharm. Biopharm.* **2011**, *79*, 566–573. [CrossRef] [PubMed]
13. Quality Guidelines. Available online: www.ich.org/products/guidelines/quality/article/quality-guidelines.html (accessed on 14 June 2018).
14. Disolution test for solid dosage forms, 2.9.3. In *European Pharmacopeia 7.0*; Council of Europe: Strasbourg, France, 2011; Volume 1, pp. 256–263.
15. Walle, T.; Hsieh, F.; DeLegge, M.H.; Oatis, J.E., Jr.; Walle, U.K. High absorption but very low bioavailability of oral resveratrol in humans. *Drug Metab. Dispos.* **2004**, *32*, 1377–1382. [CrossRef] [PubMed]
16. Baile, C.A.; Yang, J.Y.; Rayalam, S.; Hartzell, D.L.; Lai, C.Y.; Andersen, C.; Della-Fera, M.A. Effect of resveratrol on fat mobilization. *Ann. N. Y. Acad. Sci.* **2011**, *1215*, 40–47. [CrossRef] [PubMed]
17. Picard, F.; Kurtev, M.; Chung, N.; Topark-Ngarm, A.; Senawong, T.; Machado De Oliveira, R.; Leid, M.; McBurney, M.W.; Guarente, L. Sirt1 promotes fat mobilization in white adipocytes by repressing PPAR-gamma. *Nature* **2004**, *429*, 771–776. [CrossRef] [PubMed]
18. Lasa, A.; Schweiger, M.; Kotzbeck, P.; Churruca, I.; Simon, E.; Zechner, R.; Portillo, M.P. Resveratrol regulates lipolysis via adipose triglyceride lipase. *J. Nutr. Biochem* **2012**, *23*, 379–384. [CrossRef] [PubMed]
19. Kim, S.; Jin, Y.; Choi, Y.; Park, T. Resveratrol exerts anti-obesity effects via mechanisms involving down-regulation of adipogenic and inflammatory processes in mice. *Biochem. Pharmacol.* **2011**, *81*, 1343–1351. [CrossRef] [PubMed]

20. Bhatt, J.K.; Thomas, S.; Nanjan, M.J. Resveratrol supplementation improves glycemic control in type 2 diabetes mellitus. *Nutr. Res.* **2012**, *32*, 537–541. [CrossRef] [PubMed]

21. Faghihzadeh, F.; Adibi, P.; Hekmatdoost, A. The effects of resveratrol supplementation on cardiovascular risk factors in patients with non-alcoholic fatty liver disease: a randomised, double-blind, placebo-controlled study. *Br. J. Nutr.* **2015**, *114*, 796–803. [CrossRef] [PubMed]

22. Alberdi, G.; Rodríguez, V.M.; Miranda, J.; Macarulla, M.T.; Arias, N.; Andrés-Lacueva, C.; Portillo, M.P. Changes in white adipose tissue metabolism induced by resveratrol in rats. *Nutr. Metab. (Lond)* **2011**, *8*, 29. [CrossRef] [PubMed]

23. Boocock, D.J.; Faust, G.E.; Patel, K.R.; Schinas, A.M.; Brown, V.A.; Ducharme, M.P.; Booth, T.D.; Crowell, J.A.; Perloff, M.; Gescher, A.J.; et al. Phase I dose escalation pharmacokinetic study in healthy volunteers of resveratrol, a potential cancer chemopreventive agent. *Cancer Epidemiol. Biomarkers Prev.* **2007**, *16*, 1246–1252. [CrossRef] [PubMed]

24. Marier, J.F.; Vachon, P.; Gritsas, A.; Zhang, J.; Moreau, J.P.; Ducharme, M.P. Metabolism and disposition of resveratrol in rats: extent of absorption, glucuronidation, and enterohepatic recirculation evidenced by a linked-rat model. *J. Pharmacol. Exp. Ther.* **2002**, *302*, 369–373. [CrossRef] [PubMed]

25. Cho, S.J.; Jung, U.J.; Choi, M.S. Differential effects of low-dose resveratrol on adiposity and hepatic steatosis in diet-induced obese mice. *Br. J. Nutr.* **2012**, *108*, 2166–2175. [CrossRef] [PubMed]

Sample Availability: Not available

molecules

MDPI

Article

The Phytochemical and Biological Investigation of *Jatropha pelargoniifolia* Root Native to the Kingdom of Saudi Arabia

Hanan Y. Aati [1], Ali A. El-Gamal [1,2,*], Oliver Kayser [3] and Atallah F. Ahmed [1,2]

[1] Department of Pharmacognosy, Faculty of Pharmacy, King Saud University, P.O. Box 2457, Riyadh 11451, Saudi Arabia; hati@ksu.edu.sa (H.Y.A.); afahmed@ksu.edu.sa (A.F.A.)
[2] Department of Pharmacognosy, College of Pharmacy, Mansoura University, El-Mansoura 35516, Egypt
[3] Technical Biochemistry, TU Dortmund University, Emil-Figge-Strasse 66, D-44227 Dortmund, Germany; oliver.kayser@tu-dortmund.de
* Correspondence: aelgamel@ksu.edu.sa; Tel.: +966-1146-77259

Received: 29 May 2018; Accepted: 25 July 2018; Published: 28 July 2018

Abstract: Extensive phytochemical analysis of different root fractions of *Jatropha pelargoniifolia* Courb. (Euphorbiaceae) has resulted in the isolation and identification of 22 secondary metabolites. 6-hydroxy-8-methoxycoumarin-7-*O*-β-D-glycopyranoside (**15**) and 2-hydroxymethyl *N*-methyltryptamine (**18**) were isolated and identified as new compounds along with the known diterpenoid (**1**, **3**, **4**, and **7**), triterpenoid (**2** and **6**), flavonoid (**5**, **11**, **13**, **14**, and **16**), coumarinolignan (**8–10**), coumarin (**15**), pyrimidine (**12**), indole (**17**, **18**), and tyramine-derived molecules (**19–22**). The anti-inflammatory, analgesic, and antipyretic activities were evaluated for fifteen of the adequately available isolated compounds (**1–6**, **8–11**, **13**, **14**, **16**, **21**, and **22**). Seven (**4**, **6**, **10**, **5**, **13**, **16**, and **22**) of the tested compounds showed a significant analgesic effect ranging from 40% to 80% at 10 mg/kg in two in vivo models. Compound **1** could also prove its analgesic property (67.21%) when it was evaluated on a third in vivo model at the same dose. The in vitro anti-inflammatory activity was also recorded where all compounds showed the ability to scavenge nitric oxide (NO) radical in a dose-dependent manner. However, eight compounds (**1**, **4**, **5**, **6**, **10**, **13**, **16**, and **22**) out of the fifteen tested compounds exhibited considerable in vivo anti-inflammatory activity which reached 64.91% for compound **10** at a dose of 10 mg/kg. Moreover, the tested compounds exhibited an antipyretic effect in a yeast-induced hyperthermia in mice. The activity was found to be highly pronounced with compounds **1**, **5**, **6**, **10**, **13**, and **16** which decreased the rectal temperature to about 37 °C after 2 h of the induced hyperthermia (~39 °C) at a dose of 10 mg/kg. This study could provide scientific evidence for the traditional use of *J. pelargoniifolia* as an anti-inflammatory, analgesic, and antipyretic.

Keywords: *Jatropha pelargoniifolia*; alkaloids; flavonoids; coumarinolignans; diterpenes; anti-inflammatory; analgesic; antipyretic

1. Introduction

Euphorbiaceae is considered as one of the largest families of flowering plants which includes approximately 7800 species that are distributed among 300 genera and five subfamilies in tropical and subtropical regions [1,2]. Among the main genera of this family, *Jatropha* L. is represented by approximately 200 species [2]. *Jatropha* species are used in folk medicine to treat various diseases, such as skin inflammation, eye infection, chest pain, stomach pain, itching, and as a vermifuge, or as ornamental plants and energy crops in Latin America, Africa, and Asia [3]. *J. gossypiifolia*, *J. elliptica*, *J. curcas*, and *J. mollissima*, among other species of *Jatropha*, have been reported for their chemical constituents, biological activities, and medicinal uses [4]. *Jatropha glauca*, *J. curcas*, *J. spinose*, and *J. pelargoniifolia* are the only four species that are distributed in Saudi Arabia and are employed as traditional herbal medicines, owing to their anti-inflammatory, antioxidant, antiseptic, and analgesic properties [5,6].

J. pelargoniifolia Courb. of the current study is grown as a shrub and is widely known as "Obab" in Arabic. It is widely distributed in East Tropical Africa (Sudan, Eritrea, Ethiopia, Somalia, and Kenya) and the Arabian Peninsula (Yemen, Oman, and Saudi Arabia) [7]. The plant is sometimes collected from the wild for local medicinal use, especially the petiole sap which is applied to treat ulcers, severe skin inflammation, and for wound healing [7].

Previous phytochemical studies on the plants belonging to the genus *Jatropha* revealed a broad range of isolated secondary metabolites, such as diterpenoids, triterpenoids, non-conventional coumarino-lignans, alkaloids, coumarins, flavonoids, cyclic peptides, and steroids [8,9]. However, accordingly reviewed by Zhang et al. [8], the main compounds isolated from *Jatropha* genus are the terpenoids. *J. gossypiifolia* was subjected to extensive phytochemical studies that resulted in the isolation of many secondary metabolites, such as propacin, venkatasin, citlalitrione, ricinine, apigenin, jatropholones A& B, and jatrophone [4]. Moreover, curcusones A–D, taraxerol, nobiletin, curacyclines A & B, as well uracil, have been isolated from *J. curcas*. [5,6,8]. Additionally, many reported studies showed the isolation of multidione, multifidone, multifolone, and multifidol glucoside from *J. multifida*, while from *J. podagrica, there* was japodic acid, erythrinasinate, γ-sitosterol, japodagrin, and podacyclines A & B [3,8,9]. This is indeed a reflection of the versatility of the enzymatic system that is present in Euphorbiaceous plants, however nothing was reported regarding *J. pelargoniifolia*. Thus, it was of interest to explore the active constituents and their biological activity to provide evidence for the traditional use of *J. pelargoniifolia*.

2. Result and Discussion

2.1. Isolation of Compounds

The alcoholic extract of *J. pelargoniifolia* roots powder was successively partitioned with petroleum ether (60 °C), dichloromethane (DCM), ethyl acetate (EtOAc), and then *n*-butanol (*n*-BuOH) to give the correspondent organic fractions. Each fraction was subjected to chromatographic separation on normal and reversed phase (RP) silica gel to yield compounds **1**, **6**, **7**, **10**, **11**, and **16** from petroleum ether, DCM$_2$, and EtOAc fractions, respectively. Furthermore, the organic extract that was obtained after an acid-base treatment of the roots powder was isolated on a normal silica gel column which was followed by purification on RP-HPLC and/or crystallization to afford compounds **17–22** (Figure 1).

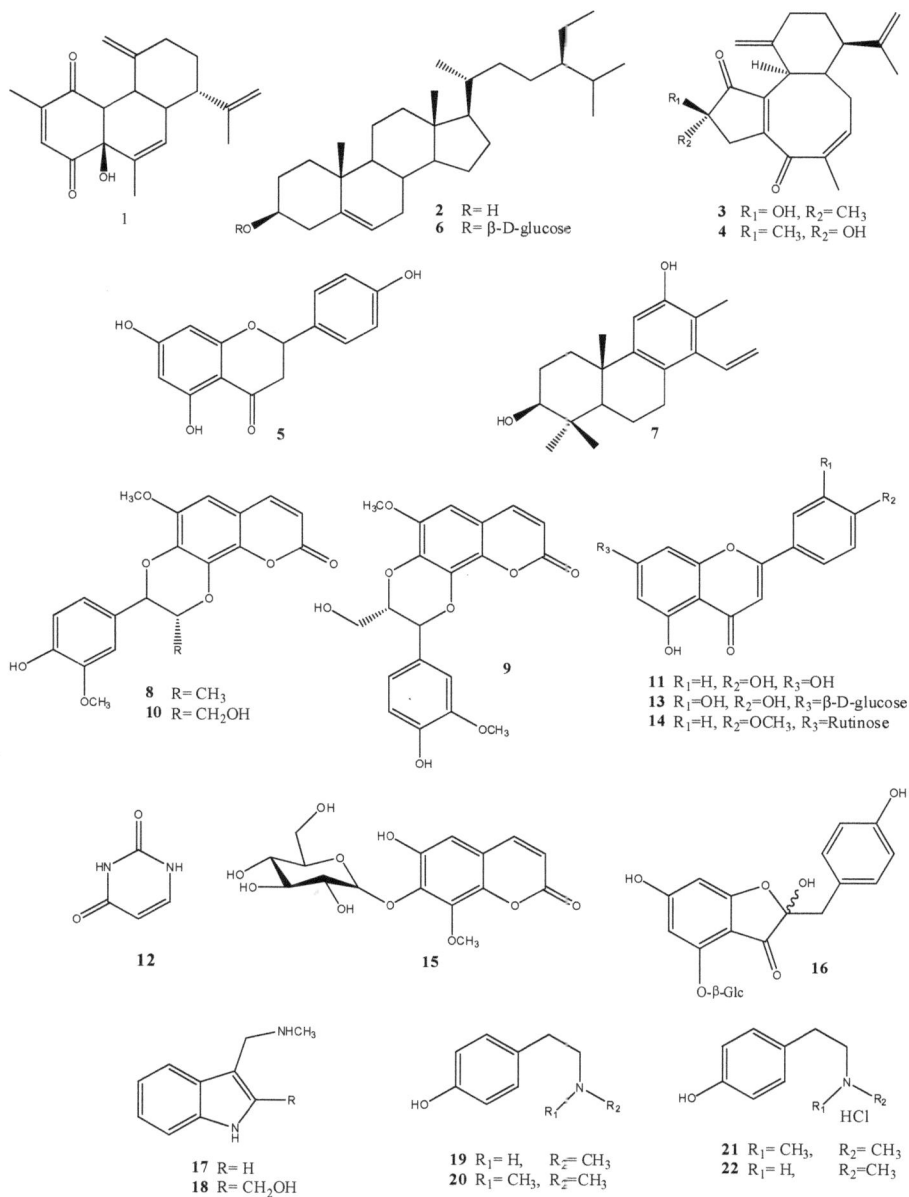

Figure 1. Chemical structures of the compounds isolated from the roots of *Jatropha pelargoniifolia*.

2.2. Structure Elucidation

The new compound **15** was obtained as white crystals. The NMR and ESIMS (Electronspray Ionization Mass Spectrometry) data established the molecular formula of **15** to be $C_{16}H_{18}O_{10}$. The IR absorption bands at max 3349, 1719, and 1625 cm^{-1} suggested the presence of hydroxyl, ester carbonyl, and aromatic functionalities, respectively. Furthermore, the ^{13}C nuclear magnetic resonance (NMR) spectrum of **15**, which was measured in deuterated methanol (CD$_3$OD, displayed sixteen signals of nine sp^2 and seven sp^3 carbons (including that of a methoxyl group). Its ^{1}H NMR spectrum showed a pair of ortho-coupled protons at δ_H 6.26 and 7.88 (each, 1H, d, J = 9.5 Hz) as was observed by ^{1}H-^{1}H correlated spectroscopy (COSY) that was assignable to H-3 and H-4 of an α-pyrone ring system of a coumarin, respectively [10]. This was further evidenced from the ^{13}C NMR carbon signals of the -pyrone at δ_C 163.5 (C, C-2), 146.5 (CH, C-4), 144.4 (C, C-8a), 116.2 (CH, C-3), and 112.7 (C, C-4a) (Table 1). Moreover, the single aromatic singlet appearing at δ_H 7.00 (1H, s) suggested **15** to be a trisubstituted coumarin. Six proton signals at δ_H 3.30–4.99 ppm, a doublet of an anomeric proton at δ_H 4.99 (1H, d, J = 7.8 Hz), and six ^{13}C NMR signals at δ_C 106.2 (CH, C-1′), 75.5 (CH, C-2′), 77.8 (CH, C-3′), 71.0 (CH, C-4′), 78.5 (CH, C-5′), and 62.2 (CH$_2$, C-6′) indicated the presence of a β-D-glucopyranosyl substituent. An aromatic methoxy substituent (δ_H/δ_C 3.91/57.0) was also revealed. The $^3J_{CH}$ correlations that were observed in the heteronuclear multiple bond correlation (HMBC) spectrum linked these two substituents to the coumarin carbons at δ_C 133.2 and 147.5, respectively (Figure 2). Thus, a hydroxy group should represent the third substituent on the coumarin carbon at δ_C 145.7. The long-range correlations (HMBC) that were found from H-5 (δ_H 7.00, 1H, s) to the carbons at δ_C 146.5 (C-4), 145.7, C-8a (δ_C 144.4), and δ_C 133.2 (C-7) indicated that C-8 (δ_C 147.5) is the position of the methoxyl group. To confirm the locations of the glucosyl and hydroxyl groups, the NMR data of **15** were further compared to those of 5-hydroxy-7-methoxycoumarin-8-*O*-β-D-glucoside and other closely related coumarin derivatives that were previously isolated from *Daphne pseudo-mezereum* [11] and *Tetraphis pellucida* [12], respectively. The structure of compound **15** was thus established as a new natural product and was identified as 6-hydroxy-8-methoxy coumarin-7-*O*-β-D-glycopyranoside.

Table 1. The ^{1}H (600 MHz, δ in ppm, J in Hz) and ^{13}C NMR (125MHz, δ in ppm) spectral data for compound **15** in deuterated methanol (CD$_3$OD).

Position	δ_H	δ_C
2	-	163.5
3	6.26 (d, J = 9.5 Hz, 1H)	116.2
4	7.88 (d, J = 9.5 Hz, 1H)	146.5
5	7.00 (s, 1H)	106.1
6	-	145.7
7	-	133.2
8	-	147.5
4a	-	112.7
8a	-	144.4
1′	4.99 (d, J = 7.8 Hz, 1H)	106.2
2′	3.57 (dd, J = 9.4, 9.4, 1H)	75.5
3′	3.46 (d, J = 1.9, 1H)	77.8
* 4′	3.47 (brs, 1H)	71.0
* 5′	3.30 (brs, 1H)	78.5
6′	3.72 (d, J = 4.9, 1H) 3.80 (d, J = 2.4, 1H)	62.2
OCH$_3$-8	3.91 (s, 3H)	57.0
OH-6	10.53	-

* Overlapped with solvent signal.

Compound **18** was isolated from the organic extract of the acid-base treated root powder as white needle-shaped crystals. It produced a positive Dragendorff's test, indicating its alkaloid nature. The IR absorption band with a spike at max 3309 cm^{-1} suggested the presence of hydroxyl and/or secondary amine functionality. The UV absorptions at max 295, 287, 279, 230 nm in MeOH were characteristic to an indole chromophore. The COSY correlations (Figure 2) disclosed the ABCD system of the aromatic protons at δ_H 7.41/7.00 (Table 2), which is consistent with 2,3-disubstituted indole alkaloids. A side chain of an ethylene and a *N*-methyl was linked to C-3 of the indole as it was manifested by 2D NMR correlations. However, comparison of ^{1}H NMR data of **18** with those of *N*-methyltryptamine

(17) revealed that the ^{1}H proton singlet at position 2 in **17** was replaced in **18** by a 2H singlet of a hydroxymethyl proton at δ_H 3.91 ppm. The ^{13}C NMR spectroscopic and ESIMS data of **18** was thus consistent with a molecular formula $C_{12}H_{16}N_2O$ of 30 mass units more than that of **17** ($C_{11}H_{14}N_2$). Furthermore, the HMBC correlation that was found from the methylene protons (δ_H 3.91, 2H, s) to C-2 and C-3 confirmed its C-2 location of the hydroxymethyl group. Finally, a full analysis of th COSY and HMBC spectral correlations (Figure 2) assigned the structure cf compound **18** to be 3-(2-(methylamino) ethyl)-1H-indol-2-yl) methanol or 2-hydroxymethyl N-methyltryptamine, a new indole alkaloid.

Table 2. The ^{1}H (700 MHz, δ in ppm, J in Hz) and ^{13}C NMR (125 MHz, δ in ppm) spectral data for compound **18** in deuterated methanol (CD_3OD).

Position	δ_H	δ_C
2	-	128.0
3	-	107.2
3a	-	130.6
4	7.29 (d, J = 7.8 Hz, 1H)	112.0
5	7.07 (t, J = 7.8 Hz, 1H)	122.4
6	7.00 (t, J = 7.8 Hz, 1H)	120.0
7	7.41 (d, J = 7.8 Hz, 1H)	118.6
7a	-	138.1
αCH_2	3.10 (t, J = 5.8 Hz, 2H)	54.1
βCH_2	3.00 (t, J = 5.8 Hz, 2H)	21.5
CH_3	2.69 (s, 3H)	44.9
CH_2OH	3.91 (s, 2H)	52.9

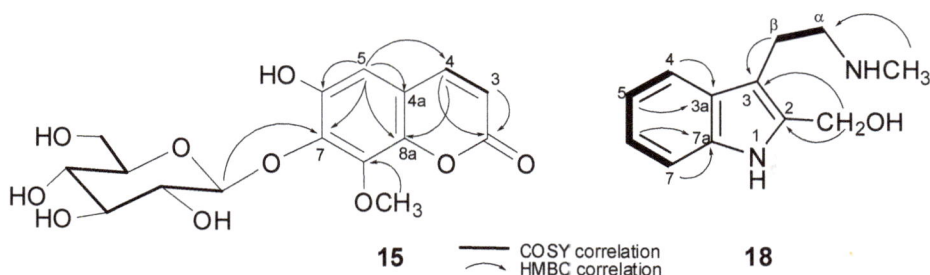

Figure 2. Selected heteronuclear multiple bond correlation ʿHMBC) and correlation spectroscopy (COSY) correlations of compounds **15** and **18**.

Compounds **1–14**, **16–17** and **19–22**, which were also isolated from *J. pelargoniifolia* roots, were found to be identical to the previously reported natural products by comparison of their spectroscopic (IR, MS, and NMR) data and were identified as jatrophadiketone (**1**) was isolated from the roots of *J. curcas* [13], β-sitosterol (**2**) isolated from *J. curcas* seed kernels and from the methanolic extract of the root bark of *Calotropis gigantean* (Linn.), [14,15], curcuscn D (**3**) and curcuson C (**4**) were isolated from *J. curcas* root extract [16], naringenin (**5**) was isolated from the root extract of *J. gossypifolia* [17,18], β-sitosterol glucoside (**6**) was isolated from the leave and tw-g extract of *J. curcas*, [19], spruceanol (**7**) was isolated from both the aerial extract of *J. divaricate* and the bark extract of *Aleurites moluccana* [20,21], propacin (**8**), cleomiscosin B (**9**), cleomiscosin A (**10**) compounds **8** and **10** were isolated from whole plant extracts of *J. gossypifolia*, while compound **9** was isolated from *Mallotus apelta* [22–26], apigenin (**11**) was identified in *J. gossypifolia* [4,27], uracil (**12**) was isolated from the leaves of *J. curcas* [28–30], cynaroside (**13**) was identified in *Scabiosa atropurpurea* aerial parts extract [31], linarin (**14**) was isolated from the extracts of aerial parts of both *Bupleurum chinense* and *Valeriana officinalis* [32,33], hovetricoside C (**16**) was separated from *Artocarpus tonkinensis* [34], N-methyltryptamine (**17**) was isolated from

Zanthoxylum arborescens [35], *N*-methyltyramine (**19**) was isolated from a beer [36], and hordenine (**20**) was separated and identified from *Ephedra aphylla* that was growing in Egypt [37]. Compounds **21** (hordenine HCl) and **22** (*N*-methyltyramine HCl) were identical to the authentic samples that were purchased from Sigma-Aldrich (St Louis, MO, USA), and on the basis of their ^{1}H NMR and TLC co-chromtographic data, they were isolated previously from *Ariocarpus kotschoubeyanus* [38].

2.3. Biological Activity

The alcoholic extract of *J. pelargoniifolia* was found to possess significant anti-inflammatory, analgesic, and antipyretic activities when it was tested on in vivo models in a dose-dependent manner [39]. This prompted us to extend the study of these activities on the isolated compounds. The anti-inflammatory, analgesic, antipyretic, and antioxidant activities for the compounds which have been isolated in good yields (**1–6**, **8–11**, **13**, **14**, **16**, **21**, and **22**) were thus evaluated for their analgesic, anti-inflammatory, antipyretic, and antioxidant activities using in vivo and in vitro models. The analgesic activities were assessed in mice via acetic acid-induced writhing, hot-plate, and tail-flick methods.

In the acetic acid-induced writhing method, compounds **1**, **4**, **6**, **9**, **11**, **13**, **14**, **16**, and **22** showed a dose-dependent analgesic activity by the reduction in the number of writhings. However, the diterpenoids (**1** and **4**), β-sitosterol glucoside (**6**), flavonoids (**5** and **13**), and tryptamine HCl (**22**) exhibited the strongest analgesic activity by inhibiting writhing in mice (49.07–65.74% inhibition) at a dose of 10 mg/kg compared with the standard antinociceptive drug (indomethacin), which showed 72.68% reduction in the number of writhings at a concentration of 4 mg/kg (Table S9). The coumarinolignan (**8**) and hordenine HCl (**21**) did not show any inhibition either at 5 or at 10 mg/kg.

In the hot plate method, the thermal responses in the mice that were treated with selected compounds after half, one, and two hours were significantly reduced ($p < 0.001$). Especially in the mice that were treated with a dose 10 mg/kg of compounds **22**, **10**, **6**, **4**, **5**, **13**, **16**, and **14**, the antinociceptive effects were reduced by 78.57, 76.19, 74.46, 73.33, 65.95, 59.57, 53.48, and 34.88%, respectively (Table S10). While in the tail-flick method, the tested animals that were treated with 10 mg/kg of compounds **22**, **1**, **10**, **4**, **13**, **16**, **5**, **6**, **21**, **3**, and **9** showed a significant ($p < 0.001$) reduction in antinociceptive activity (67.32, 67.21, 60.45, 57.59, 54.04, 53.59, 45.58, 40.30, 21.75, 20.39, and 10.87%, respectively) compared with indomethacin (95.61%), as depicted from Table S11. The obtained results confirmed that the strong analgesic activity that is exhibited by the roots of *J. pelargoniifolia* could be due to its bioactive compounds that may exert their analgesic activities through different CNS (Central Nervous System) mechanisms (peripheral and central). Therefore, further studies with purified compounds should be conducted in the future for further pharmacological and toxicological characterization in order to elucidate the mechanisms that are involved in the central analgesic effect of these compounds.

The anti-inflammatory activities of the major isolated compounds were evaluated by using the carrageenan-induced paw edema model in rats. It was found that the size of the edema was significantly reduced ($p < 0.05$–0.001) in the animals that were treated with the low doses of 5 and 10 mg/kg compared with the standard anti-inflammatory drug (phenylbutazone) at a high dose of 100 mg/kg. The rats that were treated with compounds **10**, **16**, **1**, **5**, **6**, **22**, **4**, **13**, **3**, **8**, **9**, **14**, and **21** exhibited a significant reduction in their hind paw edema in a dose-dependent manner. Therefore, at a dose of 5 mg/kg, the edema size was reduced by 20.44, 47.23, 17.95, 45.85, 33.97, 44.47, 13.53, 26.51, 7.18, 2.20, 5.52, 6.35, and 6.35%, respectively, while at 10 mg/kg, the edema size was reduced by 64.91, 55.24, 54.94, 51.38, 51.10, 50.27, 49.17, 48.61, 13.53, 12.98, 10.22, 10.22, and 8.01%, respectively relative to that reduced by phenylbutazone at 100 mg/kg (69.06%) which was almost similar to that produced by a dose of 10 mg/kg of compound **10** (64.91%; Table S12). These anti-inflammatory results were almost compatible with those of the above mentioned antinociceptive activity for the tested compounds.

In addition, the isolated compounds from the *J. pelargoniifolia* roots were tested for their antipyretic activity against yeast-induced hyperthermia in mice. All tested compounds, which were administered at doses of 5 and 10 mg/kg, showed a considerable reduction in the rectal temperature of the hyper-thermic mice, ranging between 36.73 ± 0.13 °C and 38.56 ± 0.16 °C as compared with the hypothermic effect (36.33 ± 0.11 °C) resulting from indomethacin administration (Table S13). Moreover, compounds **5, 6, 10, 13**, and **16** displayed about a 1 °C reduction in temperature less than that of the yeast-induced hyperthermia control (~38.8 °C) in the first 30 min. of the experiment.

The percentage inhibition ± SD of the nitric oxide-scavenging activity was determined for the selected compounds at concentrations of 20, 40, 60, 80, and 100 μg/mL, and the obtained results were compared with a standard antioxidant drug (ascorbic acid). Compounds **22, 4, 2, 10, 1, 14, 21, 9, 8, 11, 6, 3, 13, 5**, and **16** exhibited significant free radical-scavenging potency when compared with the free radical-scavenging activity of a strong known antioxidant drug (ascorbic acid) 87.23 ± 0.98. The ability of the tested compounds to produce antioxidant effects was found to be concentration-dependent. At a 100 μg/mL dose, the % inhibition ± SD of the tested compounds were 77.60 ± 4.22, 77.36 ± 4.22, 76.83 ± 5.01, 75.26 ± 5.54, 71.66 ± 0.70, 70.36 ± 14.73, 67.67 ± 5.75, 63.67 ± 12.85, 63.67 ± 12.85, 57.00 ± 10.21, 56.54 ± 6.03, 38.30 ± 5.63, 33.06 ± 1.86, 27.00 ± 7.85, and 25.61 ± 5.18, respectively (Table S14). The significant antioxidant activity that was associated with the administration of *J. pelargoniifolia* roots was perhaps due to its content of several phenolic and polyphenolic compounds which play an important role in free radical-scavenging activity with less cytotoxicity.

It is important to mention here that in our previous study, which was carried out on the crude alcoholic extract of *J. pelargoniifolia* roots, we observed a significant anti-inflammatory activity and analgesic potency [39], likely resulting from the presence of cleomiscosin A, hovetricoside C, jatrophadiketone, naringenin, β-sitosterol glucoside, *N*-methyltyramine HCL, curcuson C, cynaroside, curcuson D, propacin, cleomiscosin B, linarin, and hordenine HCL in good yield. Undoubtedly, a synergistic effect between these bioactive constituents produces significant antinociceptive and anti-inflammatory effects. These results justify the use of this plant in folk medicine for the treatment of pain and several inflammatory conditions. Further study will be conducted on the pure isolated compounds to investigate the exact mechanisms underlying their promising biological activities.

Our study proved that *J. pelargoniifolia* roots can be considered as a source of several biologically-active compounds such as hordenine, which exhibited various biological activities like inhibiting melanogenesis in human melanocytes, increasing the respiratory and heart rates [40], the stimulation of gastrin release, inhibition of monoamine oxidase B, and antibacterial properties [41]. Furthermore, Chrisitine et al. reported that *N*-methyltyramine increases blood pressure in an anaesthetized rat, relaxes guinea pig ileum, and increases both the force and the rate of contraction of guinea-pig right atrium by inducing the release of noradrenaline [42]. Additionally, naringenin has been reported to have several pharmacological properties, including anti-dyslipidemic, anti-obesity and antidiabetic, and antifibrotic [43]. Moreover, cleomiscosin A showed strong anti-inflammatory activity and has analgesic and antipyretic potencies [44]. Curcuson C has been reported to have antipyretic activity in vivo [45].

3. Materials and Methods

3.1. Chemicals and Analytical Instruments

The high-resolution electron spray ionization-mass spectrometry (HRESI-MS) analyses (Bruker, Bremen, Germany) were carried out on an Agilent Triple Quadrupole 6410 QQQ LC-MS mass spectrometer (Central Lab. College of Pharmacy, King Saud University (KSU)). The infra-red spectra were generally recorded in the potassium bromide pellets, unless otherwise specified, using the FTIR spectrophotometer (FT-IR Microscope Transmission, company, Waltham, MA, USA). The melting points were recorded by using a Mettler FP 80 Central Processor that was supplied with a Mettler FP 81 MBC Cell Apparatus. The spectral data for proton and carbon were measured by using Bruker AVANCE 700, 500, and 600 (College of Pharmacy, KSU and Department of Chemistry in TU Dortmund) (Bruker, Fallanden, Switzerland), resonating at either 700, 500, and 600 MHz for proton or at 125 MHz for carbon. The chemical shift values were expressed in ppm with respect to the internal standard tetramethyl silane (TMS) or residual solvent peak, and the coupling constants (J) were recorded in Hertz (Hz). The two-dimensional NMR experiments (COSY, HSQC, and HMBC) were performed using the standard Bruker program (Bruker, Fallanden, Switzerland). The silica gel 60/230–400 mesh (Qingdao Oceanic Chemical Co., Qingdao, China), RP C18 silica gel 40–63/230–400 mesh (Merck, Darmstade, Germany), and sephadex LH-20 with particle size 18–111 µm (GE Healthcare, Chicago, IL, USA) were used for column chromatography, while the silica gel and reversed phase 60 F254 (Merck, Germany) were used for thin-layer chromatography (TLC). The detection was achieved by using 10% H_2SO_4 in ethanol or ceric sulfate followed by heating. Alkaloids were tested with Mayer's reagent, Hager's reagent, and Dragendorff's reagent. All of the solvents for analytical purposes (HPLC- and analytical-grade) and the drugs for biological investigation (sodium nitroprusside, sulphanilamide, λ-carrageenan, acetic acid, ascorbic acid, and phenylbutazone) were procured from Sigma Chemical Company (Sigma-Aldrich, St Louis, MO, USA), and the solvents were distilled prior to use. The preparative and semipreparative Shimadzu HPLC were performed, characterized by Rp-18 (ODS-80 TM, TSK, Tokyo, Japan), 10 µm PS, 30 cm L × 2.15 cm i.d. fitted with a guard column (10 µm PS, 7.5 cm L × 2.15 cm i.d.) (ODS-80 TM, TSK, Tokyo, Japan), and VP 250/10 NUCLEODUR C18 HTec, 6 µm PS, 25 cm L × 2 cm i.d., respectively which both used a PDA detector.

3.2. Plant Material

The roots of *J. pelargoniifolia* were harvested from Wadi Mojasas, Jazan district (South of Saudi Arabia) in September, 2015. The plant was authenticated by Dr. Jacob Thomas, a botanist of the Science College Herbarium, KSU, where a voucher specimen (#23064) was deposited.

3.3. Animals

Male Wistar rats and white male Swiss albino mice with approximate body weights of 200 g and 20–25 g, respectively, were divided into groups of six animals. The animals were obtained from the Experimental Animal Care Center, College of Pharmacy, KSU. After a 7-day period in animal accommodation, they were divided into groups and were maintained at 12 h:12 h light-dark conditions at 55% humidity. Purina chow rat diet (UAR-Panlab, Barcelona, Spain) and drinking water were supplied to the animals ad libitum. The protocols for the present study were based on the recommendations of the Ethical Committee of the Experimental Animal Care Center of KSU (approval number CPR-7569).

3.4. Extraction, Fractionation, and Purification

The air-dried powder of the *J. pelargoniifolia* roots (2.5 kg) was divided into two parts—A and B—and 2.5 kg of part A was subjected to solvent extraction, while the remaining 500 g of the root powder (part B) was exposed to the acid-base treatment. Part A was extracted by maceration with 80% ethanol (3 L × 5) for three successive days. This process was repeated until complete exhaustion of the plant material [46]. The alcoholic extract was then concentrated to dryness under reduced pressure at 40 °C using a rotary evaporator to give 270 g of the dried alcoholic extract. The dried alcoholic extract was suspended in H_2O and was successively partitioned with petroleum ether, dichloromethane, ethyl acetate, and *n*-butanol (600–700 mL × 3) of each to obtain 13.3, 10.3, 5.1, and 33.6 g, respectively.

A part of the petroleum ether fraction (12.8 g) was chromatographed over silica gel CC (Column Chromatography) using a gradient of petroleum ether/EtOAc followed by methanol (MeOH). The 100 mL fractions of each were collected and screened by TLC, and similar fractions were combined together to give six fractions (A–F). Fraction A which was eluted by 15% EtOAc in petroleum ether (609.8 mg) was further subjected to CC and was eluted by petroleum ether/acetone gradient elution, sub fraction A1 (188.9 mg) which was eluted by 6% acetone in petroleum ether was further purified by preparative HPLC gradient elution using acetonitrile: H_2O: TFA) to yield 25.0 mg of compound **1**. Direct crystallization of fraction B, which was eluted by 20% EtOAc in petroleum ether, yielded 302.7 mg of compound **2**. Fractions C and D which were eluted with 30 and 40% EtOAc in petroleum ether, respectively, were crystallized with acetone to yield compounds **3** and **4** (14.3 and 25.4 mg, respectively). Additionally, fraction E (287.3 mg) which was eluted by 50% EtOAc was also crystallized from acetone to give 20.2 mg of compound **5**, while fraction F which was eluted with 40% MeOH in EtOAc yielded 330.4 mg of compound **6**, which was purified by crystallization with acetone.

The dichloromethane (DCM) fraction (9.8 g) was subjected to silica gel CC using a column that was packed by the wet method with petroleum ether. The polarity of the column was gradually increased by treating it with DCM, followed by MeOH to give 142 fractions, and similar fractions were pooled together depending on their TLC similarity. Fraction 48–64 which was eluted by 10% MeOH in DCM was concentrated (4.6 g) and was then subjected to repeated silica gel CC, followed by a preparative revered phase TLC using MeOH:H$_2$O (3:1) as a solvent system, leading to the isolation of white crystals of compound **7** (7.5 mg). Moreover, subtractions that were obtained using 90% acetone in petroleum ether, 100% acetone, and 10% acetone in MeOH, followed by crystallization with MeOH, afforded compounds **8** (15.1 mg), **9** (15.9 mg), and **10** (16.4 mg), respectively.

The EtOAC extract (4.6 g) was subjected to silica gel CC using a gradient of DCM/MeOH to give six fractions (I–IV). Fractions I which were eluted with 84% DCM afforded 17 mg of compound **11** after crystallization with MeOH. Fractions II which were eluted with 70% DCM afforded 8.6 mg of compound **12**. The fractions that were eluted with 35% and 40% MeOH in DCM (II and IV) were further purified by repeated acetone crystallization to give 14.9 and 13.2 mg of compounds **13** and **14**, respectively. Fractions V which were eluted with 45% MeOH were further subjected to CC using DCM/MeOH, followed by a semi-preparative HPLC (Rp-18) using MeOH:H$_2$O:TFA as a solvent system afforded 8.6 mg white crystals of compound **15**. Finally, fractions VI which were eluted with 50% MeOH in DCM were subjected to further purification over sephadex LH-20 (using water and methanol as an eluent in the gradient mode). The subfraction VI–A, which was eluted by 20% H$_2$O/MeOH was further purified over a reversed-phase column to give 12 mg of compound **16**.

Furthermore, Part B was subjected to an acid-base treatment according to the Stas-Otto method I which was described by Mandhumitha and Fowsiya [47]. The crude alkaloidal fraction was subjected to silica columns using gradient elution with solvent system DCM/MeOH:NH$_4$OH, resulting in five fractions. The first fraction which was eluted using 17% MeOH in DCM with an addition of 1% NH$_4$OH was followed by further purification by reversed-phase semipreparative HPLC using MeOH:H$_2$O:TFA to give compound **17** (6.3 mg). The second fraction which was separated by 20% MeOH in DCM to afford a subfraction, which was further purified by a semipreparative HPLC gradient elution using MeOH:H$_2$O:TFA as a solvent system, afforded white needle crystals of compound **18** (9.4 mg). The third and fourth fractions which were eluted by 23% and 26% MeOH in DCM, followed by an addition of a few drops of NH$_4$OH, afforded 6.8 and 8.8 mg of compounds **19** and **20**, respectively. The fifth fraction was eluted by 60% MeOH in DCM with an addition of a few NH$_4$OH drops to yield 86.7 mg of a mixture of two compounds, which were subjected to further purification using the reversed-phase semipreparative HPLC in gradient mode with MeOH:H$_2$O:TFA as the mobile phase, resulting in the production of the white crystals of compounds **21** and **22** (20.3 and 21.5 mg), respectively.

6-hydroxy-8-methoxycoumarin-7-*O*-β-D-glycopyranoside (compound **15**): White crystals; m.p. 219–220 °C; UV (MeOH) λ_{max} nm 325 and 250; IR (KBr) vmax (cm^{-1}): 3349, 1719, 1625, 1520, 1465, 829; ^{1}H NMR, ^{13}C NMR, and HMBC data, see Table 1 and Figure 2; HRESIMS (positive) m/z 371.0900 [M + H]$^+$ (calculated for C$_{16}$H$_{18}$O$_{10}$, 371.097825).

3-(2-(methylamino)ethyl)-1H-indol-2-yl)methanol (compound **18**): White needle crystals; m.p. 179–189 °C; UV (MeOH) λ_{max} nm: 295, 287, 279, 230; IR (KBr) v_{max} (cm^{-1}): 3309,1140, 1120, 1105, 1011, 855. ^{1}H NMR, ^{13}C NMR, and HMBC data, see Table 2 and Figure 2; HRESIMS (positive) m/z 205.1293 [M + H]$^+$ (calculated for C$_{12}$H$_{16}$N$_2$O, 205.134088).

3.5. Antinociceptive Activity Test

3.5.1. Hot-plate Method

The hot-plate method that was described by Turner was used to determine the antinociceptive activity of the compounds that were isolated from the *J. pelargoniifolia* root [48].

3.5.2. Acetic Acid-induced Writhing in Mice Test

The method of Koster et al. was used to evaluate the analgesic effect of the pure compounds that were isolated from the *J. pelargoniifolia* root [49].

3.5.3. Tail-Flick Method

Acute nociception was induced using the tail-flick apparatus (Tail flick Apparatus Harvard), following the method that was recommended by D'amour and Smith [50].

3.6. Anti-Inflammatory Activity Test

Carrageenan-Induced Edema in the Rat Paw Method

The method that was described by Winter et al. was used to evaluate the anti-inflammatory potency of the isolated compounds [51].

3.7. Antipyretic Activity Screening

Yeast-Induced Hyperthermia in Rats

Hyperthermia was induced in the mice followed by the administration of the isolated compounds, and their hypothermic activity was determined by applying the method described by Loux [52].

3.8. Antioxidant Effect

Nitric Oxide Radical-Scavenging Assay

This assay was carried out according to the procedure that was described by Green et al. [53].

3.9. Statistical Analysis

The values in the tables are given as mean $\pm$ SE. The data were analyzed by using one-way analysis of variance (ANOVA) followed by the Student's t-test. Values with $p < 0.05$ were considered significant.

4. Conclusions

The wide traditional use of *Jatropha* species as anti-inflammatory and analgesics has prompted us to investigate the chemistry and bioactivity of *J. pelargoniifolia* growing in Saudi Arabia. The phytochemical study of the plant roots resulted in the isolation of six terpenoids, five flavonoids, three coumarinolignans, two tryptamines, and two tyramines (including their HCl salts), a coumarin, and a pyrimidine. The new compounds were identified as 6-hydroxy-8-methoxy coumarin-7-*O*-β-D-glycopyranoside and 2-hydroxymethyl-*N*-methyltryptamine. To the best of our knowledge, hovetricoside C and *N*-methyltryptamine were isolated herein from the Euphorbiaceae family for the first time, while cleomiscosin B, hordenine, and *N*-methyltyramine with their salts, cynaroside, and linarin were characterized in the *Jatropha* species for the first time.

On the basis of the significant anti-inflammatory, analgesic, antipyretic, and antioxidant activities that were observed in the experimental animals for the alcoholic extract of *J. pelargoniifolia*, fifteen of the adequately isolated compounds were consequently biologically evaluated. Eleven of these compounds exhibited strong analgesic activity. Twelve out of the fifteen compounds succeeded to reduce the chemically-induced inflammatory marker in the animals in a dose-dependent manner.

Moreover, five of the compounds demonstrated an anti-pyretic effect by a reduction about a 1 °C in an induced hyperthermia model. The isolated compounds also exhibited varying degrees of nitric oxide-scavenging activity. The significant antioxidant activity that was associated with the administration the extract of *J. pelargoniifolia* roots was thus perhaps due to its phenolic content such as flavonoid, coumarins, and coumarinolignans. The synergistic effect between these bioactive constituents might explain the significant antinociceptive and anti-inflammatory effect of the alcoholic extract of *J. pelargoniifolia* roots and may scientifically justify the use of this plant in folk medicine for the treatment of pain and several inflammatory conditions.

Supplementary Materials: The following are available online. Figures S1–S4: ^{1}H-, ^{13}C-NMR, COSY and HMBC spectra of compound **15**, Figures S5–S8: ^{1}H-, ^{13}C-NMR, HMBC and HSQC spectra of compound **18**, Tables S9–S11: Analgesic effect of isolated compounds by using acetic acid-induced writhing, hot plate and tail flick methods in mice, Table S12: Effect of isolated compounds on carrageenan-induced paw edema in albino rats, Table S13: Effect of isolated compounds on yeast-induced hyperthermia in mice, Table S14: % Inhibition of nitric oxide scavenging activity for isolated compounds at different concentrations.

Author Contributions: H.Y.A., A.E.-G., and O.K. conceived and designed the experiments; H.Y.A. performed the experiments; H.Y.A., A.E.-G. analyzed the data; H.Y.A. contributed reagents/materials/analysis tools; H.Y.A. wrote the paper; A.E.-G. participated in the experiments' design and coordination and helped to draft the manuscript; O.K. participated in the experiment design and supervised. A.F.A. contributed to the structure elucidation of some of the compounds and to the partial writing of the manuscript.

Funding: The Deanship of Scientific Research at King Saud University, Project no. RG-1437-021.

Acknowledgments: The authors extend their appreciation to the Deanship of Scientific Research at King Saud University for funding the work through the Research Group, Project no. RG-1437-021.

Conflicts of Interest: The authors declare no conflict of interest.

Appendix A

Supplementary data associated with ^{1}H NMR, ^{13}C NMR, COSY, and HMBC of compounds **15** and **18** are available in Supplementary Information. In addition, Tables for all of the biological test results are provided.

References

1. Alves, M.V. Checklist das espècies de Euphorbiaceae Juss. Ocorrentes no semi-áridopernambucano, Brasil. *Acta Bot. Bras.* **1998**, *12*, 485–495. [CrossRef]
2. Webster, G.L. Classification of the Euphorbiaceae. *Ann. Mo. Bot. Gard.* **1994**, *81*, 3–143. [CrossRef]
3. Sabandar, C.W.; Ahmat, N.; Jaafar, F.M.; Sahidin, I. Medicinal property, phytochemistry and pharmacology of several *Jatropha* species (Euphorbiaceae): A review. *Phytochemistry* **2013**, *85*, 7–29. [CrossRef] [PubMed]
4. Félix-Silva, J.; Giordani, R.B.; Silva, A.A., Jr.; Zucolotto, S.M.; Fernandes-Pedrosa, M.F. *Jatropha gossypiifolia* L. (Euphorbiaceae): A review of traditional uses, phytochemistry, pharmacology, and toxicology of this medicinal plant. Evid-Based Complement. *Altern. Med.* **2014**, *2014*, 1–32.
5. Oskoueian, E.N.; Saad, W.Z.; Omar, A.; Ahmad, S.; Kuan, W.B.; Zolkifli, N.A.; Hendra, R.; Ho, Y.W. Antioxidant, anti-inflammatory and anticancer activities of methanolic extracts from *Jatropha curcas* Linn. *J. Med. Plants Res.* **2011**, *5*, 49–57.
6. Yusuf, S.O.; Maxwell, I.E. Analgesic activity of the methanolic leaf extract of *Jatropha Curcas* (Linn). *Afr. J. Biomed. Res.* **2010**, *13*, 149–152.
7. Schmelzer, G.H.; Fakim, G.A. *Jatropha pelargoniifolia*; Record from Prota4u; PROTA (Plant Resources of Tropical Africa/RessourcesVégétales de l'AfriqueTropicale): Wageningen, The Netherlands, 2007.
8. Zhang, X.P.; Zhang, M.L.; Sua, X.H.; Huoa, C.H.; Gub, Y.C.; Shi, Q.W. Chemical constituents of the plants from genus *Jatropha*. *Chem. Biodivers.* **2009**, *6*, 2166–2183. [CrossRef] [PubMed]
9. Zhu, J.Y.; Zhang, C.Y.; Dai, J.J.; Rahman, K.; Zhang, H. Diterpenoids with thioredoxin reductase inhibitory activities from *Jatropha multifida*. *Nat. Prod. Res.* **2017**, *31*, 2753–2758. [CrossRef] [PubMed]
10. Zhao, A.; Yang, X.Y. New coumarin glucopyranosides from roots of *Angelica dahurica*. *Chin. Herb. Med.* **2018**, *10*, 103–106. [CrossRef]

11. Konishi, T.; Wada, S.; Kiyosawa, S. Constituents of the leaves of *Daphne pseudo-mezereum*. *Yakugaku Zasshi: J. Pharm. Soc. Jpn.* **1993**, *113*, 670–675. [CrossRef]

12. Jung, M.; Geiger, H.; Zinsmeister, H.D. Tri- and tetrahydroxycoumarin derivatives from *Tetraphis pellucida*. *Phytochemistry* **1995**, *39*, 379–381. [CrossRef]

13. Liu, J.Q.; Yang, Y.F.; Wang, C.F.; Li, Y.; Qiu, M.H. Three new diterpenes from *Jatropha curcas*. *Tetrahedron* **2012**, *68*, 972–976. [CrossRef]

14. Habib, M.R.; Nikkon, F.; Rahman, M.; Haque, M.E.; Karim, M.R. Isolation of stigmasterol and β-sitosterol from methanolic extract of root bark of *Calotropis gigantean* (Linn). *Pak. J. Biol. Sci.* **2007**, *10*, 4174–4176. [PubMed]

15. Oskoueian, E.; Abdullah, N.; Ahmad, S.; Saad, W.Z.; Omar, A.; Ho, Y.W. Bioactive compounds and biological activities of *Jatropha curcas* L. kernel meal extract. *Int. J. Mol. Sci.* **2011**, *12*, 5955–5970. [CrossRef] [PubMed]

16. Naengchomnong, W.; Thebtaranonth, Y. Isolation and structure determination of four novel diterpenes from *Jatropha curcus*. *Tetrahedron Lett.* **1986**, *27*, 2439–2442. [CrossRef]

17. Maltese, F.; Erkelens, C.; Kooy, F.V.D.; Choi, Y.H. Identification of natural epimeric flavanone glycosides by NMR spectroscopy. *Food Chem.* **2009**, *116*, 575–579. [CrossRef]

18. Mutheeswarana, S.; Saravana, P.; Kumarb, P.; Yuvaraj, V.; Duraipandiy, N.A.; Balakrishnab, K.; Ignacimuthu, S. Screening of some medicinal plants for anticariogenic activity: An investigation on bioactive constituents from *Jatropha gossypifolia* (L.) root. *Biocatal. Agric. Biotech.* **2017**, *10*, 161–166. [CrossRef]

19. Hufford, C.D.; Oguntimein, B.O. Non-polar constituents of *Jatropha curcas*. *Lloydia* **1978**, *41*, 161–165.

20. Denton, W.R.; Harding, W.W.; Anderson, C.I.; Jacobs, H.; McLean, S.; Reynolds, W.F. New diterpenes from *Jatropha divaricata*. *J. Nat. Prod.* **2001**, *64*, 829–831. [CrossRef] [PubMed]

21. Alimboyoguen, A.B.; De Castro-Cruz, K.A.; Shen, C.C.; Li, W.T.; Ragasa, C.Y. Chemical constituents of the bark of *Aleurites moluccana* L. Willd. *J. Chem. Pharm. Res.* **2014**, *6*, 1318–1320.

22. Akbar, E.; Sadiq, Z. Coumarinolignoid rare natural product: A review. *Asian J. Chem.* **2012**, *24*, 4831–4842.

23. Begum, S.A.; Sahai, M.; Ray, A.B. Non-conventional lignans: coumarinolignans, flavonolignans, and stilbenolignans. *Prog. Chem. Org. Nat. Prod.* **2010**, *93*, 1–70.

24. Das, B.; Venkataiah, B. A minor coumarino-lignoid from *Jatropha gossypifolia*. *Biochem. Syst. Ecol.* **2001**, *2*, 213–214. [CrossRef]

25. Biswanath, D.; Kashinatham, A.; Venkataiah, B.; Srinivas, K.V.N.S.; Mahender, G.; Reddy, M.R. Cleomiscosin A, a coumarino-lignoid from *Jatropha gossypifolia*. *Biochem. Syst. Ecol.* **2003**, *31*, 1189–1191.

26. Xu, J.F.; Feng, Z.M.; L, J.; Zhang, P.C. New hepatoprotective coumarinolignoids from *Mallotus apelta*. *Chem. Biodivers.* **2008**, *5*, 591–597. [CrossRef] [PubMed]

27. Lee, D.G.; Mok, S.Y.; Choi, C.; Cho, E.J.; Kim, H.Y.; Lee, S. Analysis of apigenin in *Blumea balsamifera* Linn DC. and its inhibitory activity against aldose reductase in rat lens. *J. Agric. Chem. Environ.* **2012**, *1*, 28–33.

28. Abdelgadir, H.A.; Staden, J.V. Ethnobotany, ethnopharmacology and toxicity of *Jatropha curcas* L. (Euphorbiaceae): A review. *S. Afr. J. Bot.* **2013**, *88*, 204–218. [CrossRef]

29. Hurd, R.E.; Reid, B.R. NMR spectroscopy of the ring nitrogen protons of uracil and substituted uracils; relevance to a base pairing in the solution structure of transfer RNA. *Nucleic Acids Res.* **1977**, *4*, 2747–2756. [CrossRef] [PubMed]

30. Staubmann, R.; Schubert-Zsilavecz, M.; Hiermann, A.; Kartnig, T. A complex of 5-hydroxypyrrolidin-2-one and pyrimidine-2,4-dione isolated from *Jatropha curcas*. *Phytochemistry* **1999**, *50*, 337–338. [CrossRef]

31. Elhawary, S.S.; Eltantawy, M.E.; Sleem, A.A.; Abdallah, H.M.; Mohamed, N.M. Investigation of phenolic content and biological activities of *Scabiosa atropurpurea* L. *World Appl. Sci. J.* **2011**, *3*, 311–317.

32. Zhang, T.; Zhou, J.; Wang, Q. Flavonoids from aerial part of *Bupleurum chinense* DC. *Biochem. Syst. Ecol.* **2007**, *35*, 801–804. [CrossRef]

33. Fernández, S.; Wasowski, C.; Paladini, A.C.; Marder, M. Sedative and sleep-enhancing properties of linarin, a flavonoid-isolated from *Valeriana officinalis*. *Pharmacol. Biochem. Behav.* **2004**, *77*, 399–404. [CrossRef] [PubMed]

34. Thuy, T.T.; Kamperdick, C.; Ninh, P.T.; Lien, T.P.; Thao, T.T.P.; Sung, T.V. Immunosuppressive auronol glycosides from *Artocarpus tonkinensis*. *Pharmazie* **2004**, *59*, 297–300. [CrossRef] [PubMed]

35. Grina, J.A.; Ratcliff, M.R.; Stermitz, F.R. Old and new alkaloids from *Zanthoxylum arborescens*. *J. Org. Chem.* **1982**, *47*, 2648–2651. [CrossRef]

36. Yokoo, Y.; Kohda, H.; Kusumoto, A.; Naoki, H.; Matsumoto, N.; Amachi, T.; Suwa, Y.; Fukazawa, H.; Ishida, H.; Tsuji, K.; et al. Isolation from beer and structural determination of a potent stimulant of gastrin release. *Alcohol Alcohol.* **1999**, *34*, 161–168. [CrossRef] [PubMed]

37. Abdel-Kader, M.S.; Kassem, F.F.; Abdallah, R.M. Two alkaloids from *Ephedra aphylla* growing in Egypt. *Nat. Prod. Sci.* **2003**, *9*, 1–4.

38. Neal, J.M.; Sato, P.T.; Johnson, C.L.; McLaughlin, J.L. Cactus alkaloids X: Isolation of hordenine and *N*-methyltyramine from *Ariocarpus kotschoubeyanus*. *J. Pharm. Sci.* **1971**, *60*, 477–478. [CrossRef] [PubMed]

39. Hanan, A.; Ali, E.-G.; Oliver, K. A comparative study of the biological activities of *Jatropha pelargoniifolia* and *Jatropha glauca* native to Saudi Arabia. *Phytomedicine* **2018**, submitted.

40. Fana, Y.; Li, X.; Zhang, L.; Duana, P.; Li, F.; Zhao, D.; Wang, Y.; Wub, H. Ether-functionalized ionic liquids: Highly efficient extractants for hordenine. *Chem. Eng. Res. Des.* **2017**, *124*, 66–73. [CrossRef]

41. Ma, J.; Wang, S.; Huang, X.; Geng, P.; Wen, C.; Zhou, Y.; Yu, L.; Wang, X. Validated UPLC-MS/MS method for determination of hordenine in rat plasma and its application to pharmacokinetic study. *J. Pharm. Biomed. Anal.* **2015**, *111*, 131–137. [CrossRef] [PubMed]

42. Christine, S.; Bell, A.; Stewart-Johnsox, E. *N*-methyltyramine a biologically active amine in *Acacia* seeds. *Phytochemistry* **1979**, *18*, 2022–2023.

43. Jadeja, R.N.; Devkar, R.V. Polyphenols and flavonoids in controlling non-alcoholic steatohepatitis. In *Polyphenols in Human Health and Disease*, 1st ed.; Watson, R.R., Preedy, V.R., Zibadi, S., Eds.; Academic Press: San Diego, CA, USA, 2014; pp. 615–623.

44. Begum, S.; Saxena, B.; Goyal, M.; Ranjan, R.; Joshi, V. B.; Rao, V.; Krishnamurthy, S.; Sahai, M. Study of anti-inflammatory, analgesic and antipyretic activities of seeds of *Hyoscyamusniger* and isolation of a new coumarinolignan. *Fitoterapia* **2010**, *81*, 178–184. [CrossRef] [PubMed]

45. Picha, P.; Naengchomnong, W.; Promratanapongse, P.; Kano, E.; Hayashi, S.; Ohtsubo, T.; Zhang, S.W.; Shioura, H.; Kitai, R.; Matsumoto, H.; et al. Effect of natural pure compounds curcusones A and C from tropical herbal plant *Jatropha curcas* on thermo sensitivity and development of thermotolerance in Chinese hamster V-79 cells in vitro. *J. Exp. Clin. Cancer Res.* **1996**, *15*, 177–183.

46. Skoog, D.; Holler, F.J.; Nieman, T.A. An introduction to chromatographic separations. In *Principles of instrumental analysis*, 5th ed.; Skoog, D., Holler, F.J., Nieman, T.A., Eds.; Saunders College Publishing: Philadelphia, PA, USA, 1992; pp. 674–700.

47. Madhumitha, G.; Fowsiya, J. *A Hand Book on: Semi Micro Technique for Extraction of Alkaloids*; International E-Publication: Indore, India, 2015; p. 9.

48. Turner, R.A. *Analgesics, in Screening Methods in Pharmacology*; Academic Press: London, UK, 1965; p. 100.

49. Koster, R.; Anderson, M.; De Beer, E.J. Acetic acid for analgesic screening. *Fed. Proc.* **1959**, *18*, 412–417.

50. D'amour, F.E.; Smith, D.L. A method for determining loss of pain sensation. *J. Pharmacol. Exp. Ther.* **1941**, *72*, 74–79.

51. Winter, C.A.; Risley, E.A.; Nuss, G.W. Carregeenan-induced oedema in hind paw of the rats as an assay for anti-inflammatory drugs. *Exp. Biol. Med.* **1962**, *111*, 544–547. [CrossRef]

52. Loux, J.J.; Depalma, D.D.; Yankell, S.L. Antipyretic testing of aspirin in rats. *Toxicol. Appl. Pharmacol.* **1972**, *22*, 672–675. [CrossRef]

53. Green, L.C.; Wagner, D.A.; Glogowski, J.; Skipper, P.L.; Wishnok, J.S.; Tannenbaum, S.R. Analysis of nitrate, nitrite, and [15N] nitrate in biological fluids. *Anal. Biochem.* **1982**, *126*, 131–138. [CrossRef]

Sample Availability: Samples of the compounds **1–22** are available from the authors.

molecules

MDPI

Article

Flavonoids, Sterols and Lignans from *Cochlospermum vitifolium* and Their Relationship with Its Liver Activity

A. Berenice Aguilar-Guadarrama and María Yolanda Rios *

Centro de Investigaciones Químicas, IICBA, Universidad Autónoma del Estado de Morelos,
Avenida Universidad 1001, Col. Chamilpa, 62209 Cuernavaca, Morelos, Mexico; baguilar@uaem.mx
* Correspondence: myolanda@uaem.mx; Tel.: +52-777-329-7000 (ext. 6024)

Academic Editor: Muhammad Ilias
Received: 13 July 2018; Accepted: 3 August 2018; Published: 5 August 2018

Abstract: The sterols β-sitostenone (**1**), stigmast-4,6,8(14),22-tetraen-3-one (**2**), β-sitosterol (**3**) and stigmasterol (**4**), the aromatic derivatives antiarol (**5**) and gentisic acid (**6**), the phenylpropanes coniferyl alcohol (**7**), epoxyconiferyl alcohol (**8**) and ferulic acid (**9**), the apocarotenoid vomifoliol (**10**), the flavonoids naringenin (**11**), 7,4′-dimethoxytaxifolin (7,4′-dimethoxydihydroquercetin, **12**), aromadendrin (**13**), kaempferol (**14**), taxifolin (dihydroquercetin, **15**), prunin (naringenin-7-*O*-β-D-glucoside, **16**), populnin (kaempferol-7-*O*-β-D-glucoside, **17**) and senecin (aromadendrin-7-*O*-β-D-glucoside, **18**) and the lignans kobusin (**19**) and pinoresinol (**20**), were isolated from the dried bark of *Cochlospermum vitifolium* Spreng (Cochlospermaceae), a Mexican medicinal plant used to treat jaundice, liver ailments and hepatitis C. Fourteen of these compounds were isolated for the first time from this plant and from the *Cochlospermum* genus. Compounds **3–4**, **6–7**, **9–11**, **13–17** and **20** have previously exhibited diverse beneficial liver activities. The presence of these compounds in *C. vitifolium* correlates with the use of this Mexican medicinal plant.

Keywords: *Cochlospermum vitifolium*; Cochlospermaceae; flavonoids; lignans; aromatic compounds; carotenoids; sterols; liver activity

1. Introduction

The Cochlospermaceae family comprises seven genera: *Amoreuxia*, *Azeredia*, *Cochlospermum*, *Euryanthe*, *Lachnocistus*, *Maximilianea* and *Wittelsbachia*. In turn, the genus *Cochlospermum* (syn. *Maximilianea* and *Bixaceae*) is composed of 13 species of tropical trees ranging in height from 3 to 15 m, distributed in deciduous forests worldwide [1]. The bark and roots from the *Cochlospermum* species have been the most studied parts of these plants and only six of the 13 species of *Cochlospermum* have been chemically analyzed. Within *Cochlospermum gillivraei* the flavonoids apigenin, naringenin and afzelequin were found [2]; from *Cochlospermum gossypium* only carbohydrates were identified [3,4]; *Cochlospermum planchonii* biosynthesized the flavonoids miricetin, quercetin, aromadendrin and cianidin [5] and gallic acid, saponins, tannins, glycosides and carbohydrates [6]. From *Cochlospermum regium*, the gallic and ellagic acids, the flavonoid dihidrokaempferol-3-*O*-β-(6″-galloyl)-glucopyranoside and the lignans pinoresinol and excelsin have been isolated [7]. From *Cochlospermum tinctorium*, the triterpene arjunolic acid, along with tannins and carotenoids, β-bisabolene, 1-dodecanol, 1-hydroxy-3-octadecanone, 2-pentadecanone [8,9], alphitolic acid, cochloxantin and dihydrocochloxantin were identified [10]. Finally, the composition of the essential oils obtained from the leaves, root bark and root wood of *Cochlospermum vitifolium* has been established by GC/MS. The leaves' essential oil consist of four major components: β-caryophyllene (46.5%), α-humulene (26.0%), β-pinene (10.6%) and α-pinene (4.8%), which, together, make up 87.9% of the total oil. The essential oil derived from the root's bark is predominantly made up of β-bisabolene (29.3%), 1-hydroxy-3-hexadecanone (19.5%) and β-caryophyllene (8.2%), which corresponds to 57.0%. Furthermore, the root wood's essential oil is composed of γ-muurolene (28.4%), 1-hydroxy-3-hexadecanone (16.2%), β-caryophyllene (11.6%), β-bisabolene (11.5%) and 2-dodecanone (6.3%), which represent 74.0% of the total essential oil [11]. The ethanol extracts from the root's bark and wood are composed of gallic acid, the lignans excelsin and pinoresinol, the flavonoids naringenin and aromadendrin, and the sterols β-sitosterol, stigmasterol, 3-*O*-β-glycopyranosyl-β-sitosterol and 3-*O*-β-glycopyranosyl-stigmasterol while the root's wood also contains 1-dodecanoyl-3,5-di(tetradecanoyl)benzene [11]. A second chemical analysis yielded the apocarotenoids cochloxanthin, dihydrocochloxanthin, vitixanthin and dihydrovitixanthin [12]. On the other hand, the plant's flowers contain the flavonoids apigenin, naringenin and dihydroquercetin, and the carotenoids β-carotene, γ-carotene, lycopene, capsanthin. and zeaxanthin [13]. Finally, the stems contain naringenin and dihydroquercetin [14]. All of these studies indicate that the *Cochlospermum* species are characterized by the presence of sterols, flavonoids, carotenoids, apocarotenoids and lignans.

Cochlospermum vitifolium (common name panicua, yellow rose or pongolote) is a medicinal tree which grows up to 5 m in height and is found from México to South America. It is characterized by its attractive yellow flowers and seed pods. This plant has been used in several countries due to its medicinal properties. In Cuba, for instance, a decoction of its leaves is used in the treatment of ulcers. In Costa Rica, the sap of the leaves is used to treat jaundice, and in Guatemala it is used due to its emmenagogue effects [15]. In some Mexican states, such as Morelos, Oaxaca, Puebla, and Veracruz, a decoction of its wood and leaves is drunk as an alternative treatment for liver and kidney ailments [16]. For example, in the state of Morelos, an infusion, prepared by boiling 10 g of its dried bark in 1 L of water, is used to treat hepatitis C, jaundice, liver diseases, diabetes, metabolic syndrome, and high blood pressure [17,18].

According to different studies, *Cochlospermum vitifolium*'s lethal dose 50 (LD_{50}) in its lyophilized aqueous phase, obtained from the partitioned methanol extract, first with chloroform and then with ethyl acetate, was greater than 2000 mg/kg when administered intraperitoneally in mice [15]. The pharmacological analysis of the methanol extract from its dried bark demonstrated a decrease in noradrenaline induced vasoconstriction in rat aortic rings in a concentration and endothelium dependent manner (NO-cGMP system) [19]. The extract showed in vivo antihypertensive effects on spontaneously hypertensive rats [20] by inhibiting the [3H]-AT-II binding (angiotensin II AT1 receptor) by more than 50% [21]. In addition, hypoglycemic and antidiabetic effects were also seen in normoglycemic and STZ-nicotinamide-induced diabetic rats, both, in acute and subchronic models [22]. Additionally, the ethanol extract from the same part of the plant exhibited anti-inflammatory and immunomodulatory properties [23]. Finally, the dichloromethane extract was evaluated ex vivo using rat trachea rings to determine its relaxant activity against contractions induced by carbachol, showing a maximum effect at E_{max} = 106.58 ± 2.42% and an EC_{50} = 219.54 ± 7.61 μg/mL [24].

2. Results and Discussion

During our ongoing phytochemical research from the dichloromethane extract of the dried bark of *Cochlospermum vitifolium*, the following metabolites were characterized: the sterols β-sitostenone (**1**), stigmast-4,6,8(14),22-tetraen-3-one (**2**), β-sitosterol (**3**) and stigmasterol (**4**), the aromatic derivatives antiarol (**5**), gentisic acid (**6**), coniferyl alcohol (**7**), epoxyconiferyl alcohol (**8**) and ferulic acid (**9**), the apocarotenoid vomifoliol (**10**), and the flavonoid naringenin (**11**). On the other hand, from the methanol extract two types of metabolites were identified: the flavonoids naringenin (**11**), 7,4′-dimethoxytaxifolin (7,4′-dimethoxydihydroquercetin, **12**), aromadendrin (**13**), kaempferol (**14**), taxifolin (dihydroquercetin, **15**), prunin (naringenin-7-*O*-β-D-glucoside, **16**), populnin (kaempferol-7-*O*-β-D-glucoside, **17**) and senecin (aromadendrin-7-*O*-β-D-glucoside, **18**); along with the lignans kobusin (**19**) and pinoresinol (**20**) (Figure 1). Vomifoliol (**10**) and naringenin (**11**) were the major constituents of the dichloromethane extract, while the flavonoids naringenin (**11**) and senecin (**18**) were isolated as the major components from the methanol extract. The structures for compounds **1–20** were established by the analysis of their ^{1}H and ^{13}C NMR (1D and 2D experiments) and MS spectra parameters and its comparison with those reported in the literature. The unequivocal ^{1}H- and ^{13}C- NMR assignments for compound **2** are reported here for the first time.

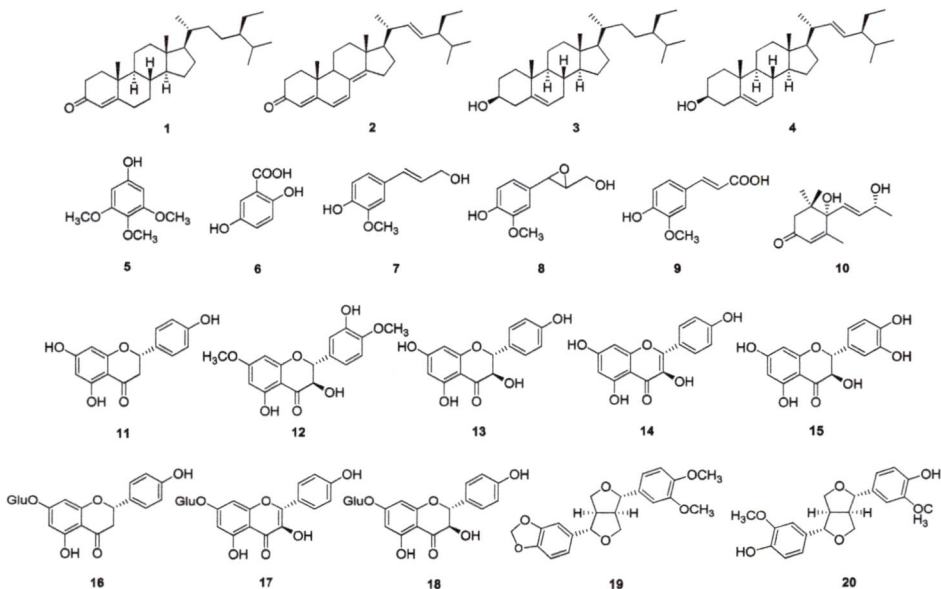

Figure 1. Chemical contents of the dried bark from *Cochlospermum vitifolium*.

Six of these 20 compounds (**3**, **4**, **11**, **13**, **15** and **20**) had been previously isolated from *Cochlospermum vitifolium*, however all other compounds were isolated here for the first time from this plant and from this genus. Compounds **1–20** belong to either the sterols; C_6, C_6–C_1 and C_6–C_3 aromatic compounds; apocarotenoids; flavonoids and lignans groups, which are the most frequently identified compounds in this genus.

Previous studies have been conducted to demonstrate the antihypertensive [19–21,23], hypoglycemic and antidiabetic [22], immunomodulatory [23] and anti-inflammatory [24] effects of *Cochlospermum vitifolium* extracts. An exhaustive revision of the existing literature indicated that several of the compounds isolated in this research (Figure 1) can be associated to its popular use as different liver treatments. Due to the fact that this plant is broadly used as a treatment for jaundice, hepatitis C and other liver ailments in Mexican traditional medicine, its methanol extract was administrated at a dose of 100 mg/kg to bile duct-obstructed rats, to determine its hepatoprotective activity, showing a statistically significant decrease of serum glutamic-pyruvic transaminase (PGT, 45%) and alkaline phosphatase (APh, 15%) [19].

The importance of such activities lie in the fact that hepatic diseases (which comprise several conditions, such as: cirrhosis, hepatitis, alcoholic liver disease, non-alcoholic fatty liver disease, cholestatic and drug-induced liver diseases and liver cancer) are extremely high-priced in terms of human suffering, loss of productivity, and medical or hospital consultations. In fact, chronic liver diseases are the major cause of mortality worldwide. In 2013, 29 million people in Europe suffered from a chronic liver condition [25] and more than 30 million Americans had hepatic disease [26]. In China, liver diseases, viral hepatitis (predominantly hepatitis B virus), non-alcoholic fatty liver and alcoholic liver disease affect approximately 300 million people [27]. From a physiopathological perspective, most chronic liver diseases begin as an inflammatory process which evolves into focal fibrosis, and afterwards, to complete fibrosis of the gland (hepatic cirrhosis), which increases the risk of liver cancer. This leads to severe hepatic injury and ultimately to liver failure and other complications.

Thirteen of the twenty compounds isolated from *Cochlospermum vitifolium* have previously exhibited in vitro and in vivo beneficial liver effects. β-Sitosterol (**3**) decreased hepatofibrosis [28], protecting against CCl$_4$-induced hepatotoxicity [29] in animal models. Stigmasterol (**4**) induced apoptosis in hepatocarcimona (HepG2) cells being a potential antineoplastic therapeutic agent [30]. Gentisic acid (**6**) showed anti-inflammatory and antimutagenic properties, demonstrating protective effects against induced genotoxicity and hepatotoxicity [31]. On the other hand, coniferyl alcohol (**7**) had a moderated anti-hepatitis B virus (HBV) activity [32]. Ferulic acid (**9**) had an in vivo hepato-protective effect against the CCl$_4$- and formaldehyde-induced hepatotoxicity [33] and also a capacity to inhibit the development of hepatic fibrosis by activation of Hepatic Stellate Cells (HSCs) in the presence of liver damage [34]. Vomifoliol (**10**) showed moderate activity against human hepatocarcinoma Hep3B cells [35]. Naringenin (**11**) had a potent lipid-lowering effect reducing the hepatic lipogenesis in rats and acting as an insulin sensitizer in vivo [36], thus, preventing rat liver damage caused by lead acetate, arsenic and high glucose. Furthermore, this same compound suppresses the metastatic potential of hepatocellular carcinoma [37]. Additionally, aromadendrin (**13**) possessed radical scavenging and activity against inflammatory, tumor and diabetic processes [38]. Kaempferol (**14**) had hepatoprotective effects in CCl$_4$-, drug- and alcoholic-induced liver injury, constituting a promising therapeutic option for patients with atherosclerotic disease [39]. Taxifolin (**15**) had antioxidant and cytoprotective effects that prevent and help treat fulminant hepatitis and hepatitis caused CCl$_4$ [40]. This natural flavonoid is licensed as Silymarin (Legalon®), a drug used for the treatment of toxic liver damage, chronic inflammatory liver disease and liver cirrhosis. Prunin (**16**) showed activity against the hepatitis B (HBV) virus with an IC$_{50}$ of 41.59 μM [41]. Administered at a dose of 25 mg/kg, populnin (**17**) exhibited in vivo hepatoprotective effects against CCl$_4$- and D-GalN-induced hepatotoxicity, preventing the development of hepatic lesions [42]. Finally, at 50 and 100 mg/kg, the lignane pinoresinol (**20**) showed hepatoprotective effects improving CCl$_4$-induced liver injury [43]. All these liver beneficial effects from the compounds isolated from *Cochlospermum vitifolium* directly correlate with its traditional use in Mexican medicine.

3. Materials and Methods

3.1. General Procedures

Compounds **1–20** were purified by successive open column chromatography (CC) using silica gel (70–230 and 230–400 mesh, Sigma-Aldrich, Toluca, México). The isolation procedures and purity of compounds were monitored by thin layer chromatography (TLC) using precoated silica gel 60 F$_{254}$ aluminium sheets, visualizing with UV-light and subsequently spraying the plates with (NH$_4$)$_4$Ce(SO$_4$)$_4$ in 2 N H$_2$SO$_4$ (Sigma-Aldrich, Toluca, México). All ^{1}H-, ^{13}C- and 2-D NMR experiments were performed in CDCl$_3$ on a Varian Unity 400 spectrometer (Varian, Inc., Palo Alto, CA, USA) equipped with a 5 mm inverse detection pulse field gradient probe at 25 °C, at 400 MHz for ^{1}H-NMR and 100 MHz for ^{13}C-NMR. Chemical shifts were referenced to tetramethylsilane as an internal standard.

3.2. Plant Material

The wood of *Cochlospermum vitifolium* was collected from "Sierra de Huautla" (20°26′10″ N, 99°05′42″ W, 1915 m above sea level), Morelos, México, in October 2011, and identified by Dr. Rolando Ramírez Rodríguez, Centro de Investigación en Biodiversidad y Conservación-UAEM. A voucher specimen (number 14628) was deposited at HUMO Herbarium from the Universidad Autónoma del Estado de Morelos, México.

3.3. Extraction and Isolation

Throughout three months the wood of this plant was dried at room temperature. The dried and ground wood (1.65 kg) was extracted with CH_2Cl_2 and MeOH. These extracts were dried under a vacuum to render 11.9 g (0.72% yield) and 25.8 g (1.56% yield) of residue, respectively.

Fractionation of the CH_2Cl_2 extract by open CC (silica gel, 70–230 mesh; 5 cm i.d. × 20 cm) was performed with a step gradient of *n*-hexane-acetone 100:0 to 0:100, collecting 330 fractions of 50 mL each. Based on TLC analysis, these fractions were pooled into nine groups, namely G-1 (fractions 1–70, *n*-hexane 100%), G-2 (fractions 71–76, 722 mg, *n*-hexane:acetone 95:5), G-3 (fractions 77–81, *n*-hexane:acetone 95:5), G-4 (fractions 82–89, 131 mg, *n*-hexane:acetone 95:5), G-5 (fractions 90–166, 780 mg, *n*-hexane:acetone 95:5), G-6 (fractions 167–188, 620 mg, *n*-hexane:acetone 9:1), G-7 (fractions 189–265, 639 mg, *n*-hexane:acetone 8:2), G-8 (fractions 266–324, 540 mg, *n*-hexane:acetone 1:1) and G-9 (fractions 325–335, acetone 100%). G-1 was made up of fatty acids, G-3 of triglycerides and G-9 of resins. The rest of the groups were subjected to column chromatography using silica gel 70–230 mesh. G-2 (2.0 cm i.d. × 30 cm, eluent *n*-hexane 100% to *n*-hexane:acetone 8:2) yielded 45 fractions of 40 mL. Fractions 23–33 (255 mg) were subjected to a second column chromatography (1.0 cm i.d. × 30 cm, eluent *n*-hexane:AcOEt 99:1) obtaining 91 fractions of 20 mL to yield β-sitosterone (**1**, 58 mg, 0.0035% with respects dry weigh of plant material). G-4 (1.5 cm i.d. × 15 cm, eluent *n*-hexane:AcOEt 99:1 to *n*-hexane:AcOEt 97:3) yielded 132 fractions of 50 mL which rendered stigmast-4,6,8(14),22-tetraen-3-one (**2**, 71 mg, 0.0043%), coniferyl alcohol (**7**, 21 mg, 0.0013%) and ferulic acid (**9**, 27 mg, 0.0016%). G-5 (2.0 cm i.d. × 30 cm, eluent *n*-hexane:AcOEt 97:3 to *n*-hexane:AcOEt 9:1) yielded 180 fractions of 50 mL resulting in the isolation of a 7:3 mixture of β-sitosterol (**3**) and stigmasterol (**4**, 280 mg, 0.0109%). G-6 (2.0 cm i.d. × 30 cm, eluent *n*-hexane:AcOEt 9:1 to *n*-hexane:AcOEt 7:3) yielded 77 fractions of 50 mL resulting in vomifoliol (**10**, 94 mg, 0.0056%). G-7 (2.0 cm i.d. × 30 cm, eluent *n*-hexane:AcOEt 9:1) yielded 50 fractions of 50 mL obtaining antiarol (**5**, 31 mg, 0.0018%) and gentisic acid (**6**, 29 mg, 0.0017%). Finally, G-8 (2.0 cm i.d. × 30 cm, *n*-hexane:acetone 8:2) yielded 187 fractions of 50 mL. Fractions 104–167 (280 mg) were subjected to a second column chromatography (1.0 cm i.d. × 30 cm, eluent *n*-hexane:AcOEt 6:4) resulting in 248 fractions of 20 mL to yield naringenin (**11**, 102 mg, 0.0061%) and epoxy-coniferyl alcohol (**8**, 36 mg, 0.0022%).

Fractionation of the MeOH extract by open CC (silica gel, 100–230 mesh; 5 cm i.d. × 30 cm) was performed with a step gradient of *n*-hexane-acetone 100:0 to 0:100, collecting 268 fractions of 50 mL each. These fractions were pooled into three groups: MG-1 (fractions 1–93, *n*-hexane to *n*-hexane:acetone 6:4), MG-2 (fractions 94–135, 2.2 g, *n*-hexane:acetone 1:1) and MG-3 (fractions 136–268, 3.29 g, *n*-hexane:acetone 1:1 to acetone). MG-1 was a complex mixture including compounds **1**, **3**, **4**, **11** and aromadendrin (**13**); MG-2 (3.0 cm *i.d.* × 30 cm, eluent *n*-hexane:AcOEt 85:15 to *n*-hexane:AcOEt 75:25) yielded 250 fractions of 50 mL which resulted in the purification of naringenin (**11**, 214 mg, 0.0129%); and MG-3 (4.0 cm *i.d.* × 30 cm, eluent *n*-hexane:AcOEt 7:3 to acetone) yielded 320 fractions of 50 mL. Fractions 33–51 (181 mg) were subjected to a second column chromatography (1.5 cm *i.d.* × 20 cm, eluent with *n*-hexane:AcOEt 7:3) collecting 56 fractions of 20 mL to yield naringenin (**11**, 41 mg, 0.0024%) and 7,4′-dimethoxy-taxifolin (**12**, 21 mg, 0.0013%). Fractions 77–83 (52 mg) were subjected to a preparative TLC (2 mm × 20 cm, eluent *n*-hexane:AcOEt 7:3 twice) to yield 30 mg (0.0018%) of a 6:4 mixture of **13** and kaempferol (**14**). Fractions 100–111 (49 mg) were subjected to a preparative TLC (2 mm × 20 cm, eluent *n*-hexane:AcOEt 7:3 twice) to yield kobusin (**19**, 19 mg, 0.0011%) and pinoresinol (**20**, 17 mg, 0.0010%). Fractions 145–182 (93 mg) were subjected to a preparative TLC (2 mm × 20 cm, eluent with CH_2Cl_2:MeOH 93:7) to yield taxifolin (**15**, 26 mg, 0.0016%). Finally, fractions 289–309 (691 mg) were subjected to a second column chromatography (2.0 cm *i.d.* × 30 cm, eluent with CH_2Cl_2:MeOH 9:1) collecting 140 fractions of 20 mL to yield prunin (**16**, 32 mg, 0.0019%), and 59 mg (0.0035%) of a 45:55 mixture of **16**, populnin (**17**), and senecin (**18**, 47 mg, 0.0028%).

Stigmast-4,6,8(14),22-tetraen-3-one (**2**). ¹H-NMR (CDCl₃): δ 6.61 (1H, d, *J* = 7.2, H-7), 6.04 (1H, d, *J* = 7.2, H-6), 5.74 (1H, s, H-4), 5.27 (1H, dd, *J* = 15.2, 6.8, H-22), 5.20 (1H, dd, *J* = 15.2, 7.6, H-23),

2.49 (1H, m, H-2a), 2.57 (1H, m, H-2b), 2.51 (1H, m, H-15b), 2.47 (1H, m, H-15a), 2.16 (1H, m, H-20), 2.15 (1H, m, H-9), 2.10 (1H, m, H-12a), 2.03 (1H, m, H-1a), 2.54 (1H, m, H-1b), 1.88 (1H, m, H-24), 1.83 (1H, m, H-16b), 1.53 (1H, m, H-16a), 1.51 (1H, m, H-25), 1.32 (1H, m, H-12b), 1.30 (1H, m, H-17), 1.06 (3H, d, *J* = 6.4, H-21), 1.00 (3H, s, H-19), 0.96 (3H, s, H-18), 0.93 (3H, d, *J* = 6.8, H-26), 0.85 (3H, d, *J* = 6.8, H-27), 0.84 (3H, t, *J* = 6.8, H-29), 0.78 (2H, m, H-28). ^{13}C-NMR (CDCl$_3$): δ 199.76 (s, C-3), 164.67 (s, C-5), 156.35 (s, C-14), 135.25 (d, C-22), 134.28 (d, C-7), 132.78 (d, C-23), 124.71 (d, C-6), 124.66 (s, C-8), 123.23 (d, C-4), 55.95 (d, C-17), 44.57 (d, C-9), 44.24 (s, C-13), 43.13 (d, C-24), 39.55 (d, C-20), 37.02 (s, C-10), 35.84 (t, C-12), 34.37 (t, C-1, C-2), 33.34 d, C-25), 27.98 (t, C-16), 25.63 (t, C-15), 22.98 (t, C-11), 21.48 (q, C-21), 20.25 (q, C-27), 19.92 (q, C-18), 19.21 (q, C-28), 17.90 (q, C-26), 16.91 (q, C-19), 14.38 (q, C-29).

4. Conclusions

Cochlospermum vitifolium biosynthesizes among other compounds the sterols **3** and **4**, the aromatic compounds **6**, **7** and **9**, the apocatrotenoid **10**, the flavonoids **11** and **13–17**, and the lignan **20**, which have demonstrated beneficial activity to alleviate different liver diseases. The presence of these compounds in the plant agrees with its traditional use in Mexican medicine and some are even included in commercial pharmaceutical formulations used in the treatment of hepatopathies. Their presence within *Cochlospermum vitifolium* extracts indicates that this plant could be an active hepatoprotective agent. However, the human consumption of this plant must be subjected to toxicity, pharmacodynamic and pharmacokinetic studies to determine how it can be a health contributor.

The isolated metabolites in this study and the chemical composition previously reported for *Cochlospermum vitifolium* agree with the metabolic content within other *Cochlospermum* species. The flavonoids, sterols, carotenoids, apocarotenoids and lignans isolated here have chemotaxonomic significance within this genus. This is the first report of compounds **1–2**, **5–10**, **12**, **14**, **16–19** from *Cochlospermum vitifolium*.

Author Contributions: A.B.A.-G. prepared the extracts, isolated, and identified the secondary metabolites. M.Y.R. designed overall research and identified the secondary metabolites.

Funding: This research was funded by CONACYT-México grant number CB-2015-241044.

Acknowledgments: All spectroscopic and spectrometric analyses were obtained from LANEM (grant numbers 279905).

Conflicts of Interest: The authors declare no conflict of interest.

References

1. Lamien-Meda, A.; Kiendrebeogo, M.; Compaoré, M.; Meda, R.N.T.; Bacher, M.; Koenig, K.; Pacher, T.; Fuehrer, H.-P.; Noedl, H.; Willcox, M.; et al. Quality assessment and antiplasmodial activity of West African *Cochlospermum* species. *Phytochemistry* **2015**, *119*, 51–61. [CrossRef] [PubMed]
2. Cook, I.F.; Knox, J.R. Flavonoids from *Cochlospermum gillivraei*. *Phytochemistry* **1975**, *14*, 2510–2511. [CrossRef]
3. Vinod, V.T.P.; Sashidhar, R.B.; Suresh, K.I.; Rama Rao, B.; Vijaya Saradhi, U.V.R.; Prabhakar Rao, T. Morphological, physico-chemical and structural characterization of gum kondagogu (*Cochlospermum gossypium*): A tree gum from India. *Food Hydrocoll.* **2008**, *22*, 899–915. [CrossRef]
4. Hongsing, P.; Palanuvej, C.; Ruangrungsi, N. Chemical compositions and biological activities of selected exudate gums. *J. Chem. Pharm. Res.* **2012**, *4*, 4174–4180.
5. Bate-Smith, E.C. Chemotaxonomie der Pflanzen. *Bull. Soc. Bot. Mem.* **1964**, 435. [CrossRef]
6. Anaga, A.O.; Oparah, N. Investigation of the methanol root extract of *Cochlospermum planchonii* for pharmacological activities in vitro and in vivo. *Pharm. Biol.* **2009**, *47*, 1027–1034. [CrossRef]
7. Solon, S.; Carollo, C.A.; Brandão, L.F.G.; Macedo, C.; Klein, A.; Dias-Junior, C.A.; Siqueira, J.M. Phenolic derivatives and other chemical compounds from *Cochlospermum regium*. *Quim. Nova* **2012**, *35*, 1169–1172. [CrossRef]
8. Diallo, B.; Vanhaelen, M.; Vanhaelen-Fastré, R.; Konoshima, T.; Kozuka, M.; Tokuda, H. Studies on inhibitors of skin-tumor promotion. Inhibitory effects of triterpenes from *Cochlospermum tinctorium* on Epstein-Barr Virus Activation. *J. Nat. Prod.* **1989**, *52*, 879–881. [CrossRef] [PubMed]

9. Diallo, B.; Vanhaelen-Fastré, R.; Vanhaelen, M. Triacylbenzenes and long-chain volatile ketones from *Cochlospermum tinctorium* rhizome. *Phytochemistry* **1991**, *30*, 4153–4156. [CrossRef]

10. Ballin, N.Z.; Traore, M.; Tinto, H.; Sittie, A.; Mølgaard, P.; Olsen, C.E.; Kharazmi, A.; Christensen, S.B. Antiplasmodial compounds from *Cochlospermum tinctorium*. *J. Nat. Prod.* **2002**, *65*, 1325–1327. [CrossRef] [PubMed]

11. Xenofonte de Almeida, S.C.; Gomes de Lemos, L.T.; Rocha Silveira, E.; Loiola Pessoa, O.D. Constituintes químicos voláteis e não-voláteis de *Cochlospermum vitifolium* (Willdenow) Sprengel. *Quim. Nov.* **2005**, *28*, 57–60. [CrossRef]

12. Achenbach, H.; Blümm, E.; Waibel, R. Vitixanthin and dihydrovitixanthin - new unusual 7′-apocarotenoic acids from. *Tetrahedron Lett.* **1989**, *30*, 3059–3060. [CrossRef]

13. Dixit, B.S.; Srivastava, S.N. Flavonoids and carotenoids of *Cochlospermum vitifolium* flowers. *Fitoterapia* **1992**, *63*, 270.

14. López, J.A. Flavonoids in *Cochlospermum vitifolium* Willd (Cochlospermaceae). *Ing. Cienc. Quim.* **1981**, *5*, 101–102.

15. Esposito-avella, M.; Brown, P.; Tejeira, I.; Buitrago, R.; Barrios, L.; Sanchez, C.; Gupta, M.P.; Cedeño, J. Pharmacological screening of Panamanian medicinal plants. Part 1. *Int. J. Crude Drug Res.* **1985**, *23*, 17–25. [CrossRef]

16. Zamora-Martinez, M.C.; de Pascual Pola, C.N. Medicinal plants used in some rural populations of Oaxaca, Puebla and Veracruz, Mexico. *J. Ethnopharmacol.* **1992**, *35*, 229–257. [CrossRef]

17. Monroy-Ortiz, C.; Castillo-España, P. *Plantas Medicinales Utilizadas en el Estado de Morelos*, 2nd ed.; Universidad Autónoma del Estado de Morelos: Cuernavaca, México, 2007; ISBN 968-878-277-7.

18. Banos, G.; Perez-Torres, I.; El Hafidi, M. Medicinal agents in the metabolic syndrome. *Cardiovasc. Hematol. Agents Med. Chem.* **2008**, *6*, 237–252. [CrossRef] [PubMed]

19. Sánchez-Salgado, J.C.; Ortiz-Andrade, R.R.; Aguirre-Crespo, F.; Vergara-Galicia, J.; León-Rivera, I.; Montes, S.; Villalobos-Molina, R.; Estrada-Soto, S. Hypoglycemic, vasorelaxant and hepatoprotective effects of *Cochlospermum vitifolium* (Willd.) Sprengel: A potential agent for the treatment of metabolic syndrome. *J. Ethnopharmacol.* **2007**, *109*, 400–405. [CrossRef] [PubMed]

20. Sánchez-Salgado, J.C.; Castillo-España, P.; Ibarra-Barajas, M.; Villalobos-Molina, R.; Estrada-Soto, S. *Cochlospermum vitifolium* induces vasorelaxant and antihypertensive effects mainly by activation of NO/cGMP signaling pathway. *J. Ethnopharmacol.* **2010**, *130*, 477–484. [CrossRef] [PubMed]

21. Caballero-George, C.; Vanderheyden, P.M.L.; Solis, P.N.; Pieters, L.; Shahat, A.A.; Gupta, M.P.; Vauquelin, G.; Vlietinck, A.J. Biological screening of selected medicinal Panamanian plants by radioligand-binding techniques. *Phytomedicine* **2001**, *8*, 59–70. [CrossRef] [PubMed]

22. Ortíz-Andrade, R.; Torres-Piedra, M.; Sánchez-Salgado, J.C.; García-Jiménez, S.; Villalobos-Molina, R.; Ibarra-Barajas, M.; Gallardo-Ortíz, I.; Estrada-Soto, S. Acute and sub-chronic effects of *Cochlospermum vitifolium* in blood glucose levels in normoglycemic and STZ-nicotinamide-induced diabetic rats. *Rev. Latinoamer. Quím.* **2009**, *37*, 122–132.

23. Deharo, E.; Baelmans, R.; Gimenez, A.; Quenevo, C.; Bourdy, G. In vitro immunomodulatory activity of plants used by the Tacana ethnic group in Bolivia. *Phytomedicine* **2004**, *11*, 516–522. [CrossRef] [PubMed]

24. Sánchez-Recillas, A.; Mantecón-Reyes, P.; Castillo-España, P.; Villalobos-Molina, R.; Ibarra-Barajas, M.; Estrada-Soto, S. Tracheal relaxation of five medicinal plants used in Mexico for the treatment of several diseases. *Asian Pac. J. Trop. Med.* **2014**, *7*, 179–183. [CrossRef]

25. HEPAMAP A Roadmap for Hepatology Research in Europe: An Overview for Policy Makers. Available online: www.easl.eu/medias/EASLimg/News/3f9dd90221ef292_file.pdf (accessed on 10 July 2018).

26. American Liver Foundation The Liver Lowdown–Liver Disease: The Big Picture. Available online: https://liverfoundation.org/for-patients/resources/liver-lowdown/ (accessed on 10 July 2018).

27. Wang, F.-S.; Fan, J.-G.; Zhang, Z.; Gao, B.; Wang, H.-Y. The global burden of liver disease: The major impact of China. *Hepatology* **2014**, *60*, 2099–2108. [CrossRef] [PubMed]

28. Kim, K.-S.; Yang, H.J.; Lee, J.-Y.; Na, Y.-C.; Kwon, S.-Y.; Kim, Y.-C.; Lee, J.-H.; Jang, H.-J. Effects of β-sitosterol derived from *Artemisia capillaris* on the activated human hepatic stellate cells and dimethylnitrosamine-induced mouse liver fibrosis. *BMC Complement. Altern. Med.* **2014**, *14*, 363. [CrossRef] [PubMed]

29. Wong, H.-S.; Chen, J.-H.; Leong, P.-K.; Leung, H.-Y.; Chan, W.-M.; Ko, K.-M. β-Sitosterol protects against carbon tetrachloride hepatotoxicity but not Gentamicin nephrotoxicity in rats via the induction of mitochondrial glutathione redox cycling. *Molecules* **2014**, *19*, 17649–17662. [CrossRef] [PubMed]

30. Kim, Y.-S.; Li, X.-F.; Kang, K.-H.; Ryu, B.; Kim, S.K. Stigmasterol isolated from marine microalgae *Navicula incerta* induces apoptosis in human hepatoma HepG2 cells. *BMB Rep.* **2014**, *47*, 433–438. [CrossRef] [PubMed]

31. Nafees, S.; Ahmad, S.T.; Arjumand, W.; Rashid, S.; Ali, N.; Sultana, S. Modulatory effects of gentisic acid against genotoxicity and hepatotoxicity induced by cyclophosphamide in Swiss albino mice. *J. Pharm. Pharmacol.* **2012**, *64*, 259–267. [CrossRef] [PubMed]

32. Wang, H.-L.; Geng, C.-A.; Ma, Y.-B.; Zhang, X.-M.; Chen, J.-J. Three new secoiridoids, swermacrolactones A–C and anti-hepatitis B virus activity from *Swertia macrosperma. Fitoterapia* **2013**, *89*, 183–187. [CrossRef] [PubMed]

33. Gerin, F.; Erman, H.; Erboga, M.; Sener, U.; Yilmaz, A.; Seyhan, H.; Gurel, A. The effects of ferulic acid against oxidative stress and inflammation in formaldehyde-induced hepatotoxicity. *Inflammation* **2016**, *39*, 1377–1386. [CrossRef] [PubMed]

34. Xu, T.; Pan, Z.; Dong, M.; Yu, C.; Niu, Y. Ferulic acid suppresses activation of hepatic stellate cells through ERK1/2 and Smad signaling pathways in vitro. *Biochem. Pharmacol.* **2015**, *93*, 49–58. [CrossRef] [PubMed]

35. Dat, N.T.; Jin, X.; Hong, Y.-S.; Lee, J.J. An isoaurone and other constituents from trichosanthes kirilowii seeds inhibit hypoxia-inducible factor-1 and nuclear factor-κB. *J. Nat. Prod.* **2010**, *73*, 1167–1169. [CrossRef] [PubMed]

36. Assini, J.M.; Mulvihill, E.E.; Burke, A.C.; Sutherland, B.G.; Telford, D.E.; Chhoker, S.S.; Sawyez, C.G.; Drangova, M.; Adams, A.C.; Kharitonenkov, A.; et al. Naringenin prevents obesity, hepatic steatosis, and glucose intolerance in male mice independent of fibroblast growth factor 21. *Endocrinology* **2015**, *156*, 2087–2102. [CrossRef] [PubMed]

37. Yen, H.-R.; Liu, C.-J.; Yeh, C.-C. Naringenin suppresses TPA-induced tumor invasion by suppressing multiple signal transduction pathways in human hepatocellular carcinoma cells. *Chem. Biol. Interact.* **2015**, *235*, 1–9. [CrossRef] [PubMed]

38. Lee, J.-W.; Kim, N.H.; Kim, J.-Y.; Park, J.-H.; Shin, S.-Y.; Kwon, Y.-S.; Lee, H.J.; Kim, S.-S.; Chun, W. Aromadendrin inhibits lipopolysaccharide-induced nuclear translocation of NF-κB and phosphorylation of JNK in RAW 264.7 macrophage cells. *Biomol. Ther.* **2013**, *21*, 216–221. [CrossRef] [PubMed]

39. Ochiai, A.; Miyata, S.; Iwase, M.; Shimizu, M.; Inoue, J.; Sato, R. Kaempferol stimulates gene expression of low-density lipoprotein receptor through activation of Sp1 in cultured hepatocytes. *Sci. Rep.* **2016**, *6*, 24940. [CrossRef] [PubMed]

40. Zhao, M.; Chen, J.; Zhu, P.; Fujino, M.; Takahara, T.; Toyama, S.; Tomita, A.; Zhao, L.; Yang, Z.; Hei, M.; et al. Dihydroquercetin (DHQ) ameliorated concanavalin A-induced mouse experimental fulminant hepatitis and enhanced HO-1 expression through MAPK/Nrf2 antioxidant pathway in RAW cells. *Int. Immunopharmacol.* **2015**, *28*, 938–944. [CrossRef] [PubMed]

41. Zhao, Y.; Geng, C.-A.; Sun, C.-L.; Ma, Y.-B.; Huang, X.-Y.; Cao, T.-W.; He, K.; Wang, H.; Zhang, X.-M.; Chen, J.-J. Polyacetylenes and anti-hepatitis B virus active constituents from *Artemisia capillaris. Fitoterapia* **2014**, *95*, 187–193. [CrossRef] [PubMed]

42. Lin, C.; Lee, H.Y.; Chang, C.H.; Namba, T.; Masao, H. Evaluation of the liver protective principles from the root of *Cudrania cochinchinensis* var. *gerontogea. Phyther. Res.* **1996**, *10*, 13–17. [CrossRef]

43. Kim, H.-Y.; Kim, J.-K.; Choi, J.-H.; Jung, J.-Y.; Oh, W.-Y.; Kim, D.C.; Lee, H.S.; Kim, Y.S.; Kang, S.S.; Lee, S.-H.; et al. Hepatoprotective effect of pinoresinol on carbon tetrachloride–induced hepatic damage in mice. *J. Pharmacol. Sci.* **2010**, *112*, 105–112. [CrossRef] [PubMed]

Sample Availability: Samples of the compounds **1**, **3**, **4**, **11** and **18** are available from the authors.

![molecules logo] *molecules*

MDPI

Article

New Chromones from a Marine-Derived Fungus, *Arthrinium* sp., and Their Biological Activity

Jie Bao [1,2], Fei He [1], Jin-Hai Yu [1], Huijuan Zhai [1], Zhi-Qiang Cheng [1], Cheng-Shi Jiang [1], Yuying Zhang [1], Yun Zhang [3], Xiaoyong Zhang [4], Guangying Chen [2,*] and Hua Zhang [1,*]

[1] School of Biological Science and Technology, University of Jinan, 336 West Road of Nan Xinzhuang, Jinan 250022, China; bio_baoj@ujn.edu.cn (J.B.); 18864838287@163.com (F.H.); yujinhai12@sina.com (J.-H.Y.); zhai18363005528@163.com (H.Z.); czq13515312897@163.com (Z.-Q.C.); jiangchengshi-20@163.com (C.-S.J.); yuyingzhang2008@163.com (Y.Z.)
[2] Key Laboratory of Tropical Medicinal Plant Chemistry of Ministry of Education, Hainan Normal University, 99 South Road of Longkun Road, Haikou 571158, China
[3] Key Laboratory of Tropical Marine Bio-Resources and Ecology, South China Sea Institute of Oceanology, Chinese Academy of Sciences, 164 West Xingang Road, Guangzhou 510301, China; zhangyun@scsio.ac.cn
[4] College of Marine Sciences, South China Agricultural University, 483 Wushan Road, Guangzhou 510642, China; zhangxiaoyong@scau.edu.cn
* Correspondence: chgying123@163.com (G.C.); bio_zhangh@ujn.edu.cn (H.Z.); Tel.: +86-898-6588-9422 (G.C.); +86-531-8973-6199 (H.Z.)

Academic Editors: Muhammad Ilias and Charles L. Cantrell
Received: 15 July 2018; Accepted: 8 August 2018; Published: 9 August 2018

Abstract: Five new chromone derivatives, arthones A–E (**1–5**), together with eight known biogenetically related cometabolites (**6–13**), were isolated from a deep-sea-derived fungus *Arthrinium* sp. UJNMF0008. Their structures were assigned by detailed analyses of spectroscopic data, while the absolute configurations of **1** and **5** were established by electronic circular dichroism (ECD) calculations and that of **2** was determined by modified Mosher ester method. Compounds **3** and **8** exhibited potent antioxidant property with DPPH and ABTS radical scavenging activities, with IC_{50} values ranging from 16.9 to 18.7 μM. Meanwhile, no compounds indicated obvious bioactivity in our antimicrobial and anti-inflammatory assays at 50.0 μM.

Keywords: *Arthrinium* sp.; chromone; polyketide; antioxidant activity

---

1. Introduction

The genus *Arthrinium* has wide geographic distribution and host range as plant pathogens, endophytes, saprobes, etc., while Poaceae and Cyperaceae are the major host plant families [1]. Although some *Arthrinium* species have been reported as phytopathogens [2–4], or even to cause cutaneous infections in humans [5], many others are known to produce diverse bioactive compounds with a variety of pharmacological applications. For instance, cytotoxic cytochalasins, pyridone alkaloids and polyketides, along with naphthalene glycosides with COX-2 inhibitory activity, were obtained from the sponge-derived fungus *A. arundinis* ZSDS1-F3 [6–9]; arundifungin with antifungal property was isolated from another *A. arundinis* species [10]; griseofulvin derivatives showing lethality against the brine shrimp *Artemia salina* were reported from a gorgonian-derived *Arthrinium* fungus [11]; antiangiogenic diterpenes were discovered from the marine species *A. sacchari* [12]; anti-parasitic dihydroisocoumarins were found in an endophytic *Arthrinium* sp. from *Apiospora montagnei* [13]; and volatile compounds in endophytic *Arthrinium* sp. MFLUCC16-0042 from *Aquilaria subintegra* were also investigated [14].

During the course of our search for new antibiotics from marine resources, an *Arthrinium* sp. UJNMF0008 from deep-sea sediment gained our interest owing to its strong inhibitory activity against *Staphylococcus aureus*. Subsequent chemical investigation on this species led to the discovery of a series of pyridone alkaloids with antibacterial and cytotoxic activities [15]. In addition to the pyridones, another class of metabolites with strong UV absorption was also revealed by chemical profiling (HPLC & ^{1}H-NMR). Further study of the remaining fractions have resulted in the isolation and structural characterization of an array of polyketide compounds, including five new chromone arthones A–E (**1–5**) and eight previously reported analogues, AGI-B4 (**6**) [16], 1,3,6-trihydroxy-8-methylxanthone (**7**) [17], 2,3,4,6,8-pentahydroxy-1-methylxanthone (**8**) [18], sydowinin A (**9**) [19], sydowinin B (**10**) [20], conioxanthone A (**11**) [21], engyodontiumone B (**12**) [22], and 8-hydroxy-3-hydroxymethyl-9-oxo-9*H*-xanthene-1-carboxylic acid methyl ester (**13**) [23]. Herein, we describe the details of the isolation, structure elucidation, and biological evaluations of compounds **1–13**.

2. Results and Discussion

2.1. Structure Elucidation

Arthone A (**1**) was isolated as a pale yellow powder and its molecular formula was established as $C_{16}H_{14}O_8$ by HR-ESIMS analysis (m/z 333.0623 [M − H]$^-$) and NMR data. Detailed analysis of the ^{1}H- and ^{13}C-NMR data (Tables 1 and 2) showed that **1** possessed the same A and B rings (5-hydroxy-7-(hydroxymethyl)-4*H*-chromen-4-one, Figure 1) moiety as those in **9–12** [19–22], which was further confirmed by the HMBC correlations from 10-O*H* to C-10, H-10 to C-2, C-3 and C-4, H-4 to C-4a, C-9a and C-10, and 1-O*H* to C-1, C-2 and C-9a (Figure 2). Two groups of olefinic signals (δ_H/δ_C 5.72 (J = 5.9 Hz)/95.1; 7.33 (J = 5.9 Hz)/157.1) including one oxygenated (δ_C 157.1, C–6) and one oxyquaternary sp^3 carbon (δ_C 84.3, C-8) revealed similar features (C ring, Figure 1) to the known compound euparvione [24] with the absence of one methyl (δ_H/δ_C 2.06/20.4 in euparvione), along with the appearance of a methyl ester moiety (δ_H/δ_C 3.70/52.8, 167.9) and one hydroxymethyl group (δ_H/δ_C 3.88, 4.12/63.6). Further HMBC correlations from H-12 to C-8, C-8a and C-11, and OC*H*$_3$ to C-11 defined the structure of C ring as shown. The planar structure of **1** was thus established with only one chiral center. The absolute configuration of **1** was established as 8*R* by comparison of its experimental and theoretical ECD spectra (Figure 3) [25].

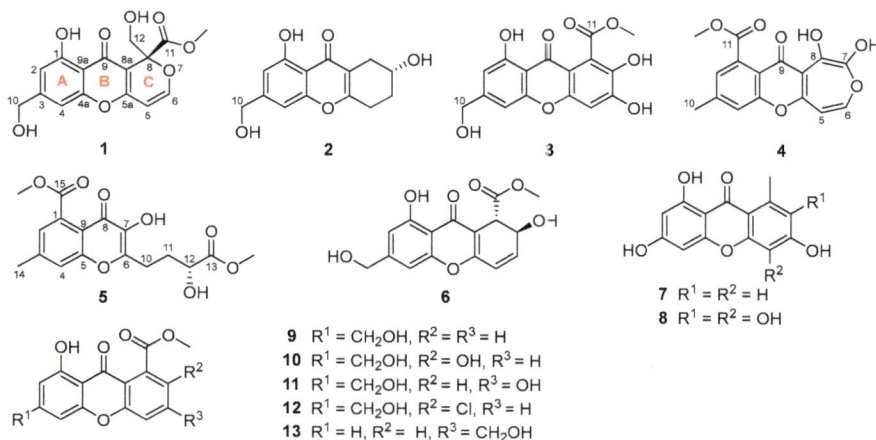

Figure 1. Chemical structures of compounds **1–13**.

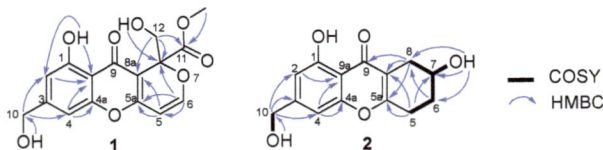

Figure 2. Key 2D-NMR correlations for **1** and **2**.

Table 1. ^{1}H-NMR (600 MHz) data for **1–5** (DMSO-d_6).

Position	1	2	3	4	Positon	5
2	6.74, brs	6.70, s	6.70, brs	7.27, d (0.9)	2	7.23, brs
4	6.96, brs	6.91, s	6.94 [a], brs	7.56, brs	4	7.54, brs
5	5.72, d (5.9)	2.79, dt (18.4, 6.4) 2.71, dt (18.4, 6.4)	6.94 [a], s	6.98, d (8.9)	10	2.84, t (7.9)
6	7.33, d (5.9)	1.87, m 1.80, m		7.34, d (8.9)	11	2.08, m 1.93, m
7		3.99, m			12	4.14, m
8		2.61, dd (16.4, 4.2) 2.32, dd (16.4, 5.8)			14	2.44, brs
10	4.56, brd (5.6)	4.54, d (5.8)	4.56, s	2.49, s	13-OCH$_3$	3.62, s
12	4.12, dd (12.5, 5.8) 3.88, dd (12.5, 7.2)				15-OCH$_3$	3.83, s
OCH$_3$	3.70, s		3.83, s	3.89, s	7-OH	9.00, brs
1-OH	12.37, s	12.67, s	12.55, s		12-OH	5.63, d (4.3)
7-OH		4.92, d (3.7)		9.45, s		
8-OH				12.05, s		
10-OH	5.50, t (5.6)	5.47, t (5.8)				
12-OH	5.29, dd (7.2, 5.8)					

[a] Interchangeable assignments.

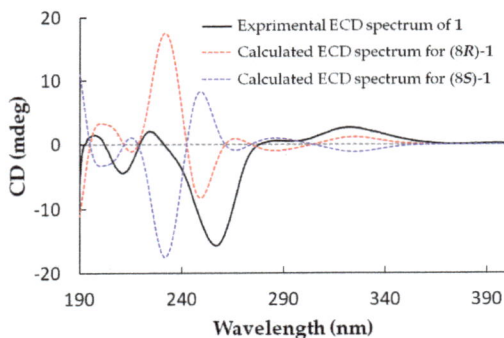

Figure 3. Experimental and theoretical ECD spectra for **1**.

Arthone B (**2**) was obtained as a pale yellow powder. The molecular formula was deduced as $C_{14}H_{14}O_5$ based on the HR-ESIMS ion at m/z 263.0916 [M + H]$^+$ (calcd. for $C_{14}H_{15}O_5$, 263.0914), indicating eight degrees of unsaturation. 1D-NMR data analysis (Tables 1 and 2) revealed that **2** also had a 5-hydroxy-7-(hydroxymethyl)-4*H*-chromen-4-one fragment but with larger chemical shifts for C-5a (δ_C 165.3) and C-8a (δ_C 114.1) compared with those of **1** (δ_C 160.6 and 103.8 for C-5a and C-8a, respectively). Correlated spectroscopy (COSY) correlations from H-7 (δ_H 3.99) to H$_2$-6 (δ_H 1.80/1.87) and H$_2$-8 (δ_H 2.32/2.61), and H-5 (δ_H 2.71/2.79) to H-6 (δ_H 1.80/1.87), revealed the spin-coupling system from H$_2$-5–H$_2$-8 (Figure 2). Key HMBC correlations from H-8 to C-5a, C-8a, and C-9 indicated the connection of C-8 to C-8a, while those from H-5 to C-5a and C-8a identified the connection of C-5 to C-5a (Figure 2). The gross structure of **2** was thus characterized, also bearing only one stereocenter.

The absolute configuration of **2** was established by modified mosher ester method [26], where analysis of the ^{1}H-NMR differences between its (R)- and (S)-MTPA esters ($\Delta\delta = \delta_S - \delta_R$) led to the assignment of 7R-configuration for **2** (Figure 4).

Table 2. ^{13}C-NMR (150 MHz) data for **1–5** (DMSO-d_6).

Position	1	2	3	4	Position	5
1	159.4, C	159.5, C	160.5, C	132.8, C	1	132.7, C
2	108.5, CH	107.6, CH	107.1, CH	124.0, CH	2	124.6, CH
3	152.0, C	151.8, C	152.9, C	147.7 ^{b}, C	3	144.4, C
4	104.1, CH	103.8, CH	103.8, CH	119.1, CH	4	119.6, CH
4a	154.7, C	155.6, C	155.3, C	156.0, C	5	155.0, C
5	95.1, CH	25.0, CH$_2$	102.5, CH	106.3, CH	6	152.7, C
5a	160.6,C	165.3, C	155.2, C	147.4 ^{b}, C	7	138.8, C
6	157.1, CH	28.80 ^{a}, CH$_2$	151.2, C	124.6, CH	8	171.1, C
7		62.9, CH	141.4, C	147.9 ^{b}, C	9	117.0, C
8	84.3, C	28.78 ^{a}, CH$_2$	117.6, C	140.5, C	10	24.8, CH$_2$
8a	103.8, C	114.1, C	108.6, C	108.7, C	11	31.2, CH$_2$
9	178.2, C	181.9, C	179.1, C	180.9, C	12	69.7, CH
9a	108.8, C	107.9, C	106.3, C	113.7, C	13	174.6, C
10	62.2, CH$_2$	62.3, CH$_2$	62.4, CH$_2$	21.3, CH$_3$	14	21.4, CH$_3$
11	167.9, C		166.8, C	168.8, C	15	169.5, C
12	63.6, CH$_2$				13-OCH$_3$	52.0, CH$_3$
OCH$_3$	52.8, CH$_3$		52.2, CH$_3$	52.7, CH$_3$	15-OCH$_3$	52.8, CH$_3$

^{a,b} Interchangeable assignments.

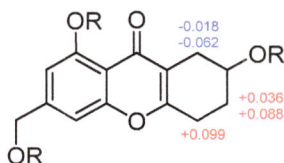

2a R = (S)-MTPA ester
2b R = (R)-MTPA ester

Figure 4. $\Delta\delta$ ($\delta_S - \delta_R$) values in ppm for MTPA eaters of **2**.

Arthone C (**3**) was yielded as a yellow powder and displayed an HR-ESIMS peak at *m/z* 333.0607 [M + H]$^+$ (calcd. for C$_{16}$H$_{13}$O$_8$, 333.0605) corresponding to the molecular formula C$_{16}$H$_{12}$O$_8$, with 16 amu more than that of its cometabolite sydowinin B (**10**) [20]. A detailed comparison of their ^{1}H- and ^{13}C-NMR data (Tables 1 and 2) revealed that **3** incorporated the same skeleton as that of sydowinin B (**10**) [20] with the downfield-shifted chemical shift of C-6 (δ_C 151.2) and upfield-shifted chemical shift of C-5 (δ_C 102.5) and C-7 (δ_C 141.4), implying the presence of a hydroxyl at C-6, and further HMBC correlations from H-5 to C-6, C-7, and C-8a confirmed this moiety. As mentioned above, the structure of compound **3** was thus elucidated, as shown in Figure 1.

Arthone D (**4**) was isolated as a yellow powder with the molecular formula of C$_{16}$H$_{12}$O$_7$ (11 degrees of unsaturation) determined by (+)-HR-ESIMS analysis at *m/z* 317.0664 ([M + H]$^+$, calcd. for C$_{16}$H$_{13}$O$_7$, 317.0656). The ^{1}H- and ^{13}C-NMR spectroscopic data for **4** (Tables 1 and 2) exhibited a similar skeleton to that of isofusidienol A [27]. However, the methoxycarbonly moiety (δ_H/δ_C 3.89/52.7, 168.8) and methyl group (δ_H/δ_C 2.49/21.3) were deduced to locate on the same benzene ring as yicathin A [28] based on the HMBC correlations from OC$\underline{H}_3$ to C-11, H-2 to C-4, C-10, and C-11, H-4 to C-2, C-10, C-4a and C-9a, and H-10 to C-2, C-3 and C-4, and the 1D-NMR data (Figure 5). A proton signal at δ_H 12.05 in the ^{1}H-NMR accounted for a hydroxyl group at C-8 due to H-bonding with C-9 carbonyl. Hence, two olefinic protons (δ_H 6.98, d, J = 8.9 Hz and 7.34, d, J = 8.9 Hz) were speculated to be located on C-5 and C-6, respectively. Finally, the hydroxyl resonanced at δ_H 9.45 could only be assigned to C-7. The seven-membered ring moiety was also supported by the HMBC correlations from H-5 to C-5a and C-8a and H-6 to C-7 and C-5a, as well as long range J^4 correlations from H-5 to C-8 and H-6 to C-8a.

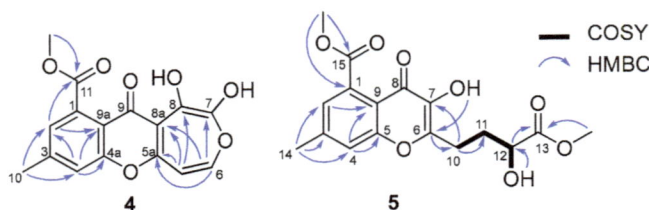

Figure 5. Key 2D-NMR correlations for **4** and **5**.

Arthone E (**5**) was isolated as a pale yellow powder and had a molecular formula of $C_{17}H_{18}O_8$ as inferred from its (+)-HR-ESIMS data at *m/z* 351.1073 [M + H]⁺. Its NMR data (Tables 1 and 2) revealed the same 7-methyl-4-oxo-4*H*-chromene-5-carboxylate moiety as arthone D (**4**). COSY correlations from H-11 (δ_H 1.93/2.08) to H-10 (δ_H 2.84) and H-12 (δ_H 4.14), and 12-OH (δ_H 5.63) to H-12 (δ_H 4.14), and HMBC correlations from OC$\underline{H}_3$ to C-13, and H-11/H-12 to C-13, revealed the moiety of C-10–C-13–OCH₃ (Figure 5). Finally, the side-chain fragment was proved to be located at C-6, as supported by the HMBC correlations from H₂-10 to C-6 and C-7, and a hydroxyl group was suggested at C-7 as indicated by the HMBC correlation from 7-OH to C-6 (Figure 5). The absolute configuration of **5** was established as 12*R* by comparison of its experimental and theoretical ECD spectra (Supplementary material Figure S31).

All the raw spectroscopic data including 1D/2D NMR and HR-ESIMS spectra for the new compounds **1** (Figures S1–S5), **2** (Figures S6–S11), **3** (Figures S12–S16), **4** (Figures S17–S21) and **5** (Figures S22–S27), and the 1D/2D NMR spectra for MTPA esters of **2** (Figures S28–S30) have been provided in the supplementary material.

2.2. Biological Activity

In order to evaluate the biological properties of **1–13**, their antioxidant activity was assayed by DPPH and ABTS radical scavenging methods with curcumin as positive control (IC$_{50}$ = 24.3 and 9.5 μM, respectively). Only compounds **3** and **8** exhibited significant antioxidant activities, with IC$_{50}$ values of 16.9 and 22.1 μM for DPPH assay, and 18.7 and 18.0 μM for ABTS assay, respectively, while others showed no significant effect at 50.0 μM. Meanwhile, the antimicrobial activity against Gram-positive bacterial strains *Mycobacterium smegmatis* ATCC 607 and *Staphylococcus aureus* ATCC 25923, Gram-negative *Escherichia coli* ATCC 8739, *Pseudomonas aeruginosa* ATCC 9027, and fungus *Candida albicans* ATCC10231, as well as anti-inflammatory activity based on the inhibition effect of NO production in lipopolysaccharide (LPS)-induced mouse macrophages RAW 264.7 cells, were evaluated for **1–13**. However, no compounds displayed obvious bioactivity in the two assays up to 50.0 μM.

3. Experimental Section

3.1. General Experimental Procedures

NMR spectra were recorded on a Bruker Avance DRX600 NMR spectrometer (Bruker BioSpin AG, Fällanden, Switzerland), with residual solvent peaks as references (DMSO-*d₆*: δ_H 2.50, δ_C 39.52). ESIMS analyses were carried out on an Agilent 1260-6460 Triple Quad LC-MS instrument (Agilent Technologies Inc., Waldbronn, Germany). HR-ESIMS data were acquired on an Agilent 6545 Q-TOF mass spectrometer (Agilent Technologies Inc., Waldbronn, Germany). UV spectra were obtained on a Shimadzu UV-2600 spectrophotometer (Shimadzu, Kyoto, Japan) with 1 cm pathway cell. Optical rotations were measured on a Rudolph VI polarimeter (Rudolph Research Analytical, Hackettstown, NJ, USA) with a 10 cm length cell. ECD spectra were acquired on a Chirascan circular-dichroism spectrometer (Applied Photophysics Ltd., Surrey, UK). IR spectra were recorded on an FT-IR VERTEX 70 (Bruker BioSpin AG, Fremont, CA, USA). All HPLC analyses and separations were carried out on Agilent 1260 series LC instruments (Agilent Technologies Inc., Waldbronn, Germany) and a YMC-Pack ODS-A column (250 × 10 mm, 5 μm) was used for HPLC separations. Column chromatography (CC) was performed on Silica gel (200–300 mesh, Yantai Jiangyou Silica Gel Development Co., Yantai, China) and Sephadex LH-20 gel (GE Healthcare Bio-Sciences AB, Uppsala, Sweden). All solvents used for CC were of analytical grade (Tianjin Fuyu Fine Chemical Co. Ltd., Tianjin, China) and solvents used for HPLC were of HPLC grade (Oceanpak Alexative Chemical Ltd., Goteborg, Sweden).

3.2. Fungal Material

The fungus strain UJNMF0008 was isolated from a marine-sediment sample collected in the South China Sea (17°55′00″ N, 115°55′31″ E; 3858 m depth, Hainan, China). This strain was identified as an *Arthrinium* sp. based on morphological traits and a molecular biological protocol by DNA amplification and comparison of its ITS region sequence with the GenBank database (100% similarity with *Arthrinium* sp. zzz1842 (HQ696050.1)). The BLAST sequenced data were deposited at GenBank (No. MG010382). The strain was deposited at the CGMCC center, Institute of Microbiology, Chinese Academy of Sciences (Beijing, China).

3.3. Fermentation and Extraction

Arthrinium sp. UJNMF0008 from a PDA culture plate was inoculated in 500 mL Erlenmeyer flasks containing 150 mL soluble-starch medium (1% glucose, 0.1% soluble starch, 1% $MgSO_4$, 0.1% KH_2PO_4, 0.1% peptone, and 3% sea salt) at 28 °C on a rotary shaker at 130 rpm for 3 days as seed cultures. Then, each of the seed cultures (20 mL) was transferred into autoclaved 1 L Erlenmeyer flasks with solid-rice medium (each flasks contained 80 g commercially available rice, 0.4 g yeast extract, 0.4 g glucose, and 120 mL water with 3% sea salt). After that, the strain was incubated statically for 30 days at 28 °C.

After fermentation, the total 4.8 kg rice culture was crushed and extracted with 15.0 L 95% EtOH three times. The EtOH extract was evaporated under reduced pressure to afford an aqueous solution and then extracted with 2.0 L ethyl acetate three times to give 80 g crude gum.

3.4. Isolation and Purification

The ethyl acetate extract (80 g) was fractionated by a silica gel column eluting with step gradient CH_2Cl_2-MeOH (v/v 100:0, 98:2, 95:5, 90:10, 80:20, 70:30, 50:50 and 0:100) to give 10 fractions (Fr.1–Fr.10) based on TLC and HPLC analysis. Fr.6 (30.2 g) was applied to CC over D101-macroporous absorption resin eluted with EtOH-H_2O (30%, 50%, 80% and 100%) to afford four subfractions (Fr.6-1–Fr.6-4). Fr.6-3 (11.6 g) was fractionated by the silica gel column with step gradient CH_2Cl_2-$(CH_3)_2CO$ (v/v 100:0–0:100) and divided into nine subfractions (Fr.6-3-1–Fr.6-3-9) and a portion of (32.5 mg) Fr.6-3-3 was further purified by HPLC eluting with MeOH-H_2O (v/v 55:45, 3.0 mL min^{-1}) to give **13** (t_R = 29.3 min, 20.6 mg). Fr.6-3-2 (0.67 g), Fr.6-3-4 (510 mg), Fr.6-3-5 (1.04 g), Fr.6-3-6 (3.9 g), and Fr.6-3-7 (2.2 g) were separated by MPLC with an ODS column eluting with gradient MeOH-H_2O (v/v 20:80–100:0) to obtain five, five, six, six and five subfractions, respectively. Fr.5-3-2-4 (37.0 mg) was further purified by HPLC eluting with MeOH-H_2O (v/v 55:45, 3.0 mL min^{-1}) to give **9** (t_R = 22.2 min, 9.8 mg). Fr.6-3-4-3 (37.0 mg) was further purified by HPLC eluting with MeOH-H_2O-CH_3CO_2H ($v/v/v$ 70:30:10^{-4}, 3.0 mL min^{-1}) to afford **4** (t_R = 14.4 min, 1.8 mg), while Fr.6-3-4-5 (50.2 mg) was further purified by HPLC eluting with MeOH-H_2O (v/v 84:16, 3.0 mL min^{-1}) to give **7** (t_R = 21.9 min, 21.0 mg). Fr.6-3-5-2 (25.0 mg) was further purified by HPLC eluting with MeOH-H_2O (v/v 48:52, 3.0 mL min^{-1}) to give **1** (t_R = 13.9, 4.5 mg), while Fr.6-3-5-5 (35.2 mg) was further purified by HPLC eluting with MeOH-H_2O (v/v 60:40, 3.0 mL min^{-1}) to give **10** (t_R = 7.9 min, 4.0 mg) and **11** (t_R = 16.4 min, 11.7 mg), and Fr.6-3-5-6

(25.1 mg) was further purified by HPLC eluting with MeOH-H$_2$O (v/v 65:35, 3.0 mL min^{-1}) to give **12** (t_R = 20.7 min, 2.4 mg). Fr.6-3-6-2 (55.2 mg) was further purified by HPLC eluting with MeOH-H$_2$O (v/v 34:66, 3.0 mL min^{-1}) to give **6** (t_R = 36.9 min, 14.3 mg). Fr.6-3-7-2 (80.1 mg) was further purified by HPLC eluting with MeOH-H$_2$O (v/v 42:58, 3.0 mL min^{-1}) to give **2** (t_R = 11.1 min, 2.7 mg). Fr.7 (15.2 g) was chromatographed on a silica gel column with step gradient CH$_2$Cl$_2$-(CH$_3$)$_2$CO (v/v 100:0–0:100) and divided into three subfractions (Fr.7-1–Fr.7-3). Fr.7-1 (3.5 g) was divided into three subfractions (Fr.7-1-1–Fr.7-1-3) by Sephadex LH-20 CC eluting with MeOH-CH$_2$Cl$_2$ (v/v 1:1), and Fr.7-1-2 (1.2 g) was fractionated by MPLC with an ODS column eluting with step gradient MeOH-H$_2$O (v/v 20:80 to 0:100) and further purified by HPLC eluting with MeOH-H$_2$O (v/v 34:66, 3.0 mL min^{-1}) to provide **5** (t_R = 46.9 min, 2.6 mg). Fr.9 (7.6 g) was subject to silica gel column with step gradient CH$_2$Cl$_2$-(CH$_3$)$_2$CO (v/v 100:0–0:100) and divided into five subfractions (Fr.9-1–Fr.9-5). Fr.9-2 (1.6 g) was separated by Sephadex LH-20 CC eluting with MeOH-CH$_2$Cl$_2$ (v/v 1:1) to obtain four subfractions (Fr.9-2-1–Fr.9-2-4), and Fr.9-2-2 (42.6 mg) was further purified by HPLC eluting with MeOH-H$_2$O-CH$_3$CO$_2$H ($v/v/v$ 54:46:10^{-4}, 3.0 mL min^{-1}) to give **3** (t_R = 17.1 min, 9.3 mg), while Fr.9-2-3 (17.0 mg) was further purified by HPLC eluting with MeOH-H$_2$O (v/v 54:46, 3.0 mL min^{-1}) to give **8** (t_R = 26.4 min, 3.7 mg).

3.4.1. Arthone A (**1**)

Pale yellow powder; $[\alpha]_D^{23}$ −6.3 (*c* 0.56, MeOH); ECD (0.20 mg mL^{-1}, MeOH) λ ($\Delta\varepsilon$) 322 (2.76), 291 (0.56), 258 (−15.78), 224 (2.50), 212 (−5.55), 204 (1.71) nm; UV (MeOH) λ_{max} (log ε) 238 (4.35), 259 (4.23), 331 (3.64) nm; IR (KBr) ν_{max} 3404, 2957, 1742, 1658, 1594, 1493, 1451, 1299, 1195, 1051, 1022, 821 cm^{-1}; ^{1}H- and ^{13}C-NMR data, Tables 1 and 2; (−)-ESIMS *m/z* 332.9 [M − H]$^-$; (−)-HR-ESIMS *m/z* 333.0623 [M − H]$^-$ (calcd. for C$_{16}$H$_{13}$O$_8$, 333.0616).

3.4.2. Arthone B (**2**)

Pale yellow powder; $[\alpha]_D^{23}$ 1.9 (*c* 0.16, MeOH); ECD (0.20 mg mL^{-1}, MeOH) λ ($\Delta\varepsilon$) 328 (0.42), 286 (0.01), 276 (0.07), 258 (−0.29), 245 (−0.01), 206 (−2.21) nm; UV (MeOH) λ_{max} (log ε) 239 (5.18), 328 (4.35) nm; IR (KBr) ν_{max} 3431, 2935, 1659, 1625, 1597, 1499, 1459, 1292, 1117, 1041 cm^{-1}; ^{1}H- and ^{13}C-NMR data, Tables 1 and 2; (−)-ESIMS *m/z* [M − H]$^-$ 260.9; (+)-HR-ESIMS *m/z* 263.0916 [M + H]$^+$ (calcd. for C$_{14}$H$_{15}$O$_5$, 263.0914).

3.4.3. Arthone C (**3**)

Yellow powder; UV (MeOH) λ_{max} (log ε) 233 (4.20), 246 (4.14), 254 (4.13), 296 (3.69), 377 (4.01) nm; IR (KBr) ν_{max} 3307, 1702, 1651, 1607, 1586, 1499, 1436, 1381, 1366, 1286, 1250, 1011, 830 cm^{-1}; ^{1}H- and ^{13}C-NMR data, Tables 1 and 2; (−)-ESIMS *m/z* 330.9 [M − H]$^-$; (+)-HR-ESIMS *m/z* 333.0607 [M + H]$^+$ (calcd. for C$_{16}$H$_{13}$O$_8$, 333.0605).

3.4.4. Arthone D (**4**)

Yellow powder; UV (MeOH) λ_{max} (log ε) 234 (4.27), 267 (4.26), 391 (3.33) nm; IR (KBr) ν_{max} 3427, 1725, 1613, 1501, 1459, 1435, 1297, 1235, 1041 cm^{-1}; ^{1}H- and ^{13}C-NMR data, Tables 1 and 2; (−)-ESIMS *m/z* 314.9 [M − H]$^-$; (+)-HR-ESIMS *m/z* 317.0664 [M + H]$^+$ (calcd. for C$_{16}$H$_{13}$O$_7$, 317.0656).

3.4.5. Arthone E (**5**)

Pale yellow powder; $[\alpha]_D^{23}$ −4.8 (*c* 0.73, MeOH); UV (MeOH) λ_{max} (log ε) 209 (4.30), 314 (3.64) nm; IR (KBr) ν_{max} 3413, 2957, 1736, 1611, 1438, 1309, 1227, 1176, 1040, 1037, 861 cm^{-1}; ^{1}H- and ^{13}C-NMR data, Tables 1 and 2; (+)-ESIMS *m/z* 351.0 [M + H]$^+$; (+)-HR-ESIMS *m/z* 351.1073 [M + H]$^+$ (calcd. for C$_{17}$H$_{19}$O$_8$, 351.1074).

3.5. Antioxidant Assay

The antioxidant activities for compounds **1–13** were determined by DPPH and ABTS methods. DPPH radical scavenging method was conducted as described formerly [29], while ABTS radical scavenging assay was performed according to the method developed by Re et al. [30] with some modifications as below. Briefly, an ABTS radical solution was prepared by mixing equal volumes of aqueous solutions of 7 mM ABTS and 4.9 mM potassium persulfate for 16 h in the dark at room temperature. Then the ABTS radical solution was diluted with EtOH to an absorbance of 0.70 ± 0.02 at 734 nm. 10 µL of the tested compounds in ethanol (final concentrations as 3.13 µM, 6.25 µM, 12.5 µM, 25.0 µM, 50.0 µM, and 100 µM) was mixed with 190 µL of the prepared diluted ABTS radical solution at room temperature, and the absorbance at 734 nm was measured after 6 min in the dark. IC_{50} values were defined as the concentrations of tested compounds resulting in 50% loss of the ABTS radical. All determinations were carried out in triplicate, and curcumin was applied as positive control.

3.6. Antimicrobial Assays

The antimicrobial activity of compounds **1–13** was assayed against the Gram-positive bacterial strains *Mycobacterium smegmatis* ATCC 607 and *Staphylococcus aureus* ATCC 25923, Gram-negative *Escherichia coli* ATCC 8739, *Pseudomonas aeruginosa* ATCC 9027, and fungus *Candida albicans* ATCC10231 by the two-fold serial dilution method in 96-well microplates as described previously [15]. Penicillin was used as positive control in the current assay.

3.7. Anti-Inflammatory Assay

Determination of nitric oxide production. Briefly, RAW 264.7 cells were plated into 96-well plates and pretreated with a series of concentrations of compounds (3.13, 6.25, 12.5, 25.0, 50.0, and 100 µM) for 1 h before treatment with 1 µg mL^{-1} LPS. After 24 h incubation, detection of accumulated nitric oxide in the cell supernatants was assayed by Griess reagent kit (Beyotime Institute of Biotechnology, Jiangsu, China) according to the manufacturer's instructions. Equal volumes of culture supernatant and Griess reagent were mixed, and the absorbance at 540 nm was measured using a Microplate Reader (Tecan, Grödig, Austria).

Cell viability assay. RAW 264.7 cells were seeded into 96-well plates at 1×10^4 cells well^{-1} and allowed to attach for 24 h. The medium was replaced with a 100 µL medium containing the indicated concentrations of compounds and further incubated for 24 h 10 µL of MTT (5 mg mL^{-1} in PBS) was added into each well and the plates were incubated for 4 h at 37 °C. Supernatants were aspirated and formed formazan was dissolved in 100 µL of dimethyl sulfoxide (DMSO). The optical density (OD) was measured at an absorbance wavelength of 490 nm using a Microplate Reader (Tecan, Switzerland).

3.8. ECD Calculations

Conformational analysis within an energy window of 3.0 kcal mol^{-1} was performed using the OPLS3 molecular mechanics force field via the MacroModel [31] panel of Maestro 10.2. The conformers were then further optimized with the software package Gaussian 09 [32] at the CAM-B3LYP/6-311G(d,p) level for **1** and B3LYP/6-311G(d,p) level for **5**, respectively, and the harmonic vibrational frequencies were also calculated to confirm their stability. Then, the 30 lowest electronic transitions for the obtained conformers in vacuum were calculated using time-dependent density functional theory (TD-DFT) method at the CAM-B3LYP/6-311G(d,p) level for **1** and B3LYP/6-311G(d,p) level for **5**, respectively. ECD spectra of the conformers were simulated using a Gaussian function with a half-bandwidth of 0.25 eV for **1** and 0.35 eV for **5**. The overall theoretical ECD spectra were obtained according to the Boltzmann weighting of each conformer.

Supplementary Materials: The following are available online. 1D/2D-NMR and HR-ESIMS spectra of compounds **1–5**, the 1D/2D NMR spectra for MTPA esters of **2**, along with experimental and theoretical ECD spectra for **5**.

Author Contributions: J.B., F.H. and H.Z. (Huijuan Zhai) carried out the microbial fermentation and the isolation of the compounds. Z.-Q.C., C.-S.J., and Y.Z. (Yuying Zhang) performed the biological tests. H.Z. (Hua Zhang), J.B., G.C., and J.-H.Y. analyzed the spectroscopic data and elucidated the structure of the compounds. Y.Z. (Yun Zhang) and X.Z. assisted with the isolation and identification of the fungal material UJNMF0008. H.Z. (Hua Zhang) and J.B. wrote the paper.

Funding: We acknowledge the financial support from the National Natural Science Foundation of China (Nos. 41506148 & 31501104), Natural Science Foundation of Shandong Province (Nos. BS2015HZ005 and JQ201721), the Young Taishan Scholars Program (tsqn20161037), Open Fund of Key Laboratory of Tropical Medicinal Plant Chemistry of Ministry of Education (rdyw2018004), Program for Innovative Research Team in University (IRT-16R19), A Project of Shandong Province Higher Educational Science and Technology Program (J18KA255), and the Shandong Talents Team Cultivation Plan of University Preponderant Discipline (No. 10027).

Conflicts of Interest: The authors declare no conflict of interest.

References

1. Wang, M.; Tan, X.-M.; Liu, F.; Cai, L. Eight new *Arthrinium* species from China. *Mycokeys* **2018**, *34*, 1–24. [CrossRef] [PubMed]

2. Li, S.-J.; Zhu, T.-H.; Zhu, H.-M.-Y.; Liang, M.; Qiao, T.-M.; Han, S.; Che, G.-N. Purification of protein AP-toxin from *Arthrinium phaeospermum* causing blight in *Bambusa pervariabilis* x *Dendrocalamopisis grandis* and its metabolic effects on four bamboo varieties. *Phytopathology* **2013**, *103*, 135–145. [CrossRef] [PubMed]

3. Li, B.J.; Liu, P.Q.; Jiang, Y.; Weng, Q.Y.; Chen, Q.H. First report of culm rot caused by *Arthrinium phaeospennum* on *Phyllostachys viridis* in China. *Plant Dis.* **2016**, *100*, 1013. [CrossRef]

4. Chen, K.; Wu, X.Q.; Huang, M.X.; Han, Y.Y. First report of brown culm streak of *Phyllostachys praecox* caused by *Arthrinium arundinis* in Nanjing, China. *Plant Dis.* **2014**, *98*, 1274. [CrossRef]

5. Crous, P.W.; Groenewald, J.Z. A phylogenetic re-evaluation of *Arthrinium*. *Ima Fungus* **2013**, *4*, 133–154. [CrossRef] [PubMed]

6. Wang, J.F.; Wang, Z.; Ju, Z.R.; Wan, J.T.; Liao, S.R.; Lin, X.P.; Zhang, T.Y.; Zhou, X.F.; Chen, H.; Tu, Z.C.; et al. Cytotoxic cytochalasins from marine-derived fungus *Arthrinium arundinis*. *Planta Med.* **2015**, *81*, 160–166. [CrossRef] [PubMed]

7. Wang, J.F.; Wei, X.Y.; Qin, X.C.; Lin, X.P.; Zhou, X.F.; Liao, S.R.; Yang, B.; Liu, J.; Tu, Z.C.; Liu, Y.H. Arthpyrones A-C, pyridone alkaloids from a sponge-derived fungus *Arthrinium arundinis* ZSDS1-F3. *Org. Lett.* **2015**, *17*, 656–659. [CrossRef] [PubMed]

8. Li, Y.L.; Wang, J.F.; He, W.J.; Lin, X.P.; Zhou, X.F.; Liu, Y.H. One strain-many compounds method for production of polyketide metabolites using the sponge-derived fungus *Arthrinium arundinis* ZSDS1-F3. *Chem. Nat. Compd.* **2017**, *53*, 373–374. [CrossRef]

9. Wang, J.-F.; Xu, F.-Q.; Wang, Z.; Lu, X.; Wan, J.-T.; Yang, B.; Zhou, X.-F.; Zhang, T.-Y.; Tu, Z.-C.; Liu, Y. A new naphthalene glycoside from the sponge-derived fungus *Arthrinium* sp. ZSDS1-F3. *Nat. Prod. Res.* **2014**, *28*, 1070–1074. [CrossRef] [PubMed]

10. Cabello, M.A.; Platas, G.; Collado, J.; Diez, M.T.; Martin, I.; Vicente, F.; Meinz, M.; Onishi, J.C.; Douglas, C.; Thompson, J.; et al. Arundifungin, a novel antifungal compound produced by fungi: Biological activity and taxonomy of the producing organisms. *Int. Microbiol.* **2001**, *4*, 93–102. [PubMed]

11. Wei, M.-Y.; Xu, R.-F.; Du, S.-Y.; Wang, C.-Y.; Xu, T.-Y.; Shao, C.-L. A new griseofulvin derivative from the marine-derived *Arthrinium* sp. fungus and its biological activity. *Chem. Nat. Compd.* **2016**, *52*, 1011–1014. [CrossRef]

12. Tsukada, M.; Fukai, M.; Miki, K.; Shiraishi, T.; Suzuki, T.; Nishio, K.; Sugita, T.; Ishino, M.; Kinoshita, K.; Takahashi, K.; et al. Chemical constituents of a marine fungus, *Arthrinium sacchari*. *J. Nat. Prod.* **2011**, *74*, 1645–1649. [CrossRef] [PubMed]

13. Ramos, H.P.; Simao, M.R.; de Souza, J.M.; Magalhaes, L.G.; Rodrigues, V.; Ambrosio, S.R.; Said, S. Evaluation of dihydroisocoumarins produced by the endophytic fungus *Arthrinium* state of *Apiospora montagnei* against *Schistosoma mansoni*. *Nat. Prod. Res.* **2013**, *27*, 2240–2243. [CrossRef] [PubMed]

14. Monggoot, S.; Popluechai, S.; Gentekaki, E.; Pripdeevech, P. Fungal endophytes: An alternative source for production of volatile compounds from agarwood oil of *Aquilaria subintegra*. *Microb. Ecol.* **2017**, *74*, 54–61. [CrossRef] [PubMed]

15. Bao, J.; Zhai, H.J.; Zhu, K.K.; Yu, J.-H.; Zhang, Y.Y.; Wang, Y.Y.; Jiang, C.-S.; Zhang, X.Y.; Zhang, Y.; Zhang, H. Bioactive pyridone alkaloids from a deep-sea-derived fungus *Arthrinium* sp. UJNMF0008. *Mar. Drugs* **2018**, *16*. [CrossRef] [PubMed]

16. Kim, H.S.; Park, I.Y.; Park, Y.J.; Lee, J.H.; Hong, Y.S.; Lee, J.J. A novel dihydroxanthenone, AGI-B4 with inhibition of VEGF-induced endothelial cell growth. *J. Antibiot.* **2002**, *55*, 669–672. [CrossRef] [PubMed]

17. Mutanyatta, J.; Matapa, B.G.; Shushu, D.D.; Abegaz, B.M. Homoisoflavonoids and xanthones from the tubers of wild and in vitro regenerated *Ledebouria graminifolia* and cytotoxic activities of some of the homoisoflavonoids. *Phytochemistry* **2003**, *62*, 797–804. [CrossRef]

18. Abdel-Lateff, A.; Klemke, C.; Konig, G.M.; Wright, A.D. Two new xanthone derivatives from the algicolous marine fungus *Wardomyces anomalus*. *J. Nat. Prod.* **2003**, *66*, 706–708. [CrossRef] [PubMed]

19. Hamasaki, T.; Sato, Y.; Hatsuda, Y. Structure of sydowinin A, sydowinin B, and sydowinol, metabolites from *Aspergillus sydowi*. *Agric. Biol. Chem.* **1975**, *39*, 2341–2345. [CrossRef]

20. Little, A.; Porco, J.A., Jr. Total syntheses of graphisin A and sydowinin B. *Org. Lett.* **2012**, *14*, 2862–2865. [CrossRef] [PubMed]

21. Wang, Y.C.; Zheng, Z.H.; Liu, S.C.; Zhang, H.; Li, E.W.; Guo, L.D.; Che, Y.S. Oxepinochromenones, furochromenone, and their putative precursors from the endolichenic fungus *Coniochaeta* sp. *J. Nat. Prod.* **2010**, *73*, 920–924. [CrossRef] [PubMed]

22. Yao, Q.F.; Wang, J.; Zhang, X.Y.; Nong, X.H.; Xu, X.Y.; Qi, S.H. Cytotoxic polyketides from the deep-sea-derived fungus *Engyodontium album* DFFSCS021. *Mar. Drugs* **2014**, *12*, 5902–5915. [CrossRef] [PubMed]

23. Wang, H.; Umeokoli, B.O.; Eze, P.; Heering, C.; Janiak, C.; Mueller, W.E.G.; Orfali, R.S.; Hartmann, R.; Dai, H.F.; Lin, W.H.; et al. Secondary metabolites of the lichen-associated fungus *Apiospora montagnei*. *Tetrahedron Lett.* **2017**, *58*, 1702–1705. [CrossRef]

24. Leon, F.; Gao, J.T.; Dale, O.R.; Wu, Y.S.; Habib, E.; Husni, A.S.; Hill, R.A.; Cutler, S.J. Secondary metabolites from *Eupenicillium parvum* and their in vitro binding affinity for human opioid and cannabinoid receptors. *Planta Med.* **2013**, *79*, 1756–1761. [CrossRef] [PubMed]

25. Berova, N.; Di, B.L.; Pescitelli, G. Application of electronic circular dichroism in configurational and conformational analysis of organic compounds. *Chem. Soc. Rev.* **2007**, *36*, 914–931. [CrossRef] [PubMed]

26. Sun, Y.-L.; Bao, J.; Liu, K.-S.; Zhang, X.-Y.; He, F.; Wang, Y.-F.; Nong, X.-H.; Qi, S.-H. Cytotoxic dihydrothiophene-condensed chromones from the marine-derived fungus *Penicillium oxalicum*. *Planta Med.* **2013**, *79*, 1474–1479. [CrossRef] [PubMed]

27. Lösgen, S.; Magull, J.; Schulz, B.; Draeger, S.; Zeeck, A. Isofusidienols: Novel chromone-3-oxepines produced by the endophytic fungus *Chalara* sp. *Eur. J. Org. Chem.* **2010**, *2008*, 698–703. [CrossRef]

28. Sun, R.-R.; Miao, F.-P.; Zhang, J.; Wang, G.; Yin, X.-L.; Ji, N.-Y. Three new xanthone derivatives from an algicolous isolate of *Aspergillus wentii*. *Magn. Reson. Chem.* **2013**, *51*, 65–68. [CrossRef] [PubMed]

29. Cao, Y.-K.; Li, H.-J.; Song, Z.-F.; Li, Y.; Huai, Q.-Y. Synthesis and biological evaluation of novel curcuminoid derivatives. *Molecules* **2014**, *19*, 16349–16372. [CrossRef] [PubMed]

30. Re, R.; Pellegrini, N.; Proteggente, A.; Pannala, A.; Yang, M.; Riceevans, C. Antioxidant activity applying an improved ABTS radical cation decolorization assay. *Free Radic. Biol. Med.* **2013**, *26*, 1231–1237. [CrossRef]

31. *MacroModel*, 9.7.211; Schrödinger: New York, NY, USA, 2009.

32. *Gaussian* 09, Revision B.01; Gaussian, Inc.: Wallingford, CT, USA, 2010.

Sample Availability: Samples of the compounds **1–13** are available from the authors.

molecules

MDPI

Article

Isolation and Identification of the Five Novel Flavonoids from *Genipa americana* Leaves

Larissa Marina Pereira Silva [1], Jovelina Samara Ferreira Alves [1],
Emerson Michell da Silva Siqueira [1], Manoel André de Souza Neto [1], Lucas Silva Abreu [2],
Josean Fechine Tavares [2], Dayanne Lopes Porto [3], Leandro de Santis Ferreira [3],
Daniel Pecoraro Demarque [4], Norberto Peporine Lopes [4], Cícero Flávio Soares Aragão [3]
and Silvana Maria Zucolotto [1,*]

[1] Research Group on Bioactive Natural Products (PNBio), Laboratory of Pharmacognosy,
     Department of Pharmacy, Federal University of Rio Grande do Norte (UFRN), Natal 59010-180, Brazil;
     larissamarinaps@gmail.com (L.M.P.S.); jsfa.farma@gmail.com (J.S.F.A.);
     siqueira_emerson@hotmail.com (E.M.d.S.S.); rufinim@gmail.com (M.A.d.S.N.)
[2] Multiuser Laboratory of Characterization and Analysis (LMCA), Federal University of Paraíba,
     João Pessoa 58051-900, Brazil; lucas.abreu@ltf.ufpb.br (L.S.A.); josean@ltf.ufpb.br (J.F.T.)
[3] Laboratory of Quality Control of Medications (LCQMed), Department of Pharmacy,
     Federal University of Rio Grande do Norte (UFRN), Natal 59010-180, Brazil;
     daylopesporto@hotmail.com (D.L.P.); lean_sf@yahoo.com.br(L.d.S.F.);
     cicero.aragao@yahoo.com.br (C.F.S.A.)
[4] Nucleus Research in Natural and Synthetic Products (NPPNS), Department of Physics and Chemistry,
     Faculty of Pharmaceutical Sciences of Ribeirão Preto, University of São Paulo, São Paulo 14040-903, Brazil;
     dpdemarque@gmail.com (D.P.D.); npelopes@fcfrp.usp.br (N.P.L.)
*   Correspondence: silvanazucolotto@ufrnet.br; Tel.: +55-084-33429818

Received: 15 August 2018; Accepted: 5 September 2018; Published: 2 October 2018

Abstract: *Genipa americana* is a medicinal plant popularly known as "jenipapo", which occurs in Brazil and belongs to the Rubiaceae family. It is a species widely distributed in the tropical Central and South America, especially in the Cerrado biome. Their leaves and fruits are used as food and popularly in folk medicine to treat anemias, as an antidiarrheal, and anti-syphilitic. Iridoids are the main secondary metabolites described from *G. americana*, but few studies have been conducted with their leaves. In this study, the aim was to chemical approach for identify the main compounds present at the extract of *G. americana* leaves. The powdered leaves were extracted by maceration with EtOH: water (70:30, *v/v*), following liquid-liquid partition with petroleum ether, chloroform, ethyl acetate and *n*-butanol. A total of 13 compounds were identified. In addition three flavonoids were isolated from the ethyl acetate fraction: *quercetin-3-O-robinoside* (**GAF 1**), *kaempferol-3-O-robinoside* (**GAF 2**) and *isorhamnetin-3-O-robinoside* (**GAF 3**) and, from *n*-butanol fraction more two flavonoids were isolated, *kaempferol-3-O-robinoside-7-O-rhamnoside* (robinin) (**GAF 4**) and *isorhamnetin-3-O-robinoside-7-rhamnoside* (**GAF 5**). Chemical structures of these five flavonoids were elucidated using spectroscopic methods (MS, ^{1}H and ^{13}C-NMR 1D and 2D). These flavonoids glycosides were described for the first time in *G. americana*.

Keywords: Rubiaceae; jenipapo; HPLC-ESI-IT-MS/MS; flavonoids glycosides

---

1. Introduction

Genipa americana L. belongs to Rubiaceae family, which contains approximately 140 genera and 2120 species occurring and in various places around the world, mainly in tropical and subtropical regions [1,2]. *Genipa americana* L. is a native Brazilian plant, with perennial foliage and widely distributed in tropical Central and South America, including the Cerrado biome [3–5]. It is popularly known as "jenipapo", which comes from the indigenous language Tupi-Guarani and means "fruit used for painting" [6]. Its fruits are usually processed for liqueurs, candies, and ice cream preparations [7], and are used as tonic against anemia [8]. Ethnopharmacological studies report that different parts of the species have been used in folk medicine, including leaves as an antidiarrheal and anti-syphilitic (decoction) [9]; leaf preparation for treatment of fevers (macerated) [10]; and to treat liver diseases [8].

Most of the phytochemical and pharmacological studies reported to *G. americana* were conducted with its fruits [4]. Regarding pharmacological studies especially with its leaves there are reports about anti-inflammatory [11], antiangiogenic [12], antidiarrheal, and anti-syphilis activity [9], inhibition of larval development and eclosion of sheep gastrointestinal nematodes [13], antidiabetic [14], and anticonvulsant effects [15].

Concerning the phytochemical composition, it has been reported to *G. americana* leaves the presence of secondary metabolites such as tannins [13], monoterpenes [16], and flavonoids [17], being iridoids the main secondary metabolites described until this moment [4,16,18–21]. Mainly for the leaves, the presence of following iridoids was already described: geniposidic acid [22], genipatriol [23], genamesides A–D [19], and our group recently isolated and identified the iridoids 1-hydroxy-7-(hydroxymethyl)-1,4aH,5H,7aH-cyclopenta[c]pyran-4-carbaldehyde, and 7-(hydroxymethyl)-1-methoxy-1H,4aH,5H,7aH-cyclopenta[c]pyran-4-carbaldehyde [24].

Meanwhile, although flavonoids are a large class of plant secondary metabolites widely distributed throughout the plant kingdom [24], that is structurally resembling in that most have a C_{15} phenyl-benzopyrone skeleton [25], the reports about its presence in *G. americana* are scarce, making it a topic that needs further investigation. Our research group detected the presence of flavonoids for the first time in *G. americana* leaves through a fingerprint by thin layer chromatography (TLC) using a specific reagent to detect flavonoids (NP reagent) [24].

In this context, the present study aimed to study the hydroethanolic extract (HE) obtained from leaves of *Genipa americana*, in terms of its phytochemical composition, especially with respect to flavonoid profiles obtained by UHPLC-DAD and LC-IT-ESI-MS/MS.

2. Results and Discussion

2.1. Extraction and TLC Analysis

Despite being a class of secondary metabolites that have shown promising biological activities in various studies [26], only our study described the presence of flavonoids in *G. americana* [24]. Therefore, the phytochemical investigation of this work was conducted to isolate and purify the flavonoids of *G. americana* leaf extract. Initially, the extract (HE) was fractioned and TLC analysis of the HE and fractions, with specific spray reagents, indicated the presence of flavonoids and iridoids [7,27]. The phytochemical screening of the HE, revealed with vanillin sulfuric acid, showed characteristics yellow spots (R_f = 0.33 until 0.90), suggesting the presence of flavonoids as observed in our previous work [24]. Afterwards, a liquid-liquid extraction was carried out, affording the following fractions: petroleum ether (PE), chloroform (CHCl$_3$), ethyl acetate (EtOAc), n-butanol (n-BuOH) and the residual aqueous fraction (RA). The PE fraction showed no spot in the TLC plate after spraying with vanillin sulfuric acid; CHCl$_3$ fraction showed two purple spots (R_f = 0.93 and 0.83), suggesting the presence of iridoids; the EtOAc fraction showed only one zone more evident (R_f = 0.81 —purple), that indicated the presence of iridoids. When EtOAc fraction was revealed with Natural Product (NP) Reagent (UV 365 nm) a yellow zone was observed (R_f = 0.55), suggesting the presence of flavonoids. TLC analysis of the n-BuOH fraction showed four yellow and orange spots (R_f = 0.21 until 0.50—NP Reagent/UV 365 nm), suggestive of flavonoids [8].

2.2. UHPLC-DAD Characterization

The Research Group on Bioactive Natural Products (PNBio) has been studying the leaves of *G. americana* and in a recent publication [24] the presence of two new iridoids for its leaves has been identified, showing the chemical potential of the species.

As described earlier, most of the studies have described the presence of iridoids in the genus *Genipa*, being considered a chemotaxonomic marker [2,24]. Only our study identified flavonoids in leaves [24]. Furthermore, only one article described the presence of leucoanthocyanidins, catechins, and flavanones in the fruits of *G. americana* by HPLC analysis [28], and another study described its presence in fruits by colorimetric methods [13], demonstrating that the presence of flavonoids in leaf extract was not described previously.

Extract and fractions were analyzed by ultra-performance liquid chromatography (UHPLC). The chromatogram of the HE, recorded in 254 nm, showed four main peaks (3.5 min, UV: 246, 265 and 350 nm; 3.9 min, UV: 255 and 354 nm; 16.6 min, UV: 242 and 287 nm; 17.2 min, UV: 243 nm). Accordingly, the UV spectra of each peak observed in the chromatogram can suggest the presence the flavonoids and iridoids in HE of *G. americana* leaves (Figure 1).

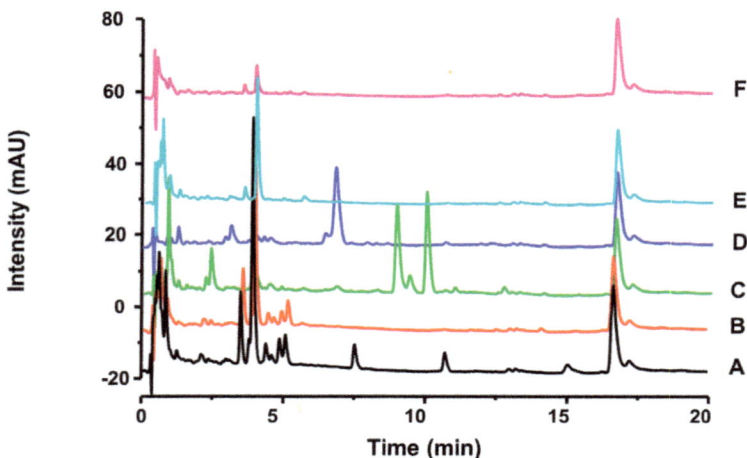

Figure 1. UHPLC-DAD chromatogram of *G. americana* leaf extract and fractions. Solvent system: mix of acetonitrile: water (with 0.3% of acetic acid); Flow rate: 0.3 mL/min; *Software* LC *Solution*; Wavelength: 254 nm. column C_{18} Shim-pack XR-ODS (Shimadzu®, Japan) (30 × 2 mm, 2.2 μm), the temperature 25 ± 2 °C. **A**: EtOAc fraction; **B**: *n*-BuOH fraction; **C**: CHCl₃ fraction; **D**: PE fraction; **E**: Aqueous residual fraction; **F**: Extract HE.

2.3. HPLC-ESI-IT-MS/MS Characterization

Through the TLC and UHPLC-DAD, we can suggest the presence of iridoids and flavonoids in leaves extract from *G. americana*. Therefore, it was decided to perform an HPLC-MS/MS study to allow the identification these compounds.

In this way, we selected of HE, Fr.EtOAc, and Fr.*n*-BuOH to analyze. The HPLC-MS/MS revealed the presence of several majority peaks and allowed suggest the structures of 13 compounds—including five flavonoids (subsequently isolated), using the program DataAnalysis 4.2 (Bruker, Billerica, MA, USA) (Table 1).

Table 1. Phytochemical profile of leaf extract from *G. americana* obtained by HPLC-IT-ESI-MS/MS. For conditions, please see Section 3.4.

Peak n°	Rt (min)	m/z Pos/Neg	m/z MS² Pos/Neg	m/z MS³ Pos/Neg	LC-DAD λ$_{max}$ [nm]	Molecular Formula	Compound (MM)
1	5.2 [a]	-/342	-/179	-/-	-	$C_{16}H_{22}O_8$	Coniferin (342)
2	24 [b]	559/535, 571⁻	-/535	373, 210	-	$C_{22}H_{32}O_{15}$	Asystasioside D (536)
3	25.1 [a]	397/373, 409	379, 217/210	172/166, 148, 122	-	$C_{16}H_{22}O_{10}$	Geniposidic acid (374)
4	26.1 [a]	-/357, 393	-/194	-/-	-	$C_{16}H_{25}O_9$	Tarenoside (358)
5	26.9 [b]	-/375	-/213	-/169, 125	-	$C_{16}H_{24}O_{10}$	Loganic acid (376)
6	33.4 [a,b]	355/353	162/190	-/-	-	$C_{16}H_{18}O_9$	Chlorogenic acid (354)
7	48.8 [a,b,c]	741/739⁻	595/593	433, 287/285	266, 346	$C_{33}H_{39}O_{20}$	Kaempferol-3-O-hexoside-deoxyhexoside-7-O-deoxyhexoside (740)
8	50.1 [a,b]	771/769	625, 463/623	317/315	255, 354	$C_{34}H_{42}O_{21}$	Isorhamnetin-3-O-hexoside-deoxyhexoside-7-O-deoxyhexoside (770)
9	51.4 [a,b]	611/609⁻	465/301	303/-	255, 366	$C_{27}H_{30}O_{16}$	Quercetin-3-O-hexoside-deoxyhexoside (610)
10	51.7 [b]	-/515	-/353, 190, 178	-/-	-	$C_{25}H_{24}O_{12}$	1,3-Di-O-caffeoylquinic acid (516)
11	52	519, 373, 211/517, 371⁻	-/-	-/-	-	$C_{25}H_{30}O_{14}$	Teneoside A (518)
12	53.9 [a,b]	595/593	2834.7/449, 288	-/-	265, 355	$C_{27}H_{30}O_{16}$	Kaempferol-3-O-hexoside-deoxyhexoside (594)
13	55.2 [a,b]	625/623	479/315	317/-	255, 370	$C_{28}H_{33}O_{16}$	Isorhamnetin-3-O-hexoside-deoxyhexoside (624)

[a] = HE; [b] = Fr. EtOAc; [c] = Fr. *n*-BuOH; Rt: retention time; MM: molecular mass.

Coniferin (MM = 342, $C_{16}H_{22}O_8$, Table 1), showed signals at 342 $[M - H]^-$; 179 $[M - H - Glc]^-$. Already identified in *Ginkgo biloba* L. (Ginkgoaceae) [29].

Asystasioside D (MM = 536, $C_{22}H_{32}O_{15}$, Table 1): showed signals at 559 $[M + Na]^+$, 535 $[M - H]^-$; 571 $[M + Cl]^-$, 535, 373 $[M - H - Glc]$; 210 $[M - H - Glc - Glc]$ [aglycone genipinic acid]. This compound is a iridoid glucoside and had already been described in *Asystasia bella* (Harv.) [30].

Geniposidic acid (MM = 374, $C_{16}H_{22}O_{10}$, Table 1), showed signals at 397 $[M + Na]^+$, 373 $[M - H]^-$, 409 $[M + Cl]^-$, 379 $[M + Na - H_2O]^+$, 217 $[M + Na - H_2O - Glc]^+$, 210 $[M - H - Glc]^-$, 172 $[M + Na - H_2O - Glc - CO_2]^+$, 166 $[M - H - Glc - CO_2]^-$, 148 $[M - H - Glc - CO_2 - H_2O]^-$, 122 $[M - H - Glc - C_3H_4O_3]^-$. It is a iridoid already identified in fruit of *Gardenia jasminoides* Ellis (Rubiaceae) [19,22,31,32] and *Wendlandia formosana* Cowan leaves [33].

Tarenoside (MM = 358, $C_{16}H_{25}O_9$, Table 1): showed signals at 357 $[M - H]^-$; 393 $[M + Cl]^-$, 194 $[M - H - Glc]^-$. This compound is an iridoid already identified in *Wendlandia formosana* Cowan leaves and *Genipa americana* [2,33].

Loganic acid (MM = 376, $C_{16}H_{24}O_{10}$, Table 1): showed signals at 375 $[M - H]^-$; 213 $[M - H - Glc]$; 169, 125. It had already been described in *Anthocephalus chinensis* (Rubiaceae) and *Ophiorrhiza liukiuensis* (Rubiaceae) [2]. Loganic acid is described for the first time to *Genipa* genus.

Chlorogenic acid (MM = 354, $C_{16}H_{18}O_9$, Table 1), showed signals at 355 $[M + H]^+$, 353 $[M - H]^-$, 162 [caffeic acid], 190 [quinic acid]. It had already been isolated in fruit of *Gardenia jasminoides* Ellis (Rubiaceae) [31,34], but the first time is described in the *Genipa* genus.

Kaempferol-3-O-hexoside-deoxyhexoside-7-O-deoxyhexoside (**GAF 4**, MM = 740, $C_{33}H_{39}O_{20}$, Table 1), showed signals at 741 $[M + H]^+$, 739 $[M - H]^-$, 595, 593 $[M - Glc]^-$, by scission of the robinose glycan residue to yield the radical anion 433 $[M - Glc - Glc]^-$, 287; radical anion fragments by loss of the rhamnose glycan residue to yield the radical anion 285 $[M - Glc - Glc - Glc]^-$ [25].

Isorhamnetin-3-O-hexoside-deoxyhexoside-7-O-deoxyhexoside (**GAF 5**, MM = 770, $C_{34}H_{42}O_{21}$, Table 1), showed signals at 769 $[M - H]^-$; 771 $[M - H]^+$; 625 $[M - Rha + H]^+$; 463; 623 $[M - Rha - H]^-$; 317; 315 $[M - Rha - Rha - Glc-H]^-$ [35].

Quercetin-3-O-hexoside-deoxyhexoside (**GAF 1**, MM = 610, $C_{27}H_{30}O_{16}$, Table 1), showed signals at m/z 611 $[M + H]^+$, 609 $[M - H]^-$, 465, corresponding to the loss of rhamnose unit, 301 $[M - Glc - Glc]^-$, corresponding to the loss of the rhamnosylgalactose unit, and 303. It is suggested that the sugar is of galactose isomerism, since this form was isolated, and plants usually produces a conformation of sugar glucose or galactose, not being the rutin.

1,3-Di-O-caffeoylquinic acid (cynarine, MM = 516, $C_{25}H_{24}O_{12}$, Table 1): showed signals at 515 $[M - H]^-$, 353 [chlorogenic acid]; 190 [acid quinic]; 178. This compound is a chlorogenic acid derivatives. It had already been isolated in *Cynara cardunculus* leaves [36].

Teneoside A (MM = 518, $C_{22}H_{30}O_{14}$, Table 1): showed signals at 519 $[M + H]^+$, 373 $[M - Ram + H]^+$, 211 $[M - Ram - Glc + H]^+$, 517 $[M - H]^-$, 371 $[M - Ram - H]^-$. This is an iridoid glucoside, and it has already been isolated in *Hedyotis tenelliflora* Blume (Rubiaceae) [2], but is also described for the first time in the *Genipa* genus.

Kaempferol-3-O-hexoside-deoxyhexoside (**GAF 2**, MM = 594, $C_{27}H_{30}O_{16}$, Table 1) showed signals at 595 $[M + H]^+$, 593 $[M - H]^-$, and 449 $[M - Glc]^-$ due to the loss of rhamnose, and 288 $[M - Glc]^-$ due to the loss of the rhamnosylgalactose unit. Likewise, it is suggested that the sugar is galactose, not being, therefore the *Kaempferol-3-O-rutinosideo*.

Isorhamnetin-3-O-hexoside-deoxyhexoside (**GAF 3**, MM = 624, $C_{28}H_{33}O_{16}$, Table 1) showed signals at 625 $[M + H]^+$, 623 $[M - H]^-$, 479, 315, and 317 [37].

The present study contributed to describe a new phytochemical approach for *Genipa americana*, considering that until this moment most of the studies reported only the presence of iridoids [2,4,24].

2.4. Isolation of Major Compounds

In order to verify the positions of the sugars of the O-glycosides present in the flavonoids identified by HPLC-MS/MS (Section 2.3), the major constituents of the EtOAc and *n*-BuOH fractions were isolated and identified.

Based on the chromatographic profile observed by TLC and UHPLC analysis, EtOAc fraction (3.27 g) was chosen initially to be fractionated by vacuum liquid chromatography (VLC) and then fractions 8 and 9 were gathered and chromatographed by classical column using silica gel (0.063–0.200 mm) and Sephadex LH-20 gel, which yielded subfractions 134. Then subfraction (61–111) was gathered and isolated by preparative HPLC and three well-separated peaks were obtained in the chromatogram. Thereby, the following flavonoids were purified: *quercetin-3-O-robinoside* (**GAF 1**, 4.2 mg), *kaempferol-3-O-robinoside* (**GAF 2**, 4.0 mg) and *isorhamnetin-3-O-robinoside* (**GAF 3**, 6.0 mg).

Since the *n*-BuOH fraction (5.0 g) also had a flavonoid rich phytochemical profile (Figure 1) and higher yield, it was submitted to fractionation by VLC with silica gel, which yielded 11 subfractions. The fraction 4 (1.07 g) was isolated by classical column chromatography, and its sub-fraction 4-E of chromatography (1.0 g) was dissolved in methanol and chilled to 2 °C, for five days, which promoted the crystallization of a part of the sample, allowed its separation. These fractions were named fraction 4-E.1 (crystallized) and fraction 4-E.2 (not crystallized). The fraction 4-E.1 was submitted to preparative HPLC and two flavonoids were further separated: *kaempferol-3-O-robinoside-7-O-rhamnoside* (robinin) (**GAF 4**, 10.0 mg) and *isorhamnetin-3-O-robinoside-7-O-rhamnoside* (**GAF 5**, 9.0 mg). Under the above conditions, a satisfactory separation of targeted compounds was obtained.

The chemical structures of these five compounds were elucidated by MS and 1D and 2D ^{1}H- and ^{13}C-NMR spectroscopic analyses. Comparison with literature data allowed to confirm the structures of the compounds **GAF 1, 2, 3, 4,** and **5** as flavonoids (Figure 2).

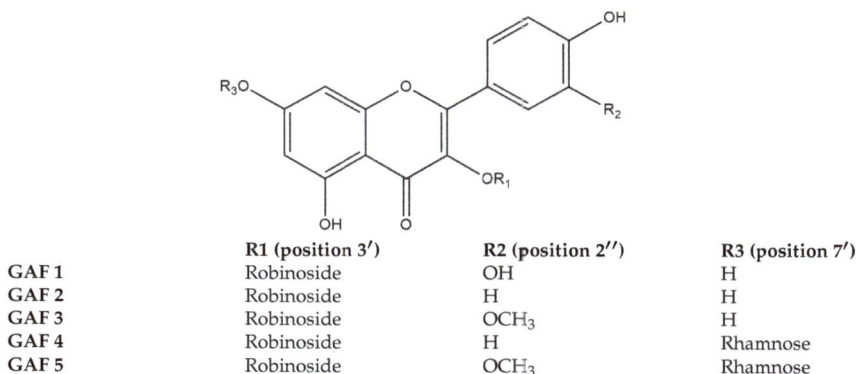

	R1 (position 3′)	R2 (position 2″)	R3 (position 7′)
GAF 1	Robinoside	OH	H
GAF 2	Robinoside	H	H
GAF 3	Robinoside	OCH$_3$	H
GAF 4	Robinoside	H	Rhamnose
GAF 5	Robinoside	OCH$_3$	Rhamnose

Figure 2. Chemical structures of flavonoids **GAF 1–5** isolated from the leaves of *Genipa americana*.

In the study, a simple and effective procedure allowed the isolation of flavonoids from the leaves of *G. americana*. Previously only two papers described the detection of flavonoids in fruits [13,28] and one study published by our group identified these compounds in leaves [24] from *G. americana*. It is important to mention that this is a new phytochemical approach concerning the leaves of this species. Flavonoids are products of secondary metabolism in plants and are of interest to the pharmaceutical and food industries because of their reported wide range of biological effects [26], and they have long been associated with good health benefits, which could be attributed to their antioxidant capabilities [26].

2.5. Identification of the Isolated Compounds

To confirm the positions of the sugars of the O-glycosylated flavonoids identified by HPLC-MS/MS, major compounds from EtOAc and *n*-BUOH fractions were isolated, obtaining five flavonoids. Their structures were identified by comparison of their spectroscopic data reported, including ESI-MS and NMR data.

Quercetin 3-O-robinoside [38] (**GAF 1**, Figure 2): λ_{max} 255, 366 nm. It was assigned with a molecular formula $C_{27}H_{30}O_{16}$ of ESI-MS, *m/z* 609.1403 [M − H]$^-$. ^{1}H-NMR (400 MHz, DMSO-d_6) δ: 7.66 (1H, dd, *J* = 8.5, 2. 2 Hz, H-6′), 7.53 (1H, d, *J* = 2.2 Hz, H-2′), 6.82 (1H, d, *J* = 8.5 Hz, H-5′), 6.39 (1H, s, H-8), 6.19 (1H, s, H-6), 5.32 (1H, d, *J* = 7.7 Hz, galactosyl H-1″), 4.42 (1H, d, *J* = 1.3 Hz, rhamnosyl H-1‴), 3.10 (2H, t, *J* = 9.5 Hz, H-4‴), 1.07 (3H, d, *J* = 6.2 Hz, H-6‴, rhamnosyl-Me). ^{13}C-NMR (100 MHz, DMSO-d_6) δ: 177.35 (C-4), 164.25 (C-7), 161,20 (C-5), 156.38 (C-9), 156.26 (C-2), 148.56 (C-4′), 144.85 (C-3′), 133.47 (C-3), 121.93 (C-6′), 121.03 (C-1′), 115.96 (C-2′), 115.19 (C-5′), 103.77 (C-10), 102.07 (C-1″), 99.99 (C-1‴), 98.80 (C-6), 93.58 (C-8), 73.54 (C-5″), 73.06 (C-3″), 71.92 (C-4‴), 71.10 (C-2″), 70.62 (C-3‴), 70.43 (C-2‴), 68.28 (C-5‴), 68.04 (C-4″), 65.09 (C-6″), 17.92 (C-6‴). The ^{13}C-NMR spectrum showed the sugars to be in the β-D-galactopyranose and α-L-rhamnopyranose forms.

The compounds **GAF 1** has already been identified in: *Alternanthera brasiliana* (L.) Kuntze (Amaranthaceae) [37]; *Strychnos variabilis* (Loganiaceae) leaves; *Robinia pseudacacia* (Fabaceae) fruits; *Lespedeza hedysaroides* (Fabaceae) epigeal part; *Costus sanguineus* (Costaceae)leaves; *Crataegus pinnatifida* (Rosaceae) flowers; *Brickellia chlorolepis* (Asteraceae) leaves [38]; *Lysimuchia vulgaris* (Primulaceae) [39]; and *Aspalathus linearis* (Fabaceae) [40]. No reports were found for the Rubiaceae family.

Kaempferol-3-O-robinoside [38,41] (**GAF 2**, Figure 2): λ_{max} 265, 355 nm. It was assigned with a molecular formula $C_{27}H_{30}O_{16}$ by ESI-MS, *m/z* 593.1508 [M − H]$^-$. ^{1}H-NMR (400 MHz, DMSO-d_6) δ: 8.05 (2H, d, *J* = 8.8 Hz, H-2′, H-6′), 6.86 (2H, d, *J* = 8.8 Hz, H-3′, H-5′), 6.41 (1H, s, H-8), 6.19 (1H, s, H-6), 5.31 (1H, d, *J* = 7.7 Hz, galactosyl H-1″), 4.40 (1H, s, rhamnosyl H-1‴), 3.09 (1H, t, *J* = 9.5 Hz, H-4‴), 1.06 (2H, d, *J* = 6.1 Hz, H-6‴). ^{13}C-NMR (100 MHz, DMSO-d_6) δ: 177.39 (C-4), 164.58 (C-7), 161.18 (C-5), 159.97 (C-3′), 156.54 (C-2), 156.47 (C-9), 149.41 (C-4′), 133.29 (C-3), 130.96 (C-6′), 121.2 (C-2′), 120.85 (C-1′), 115.07 (C-5′), 103.79 (C-10), 100.06 (C-1‴), 102.07 (C-1″), 98.85 (C-6), 93.79 (C-8), 73.52 (C-5″), 72.97 (C-3″), 71.90 (C-4‴), 71.10 (C-2″), 70.61 (C-2‴), 70.42 (C-3‴), 68.29 (C-5‴), 68.00 (C-4″), 65.30 (C-6″), 17.93 (C-6‴).

The compounds **GAF 2** has already been identified in: *Strychnos variabilis* (Loganiaceae) leaves; *Atropa beladona* (Solanaceae) leaves [38]; *Blackstonia perfoliata* (L.) (Gentianaceae) aerial parts [42]; *Gynura formosana* Kitam. (Asteraceae) [43]; *Astragalus tana* Sosn. (Fabaceae) [37]; *Caragana chamlagu* Lam. (Fabaceae) [37]; *Alternanthera brasiliana* (L.) Kuntze (Amaranthaceae) [37]; and *Rumex chalepensis* Mill (Polygonaceae) [41]. No reports were found for the *Genipa* genus.

Isorhamnetin-3-O-robinoside [37] (**GAF 3**, Figure 2): λ_{max} 255, 370 nm. It was assigned with a molecular formula $C_{28}H_{33}O_{16}$ by ESI-MS, *m/z* 623.1596 [M − H]$^-$. ^{1}H-NMR (400 MHz, DMSO-d_6) δ: 8.00 (1H, d, *J* = 2.3 Hz, H-2′), 7.51 (1H, dd, *J* = 8.4, 2.1 Hz; H-6′), 6.90 (1H, d, *J* = 8.3 Hz, H-5′), 6.42 (1H, s, H-8), 6.19 (1H, d, *J* = 1.8 Hz, H-6), 5.45 (1H, d, *J* = 7.7 Hz; H-1″), 4.42 (1H, d, *J* = 1.7 Hz; H-1‴), 3.85 (3H, s, H-3′-O-CH$_3$), 3.08 (2H, t, *J* = 9.4 Hz, H-4‴), 1.05 (3H, d, *J* = 6.1 Hz, H-6‴). ^{13}C-NMR (100 MHz, DMSO-d_6) δ: 177.29 (C-4), 164.65 (C-7), 161.19 (C-5), 156.46 (C-9), 156.33 (C-2), 149.45 (C-4′), 146.99 (C-3′), 133.08 (C-3), 121.96 (C-6′), 121.05 (C-1′), 115.16 (C-5′), 113.44 (C-2′), 103.84 (C-10), 101.85 (C-1″), 100.06 (C-1‴), 98.92 (C-6), 93.82 (C-8), 73.56 (C-5″), 72.94 (C-3″), 71.14 (C-2″), 70.6 (C-4‴), 70.4 (C-2‴), 68.3 (C-5‴), 67.98 (C-4″), 65.18 (C-6″), 55.94 (C-3′-O-CH$_3$), 17.90 (C-6‴).

The compounds **GAF 3** is a diglycoside that has already been identified in: *Gomphrena martiana* Moquin (Amaranthaceae) [39]; *Blackstonia perfoliata* (L.) (Gentianaceae) aerial parts [42]; *NitrariaRetusa* (Nitrarioideae) leaves [40]; and *Calotropis procera* R. Br. (Asclepiadaceae) leaves [44]. No reports were found for the Rubiaceae family.

Kaempferol-3-O-robinoside-7-O-rhamnoside (robinin) [45] (**GAF 4**, Figure 2): λ_{max} 266, 346 nm. It was assigned with a molecular formula $C_{33}H_{39}O_{20}$ by ESI-MS, m/z 739.2062. $[M - H]^-$. ^{1}H-NMR (400 MHz, DMSO-d_6) δ: 8.10 (2H, d, J = 9.0 Hz, H-2′, 6′), 6.88 (2H, d, J = 9.0 Hz, H-3′, 5′), 6.81 (1H, d, J = 2.1 Hz, H-8), 6.46 (1H, d, J = 2.1 Hz, H-6), 5.56 (1H, d, J = 1.4 Hz, H-1′), 5.36 (1H, d, J = 7.6 Hz, H-1″), 4.40 (1H, d, J = 1.7 Hz, H-1‴), 3.09 (1H, dd, J = 12 Hz, 6.5, H-4‴), 1.12 (2H, d, J = 6.1 Hz, H-6‴′), 1.05 (2H, d, J = 6.2 Hz, H-6‴). ^{13}C-NMR (100 MHz, DMSO-d_6) δ: 177.66 (C-4), 161.64 (C-5), 160.19 (C-4′), 157.08 (C-2), 156.04 (C-9), 133.57 (C-3), 131.11 (C-6′), 120.71 (C-1′), 115.14 (C-5′), 105.60 (C-10), 103.29 (C-1″), 100.03 (C-1‴), 99.73 (C-6), 98.39 (C-1‴′), 94.88 (C-8), 73.61 (C-5″), 72.96 (C-3″), 71.92 (C-4′), 71.62 (C-4‴), 71.10 (C-2″, 2‴′), 70.60 (C-3‴), 70.43 (C-2‴), 70.27 (C-3‴′), 69.85 (C-5‴), 68.28 (C-5‴′), 68.00 (C-4″), 17.93 (C-6‴).

The compounds **GAF 4** is a flavone triglycoside of the well-studied has already been identified in: *Alternanthera brasiliana* (Amaranthaceae) leaves [37]; *Strychnos variabilis* (Loganiaceae) leaves; *Atropa beladona* (Solanaceae) leaves [38]; *Astragalus falcatus* Lam. (Leguminosae) flowers [45]; *Robinia pseudoacacia* (Leguminosae) leaves [46]; and *Melilotus elegans* Salzm. ex Ser. (Leguminosae) leaves [47]. No reports were found for the Rubiaceae family.

Isorhamnetin-3-O-robinoside-7-O-rhamnoside [35] (**GAF 5**, Figure 2): λ_{max} 255, 354 nm. It was assigned with a molecular formula $C_{34}H_{42}O_{21}$ by ESI-MS, m/z 769.2220 $[M - H]^-$. ^{1}H-NMR (400 MHz, DMSO-d_6) δ: 8.02 (1 H, d, J = 2.1 Hz, H-2′), 7.60–7.57 (1H, m, H-6′), 6.93–6.90 (1H, m, H-5′), 6.46 (1H, d, J = 2.2 Hz, H-8), 5.57 (1H, d, J = 2.1 Hz, H-1‴′), 5.48 (1H, d, J = 7.7 Hz, H-1″), 3.87 (3H, s, H-3′-O-CH₃), 3.12–3.05 (1H, m, H-4‴), 1.13 (3H, d, J = 6.1 Hz, H-6‴′), 1.06 (3H, d, J = 6.1 Hz, H-6‴). ^{13}C-NMR (100 MHz, DMSO-d_6) δ: 178.01 (C-4), 162.08 (C-7), 161.34 (C-5), 157.35 (C-9), 156.46 (C-2), 150.12 (C-4′), 147.50 (C-3′), 133.84 (C-3), 122.69 (C-6′), 121.33 (C-1′), 115.64 (C-5′), 113.87 (C-2′), 106.08 (C-10), 102.16 (C-1″), 98.81 (C-6), 95.18 (C-8), 73.38 (C-5″), 72.34 (C-4‴), 71.58 (C-4‴′), 71.05 (C-2″), 70.73 (C-3‴′), 70.30 (C-3‴), 56.40 (C-3′-O-CH₃), 18.38 (C-6‴′), 18.34 (C-6‴)

The compounds **GAF 5** is a triglycoside that has already been identified in *Rhazya stricta* Decaisne (Apocynaceae) leaves [35]; and *Blackstonia perfoliata* (L.) (Gentianaceae) aerial parts [42].

Only after the isolation was it possible to analyze the compounds by ^{1}H- and ^{13}C-NMR and to differentiate the types of sugar present, being possible to identify the final portion as galactose. In addition, it can be confirmed that the sugars are present in positions-3 (**GAF 1–5**) and-7 (**GAF 4–5**).

Compound **GAF 1** exhibited in vitro inhibitory activities against leukemia K562 cells in different extents [47], and cytotoxic activity against HepG-2cells.The hydroxylation pattern of the C-rings of the flavonoid compounds like quercetin aglycone, play an essential role in their cytotoxic activities, especially the inhibition of protein kinase antiproliferation activity [48].

Kaempferol-3-O-robinoside (**GAF 2**) significantly inhibited the human lymphocyte proliferation *in vitro*, to a greater extent (IC50 $\cong$ 25 μg mL^{-1}) and were twice more active than crude extract of *Alternanthera brasiliana* [37]. The compound **GAF 2** showed radical scavenging activities when evaluated using the DPPH method. The IC50 values of the DPPH radical were 286.7 mM [43].

The isorhamnetin 3-O-robinoside (**GAF 3**) showed a protective effect against lipid peroxidation induced by H_2O_2 and antigenotoxic potential on human chronic myelogenous leukemia cell line K562 [49].

Flavonoids are considered as a class of natural products of high pharmacological potency but, unfortunately, many of them have a low solubility in water. We have also isolated one flavonol glycosides very soluble in water: robinin (**GAF 4**). Similar compound have been previously isolated from leaves of *Atropa belladonna* [38] but your structure have not been fully elucidated. Robinin (**GAF 4**) displayed a marked activity, inhibiting edema (38.8%) at a concentration of 0.0027 mmol/kg of body weight, four hours after injection of carrageenin [47]. **GAF 4** was also able to inhibit lymphocyte proliferation to a greater extent (IC50 $\cong$ 25 μg mL^{-1}) and were twice more active than crude extract of *Alternanthera brasiliana* [37].

No reports about pharmacological activities were found for **GAF 5**.

This work described the first time the isolation and identification of flavonoids in leaves at *G. americana*. This is an important finding for subsequent studies aimed at the standardization of leaf extracts.

3. Materials and Methods

3.1. Plant Material

The leaves of *Genipa americana* L., Rubiaceae, were collected in Natal City, Rio Grande do Norte State, Brazil, at coordinates lat: −6.1278 long: −35.1115 WGS22, in May 2012. The plant was identified by botanist Alan de Araújo Roque (UFRN) and a voucher specimen was deposited at the Herbarium of the Federal University of Rio Grande do Norte (UFRN), Brazil, with an identification number 12251.

The collection of the plant material was conducted under authorization of Brazilian Authorization and Biodiversity Information System (SISBIO) (process number 35017) and National System for the Management of Genetic Heritage and Associated Traditional Knowledge (SISGEN) (process number A618873).

3.2. Extraction and TLC Characterization

The leaves of *G. americana* was evaporated at 40 °C in a circulating air oven and powdered leaves (600 g) were extracted by maceration with ethanol:water (70:30, v/v) for five days (plant:solvent, 1.5:10, w/v; 4 L; at room temperature). Then, the organic solvent was evaporated under reduced pressure in rotary evaporator (temperature below 45 °C) and water residue was freeze-dried, obtaining the hydroethanolic extract (HE).

The HE was submitted to a liquid-liquid extraction with organic solvents in order of increasing polarity: petroleum ether (PE) (3 × 200 mL), was sequentially partitioned with chloroform (3 × 200 mL), ethyl acetate (3 × 200 mL) and *n*-butanol (3 × 200 mL). The fractions were evaporated under reduced pressure (temperature below 45 °C) and respectively afforded the PE (0.70 g), CHCl$_3$ (2.9589 g), EtOAc (3.27 g), and *n*-BuOH (14.51 g) fractions of the leaves of *G. americana*.

In the phytochemical screening the HE and fractions were analyzed by TLC using aluminum sheets, coated with silica gel F254 as absorbent and chromatographed with ethyl acetate:formic acid:water: methanol (10:1.6:1.5:0.6, $v/v/v/v$) as mobile phase. The TLC was analyzed under 254 and 365 nm ultraviolet (UV) light and then sprayed with vanillin sulfuric acid (4%) or natural product reagent (0.5%)—NP reagent.

3.3. UHPLC Characterization

The samples were analyzed by ultra-high performance liquid chromatography coupled with a diode array detector (UHPLC/HPLC-DAD), model UFLC (Shimadzu, Kyoto, Japan), containing a quaternary pump system (LC-20A$_3$ XR), equipped with a degasser (DGU-20A$_3$), auto-sampler (SII-20AC XR), column oven (CTO-20AC), and diode-array detectors (SPD-M20A), with *Software* LC *Solution* (Shimadzu, Kyoto, Japan) controlled system. A C$_{18}$ column Shim-pack XR-ODS 30 × 2 mm, 2.2 μm (Shimadzu, Kyoto, Japan); a temperature of 25 ± 2 °C was used for the analysis and separation of the compounds and was achieved using a solvent system mixture of acetonitrile: acidified water (with 0.3% formic acid) as the mobile phase with a flow rate of 0.3 mL/min, and a detector was set at 254 nm.

3.4. HPLC-ESI-IT-MS/MS Characterization

The hydroethanolic extracts and EtOAc and BuOH fractions were analyzed by HPLC-IT-MS/MS. UFLC (Shimadzu, Kyoto, Japan) containing two LC20AD solvent pumps, a SIL20AHT auto sampler, a SPD-M20A detector and a CBM20A system controller, coupled with an ion-trap mass spectrometer (AmaZon X, Bruker, Billerica, MA, USA). LC experiments were performed using a C_{18} column (Kromasil—250 mm × 4.6 mm × 5 μm) and the following gradient elution: solvent A: water and formic acid (0.1%, v/v); solvent B: acetonitrile; injection volume of 20 μL, and flow rate of 0.6 mL/min.

The ion-trap analysis parameters are as follows: capillary 4.5 kV, ESI in positive mode, final plate offset 500 V, 40 psi nebulizer, dry gas (N_2) with flow rate of 8 mL/min and a temperature of 300 °C. CID fragmentation was achieved in auto MS/MS mode using advanced resolution mode for MS and MS/MS mode. The spectra (m/z 50–1000) were recorded every two seconds.

The data obtained were interpreted with the help of the following: *Metlin, MassBank* and *Scienfinder*.

3.5. Isolation of Major Compounds

According to their profiles by TLC, the EtOAc and *n*-BuOH fractions were gathered for subsequent isolation. Fractionation of the compounds of *G. americana* started with the EtOAc fraction (3.27 g), which was submitted to vacuum liquid chromatography (VLC) (10 × 15 cm) on a sintered funnel filled with silica gel 60 (0.063–0.200 mm) and eluted with *n*-hexane (50:10):methanol (50:50):water (10:100) v/v, 150 mL. This procedure yielded in 10 fractions. In this step, the fractions that showed the presence of flavonoids in TLC analysis were chosen for isolation. Fractions 8 and 9 of VLC were gathered (1.5 g) and further subjected to classical column chromatography (25 × 3 cm) in a Sephadex LH-20(GE Healthcare Bio-Science AB, Uppsala, Sweden), eluted with chloroform:water:methanol (9:0.1:0.9; v/v; 2.0 mL/min) affording 134 fractions. Fractions 61–111 were gathered (Fr. 61–111, 153 mg) and submitted to purification by HPLC (mobile phase: acetonitrile (21–23%) using a Shimadzu Shimpack ODS (H) kit, C_{18} column (200 mm × 20 mm, 5 μm), flow rate 10 mL/min and UV 254 nm, 270 nm and 340 nm) (for details, see Section 3.6). This procedure yielded compounds **GAF 1** (4.2 mg), **GAF 2** (4.0 mg) and **GAF 3** (6.0 mg).

The *n*-BuOH fraction (5.0 g) was submitted to VLC (10 × 15 cm) on silica gel 60 (0.063–0.200 mm) and eluted with gradient-mode methanol:EtOAc (0–100%; v/v). This procedure resulted in 11 fractions. Fraction 4 (1.07 g) was subjected to classical column chromatography (25 × 3 cm) on silica gel 60 (0.063–0.200 mm) and eluted with EtOAc:formic acid:water:methanol (6.5:1.4:1:1; v/v), sub-fraction 4-E showed greater intensity of phenolic compounds in analysis by TLC, being selected for isolation steps. The fraction 4-E of chromatography (1.0 g) was dissolved in methanol and chilled to 2 °C, for five days, which promoted the crystallization of a part of the sample, which was separated and named fraction 4-E.1 (crystallized, 60 mg) and fraction 4-E.2 (not crystallized, 440 mg). The sub-fraction 4-E.2 was submitted to preparative HPLC developed with water:acetonitrile (0–30 min, 18%; 10 mL/min) (for details, see Section 3.6) using C_{18} column (200 mm × 20 mm, 5 μm), and UV 254 nm, 270 nm, and 340 nm. This procedure yielded compounds **GAF 4** (10.0 mg) and **GAF 5** (9.0 mg).

3.6. Preparative HPLC Optimization and Analyses

In order to obtain the isolate compounds, these fractions were subjected to preparative HPLC for further purification. The preparative HPLC separations were used with a C_{18} column (200 mm × 20 mm, 5 μm); and mobile phase was selected based on the polarity of the likely compounds and the analytical HPLC conditions. Several mobile phases composed of acetonitrile (B)–water(A) in various concentrations of acetonitrile (15%, 18%, 20%, 21%, 23%, 24%, 25%, 30%) were tested. The results indicated that the best separation conditions were achieved using acetonitrile in a gradient mode (0–28 min, 21–23%) for the ethyl acetate fraction and isocratic mode (0–30 min, 18%) for *n*-BuOH fraction and a flow rate of 10 mL/min and a monitoring wavelength of 254 nm, 270 nm and 340 nm,

Molecules **2018**, *23*, 2521

with LabSolutions software (Shimadzu, Kyoto, Japan). Under the above conditions, a satisfactory separation of each the targeted compounds was achieved.

3.7. MS/MS of Isolated Compounds

The isolated compounds were analysis by MS-MS in positive and negative modes were by mass of direct infusion the microTOF II-ESI-TOF—Bruker Daltonics (Bruker, Billerica, MA, USA), with a drying gas flow rate o f4 L/min at 180 °C, nebulizer gas 0.4 bar (pressure), internal calibration standard: TFA, and syringe flow: 10 µL/min.

3.8. Nuclear Magnetic Resonance Spectroscopy (NMR) of Isolated Compounds

The nuclear magnetic resonance (NMR) spectra were performed on a (Bruker, Billerica, MA, USA) (400 MHz for ^{1}H and 100 MHz for ^{13}C) and chemical shifts are given in ppm relative to residual DMSO-d_6 (2.5), and to the central peak of the triplet related to DMSO-d_6 carbon (39.5 ppm).

4. Conclusions

Through characteristic fragmentation patterns of substances obtained by MS/MS data, 13 compounds were identified. The flavonoids were isolated and identified.

In contrast to literature, which describes mainly the presence of iridoids for the fruit and leaf extracts of *G. americana*, in this paper the leaves showed to be rich in *O-glycosidic* flavonoids. The isolation was carried out in few steps and allowed the identification of five flavonoids glycosides not described until this moment to the *G. americana* leaf extract. These flavonoids **GAF 1, 3, 4,** and **5** were identified for the first time in the Rubiaceae family and flavonoid **GAF 2** is unknown for genus *Genipa*. All flavonol aglycones have sugars units, attached only at the 3-position. The similarities of the compounds of five chemical structures suggest a common biosynthetic pathway in this species. The present article was conducted to evaluate the chromatograph profiles of the leaf extract from *Genipa americana*, to be used in future to the quality control for this species. It can also be suggesting that flavonoids identified for this species may be associated, at least in part, to pharmacological properties of the plant.

Author Contributions: L.M.P.S. performed the experiments, analyzed the results, identified the compounds, and wrote the manuscript; J.S.F.A. performed the isolation of the chemical compounds; D.L.P. performed the characterization by UHPLC-DAD; C.F.S.A. and J.F.T. contributed the reagents/materials/analysis tools; L.S.A., E.M.d.S.S., M.A.d.S.N., N.P.L., L.d.S.F. and D.P.D. helped in the identification of the compounds; S.M.Z. conceived and designed the experiments, analyzed the data, interpreted the results, and wrote the manuscript. All the authors read and approved the final manuscript.

Funding: We thank the Conselho Nacional de Desenvolvimento Científico e Tecnológico-CNPq 478661/2010-e *Coordenação de Aperfeiçoamento de Pessoal de Nível Superior*—CAPES for funding this project and UFRN—*Universidade Federal do Rio Grande do Norte* for support.

Acknowledgments: The authors acknowledge all contributors for their valuable time and commitment to the study.

Conflicts of Interest: The authors declare no conflict of interest.

References

1. Barbosa, M.R.; Zappi, D.; Taylor, C.; Cabral, E.; Jardim, J.G.; Pereira, M.S.; Calió, M.F.; Pessoa, M.C.R.; Salas, R.; Souza, E.B.; et al. Rubiaceae em Flora do Brasil 2020 em construção.Jardim Botânico do Rio de Janeiro. Available online: http://floradobrasil.jbrj.gov.br/jabot/FichaPublicaTaxonUC/FichaPublicaTaxonUC.do?id=FB14045 (accessed on 8 March 2018).
2. Martins, D.; Nunez, C.V. Secondary metabolites from rubiaceae species. *Molecules* **2015**, *20*, 13422–13495. [CrossRef] [PubMed]
3. Santos, O.A.; Couceiro, S.R.M.; Rezende, A.C.C.; Silva, M.D.S. Composition and richness of woody species in riparian forests in urban areas of Manaus, Amazonas, Brazil. *Landsc. Urban Plan* **2016**, *150*, 70–78. [CrossRef]

4.   Ueda, S.; Iwahashi, Y. Production of anti-tumor-promoting iridoid glucosides in *Genipa americana* and its cell cultures. *J. Nat. Prod.* **1991**, *54*, 1677–1680. [CrossRef] [PubMed]

5.   Zappi, D. Genipa in Lista de Espécies da Flora do Brasil. Jardim Botânico do Rio de Janeiro. Available online: http://floradobrasil.jbrj.gov.br/jabot/floradobrasil/FB14045 (accessed on 22 November 2017).

6.   Almeida, E.R.; Almeida, E.; Almeida, E. *Plantas Medicinais Brasileiras: Conhecimentos Populares e Científicos*; Hemus: São Paulo, Brasil, 1993; pp. 215–216.

7.   Gomes, R.P. *Fruticultura Brasileira*; Nobel: São Paulo, Brasil, 1982; pp. 278–281.

8.   Agra, M.F.; Silva, K.N.; Basílio, I.J.L.D.; de Freitas, P.F.; Barbosa-Filho, J.M. Survey of medicinal plants used in the region Northeast of Brazil. *Rev. Bras. Farmacogn.* **2008**, *18*, 472–508. [CrossRef]

9.   Mors, W.B.; Rizzini, C.T.; Pereira, N. *Medicinal Plants of Brazil*; Reference Publications, Inc.: Algonac, MI, USA, 2000.

10.   Delprete, P.G.; Smith, L.B.; Klein, R.M. Rubiáceas. In *Flora Ilustrada Catarinense*; Reis, A., Ed.; Herbário Barbosa Rodrigues: Itajaí, Santa Catarina, Brizal, 2005.

11.   Koo, H.J.; Song, Y.S.; Kim, H.J.; Lee, Y.H.; Hong, S.M.; Kim, S.J.; Kim, B.C.; Jin, C.; Lim, C.J.; Park, E.H. Antiinflammatory effects of genipin, an active principle of gardenia. *Eur. J. Pharmacol.* **2004**, *495*, 201–208. [CrossRef] [PubMed]

12.   Kim, B.C.; Kim, H.G.; Lee, S.A.; Lim, S.; Park, E.H.; Kim, S.J.; Lim, C.J. Genipin-induced apoptosis in hepatoma cells is mediated by reactive oxygen species/c-Jun NH 2-terminal kinase dependent activation of mitochondrial pathway. *Biochem. Pharmacol.* **2005**, *70*, 1398–1407. [CrossRef] [PubMed]

13.   Nogueira, F.A.; Nery, P.S.; Morais-Costa, F.; de Faria Oliveira, N.J.; Martins, E.R.; Duarte, E.R. Efficacy of aqueous extracts of *Genipa americana* L. (Rubiaceae) in inhibiting larval development and eclosion of gastrointestinal nematodes of sheep. *J. Appl. Anim. Res.* **2014**, *42*, 356–360. [CrossRef]

14.   De Souza, P.M.; De Salesi, P.M.; Simeoni, L.A.; Silva, E.C.; Silveira, D.; Magalhães, P.O. Inhibitory activity of α-amylase and α-glucosidase by plant extracts from the Brazilian cerrado. *Planta Med.* **2012**, *78*, 393–399. [CrossRef] [PubMed]

15.   Nonato, D.T.T.; Vasconcelos, S.M.M.; Mota, M.R.L.; de Barros Silva, P.G.; Cunha, A.P.; Ricardo, N.M.P.S.; Pereira, M.G.; Assreuy, A.M.S.; Chaves, E.M.C. The anticonvulsant effect of a polysaccharide-rich extract from *Genipa americana* leaves is mediated by GABA receptor. *Biomed. Pharmacother.* **2018**, *101*, 181–187. [CrossRef] [PubMed]

16.   Ono, M.; Ishimatsu, N.; Masuoka, C.; Yoshimitsu, H.; Tsuchihashi, R.; Okawa, M.; Kinjo, J.; Ikeda, T.; Nohara, T. Three new monoterpenoids from the fruit of *Genipa americana*. *Chem. Pharm. Bull.* **2007**, *55*, 632–634. [CrossRef] [PubMed]

17.   De Bentes, A.S.; Mercadante, A.Z. Influence of the stage of ripeness on the composition of iridoids and phenolic compounds in genipap (*Genipa americana* L.). *J. Agric. Food Chem.* **2014**, *5*, 62. [CrossRef] [PubMed]

18.   Djerassi, C.; Gray, J.D.; Kincl, F. Isolation and characterization of genipin. *J. Org. Chem.* **1960**, *25*, 2174–2177. [CrossRef]

19.   Ono, M.; Ueno, M.; Masouka, C.; Ikeda, T.; Nohara, T. Iridoid glucosides from the fruit of *Genipa americana*. *Chem. Pharm. Bull.* **2005**, *53*, 1342–1344. [CrossRef] [PubMed]

20.   Tallent, W.H. Two new antibiotic cyclopentanoid monoterpenes of plant origin. *Tetrahedron Lett.* **1964**, *20*, 1781–1787. [CrossRef]

21.   Hsua, H.; Yang, J.; Lin, S.; Linb, C. Comparisons of geniposidic acid and geniposide on antitumor and radioprotection after sublethal irradiation. *Cancer Lett.* **1997**, *113*, 31–37. [CrossRef]

22.   Guarnaccia, R.; Madyastha, K.M.; Tegtmeyer, E.; Coscia, C.J. Geniposidic acid, an iridoid glucoside from *Genipa americana*. *Tetrahedron Lett.* **1972**, *50*, 5125–5127. [CrossRef]

23.   Hossain, C.F.; Jacob, M.R.; Clark, A.M.; Walker, L.A.; Nagle, D.G. Genipatriol, a new cycloartane triterpene from *Genipa spruceana*. *J. Nat. Prod.* **2003**, *66*, 398–400. [CrossRef] [PubMed]

24.   Alves, J.S.F.; Medeiros, L.A.; Fernandes-Pedrosa, M.F.; Araújo, R.M.; Zucolotto, S.M. Iridoids from leaf extract of *Genipa americana*. *Rev. Bras. Farmacogn.* **2017**, *27*, 641–644. [CrossRef]

25.   March, R.E.; Miao, X.-S.; Metcalfe, C.D. A fragmentation study of a flavone triglycoside, kaempferol-3-O-robinoside-7-O-rhamnoside. *Rapid Commun. Mass Spectrom.* **2004**, *18*, 931–934. [CrossRef] [PubMed]

26.   Panche, A.N.; Diwan, A.D.; Chandra, S.R. Flavonoids: An overview. *J. Nutr. Sci.* **2016**, *5*, 1–15. [CrossRef] [PubMed]

27.   Wagner, H.; Bladt, S. *Plant Drug Analysis: A Thin Layer Chromatography Atlas*; Springer: Berlin, Germany, 2001.

28. Omena, C.M.B.; Valentim, I.V.; Guedes, G.S.; Rabelo, L.A.; Mano, C.M.; Bechara, E.J.H.; Sawaya, A.C.H.F.; Trevisan, M.T.S.; Da Costa, J.G.; Ferreira, R.C.S.; et al. Antioxidant, anti-acetylcholinesterase and cytotoxic activities of ethanol extracts of peel, pulp and seeds of exotic Brazilian fruits. *Food Res. Int.* **2012**, *49*, 334–344. [CrossRef]

29. Kuroda, K.; Yagami, S.; Takama, R.; Fukushima, K. Distribution of coniferin in freeze-fixed stem of *Ginkgo biloba* L. by cryo-TOF-SIMS/SEM. *Sci. Rep.* **2016**, *6*, 31525. [CrossRef]

30. Demuth, H.; Jensen, S.R.; Nielsen, B.J. Iridoid glucosides from *Asystasia bella*. *Phytochemistry* **1989**, *28*, 3361–3364. [CrossRef]

31. Wu, X.; Zhou, Y.; Yin, F.; Mao, C.; Li, L.; Cai, B.; Lu, T. Quality control and producing areas differentiation of Gardeniae Fructus for eight bioactive constituents by HPLC–DAD–ESI/MS. *Phytomedicine* **2014**, *21*, 551–559. [CrossRef] [PubMed]

32. Fu, Z.; Xue, R.; Li, Z.; Chen, M.; Sun, Z.; Hu, Y.; Huang, C. Fragmentation patterns study of iridoid glycosides in Fructus Gardeniae by HPLC-Q/TOF-MS/MS. *Biomed. Chromatogr.* **2014**, *28*, 1795–1807. [CrossRef] [PubMed]

33. Choze, R.; Delprete, P.G.; Lião, L.M. Chemotaxonomic significance of flavonoids, coumarins and triterpenes of *Augusta longifolia* (Spreng.) Rehder, Rubiaceae-Ixoroideae, with new insights about its systematic position within the family. *Braz. J. Pharmacogn.* **2010**, *3*, 20. [CrossRef]

34. Tauchen, J.; Bortl, L.; Huml, L.; Miksatkova, P.; Doskocil, I.; Marsik, P.; Villegas, P.P.P.; Flores, Y.B.; Damme, P.V.; Lojka, B.; et al. Phenolic composition, antioxidant and anti-proliferative activities of edible and medicinal plants from the Peruvian Amazon. *Rev. Bras. Farmacogn.* **2016**, *26*, 728–737. [CrossRef]

35. Andersen, W.K.; Omar, A.A.; Christensen, S.B. Isorhamnetin3-(2,6-dirhamnosylgalactoside)-7-rhamnoside and 3-(6-rhamnosylgalactoside)-7-rhamnoside from *Rhazya stricta*. *Phytochemistry* **1987**, *26*, 291–294. [CrossRef]

36. Zhu, X.; Zhang, H.; Lo, R. Phenolic Compounds from the Leaf Extract of Artichoke (*Cynara scolymus* L.) and Their Antimicrobial Activities. *J. Agric. Food Chem.* **2004**, *52*, 7272–7278. [CrossRef] [PubMed]

37. Brochado, C.D.O.; Almeida, A.P.; Barreto, B.P.; Costa, L.P.; Ribeiro, L.S.; Pereira, R.L.C.; Koatz, V.L.G.; Costa, S.S. Flavonol robinobiosides and rutinosides from *Alternanthera brasiliana* (Amaranthaceae) and their effects on lymphocyte proliferation in vitro. *J. Braz. Chem. Soc.* **2003**, *14*, 449–451. [CrossRef]

38. Brasseur, T.; Angenot, L. Six flavonol glycosides from leaves of *Strychnos variabilis*. *Phytochemistry* **1988**, *27*, 1487–1490. [CrossRef]

39. Buschi, C.A.; Pomilio, A.B. Isorhamnetin 3-*O*-robinobioside from *Gomphrena martiana*. *J. Nat. Prod.* **1982**, *45*, 557–559. [CrossRef]

40. Halim, A.F.; Saad, H.E.A.; Hashish, N.E. Flavonol glycosides from *Nitraria retusa*. *Phytochemistry* **1995**, *40*, 349–351. [CrossRef]

41. Hasan, A.; Ahmed, I.; Jay, M.; Voirin, B. Flavonoid glycosides and an anthraquinone from *Rumex chalepensis*. *Phytochemistry* **1995**, *39*, 1211–1213. [CrossRef]

42. Kaouadji, M. Flavonol diglycosides from *Blackstonia perfoliata*. *Phytochemistry* **1990**, *29*, 1345–1347. [CrossRef]

43. Hou, W.-C.; Lin, R.-D.; Lee, T.-H.; Huang, Y.-H.; Hsu, F.-L.; Lee, M.-H. The phenolic constituents and free radical scavenging activities of *Gynura formosana* Kiamnra. *J. Sci. Food Agric.* **2005**, *85*, 615–621. [CrossRef]

44. Gallegos-Olea, R.S.; Borges, M.O.R.; Borges, A.C.R.; Freire, S.M.F.; Silveira, L.M.S.; Vilegas, W.; Rodrigues, C.M.; Oliveira, A.V.; Costa, J.L. Flavonoides de *Calotropis procera* R. Br. (Asclepiadaceae). *Rev. Bras. Plantas Med.* **2008**, *10*, 29–33.

45. Tsiklauri, L.K.; An, G.; Alania, M.D.; Kemertelidze, E.P.; Morris, M.E. Optimum HPLC parameters for simultaneous determination of robinin and kaempferol. *Pharm. Chem. J.* **2012**, *46*, 64–67. [CrossRef]

46. Veitch, N.C.; Elliott, P.C.; Kite, G.C.; Lewis, G.P. Flavonoid glycosides of the black locust tree, *Robinia pseudoacacia* (Leguminosae). *Phytochemistry* **2010**, *71*, 479–486. [CrossRef] [PubMed]

47. Asres, K.; Eder, U.; Bucar, F. Studies on the anti-inflammatory activity of extracts and compounds from the leaves of *Melilotus elegans*. *Ethiop. Pharm. J.* **2000**, *18*, 15–23.

48. Ghareeb, M.A.; Shoeb, H.A.; Madkour, H.M.F.; Refahy, L.A.; Mohamed, M.A.; Saad, A.M. Antioxidant and cytotoxic activities of flavonoidal compounds from *Gmelina arborea* Roxb. *Global J. Pharm.* **2014**, *8*, 87–97. [CrossRef]

49. Boubaker, J.; Sghaier, M.B.; Skandrani, I.; Ghedira, K.; Chekir-Ghedira, L. Isorhamnetin 3-*O*-robinobioside from *Nitraria retusa* leaves enhance antioxidant and antigenotoxic activity in human chronic myelogenous leukemia cell line K562. *BMC Complement. Altern. Med.* **2012**, *12*, 135. [CrossRef] [PubMed]

Sample Availability: Samples of the compounds **GAF 1**, **GAF 2**, **GAF 3**, **GAF 4** and **GAF 5** are available from the authors.

molecules

MDPI

Article

Essential Oils of Five *Baccharis* Species: Investigations on the Chemical Composition and Biological Activities

Jane M. Budel [1,2,*], Mei Wang [2], Vijayasankar Raman [2], Jianping Zhao [2], Shabana I. Khan [2], Junaid U. Rehman [2], Natascha Techen [2], Babu Tekwani [2], Luciane M. Monteiro [1], Gustavo Heiden [3], Inês J. M. Takeda [4], Paulo V. Farago [1] and Ikhlas A. Khan [2]

1   Departamento de Ciências Farmacêuticas, Universidade Estadual de Ponta Grossa (UEPG), Ponta Grossa, PR 84030-900, Brasil; lmmonteiro@hotmail.com (L.M.M.); pvfarago@gmail.com (P.V.F.)
2   National Center for Natural Products Research, School of Pharmacy, University of Mississippi, Mississippi, MS 38677, USA; meiwang@olemiss.edu (M.W.); vraman@olemiss.edu (V.R.); jianping@olemiss.edu (J.Z.); skhan@olemiss.edu (S.I.K.); jurehman@olemiss.edu (J.U.R.); ntechen@olemiss.edu (N.T.); btekwani@olemiss.edu (B.T.); ikhan@olemiss.edu (I.A.K.)
3   Embrapa Clima Temperado, Pelotas, RS 70770-901, Brazil; gustavo.heiden@embrapa.br
4   Departamento de Meio Ambiente, Universidade Estadual de Maringá (UEM), Umuarama, PR 87020-900, Brazil; takedaines@bol.com.br
*   Correspondence: janemanfron@hotmail.com; Tel.: +55-42-3220-3000

Academic Editors: Charles L. Cantrell and Muhammad Ilias
Received: 7 August 2018; Accepted: 10 October 2018; Published: 12 October 2018

Abstract: This paper provides a comparative account of the essential oil chemical composition and biological activities of five Brazilian species of *Baccharis* (Asteraceae), namely *B. microdonta*, *B. pauciflosculosa*, *B. punctulata*, *B. reticularioides*, and *B. sphenophylla*. The chemical compositions of three species (*B. pauciflosculosa*, *B. reticularioides*, and *B. sphenophylla*) are reported for the first time. Analyses by GC/MS showed notable differences in the essential oil compositions of the five species. α-Pinene was observed in the highest concentration (24.50%) in *B. reticularioides*. Other major compounds included α-bisabolol (23.63%) in *B. punctulata*, spathulenol (24.74%) and kongol (22.22%) in *B. microdonta*, β-pinene (18.33%) and limonene (18.77%) in *B. pauciflosculosa*, and β-pinene (15.24%), limonene (14.33%), and spathulenol (13.15%) in *B. sphenophylla*. In vitro analyses for antimalarial, antitrypanosomal, and insecticidal activities were conducted for all of the species. *B. microdonta* and *B. reticularioides* showed good antitrypanosomal activities; *B. sphenophylla* showed insecticidal activities in fumigation bioassay against bed bugs; and *B. pauciflosculosa*, *B. reticularioides*, and *B. sphenophylla* exhibited moderate antimalarial activities. *B. microdonta* and *B. punctulata* showed cytotoxicity. The leaves and stems of all five species showed glandular trichomes and ducts as secretory structures. DNA barcoding successfully determined the main DNA sequences of the investigated species and enabled authenticating them.

Keywords: *Baccharis*; antimalarial activity; antitrypanosomal activity; insecticidal activity; GC/MS; DNA barcoding; microscopy

1. Introduction

Baccharis L. (Asteraceae) is an important genus comprising 435 species distributed from Argentina to the United States (USA). In Brazil, the genus is represented by 179 species [1]. Several species of *Baccharis* are frequently used in traditional medicine as analgesic, antidiabetic, anti-inflammatory, digestive, diuretic, and spasmolytic agents [2,3]. These properties provide an excellent rationale for systematically studying their therapeutic properties. However, the correct identification of different

Baccharis species is challenging due to their morphological similarities. Moreover, many different species of *Baccharis* are collectively called "vassouras" in Brazil, causing further confusion [4].

Baccharis species have also provided valuable biomolecules in the discovery of new medicinal natural products [5]. Species of the genus produce essential oils (EOs) that are composed mainly of monoterpenoids and sesquiterpenoids. Several such EOs are used in the fragrance industry and for pharmaceutical purposes [3,6–8].

Many of the medicinal properties described for *Baccharis* are attributed to their EOs [3]. The EOs of the genus have been reported to have several biological activities, including antibacterial, antifungal [9,10], antiprotozoal, antiviral, antioxidant, anti-inflammatory, antimutagenic, antiulcer, chemopreventive, repellent, sedative [7,11–13], schistosomicidal [14], and larvicidal against *Aedes aegypti* [15] and cytotoxic properties [16]. *B. microdonta* DC. has shown antibacterial activity against *Salmonella typhi* [17] and anti-inflammatory properties [18]. Another species, *B. pauciflosculosa* DC., has exhibited antimicrobial activities [17,19].

However, few studies have been carried out on the chemical analysis of the EOs of *Baccharis* species [20]. Thus, the major aims of this study were to identify the marker compounds and compare the chemical profiles of the EOs of five *Baccharis* species, namely *B. microdonta*, *B. pauciflosculosa*, *B. punctulata* DC., *B. reticularioides* Deble and A.S.Oliveira, and *B. sphenophylla* Dusén ex Malme, collected from Brazil. To the best of our knowledge, no previous reports exist on the antitrypanosomal, antimalarial, or insecticidal activities of these five species. The study also analyzes the micromorphology of secretory structures, in which the EOs are stored by microscopy and discriminate the species by DNA barcoding.

2. Results and Discussion

2.1. Yield and Chemical Composition of Essential Oils

Essential oils are liquid, volatile, clear, and rarely colored substances that are soluble in organic solvents and usually of lesser density than water. EOs extracted by hydrodistillation from the vegetative aerial parts of the *Baccharis* species presented a strong and characteristic aroma. The EO of *B. punctulata* was green, whereas it was yellow in *B. microdonta* and light yellow in the other species. The yield of EO was 0.58% (v/w) in *B. microdonta*, 0.93% (v/w) in *B. pauciflosculosa*, 0.29% (v/w) in *B. punctulata*, 0.59% (v/w) in *B. reticularioides*, and 0.53% (v/w) in *B. sphenophylla*. As per the literature, the yield of EOs in the *Baccharis* species ranged between 0.08–2.82%. *B. obovata* Hook. and Arn. presented the highest yield [21], whereas the lowest content was achieved for *B. lateralis* Baker (syn. *B. schultzii* Baker) [22], which were collected in Argentina and Brazil, respectively.

The yield of EOs can be influenced by the physiological variations inherent in the plant, environmental conditions, phenological factors, genetic characteristics of the cultivars, drying conditions applied to the plant material prior to extraction, the process employed for grinding, the storage conditions, and the EO extraction methods [23–25]. Different yields of EOs extracted from the leaves of *B. microdonta* collected in São Paulo, Brazil were reported. Sayuri et al. [26] showed yields between 0.06–0.35%, whereas Lago et al. [22] presented yields of 0.08–0.21% for *B. microdonta*. However, these authors extracted the EOs only from the leaves of *B. microdonta*. In the present work, mixtures of leaves and stems were used for extraction. The anatomical study of *Baccharis* species revealed several large secretory cavities in the cortex of the stems, which may have contributed to the increased yield, as observed by Saulle et al. [27] for *Eucalyptus saligna* Sm. (Myrtaceae).

Chemically, EOs are complex mixtures that can contain 20–60 compounds in different concentrations, and are characterized by two or three main components with higher concentrations compared to others present in lower concentrations. In general, the major compounds that are present in EOs are responsible for the biological activities [28,29].

The chemical compositions of EOs from the five species of *Baccharis* were investigated, and their profiles (Figure 1) were compared. Table 1 compares relative retention indices (RRIs) using non-polar and polar columns, the literature-reported retention indices (RI lit), chemical identity, and relative peak area percentage (%) concentration of the chemical constituents of the five species.

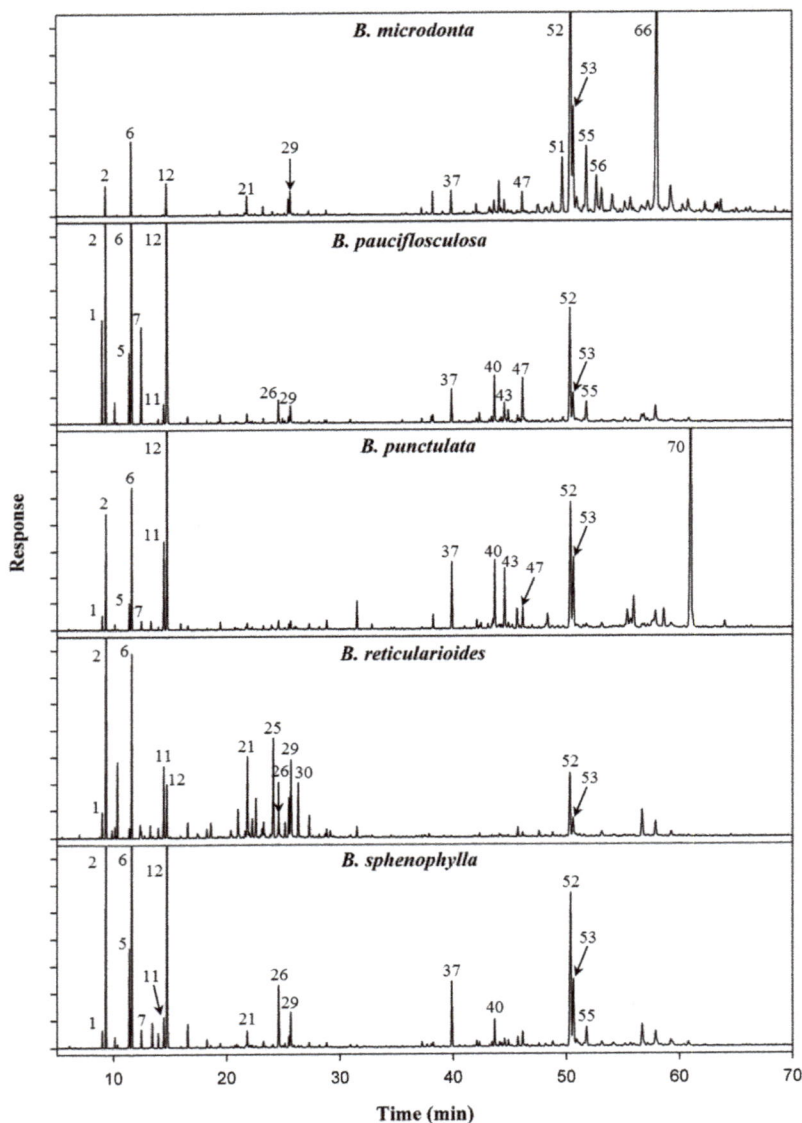

Figure 1. GC chromatograms of *Baccharis* species obtained for non-polar column. Compound identification is consistent with Table 1.

Monoterpenoids and sesquiterpenoids are frequently found in the EOs of *Baccharis* [3,30]. Sesquiterpenoids seemed to be more abundant in the majority of the species [3,6,7,16,22,30–33]. However, the EO of some species contained more monoterpenoids than sesquiterpenoids, such as in *B. obovata* [21], *B. schultzii*, *B. regnelli* Sch. Bip. ex Baker, *B. uncinella* DC. [22], *B. darwinii* Hook. & Arn. [34], *B. tridentata* Vahl. [33] and *B. trimera* (Less.) DC. [35]

Considering that raw percentages of volatile compounds are usually reported by the use of the non-polar GC column, in the present work, only these percentages were discussed. Both monoterpenoids and sesquiterpenoids were found in the EOs of all of the *Baccharis* species that were analyzed, although in different concentrations. Higher concentrations of monoterpenoids (60.78%) were found in *B. reticularioides*, which was formed by 27.45% of monoterpenoids hydrocarbons and 33.33% of oxygenated monoterpenoids. In the case of sesquiterpenoids, *B. microdonta* showed the highest concentrations (41.66%), comprising 18.33% of sesquiterpenoids hydrocarbons and 23.33% of oxygenated sesquiterpenoids. The sesquiterpenoid cyclic alcohols found in the present study, such as ledol, spathulenol, viridiflorol, and palustrol, are not only important in the perfume industry due to their agreeable aromatic notes, but also have taxonomic value [33].

A recent bibliographic review reported around 60 compounds identified in the EOs of 16 species of *Baccharis* that showed biological activities. The main components that were found in these species were α-thujene, β-caryophyllene, β-pinene, camphor, caryophyllene, caryophyllene oxide, limonene, nerolidol, thymol, thymol acetate, thymol methyl ether, sabinene, and spathulenol [3]. In the present study, α-pinene, β-pinene, limonene, trans-pinocarveol, pinocarvone, myrtenal, α-terpineol, cis-carveol, carvone, β-selinene, δ-cadinene, spathulenol, caryophyllene oxide, and δ-cadinol were found in all of the species of *Baccharis* studied. However, significant differences in their concentrations were observed.

The compounds' identification is given in Table 1. *B. microdonta* contained spathulenol (22.74%) (Figure S2) and kongol (22.22%) (Figure S1), *B. pauciflosculosa* showed β-pinene (18.33%) and limonene (18.77%), *B. punctulata* contained α-bisabolol (23.63%), *B. reticularioides* presented α-pinene (24.50%), and *B. sphenophylla* showed α-pinene (10.74%), β-pinene (15.24%), limonene (14.33%), and spathulenol (13.15%) as the major compounds. It is important to highlight that kongol and α-bisabolol were found only in *B. microdonta* and *B. punctulata*, respectively. These compounds can be considered chemical markers for these species.

Differences in the EO chemical compositions have been reported for *B. punctulata* collected from different geographical locations. The EOs of the samples collected from Uruguay have shown β-phellandrene (5.2%), bornyl acetate (5.2%), α-cadinol (4.2%), δ-elemene (3.7%), and the ketone shyobunone (3.5%) as the major compounds [36], whereas the EOs sourced in Guaíba, Brazil have comprised bicyclogermacrene (9.73%), cis-cadin-4-en-7-ol (6.77%), and (Z)-ocimene (6.33%) [23]. Of these compounds, only bornyl acetate (1.32%) and bicyclogermacrene (3.10%) were found in the present study in *B. punctulata* and in low concentrations. α-Bisabolol was not reported in the previous studies, but it was found in higher concentration (23.63%) in the present work.

For *B. microdonta*, Lago et al. [22] reported 24% of caryophyllene oxide, whereas Sayuri et al. recorded 31% of elemol, 34% of spathulenol, 19% of β-caryophyllene, and 24% of germacrene D as the major compounds. Both of these studies used samples collected from Campos do Jordão, Brazil. None of the earlier studies have reported kongol from any of the five *Baccharis* species, whereas this compound was found in high quantity (22.22%) in *B. microdonta* in the present study. Even though the chemical composition of EOs is frequently associated to environmental and phenological influences, it is necessary to investigate whether these variations in *B. punctulata* and *B. microdonta* are possibly linked to different chemotypes. Not only the compositions of EOs, but also the quantities of the compounds vary throughout the life of the plant. This is related to the circadian rhythms, seasonal conditions, and environmental influences that impact the development of the species [24]. To avoid some of these factors, in the present work, all five species were grown in the same locality and collected on the same day and at the same time. Additionally, the sample preparation, hydrodistillation, and characterization of the EOs by GC/MS analysis of all of the materials were performed under the same experimental conditions.

Table 1. Chemical compositions of the essential oils of *Baccharis* species. RRI: relative retention indices.

No.	RRI [a]	RI Lit [b]	RRI [c]	RI Lit [d]	Compound Name	Peak Area % [e]										ID
						B. microdonta		*B. pauciflosculosa*		*B. punctulata*		*B. reticularioides*		*B. sphenophylla*		
						NC	PC	NC	PC	NC	PC	NC	PC	NC	PC	
1	937	924	1019	1038	α-Thujene	-	-	3.42	2.94	0.41	0.43	0.89	1.18	0.54	0.37	MS, RI
2	942	932	1014	1036	α-Pinene	0.72	0.73	10.45	9.44	3.55	3.15	24.50	24.78	10.74	8.04	tR, MS, RI
3	954	946	1050	1083	Camphene [f]	-	-	0.78	0.75	0.19	0.18	0.43	0.42	0.39	0.30	tR, MS, RI
4	950	953	1109	-	Thuja-2,4(10)-diene	-	-	-	-	-	-	2.91	3.65	0.12	0.09	MS, RI
5	973	969	1105	1130	Sabinene [f]	-	-	2.75	2.62	0.89	0.79	0.39	0.85	3.82	3.18	tR, MS, RI
6	976	974	1089	1124	β-Pinene	2.24	2.33	18.33	16.50	4.95	4.41	7.68	9.24	15.24	13.17	tR, MS, RI
7	988	988	1156	1156	β-Myrcene	-	-	3.65	2.76	0.30	0.23	0.27	0.29	0.67	0.57	tR, MS, RI
8	1001	1002	1151	1177	α-Phellandrene	-	-	-	-	0.33	0.11	-	-	-	-	tR, MS, RI
9	1003	1008	-	1141	δ-(3)-Carene	-	-	-	-	-	-	-	-	0.99	0.82	tR, MS, RI
10	1010	1014	1198	1188	α-Terpinene	-	-	0.19	-	0.10	-	0.43	0.39	0.58	0.64	tR, MS, RI
11	1018	1020	1271	1272	p-Cymene	-	-	0.82	1.24	3.44	1.94	3.18	3.27	1.31	1.79	tR, MS, RI
12	1021	1024	1189	1206	Limonene [f]	1.12	1.14	18.77	14.99	11.35	9.77	2.47	2.75	14.33	11.81	tR, MS, RI
13	1040	1044	1254	1250	trans-β-Ocimene	-	-	0.31	-	0.25	0.14	0.14	0.08	1.03	0.41	MS, RI
14	1049	1054	1243	1251	γ-Terpinene	-	-	-	-	0.16	0.06	0.70	0.38	0.37	0.18	tR, MS, RI
15	1074	1086	1267	1287	Terpinolene [f]	-	-	-	-	-	-	0.44	0.20	0.11	0.12	tR, MS, RI
16	1079	1089	1462	-	p-Cymenene	-	-	-	-	-	-	0.85	1.01	0.15	0.22	MS, RI
17	1092	1095	1593	1506	Linalool [f]	0.14	-	0.41	0.57	0.31	0.37	-	-	-	-	tR, MS, RI
18	1106	1101	1449	-	Thujone [f]	-	-	-	-	-	-	0.52	0.55	-	-	tR, MS, RI
19	1115	1122	1518	-	α-Campholenal	-	-	-	-	-	-	1.63	2.38	-	-	MS, RI
20	1124	1135	1607	-	Nopinone	-	-	-	-	-	-	0.38	0.42	-	-	MS, RI
21	1127	1135	1695	-	trans-Pinocarveol	0.73	1.23	0.41	0.92	0.27	0.37	4.44	6.84	0.79	1.23	tR, MS, RI
22	1133	1140	1700	-	trans-Verbenol	-	-	0.10	0.34	0.11	0.18	0.94	0.61	0.11	-	MS, RI
23	1138	-	1706	-	Unknown 1 [m/z 94 (100%), 79 (89.9%), 59 (79.7%), 91 (51.7%)]	-	-	-	-	-	-	2.12	2.10	-	-	MS
24	1148	1160	1597	-	Pinocarvone [f]	0.32	0.29	0.26	0.16	0.16	0.09	0.80	1.03	0.27	0.34	tR, MS, RI
25	1161	-	1769	-	α-Phellandren-8-ol	-	-	-	-	-	-	5.60	3.72	0.15	0.03	MS
26	1168	1174	1643	1628	Terpinene-4-ol [f]	-	-	1.19	1.23	0.47	0.44	1.34	1.29	3.10	3.45	tR, MS, RI
27	1176	1179	1882	1846	p-Cymene-8-ol	-	-	0.11	0.17	-	-	0.83	1.16	0.21	0.32	MS, RI
28	1181	1195	1659	-	Myrtenal [f]	0.59	0.67	0.34	0.41	0.24	0.27	2.14	3.25	0.53	0.62	tR, MS, RI
29	1184	1186	1740	1731	α-Terpineol	0.97	0.65	0.92	0.74	0.42	0.26	4.82	0.61	1.85	1.42	tR, MS, RI
30	1193	1204	1737	1733	Verbenone [f]	-	-	-	-	-	-	2.95	2.65	-	-	tR, MS, RI

Table 1. Cont.

No.	RRI [a]	RI Lit [b]	RRI [c]	RI Lit [d]	Compound Name	Peak Area % [e]										ID
						B. microdonta		B. pauciflosculosa		B. punctulata		B. reticularioides		B. sphenophylla		
						NC	PC	NC	PC	NC	PC	NC	PC	NC	PC	
31	1208	1225	1850	1820	cis-Carveol	0.15	-	0.13	0.34	0.27	0.30	1.19	1.15	0.27	0.38	MS, RI
32	1231	1239	1765	1715	Carvone [f]	0.17	-	0.13	0.22	0.42	0.46	0.43	0.46	0.22	0.31	MS, RI
33	1235	-	1867	-	2-Carene-4-ol	-	-	-	-	-	-	0.32	0.17	-	-	MS
34	1271	1284	1615	1599	Bornyl acetate [f]	-	-	-	-	1.32	1.04	0.60	0.51	0.12	0.09	t_R, MS, RI
35	1356	1374	1515	1493	α-Copaene	0.26	0.28	0.21	0.33	0.67	0.53	-	-	0.26	0.34	MS, RI
36	1370	1389	1621	1591	β-Elemene	0.87	1.10	0.41	0.51	3.35	2.76	-	-	0.23	-	MS, RI
37	1395	1417	1624	1617	β-Caryophyllene	0.97	0.92	1.80	1.55	0.42	0.41	-	-	3.61	3.46	t_R, MS, RI
38	1427	1437	1707	1672	α-Humulene	0.39	0.26	0.19	-	0.17	0.54	-	-	0.31	0.12	t_R, MS, RI
39	1447	1478	1720	1692	γ-Muurolene	0.15	0.28	0.26	0.56	3.63	2.48	-	-	0.27	0.47	MS, RI
40	1451	1484	1737	1712	Germacrene-D	0.64	0.41	2.56	0.68	0.12	0.16	-	-	1.44	0.80	MS, RI
41	1458	1489	1746	1756	β-Selinene	1.35	1.23	0.15	0.11	0.12	0.13	0.18	-	0.24	0.21	MS, RI
42	1460	1496	1725	-	Ledene [f]	-	-	0.29	0.10	3.10	2.85	-	0.07	0.17	0.14	t_R, MS, RI
43	1464	1500	1761	1744	Bicyclogermacrene [f]	0.58	0.48	1.25	0.01	-	-	-	-	0.48	0.05	MS, RI
44	1464	1498	1751	1729	α-Selinene	0.14	-	-	-	0.26	0.27	-	-	-	-	MS, RI
45	1469	1500	1756	1730	α-Muurolene	-	-	0.63	0.73	1.18	0.78	-	-	0.29	0.26	MS, RI
46	1480	1505	1705	1745	β-Bisabolene	1.00	1.17	-	-	1.13	1.35	0.28	1.04	-	-	MS, RI
47	1488	1522	1787	1761	δ-Cadinene	0.78	0.37	2.74	1.96	-	-	0.43	0.18	0.79	1.52	MS, RI
48	1509	1544	1930	1916	α-Calacorene	-	-	-	-	-	-	-	-	0.18	0.21	MS, RI
49	1520	1548	2080	2078	Elemol [f]	-	-	0.12	-	1.02	1.09	-	-	-	-	MS, RI
50	1537	1561	2051	2044	(E)-Nerolidol	3.23	3.22	0.43	0.60	0.13	0.30	-	-	0.17	0.17	MS, RI
51	1540	1567	1942	1931	Palustrol [f]	22.74	24.19	-	-	0.13	-	-	-	-	-	MS, RI
52	1551	1577	2114	2153	Spathulenol [f]	6.84	7.47	9.53	12.18	9.96	11.66	5.52	2.70	13.15	14.92	t_R, MS, RI
53	1555	1582	1987	1966	Caryophyllene oxide [f]	0.59	0.69	2.11	3.44	5.30	6.01	1.37	1.34	5.34	6.78	t_R, MS, RI
54	1559	1590	2128	-	Globulol	4.36	4.90	-	-	-	-	-	-	0.36	0.32	MS, RI
55	1571	1592	2080	2112	Viridiflorol [f]	2.38	2.55	1.81	2.69	-	-	-	-	1.66	2.01	t_R, MS, RI
56	1585	1607	2081	-	Ledal [f]	1.03	0.57	-	-	-	-	-	-	-	-	MS, RI
57	1622	1627	2063	2037	1-epi-Cubenol	-	-	0.38	0.39	-	-	-	-	-	-	MS, RI
58	1625	1630	-	-	γ-Eudesmol	-	-	-	-	1.09	1.65	-	-	-	-	MS, RI
59	1630	-	-	-	Unknown 2 [m/z 119 (100%), 105 (92.9%), 91 (90.2%), 93 (79.5%)]	1.37	1.22	-	-	-	-	-	-	-	-	MS
60	1633	1638	2094	-	epi-α-Cadinol	0.18	0.38	-	-	1.95	2.09	-	-	-	-	MS, RI
61	1644	1644	2151	2150	δ-Cadinol	0.42	0.33	0.77	0.65	0.26	0.64	2.64	1.56	2.27	2.03	MS, RI
62	1647	1640	2164	-	epi-α-Muurolol	0.19	-	0.65	0.75	0.27	0.33	-	-	-	-	MS, RI
63	1659	1649	2196	2248	β-Eudesmol	-	-	-	-	0.64	1.65	-	-	-	-	t_R, MS, RI

Table 1. Cont.

No.	RRI[a]	RI Lit[b]	RRI[c]	RI Lit[d]	Compound Name	B. microdonta NC	PC	B. pauciflosculosa NC	PC	B. punctulata NC	PC	B. reticularioides NC	PC	B. sphenophylla NC	PC	ID
64	1661	1656	2120	-	α-Bisabolol oxide B	-	-	-	-	1.17	0.47	-	-	-	-	MS, RI
65	1662	1652	2200	2224	α-Cadinol	-	-	1.44	2.08	-	-	1.36	0.46	1.49	2.10	MS, RI
66	1665	-	2214	-	**Kongol**	22.22	20.09	-	-	-	-	-	-	-	-	tR, MS
67	1672	1675	2209	2203	Cadalene	-	-	-	-	1.34	1.32	-	-	-	-	MS, RI
68	1697	-	2278	-	Murolan-3,9(11)-diene-10-peroxy	0.54	0.64	-	-	-	-	-	-	-	-	MS
69	1705	-	2292	-	(1R,7S,E)-7-Isopropyl-4,10-dimethylenecyclodec-5-enol	0.83	0.94	0.39	0.28	-	-	-	-	0.42	0.46	MS
70	1708	1685	2190	2022	**α-Bisabolol**	-	-	-	-	23.63	20.72	74.50	66.66	76.66	69.84	tR, MS, RI
					Compounds identified (%)	58.33	52.63	77.36	67.27	77.96	73.33	74.50	66.66	76.66	69.84	
					Monoterpenoids hydrocarbons	6.67	5.26	20.75	16.36	22.03	20.00	27.45	24.56	25.00	23.81	
					Oxygenated monoterpenoids	10.00	7.02	16.98	16.36	15.26	15.00	33.33	29.82	18.33	15.87	
					Sesquiterpenoids hydrocarbons	18.33	17.54	20.75	18.19	18.64	18.33	5.88	5.26	20.00	17.46	
					Oxygenated sesquiterpenoids	23.33	22.81	18.88	16.36	22.03	20.00	7.84	7.02	13.33	12.70	

RRI[a], relative retention indices calculated against n-alkanes on the DB-5MS column; RI lit[b], retention index literature (DB-5 column) [37]; RRI[c], relative retention indices calculated against n-alkanes on the DB-WAX column; RI lit[d], retention index literature (CW20M column) [38], Peak Area%[e]; stereoisomers not identified[f]; NC, non-polar column; PC, polar column; tR, identification based on the retention times (tR) of genuine compounds on the DB-5MS column; MS, identified on the basis of computer matching of the mass spectra with those of the Wiley and NIST libraries and comparison with literature data. The compounds in bold represent the major compounds.

Retta et al. [8] analyzed five species of *Baccharis*, namely *B. gaudichaudiana* DC., *B. microcephala* (Less.) DC., *B. penningtonii* Heering, *B. phyteumoides* (Less.) DC., and *B. spicata* (Lam.) Baill., and reported that they were qualitatively, but not quantitatively, similar. In the present study, although the five species presented similar qualitative patterns, some compounds were found only in one specific species. Qualitative similarities in these species were expected, as they belonged to the same genus. However, they were classified into two different taxonomic groups: *B. punctulata* belonged to subgenus *Molina*, whereas the other four species belonged to the subgenus *Baccharis*.

By comparing the GC chromatograms of the EOs of the five species of *Baccharis*, it is possible to distinguish them by the quality and quantity of their major constituents (Figure 1).

2.2. Antimalarial Activity

In order to explore the antimalarial properties of the five species of *Baccharis*, their EOs were investigated against chloroquine-sensitive (D6) and chloroquine-resistant (W2) strains of *Plasmodium falciparum* (Table 2). The EOs of *B. microdonta* and *B. punctulata* were cytotoxic to Vero cells (selectivity control), and because of this result, they were not indicated to be used in cellular media as an antimalarial. These two EOs differed from other studied species due to the presence of the chemical markers spathulenol (22.74%) and kongol (22.22%) for *B. microdonta* and α-bisabolol for *B. punctulata* (23.63%). Due to their cytotoxic properties, these EOs can be further explored in other cytotoxicity or anticancer studies, as previously reported by Pereira et al. for *B. milleflora* DC. [16]. Otherwise, the EO of *B. pauciflosculosa* showed moderate antimalarial activity against both *P. falciparum* clones (lower than 15 μg/mL), while *B. reticularioides* and *B. sphenophylla* demonstrated discrete antimalarial effects. Significant differences in the quality and quantity of the chemical components of the five EOs can be strongly related to these data. In particular, the variation in the quantities of the main components e.g., monoterpenes β-pinene (18.33%) and limonene (18.77%), might be responsible for the antimalarial effect of the EO of *B. pauciflosculosa*.

Table 2. Activities of essential oils of *Baccharis* species against *Plasmodium falciparum*.

Sample Name	P. falciparum (D6 Clone)		P. falciparum (W2 Clone)		Cytotoxicity (Vero Cells)
	IC_{50} (μg/mL)	SI	IC_{50} (μg/mL)	SI	IC_{50} (μg/mL)
B. microdonta	14.75 ± 3.80	2.4	23.93 ± 4.64	1.5	35.80 ± 7.29
B. pauciflosculosa	10.90 ± 0.98	>4.3	14.20 ± 1.08	>3.3	NC
B. punctulata	17.26 ± 0.83	2.2	19.73 ± 4.11	1.9	37.81 ± 6.36
B. reticularioides	20.32 ± 4.37	>2.3	34.35 ± 4.11	>1.4	NC
B. sphenophylla	27.58 ± 1.64	>1.7	32.53 ± 16.5	>1.5	NC
Chloroquine	0.014	>17	0.117	>2	NC
Artemisinin	0.004	>31.8	0.003	>71.3	NC

NC: No cytotoxicity up to 47.6 μg/mL of essential oils and 0.238 μg/mL for chloroquine and artemisinin; SI: selectivity index (IC_{50} for cytotoxicity/IC_{50} for antimalarial activity) values were measured in triplicate ($n = 3$). They are presented as mean ± SD.

In the genus *Baccharis*, antimalarial studies were carried out for a few species using their plant extracts or isolated compounds. *B. dracunculifolia* DC. is the most important plant source of the Brazilian green propolis, and showed antimalarial activities against *P. falciparum* (D6) using crude hydroalcoholic green propolis extract (13 μg/mL) and hautriwaic acid lactone with IC_{50} values of 0.8 μg/mL (D6 clone) and 2.2 μg/mL (W2 clone) [39]. The extracts of leaves from *B. rufescens* Spreng. and *B. genistelloides* (Lam.) Pers. also showed in vitro antimalarial activity, achieving 100% of inhibition at 100 μg/mL against a *P. falciparum* chloroquine-resistant strain [40]. In spite of the antiplasmodial activity of plant EOs widely reported in the literature [28], the present work represents the first study involving the antimalarial effect using the EOs of the *Baccharis* species.

Additionally, the selectivity index (SI) was calculated to predict how toxic the samples were to normal cells. The calculated selectivity indices showed that *E. pauciflosculosa* had better selectivity to

P. falciparum clones than for Vero cells. This EO was safer than other examined samples, making it a candidate for further development as an antimalarial agent, mainly via the inhalation route.

2.3. Antitrypanosomal Activity

Trypanosoma brucei is a protozoan that causes human African trypanosomiasis (HAT). There are currently only four drugs available for its treatment, namely pentamidine, melarsoprol, suramin, and eflornithine. Considering the lack of phytochemical and pharmacological data available for plants with efficacy against trypanosomes and aiming at proposing alternative treatments for HAT, an initial screening of the EOs of the five species of *Baccharis* were carried out against *T. brucei* (Table 3).

All five species presented remarkable antitrypanosomal activities at concentrations ranging from 0.31–1.69 µg/mL (IC_{50}) and 0.52–2.68 µg/mL (IC_{90}). *B. pauciflosculosa* demonstrated the highest effect, 0.31 µg/mL (IC_{50}) and 0.52 µg/mL (IC_{90}), followed by *B. reticularioides*, which showed 0.96 µg/mL (IC_{50}) and 2.49 µg/mL (IC_{90}), and *B. sphenophylla*, which presented 1.14 µg/mL (IC_{50}) and 2.38 µg/mL (IC_{90}). This is the first report on the antitrypanosomal activities for these species. Good EO activity was also reported for other species, such as *Cymbopogon giganteus* Chiov. (Poaceae) (IC_{50} of 0.25 µg/mL) [41] and *Juniperus oxycedrus* L. (Cupressaceae) (IC_{50} of 0.9 µg/mL) [42].

Costa et al. [42] investigated some isolated monoterpenoids (1,8-cineole, borneol, camphor, carvacrol, citral, eugenol, linalool, thymol, and α-pinene) against *T. brucei*. α-Pinene and citral exhibited the highest activities with IC_{50} values of 2.9 µg/mL and 18.9 µg/mL, respectively. Therefore, the activities observed in this work could be attributed to the presence of α-pinene in the EOs of *B. pauciflosculosa* (10.45%), *B. reticularioides* (24.50%), and *B. sphenophylla* (10.74%). *B. microdonta* and *B. punctulata* demonstrated a lower antitrypanosomal effect than other species, and contained low concentrations of α-pinene (0.72% and 3.55%, respectively).

Table 3. In vitro antitrypanosomal activity of essential oils of *Baccharis* species against *T. brucei*.

Sample Name	IC_{50} (µg/mL) *	IC_{90} (µg/mL) *
B. microdonta	1.688 ± 0.354	2.683 ± 0.123
B. pauciflosculosa	0.306 ± 0.056	0.516 ± 0.043
B. punctulata	1.054 ± 0.211	1.969 ± 0.201
B. reticularioides	0.955 ± 0.121	2.484 ± 0.165
B. sphenophylla	1.143 ± 0.113	2.378 ± 0.201
Pentamidine	0.007 ± 0.001	0.011 ± 0.002
α-Difluoromethylornithine (DFMO)	5.506 ± 0.412	12.052 ± 0.613

* Values were measured in triplicate ($n = 3$). They are presented as mean $\pm$ SD.

2.4. Insecticidal Studies with Bed Bugs

Most of the plant-based insecticides and repellents are derived from plants containing EOs [43]. In addition, receptors responding to DEET (*N,N*-diethyl-3-methylbenzamide) can also respond to volatile terpenes [44]. Therefore, this exploratory study was aimed at investigating the insecticidal potential of EOs of *Baccharis* species against bed bugs because of the increasing demands for information about effective control tactics and their public health risks.

The results of fumigation studies involving bed bugs are illustrated in Figure 2. Out of the five EOs analyzed, only *B. sphenophylla* produced 66.67 ± 3.33% mortality in the insecticide-resistant strain 'Bayonne', while producing 83.33 ± 3.33% mortality in the susceptible strain 'Ft.Dix', 24 h after treatment. All of the other EOs showed less than 15% mortality. In particular, *B. sphenophylla* EO exhibited a wide range of volatile compounds with no specific chemical markers. In that sense, its fumigation effect could be attributed to the synergic effects of terpenes, as previously reported in the literature [45]. Several monoterpenes were isolated and demonstrated fumigation effects on different insects, e.g., α-pinene, β-pinene, 3-carene, limonene, myrcene, α-terpinene, and camphene [43]. Of these compounds described in the literature, only camphene was not present in the EO of *B. sphenophylla*, which reinforces the hypothesis that its effect was based on a synergism of various volatile compounds present in the EO.

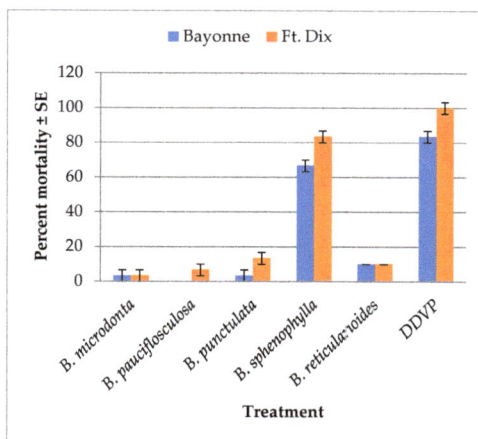

Figure 2. Mean percent mortality (±SE) caused by the essential oils of five species of *Baccharis* against two strains of bed bugs (*Cimex lectularius*) 24 h after treatment in a fumigation bioassay. Essential oil dose: 250 µg/125 mL of air, 2,2-dichlorovinyl dimethyl phosphate (DDVP) used as standard (2 µg/125 mL of air). Mean and standard error were calculated in John´s Macintosh Project (JMP) 10.0.

None of the EOs showed high mortality when applied topically at 50 µg/bug. Only in *B. punctulata* did the mortality reach 20% seven days after the treatment (Figure 3). In residual study, none of the EOs produced mortality in bed bugs seven days of exposure at 100 µg/cm^2.

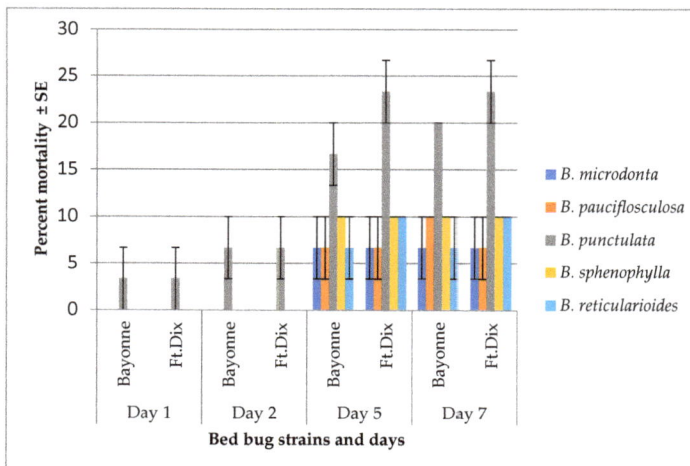

Figure 3. Mean percent mortality (±SE) caused by the essential oils (EOs) of five species of *Baccharis* applied topically on bed bug *Cimex lectularius*. Essential oil dose: 50 µg/bug, deltamethrin (standard) produced 100% (Ft. Dx.) and 56.67% (Bayonne) mortality at 2.4 ng/bug 24 h after treatment. Mean and standard error were calculated in JMP 10.0.

2.5. Secretory Structures

In Asteraceae, EOs are biosynthesized and accumulated in various secretory structures, such as idioblast oil cells, oil cavities, secretory ducts, and glandular trichomes [46]. In *Baccharis*, EOs can be found in roots, stems, leaves and flowers [35,47,48], and are stored in secretory ducts and glandular trichomes [4].

In the present study, the leaves and stems of all of the *Baccharis* species showed glandular trichomes, either isolated or in clusters (Figure 4a–d), and frequently inserted in small epidermal depressions. There were three types of glandular trichomes, namely biseriate (Figure 4a,c,d), flagelliform with straight body (Figure 4a,b,d), and flagelliform C-shaped (Figure 4c). Biseriate glandular trichomes were present in all of the species, except for *B. pauciflosculosa*. The flagelliform trichomes with straight body were found in all of the species except *B. punctulata*, and only this species had flagelliform C-shaped trichomes (Figure 4c).

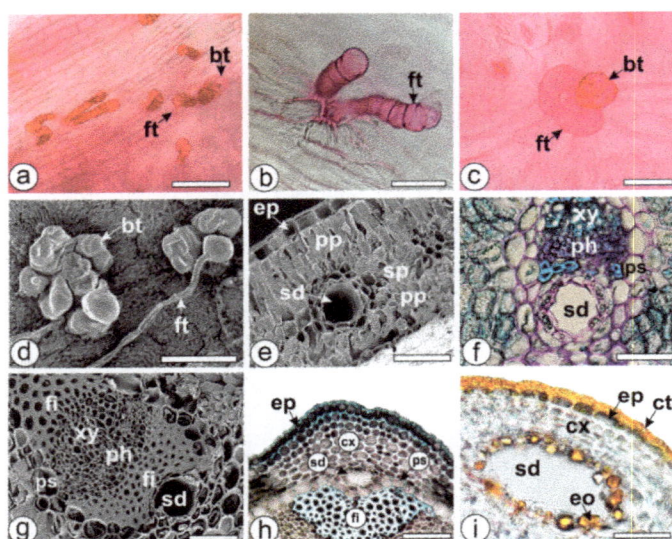

Figure 4. Anatomy of *Baccharis* [light (**a,b,c,f,h,i**) and scanning electron microscopy (**d,e,g**)]. Leaf epidermis in surface view (**a–d**), cross-sections of the leaf (**e–g**) and of the stem (**h,i**). *B. microdonta* (**a**), *B. pauciflosculosa* (**b,e,f**), *B. punctulata* (**c,h**), *B. reticularioides* (**d,g**), *B. sphenophylla* (**i**). [bt–biseriate glandular trichome, ct—cuticle, cx—cortex, eo—essential oil, ep—epidermis, fi—fibers, ft—flagelliform trichome, ph—phloem, pp—palisade parenchyma, sd—secretory ducts, sp—spongy parenchyma, xy—xylem]. Scale bars: **b, c, d, f, g,** *i* = 50 μm; *e* = 100 μm; **a,** *h* = 200 μm.

All the studied *Baccharis* species presented secretory ducts in the mesophyll (Figure 4e,f) and midrib of the leaves (Figure 4g), and in the cortex of the stems (Figure 4h,i). They showed a uniseriate epithelium formed by four to 20 cells with large nuclei and dense cytoplasm containing EO droplets, and were found next to the parenchyma sheath near the phloem (Figure 4e–i). These secretory ducts could also sometimes release other chemical compounds such as resins and tannins beside EOs [49]. Essential oils in the trichomes and ducts, and lipophilic compounds in the cuticle reacted positively with Sudan III in the histochemical tests (Figure 4i).

2.6. Identification of the Samples by DNA

Classical methods for the identification of medicinal plants include organoleptic, macroscopic, and microscopic methods and chemical profiling. Modern techniques, such as DNA barcoding, have emerged recently and are often used in plant identification [50]. Considering the morphological similarities among *Baccharis* species [4], all four genomic regions, namely ITS, ETS, psbA-trnH, and trnL-trnF were subjected to amplification and sequencing in order to provide molecular data for the differentiation of the species. Only two samples (ECT0000641, ECT0000642) resulted in an ETS PCR product and consequently sequence data. Only a single sequence per authenticated species was available for sequence comparison. Table 4 shows KP2 distances between five *Baccharis* samples and authenticated species. The KP2 value determines the genetic distance between samples and the lowest KP2 value of 0.000 indicates a 100% match. The psbA-trnH sequences of samples *B. reticularioides* (ECT0000642) and *B. sphenophylla* (ECT0000647) matched 100% to three different species, indicating that the genomic region has not much variation in its sequence to be helpful to distinguish between species.

Table 4. KP2 distances between five *Baccharis* samples and authenticated species.

Sample	Lowest Kp Distance (ITS)		Lowest Kp Distance (trnL-trnF)		Lowest Kp Distance (ETS)		Lowest Kp Distance (psbA-trnH)				
	Species Match	Value	Species Match	Value	Species Match	Value	Species Match(es)				Value
B. pa.	B. pa	0.000	B. il	0.000	B. pa	0.003	B. pa	B. il	B. re	B. sp	0.004
B. re	B. re	0.000	B. il	0.001	B. re	0.001	B. pa	B. il	B. re		0.000
B. mi	B. mi	0.001	B. mi	0.000	NA		B. mi				0.000
B. pu	B. pu	0.006	B. pu	0.000	NA		B. pu				0.000
B. sp	B. sp	0.000	B. il	0.004	NA		B. pa	B. il	B. re		0.000

A KP2 distance of 0.000 represents a 100% identity match of a sample with a species. Authenticated samples B. mi: *B. microdonta* (GH1599), B. pa: *B. pauciflosculosa* (GH1558), B. il: *B. illirita* (GH1586), B. pu: *B. punctulata* (GH1892), B. re: *B. reticularioides* (GH1426), B. sp: *B. sphenophylla* (GH1438), NA: not analyzed due to low sequence quality.

The samples analyzed were morphologically identified. The ITS sequences of samples *B. pauciflosculosa*, *B. reticularioides*, *B. sphenophylla*, and the trnL-trnF sequences of samples *B. microdonta* and *B. punctulata* support the species identification based on morphology as the sequences 100% matched the sequences from previously authenticated samples.

Only single sequences of authenticated samples were available for the sequence alignments. To achieve a more reliable way of sample identification based on genomic regions, more authenticated samples should be analyzed to get a better representation of the intraspecific sequence variations of a certain species.

3. Materials and Methods

3.1. Plant Material

Fresh samples of vegetative aerial parts were collected, in triplicate, from *B. microdonta*, *B. pauciflosculosa*, *B. punctulata*, *B. reticularioides*, and *B. sphenophylla* in March 2016 from open and sunny habitats in Campos Gerais, Ponta Grossa, Paraná, Southern Brazil (coordinates 25°5′11″ S and 50°6′23″ W). The specimens were registered as ECT0000644 (*B. microdonta*), ECT0000641 (*B. pauciflosculosa*), ECT0000645 (*B. punctulata*), ECT0000642 (*B. reticularioides*), and ECT0000647 (*B. sphenophylla*), and deposited in the Herbarium of Embrapa Clima Temperado (ECT) in Rio Grande do Sul, Brazil. The access to the botanical material was authorized and licensed by the Conselho de Gestão do Patrimônio Genético (CGEN/SISGEN) registered under number A429DA6. The collected plant materials were selected and standardized in order to obtain leaves and stems with the same pattern. Then, the materials were dried in the shade at room temperature and cut into small pieces (~1 cm).

3.2. Extraction of Essential Oil (EO)

Dried plant material (100 g) was subjected to hydrodistillation for 3 h, in triplicate, using a Clevenger-type apparatus for the extraction of EOs. The EOs obtained were dried using anhydrous Na_2SO_4, stored in glass vials with Teflon-sealed caps, and kept under -4 ± 0.5 °C with no light until analysis. The yield of EO was calculated in volume/mass % [51].

3.3. Chemicals

GC-grade *n*-hexane (>99%) was purchased from Sigma Aldrich (St. Louis, MO, USA). The reference standards of α-pinene, β-pinene, camphene, sabinene, α-thujone, β-myrcene, *p*-cymene, limonene, γ-terpinene, α-terpineol, terpinolene, *trans*-pinocarveol, terpinen-4-ol, verbenone, β-elemene, carveol, bornyl acetate, caryophyllene oxide, viridiflorol, α-bisabolol, and myrtenal were also purchased from Sigma-Aldrich.

Kongol (Figure S1) and spathulenol (Figure S2) were isolated from EO of *B. microdonta* and identified by NMR spectroscopy in the present study. Briefly, 180 mg of EO of *B. microdonta* was subjected to a Biotage ZIP KP-SIL 45-g cartridge, and the isolation was performed on a Biotage Isolera™ system (Biotage, Charlotte, NC). Hexanes-ethyl acetate was used for eluting with increasing proportions of ethyl acetate from 0% to 20%. The eluted fractions (12 mL each) were collected and detected by using thin layer chromatography. Kongol (23.8 mg) and spathulenol (25.3 mg) were obtained from the fractions 102–106 and 95–99, respectively. The proton and carbon NMR spectra of the two isolates were recorded using an Agilent DD2-500 NMR spectrometer (Agilent, Santa Clara, CA) equipped with a One NMR probe operating at 499.79 MHz for ^{1}H and 125.67 MHz for ^{13}C. The spectrum of spathulenol was identical to that of the reference standard. The spectral data of isolated kongol was also in agreement with that reported in the literature [52].

3.4. Gas Chromatography-Mass Spectrometry (GC/MS) Analysis

The EOs of *B. microdonta*, *B. pauciflosculosa*, *B. punctulata*, *B. reticularioides*, and *B. sphenophylla* were analyzed by GC/MS using an Agilent 7890A GC system equipped with a 5975C quadrupole mass spectrometer and a 7693 autosampler (Agilent Technologies, Santa Clara, CA, USA). Ten microliters of EOs were dissolved in 1 mL of *n*-hexane for each oil sample, and 1 μL of the sample solution was injected. Helium was used as the carrier gas at a flow rate of 1 mL/min. The inlet temperature was set to 250 °C with a split injection mode for a split ratio of 50:1. Separation was performed on two columns with different polarity, non-polar DB-5MS (column 1) and polar DB-WAX (column 2) capillary columns (Agilent J&W Scientific, Folsom, CA, USA) with the same dimensions of 30 m × 0.25 mm i.d. × 0.25 μm film thickness. The oven temperature program was as follows: (1) Column 1: the initial temperature was 45 °C (held for 2 min); it then increased to 130 °C at a rate of 2 °C/min (held for 10 min), to 150 °C at a rate of 2 °C/min, and finally to 250 °C at a rate of 2 °C/min and isothermal for 10 min at 280 °C with a total experiment time of 70 min; (2) Column 2: the initial temperature was 40 °C (held for 4 min); it then increased to 200 °C at a rate of 3 °C/min, and to 240 °C at a rate of 20 °C/min. Triplicate injections were made for each sample.

Mass spectra were recorded at 70 eV at a scan mode from m/z 35 to 500. The transfer line temperature was 260 °C. The ion source and quadrupole temperatures were 230 °C and 130 °C, respectively. Data acquisition was performed with Agilent MSD Chemstation (F.01.03.2357).

Compound identification involved the comparison of the mass spectra with the databases (Wiley and the National Institute of Standards and Technology (NIST) using a probability-based matching algorithm. Further identification was based on the relative retention indices compared with the literature [38] and the reference standards purchased from commercial sources or isolated in-house.

The raw percentage from the peak area of each compound was obtained in full-scan GC/MS analyses (DB-5MS and DB-WAX columns). Further standardization was not carried out, since our aim was focused on identifying the essential oil compounds for species differentiation.

3.5. Antimalarial Activity

The antimalarial activity of EOs from *Baccharis* species was determined using a colorimetric assay based on plasmodial lactate dehydrogenase (LDH) activity as described by Kumar et al. [53]. A suspension of red blood cells infected with D6 or W2 strains of *P. falciparum* was added to the wells of a 96-well plate containing test samples diluted in medium at several concentrations. Parasitic LDH activity was determined according to the method described by Makler and Hinrichs [54]. Chloroquine and artemisinin were included as the drug controls. IC_{50} values were calculated from the dose-response curves using Excelfit®. DMSO (0.25%) was used as the vehicle control. For calculating the selectivity index of the antimalarial activity of EOs, their toxicity to Vero cells (monkey kidney fibroblasts) was also determined. Essential oils at different concentrations were added, and plates were again incubated for 48 h. The number of viable cells was determined using a vital dye (WST-8). Doxorubicin was used as a positive control.

3.6. Antitrypanosomal Activity

The screening that was employed to test the antitrypanosomal activity of the EOs of *Baccharis* species against *T. brucei* was detailed in a previous paper by Jain et al. [55]. Briefly, the samples were tested against trypomastigotes cultures of *T. brucei*. The cell cultures of *T. brucei* were treated with varying concentrations of the samples, and the growth of the parasite cells were monitored with Alamar blue assay. The results were analyzed with ExcelFit® to determine the IC_{50} and IC_{90} values.

3.7. Insecticidal Studies against Bed Bugs

The bed bug strains (Bayonne 'Insecticide resistant' and Ft. Dix 'Susceptible') were provided by Dr. Changlu Wang, Department of Entomology, Rutgers University, New Brunswick, NJ, and their colony was raised as explained by Montes et al. [56] using blood feeders (CG-1836-75 ChemGlass). The insecticidal activity of EOs against bed bugs was evaluated by fumigation, topical application, and residual studies. For fumigation test, the bed bugs were subjected to vapor toxicity in 125-mL clear glass jars using two microliter aliquots of 125 µg/µL EO stock solution that was injected directly onto inner bottle wall ~4 cm from the bottom. The jars were covered immediately with a screw cap and then sealed with parafilm 'M'. The jars were then placed in the growth chamber, and data for mortality was recorded 24 h after treatment. Solutions were made in acetone, and the control treatment received acetone only. 2,2-Diclorovinil-dimetilfosfato (DDVP) was used as the standard.

Studies in topical application were performed with adult bugs, which were separated in the Petri dishes and anesthetized with CO_2. Using a hand-held repeating dispenser, 1 µL of treatment solution (50 µg/bug) in acetone was delivered onto the dorsal surface of the abdomen. Control bugs received 1 µL of acetone alone. Data for the mortality of the bed bugs was recorded for seven days after treatment. There were three replicates with 10 bugs (mixed sex)/replicate. Deltamethrin was used as the standard (2.4 ng/bug).

For residual studies, the method described by Campbell and Miller [57] was used with minor modifications. A 100-µL aliquot of treatment (diluted in acetone) was applied on 20-cm² Whatman #1 filter paper achieving 100 µg/cm² of residues. The treated filter papers were then placed in the Petri dish. Control treatments received only acetone. Ten adult bugs were released on the filter paper and mortality was recorded as mentioned in topical application. Deltamethrin was used as standard. Data of insecticidal investigations were analyzed for means, standard error, and one-way ANOVA in JMP 10.0.

3.8. Microscopic Procedure

The methods employed for light and scanning electron microscopy analysis of leaves and stems of *Baccharis* species are fully detailed in a previous paper by Budel et al. [4].

3.9. DNA Extraction, PCR, Sequencing

To extract genomic DNA from *Baccharis*, 100 mg of freeze-dried leaves were ground to fine powder. Genomic DNA from *Baccharis* samples was extracted using the DNeasy Plant Mini Kit (Qiagen Inc., Valencia, Spain). Four genomic regions—namely ITS, ETS, *psbA-trnH*, and *trnL-trnF*—were amplified in 25-μL reactions.

The PCR consisted of a 25-μL reaction mixture containing 2 μL of the DNA solution, 1x PCR reaction buffer, 0.2 mM of dNTP mixture, 0.2 μM of each forward and reverse primers (Table 5), 1.5 mM of MgCl$_2$ and 1 U of Platinum Taq DNA Polymerase (Invitrogen, Carlsbad, CA, USA). The program comprised of one initial denaturation step at 94 °C for 3 min, followed by 35 cycles at 94 °C for 30 s, X °C for 30 s, and 72 °C for X s (see Table 5 for annealing temperature and extension time).

Table 5. List of primers, T_M, and extension time.

Genomic Regions	Sequence in 5′-3′	Source	T_M	Extension Time at 72 °C
ETS1f 18S-2L	CTTTTTGTGCATAATGTATATATAGGGGG TGACTACTGGCAGGATCAACCAG	Linder et al. [58]	45 °C	60 s
ITS4 ITS5	TCCTCCGCTTATTGATATGC GGAAGTAAAAGTCGTAACAAGG	White et al. [59]	52 °C	30 s
trnL-F-trnC trnL-F-trnF	CGAAATCGGTAGACGCTACG ATTTGAACTGGTGACACGAG	Taberlet et al. [60]	52 °C	60 s
psbA trnH (GUG)	CGAAGCTCCATCTACAAATGG ACTGCCTTGATCCACTTGGC	Hamilton et al. [61]	56 °C	30 s

T_M = Melting temperature.

After amplification, each PCR reaction was analyzed by electrophoresis on a 1.5% borate agarose gel and visualized under UV light. The sizes of the PCR products were compared to the molecular size standard 1 kb plus DNA ladder (cat no.: 10787-018, Invitrogen, Carlsbad, CA, USA).

Successfully amplified PCR products were isolated with NucleoSpin® Gel and a PCR Clean-up kit (MACHEREY-NAGEL, cat no. 740609.50) and eluted with 30 μL of Buffer AE from the DNeasy Plant Mini Kit (Qiagen Inc., Valencia, Spain). PCR products were sequenced in both directions at GeneWiz (South Plainfield, NJ, USA). Sequences were analyzed with DNASTAR (DNASTAR, Madison, WI, USA) and Clone manager 9 (Scientific and Educational Software, Cary, NC, USA) and visually inspected. Contiguous sequences were screened against previously sequenced authenticated samples.

4. Conclusions

In the present work, profiles of EOs from five species of *Baccharis* were analyzed and compared. The chemical compositions of EOs of *B. pauciflosculosa*, *B. reticularioides*, and *B. sphenophylla* are reported for the first time. Although the qualitative compositions of the EOs of these species were more or less similar, they showed distinctive differences in the quantity of the components. Some compounds were unique to these species, and hence can be used as chemical markers for species identification and authentication. *B. microdonta* differed from the other species by having kongol and spathulenol in high concentrations. *B. pauciflosculosa* showed β-pinene and limonene as major compounds. α-Bisabolol was found only in *B. punctulata*. *B. reticularioides* showed α-pinene, while *B. sphenophylla* presented α-pinene, β-pinene, limonene, and spathulenol as major compounds.

B. microdonta and *B. punctulata* exhibited cytotoxicity, whereas *B. pauciflosculosa*, *B. reticularioides*, and *B. sphenophylla* showed moderate antimalarial activities. Only *B. sphenophylla* EO showed strong toxicity to bed bug *viz.*, and 66.67% and 83.33% mortality in 'Bayonne' and 'Ft.Dix' in the fumigation bioassay. *B. pauciflosculosa* and *B. reticularioides* showed good antitrypanosomal activities.

The leaves and stems of all five *Baccharis* species possessed glandular trichomes and ducts as secretory structures. All three types of glandular trichomes that were observed in this study contained

Molecules **2018**, *23*, 2620

EOs. DNA barcoding using ITS and trnL-trnF sequences were useful for the authentication of the studied *Baccharis* species.

Supplementary Materials: The supplementary materials are available online, Figure S1: Proton and Carbon NMR spectra of kongol, Figure S2: Proton and Carbon NMR spectra of spathulenol.

Author Contributions: Collection of plant materials, I.J.M.T., P.V.F. and J.M.B.; Identification of plant materials, G.H.; Extraction of EOs, L.M.M. and J.M.B.; Microscopy analyses, J.M.B. and V.R.; Chemical profile, J.M.B., M.W. and J.Z.; DNA barcoding, N.T.; Antimalarial activity, S.I.K.; Antitrypanosomal activity, B.T.; Insecticidal activity, J.U.R.; Supervision of the laboratory work, I.A.K.

Funding: This research was funded by CAPES–Coordenação de Aperfeiçoamento de Pessoal de Nível Superior, scholarship number (88881.119611/2016-01) and UEPG–Universidade Estadual de Ponta Grossa. The bed bug research was supported by USDA-Discovery & Development of Natural Products based insect management for medical, veterinary & Urban (58-6066-6-043).

Conflicts of Interest: The authors declare no conflict of interests.

References

1. Heiden, G.; Schneider, A. Baccharis. In *Lista de Espécies da Flora do Brasil*; Jardim Botânico do Rio de Janeiro: Rio de Janeiro, Brazil, 2015.
2. Budel, J.M.; Matzenbacher, N.I.; Duarte, M.R. *Genus Baccharis (Asteraceae): A Review of Chemical and Pharmacological Studies*; Studium Press LLC: Houston, TX, USA, 2008; pp. 1–18.
3. Ramos Campos, F.; Bressan, J.; Godoy Jasinski, V.C.; Zuccolotto, T.; da Silva, L.E.; Bonancio Cerqueira, L. Baccharis (Asteraceae): Chemical Constituents and Biological Activities. *Chem. Biodivers.* **2016**, *13*, 1–17. [CrossRef] [PubMed]
4. Budel, J.M.; Raman, V.; Monteiro, L.M.; Almeida, V.P.; Bobek, V.B.; Heiden, G.; Takeda, I.J.M.; Khan, I.A. Foliar anatomy and microscopy of six Brazilian species of Baccharis (Asteraceae). *Microsc. Res. Tech.* **2018**. [CrossRef] [PubMed]
5. Abad, M.J.; Bermejo, P. Baccharis (Compositae): A review update. *Arkivoc* **2007**, *7*, 76–96.
6. Simões-Pires, C.A.; Debenedetti, S.; Spegazzini, E.; Mentz, L.A.; Matzenbacher, N.I.; Limberger, R.P.; Henriques, A.T. Investigation of the essential oil from eight species of Baccharis belonging to sect. Caulopterae (Asteraceae, Astereae): A taxonomic approach. *Plant Syst. Evol.* **2005**, *253*, 23–32. [CrossRef]
7. Budel, J.M.; Duarte, M.R.; Döll-Boscardin, P.M.; Farago, P.V.; Matzenbacher, N.I.; Sartoratto, A.; Sales Maia, B.H.L.N. Composition of essential oils and secretory structures of Baccharis anomala, B. megapotamica and B. ochracea. *J. Essent. Oil Res.* **2012**, *24*, 19–24. [CrossRef]
8. Retta, D.; Gattuso, M.; Gattuso, S.; Di Leo Lira, P.; van Baren, C.; Bandoni, A. Volatile constituents of five Baccharis Species from Northeastern Argentina. *J. Braz. Chem. Soc.* **2009**, *20*, 1379–1384. [CrossRef]
9. Negreiros, M.O.; Pawlowski, Â.; Zini, C.A.; Soares, G.L.G.; Motta, A.S.; Frazzon, A.P.G. Antimicrobial and antibiofilm activity of Baccharis psiadioides essential oil against antibiotic-resistant Enterococcus faecalis strains. *Pharm. Biol.* **2016**, *54*, 3272–3279. [CrossRef] [PubMed]
10. Perera, W.H.; Bizzo, H.R.; Gama, P.E.; Alviano, C.S.; Salimena, F.R.G.; Alviano, D.S.; Leitão, S.G. Essential oil constituents from high altitude Brazilian species with antimicrobial activity: Baccharis parvidentata Malag., Hyptis monticola Mart. ex Benth. and Lippia origanoides Kunth. *J. Essent. Oil Res.* **2017**, *29*, 109–116. [CrossRef]
11. Florão, A.; Budel, J.M.; Duarte, M.R.; Marcondes, A.; Rodrigues, R.A.F.; Rodrigues, M.V.N.; Santos, C.A.M.; Weffort-Santos, A.M. Essential oils from Baccharis species (Asteraceae) have anti-inflammatory effects for human cells. *J. Essent. Oil Res.* **2012**, *24*, 561–570. [CrossRef]
12. Valarezo, E.; Rosales, J.; Morocho, V.; Cartuche, L.; Guaya, D.; Ojeda-Riascos, S.; Armijos, C.; González, S. Chemical composition and biological activity of the essential oil of Baccharis obtusifolia Kunth from Loja, Ecuador. *J. Essent. Oil Res.* **2015**, *27*, 212–216. [CrossRef]
13. Sobrinho, A.C.N.; de Souza, E.B.; Rocha, M.F.G.; Albuquerque, M.R.J.R.; Bandeira, P.N.; dos Santos, H.S.; de Paula Cavalcante, C.S.; Oliveira, S.S.; Aragão, P.R.; de Morais, S.M.; et al. Chemical composition, antioxidant, antifungal and hemolytic activities of essential oil from Baccharis trinervis (Lam.) Pers. (Asteraceae). *Ind. Crops Prod.* **2016**, *84*, 108–115. [CrossRef]

14. De Oliveira, R.N.; Rehder, V.L.; Santos Oliveira, A.S.; Junior, I.M.; de Carvalho, J.E.; de Ruiz, A.L.; Jeraldo Vde, L.; Linhares, A.X.; Allegretti, S.M. Schistosoma mansoni: In vitro schistosomicidal activity of essential oil of *Baccharis trimera* (less) DC. *Exp. Parasitol.* **2012**, *132*, 135–143. [CrossRef] [PubMed]

15. Botas, G.; Cruz, R.; de Almeida, F.; Duarte, J.; Araújo, R.; Souto, R.; Ferreira, R.; Carvalho, J.; Santos, M.; Rocha, L.; et al. *Baccharis reticularia* DC. and Limonene Nanoemulsions: Promising Larvicidal Agents for *Aedes aegypti* (Diptera: Culicidae) Control. *Molecules* **2017**, *22*, 1990. [CrossRef] [PubMed]

16. Pereira, C.B.; Kanunfre, C.C.; Farago, P.V.; Borsato, D.M.; Budel, J.M.; de Noronha Sales Maia, B.H.L.; Campesatto, E.A.; Sartoratto, A.; Miguel, M.D.; Miguel, O.G. Cytotoxic mechanism of *Baccharis milleflora* (Less). DC. essential oil. *Toxicol. In Vitro* **2017**, *42*, 214–221. [CrossRef] [PubMed]

17. Perez, C.; Anesini, C. Inhibition of *Pseudomonas aeruginosa* by Argentinean medicinal plants. *Fitoterapia* **1994**, *65*, 169–172.

18. Soares, V.C.G.; Bristot, D.; Pires, C.L.; Toyama, M.H.; Romoff, P.; Pena, M.J.; Favero, O.A.; Toyama, D.O. Evaluation of Extracts and Partitions from Aerial Parts of *Baccharis microdonta* on Enzymatic Activity, Pro-Inflammatory and Myotoxic Activities Induced by Secretory Phospholipase A2 from *Bothrops jararacussu*. *Toxicon* **2012**, *60*, 208–208. [CrossRef]

19. Anesini, C.; Perez, C. Screening of plants used in Argentine folk medicine for antimicrobial activity. *J. Ethnopharmacol.* **1993**, *39*, 119–128. [CrossRef]

20. Budel, J.M.; Wang, M.; Raman, V.; Zhao, J.; Khan, S.I.; Rehman, J.U.; Monteiro, L.M.; Heiden, G.; Farago, P.V.; Khan, I.A. Chemical Composition of Essential oils, biological activity and secretory structures of species of Baccharis from Brazil. In Proceedings of the American Society of Pharmacognosy Annual Meeting, Lexington, KY, USA, 21–25 July 2018; pp. 78–79.

21. Malizia, R.A.; Cardell, D.A.; Molli, J.S.; González, S.; Guerra, P.E.; Grau, R.J. Volatile Constituents of Leaf Oils from the Genus *Baccharis*. Part II: *Baccharis obovata* Hooker et Arnott and *B. salicifolia* (Ruiz and Pav.) Pers. Species from Argentina. *J. Essent. Oil Res.* **2005**, *17*, 194–197. [CrossRef]

22. Lago, J.H.G.; Romoff, P.; Fávero, O.A.; Soares, M.G.; Baraldi, P.T.; Corrêa, A.G.; Souza, F.O. Composição química dos óleos essenciais das folhas de seis espécies do gênero *Baccharis* de "Campos de Altitude" da mata atlântica paulista. *Quím. Nov.* **2008**, *31*, 727–730. [CrossRef]

23. Schossler, P.; Schneider, G.L.; Wunsch, D.; Soares, G.L.G.; Zini, C.A. Volatile compounds of *Baccharis punctulata*, *Baccharis dracunculifolia* and *Eupatorium laevigatum* obtained using solid phase microextraction and hydrodistillation. *J. Braz. Chem. Soc.* **2009**, *20*, 277–287. [CrossRef]

24. Gobbo-Neto, L.; Lopes, N.P. Plantas medicinais: Fatores de influência no conteúdo de metabólitos secundários. *Quím. Nov.* **2007**, *30*, 374–381. [CrossRef]

25. Tischer, B.; Vendruscolo, R.G.; Wagner, R.; Menezes, C.R.; Barin, C.S.; Giacomelli, S.R.; Budel, J.M.; Barin, J.S. Effect of grinding method on the analysis of essential oil from *Baccharis articulata* (Lam.) Pers. *Chem. Pap.* **2017**, *71*, 753–761. [CrossRef]

26. Sayuri, V.A.; Romoff, P.; Fávero, O.A.; Ferreira, M.J.P.; Lago, J.H.G.; Buturi, F.O.S. Chemical Composition, Seasonal Variation, and Biosynthetic Considerations of Essential Oils from *Baccharis microdonta* and *B. elaeagnoides* (Asteraceae). *Chem. Biodivers.* **2010**, *7*, 2771–2782. [CrossRef] [PubMed]

27. Saulle, C.C.; Raman, V.; Oliveira, A.V.G.; Maia, B.H.L.N.S.; Meneghetti, E.K.; Flores, T.B.; Farago, P.V.; Khan, I.A.; Budel, J.M. Anatomy and volatile oil chemistry of *Eucalyptus saligna* cultivated in South Brazil. *Rev. Bras. Farmacogn.* **2018**, *28*, 125–134. [CrossRef]

28. Raut, J.S.; Karuppayil, S.M. A status review on the medicinal properties of essential oils. *Ind. Crops Prod.* **2014**, *62*, 250–264. [CrossRef]

29. Guimarães, A.G.; Oliveira, A.P.; Ribeiro, E.A.N.; Claudino, F.S.; Almeida, J.R.G.S.; Lima, J.T.; Bonjardin, L.R.; Ribeiro, L.A.A.; Quintas-Júnior, L.J.; Santos, M.R.V. Atividade farmacológica de monoterpenos. In *Farmacognosia: Coletânea Científica*; Souza, G.H.B., Mello, J.C.P., Lopes, N.P., Eds.; UFOP: Ouro Preto, Brazil, 2012; pp. 219–250.

30. Bogo, C.A.; de Andrade, M.H.; de Paula, J.P.; Farago, P.V.; Döll-Boscardin, P.M.; Budel, J.M. Comparative analysis of essential oils of *Baccharis* L.: A review. *Revis. Strict. Sensu* **2016**, *1*, 1–11. [CrossRef]

31. Xavier, V.B.; Vargas, R.M.F.; Minteguiaga, M.; Umpiérrez, N.; Dellacassa, E.; Cassel, E. Evaluation of the key odorants of *Baccharis anomala* DC. essential oil: New applications for known products. *Ind. Crops Prod.* **2013**, *49*, 492–496. [CrossRef]

32. De Assis Lage, T.C.; Montanari, R.M.; Fernandes, S.A.; de Oliveira Monteiro, C.M.; de Oliveira Souza Senra, T.; Zeringota, V.; da Silva Matos, R.; Daemon, E. Chemical composition and acaricidal activity of the essential oil of *Baccharis dracunculifolia* De Candole (1836) and its constituents nerolidol and limonene on larvae and engorged females of *Rhipicephalus microplus* (Acari: Ixodidae). *Exp. Parasitol.* **2015**, *148*, 24–29. [CrossRef] [PubMed]

33. Minteguiaga, M.; Umpiérrez, N.; Xavier, V.; Lucas, A.; Mondin, C.; Fariña, L.; Cassel, E.; Dellacassa, E. Recent Findings in the Chemistry of Odorants from Four *Baccharis* Species and Their Impact as Chemical Markers. *Chem. Biodivers.* **2015**, *12*, 1339–1348. [CrossRef] [PubMed]

34. Kurdelas, R.R.; López, S.; Lima, B.; Feresin, G.E.; Zygadlo, J.; Zacchino, S.; López, M.L.; Tapia, A.; Freile, M.L. Chemical composition, anti-insect and antimicrobial activity of *Baccharis darwinii* essential oil from Argentina, Patagonia. *Ind. Crops Prod.* **2012**, *40*, 261–267. [CrossRef]

35. Minteguiaga, M.; Mercado, M.I.; Ponessa, G.I.; Catalán, C.A.N.; Dellacassa, E. Morphoanatomy and essential oil analysis of *Baccharis trimera* (Less.) DC. (Asteraceae) from Uruguay. *Ind. Crops Prod.* **2018**, *112*, 488–498. [CrossRef]

36. Minteguiaga, M.; González, A.; Cassel, E.; Umpierrez, N.; Fariña, L.; Dellacassa, E. Volatile Constituents from *Baccharis* spp. L. (*Asteraceae*): Chemical Support for the Conservation of Threatened Species in Uruguay. *Chem. Biodivers.* **2018**, *15*, e1800017. [CrossRef] [PubMed]

37. Adams, R.P. *Identification of Essential Oil Components by Gas Chromatography/Mass Spectroscopy*, 4th ed.; Allured Publishing Corporation: Carol Stream, IL, USA, 2007.

38. Davies, N. Gas chromatographic retention indices of monoterpenes and sesquiterpenes on methyl silicon and Carbowax 20M phases. *J. Chromatogr. A* **1990**, *503*, 1–24. [CrossRef]

39. Da Silva Filho, A.A.; Resende, D.O.; Fukui, M.J.; Santos, F.F.; Pauletti, P.M.; Cunha, W.R.; Silva, M.L.; Gregorio, L.E.; Bastos, J.K.; Nanayakkara, N.P. In vitro antileishmanial, antiplasmodial and cytotoxic activities of phenolics and triterpenoids from *Baccharis dracunculifolia* DC. (Asteraceae). *Fitoterapia* **2009**, *80*, 478–482. [CrossRef] [PubMed]

40. Munoz, V.; Sauvain, M.; Bourdy, G.; Arrazola, S.; Callapa, J.; Ruiz, G.; Choque, J.; Deharo, E. A search for natural bioactive compounds in Bolivia through a multidisciplinary approach. Part III. Evaluation Of the antimalarial activity of plants used by Altenos Indians. *J. Ethnopharmacol.* **2000**, *71*, 123–131. [CrossRef]

41. Kpoviessi, S.; Bero, J.; Agbani, P.; Gbaguidi, F.; Kpadonou-Kpoviessi, B.; Sinsin, B.; Accrombessi, G.; Frederich, M.; Moudachirou, M.; Quetin-Leclercq, J. Chemical composition, cytotoxicity and in vitro antitrypanosomal and antiplasmodial activity of the essential oils of four *Cymbopogon* species from Benin. *J. Ethnopharmacol.* **2014**, *151*, 652–659. [CrossRef] [PubMed]

42. Costa, S.; Cavadas, C.; Cavaleiro, C.; Salgueiro, L.; do Céu Sousa, M. In vitro susceptibility of *Trypanosoma brucei brucei* to selected essential oils and their major components. *Exp. Parasitol.* **2018**, *190*, 34–40. [CrossRef] [PubMed]

43. Viegas Júnior, C. Terpenos com atividade inseticida: Uma alternativa para o controle químico de insetos. *Quím. Nov.* **2003**, *26*, 390–400. [CrossRef]

44. Maia, M.F.; Moore, S.J. Plant-based insect repellents: A review of their efficacy, development and testing. *Malar. J.* **2011**, *10*, S11. [CrossRef] [PubMed]

45. Lima, R.K.; Cardoso, M.G.; Moraes, J.C.; Carvalho, S.M.; Rodrigues, V.G.; Guimarães, L.G.L. Chemical composition and fumigant effect of essentialoil of *Lippia sidoides* Cham. and monoterpenes against *Tenebrio molitor* L. (Coleoptera: Tenebrionidae). *Ciênc. Agrotecnol.* **2011**, *35*, 664–671. [CrossRef]

46. Gottlieb, O.; Salatino, A. Função e evolução de óleos essenciais e de suas estruturas secretoras. *Ciênc. Cul.* **1987**, *39*, 707–716.

47. Budel, J.M.; Paula, J.P.; Santos, V.L.P.; Franco, C.R.C.; Farago, P.V.; Duarte, M.R. Pharmacobotanical study of *Baccharis pentaptera. Rev. Bras. Farmacogn.* **2015**, *25*, 314–319. [CrossRef]

48. Bobek, V.B.; Heiden, G.; Oliveira, C.F.; Almeida, V.P.; Paula, J P.; Farago, P.V.; Nakashima, T.; Budel, J.M. Comparative analytical micrographs of vassouras (*Baccharis*, Asteraceae). *Rev. Bras. Farmacogn.* **2016**, *26*, 665–672. [CrossRef]

49. Espinar, L.A. Las especies de *Baccharis* (Compositae) de Argentina Central. In *Facultad de Ciencias Exactas, Fisicas y Naturales*; Universidad Nacional de Córdoba: Córdoba, Argentina, 1974.

50. Techen, N.; Parveen, I.; Pan, Z.; Khan, I.A. DNA barcoding of medicinal plant material for identification. *Curr. Opin. Biotechnol.* **2014**, *25*, 103–110. [CrossRef] [PubMed]

51. ANVISA. *Farmacopeia Brasileira*; ANVISA (Agência Nacional de Vigilância Sanitária): Brasília, Brasil, 2010.
52. Kesselmans, R.P.W.; Wijnberg, J.B.P.A.; de Groot, A.; van Beek, T.A. Chromatographic and Spectroscopic Data of All Stereoisomers of Eudesm-11-en-4-ol. *J. Essent. Oil Res.* **1992**, *4*, 201–217. [CrossRef]
53. Kumar, N.; Khan, S.I.; Beena; Rajalakshmi, G.; Kumaradhas, P.; Rawat, D.S. Synthesis, antimalarial activity and cytotoxicity of substituted 3,6-diphenyl-[1,2,4,5]tetraoxanes. *Bioorg. Med. Chem.* **2009**, *17*, 5632–5638. [CrossRef] [PubMed]
54. Makler, M.T.; Hinrichs, D.J. Measurement of the Lactate Dehydrogenase Activity of *Plasmodium falciparum* as an Assessment of Parasitemia. *Am. J. Trop. Med. Hyg.* **1993**, *48*, 205–210. [CrossRef] [PubMed]
55. Jain, S.; Jacob, M.; Walker, L.; Tekwani, B. Screening North American plant extracts in vitro against *Trypanosoma brucei* for discovery of new antitrypanosomal drug leads. *BMC Complement. Altern. Med.* **2016**, *16*, 131. [CrossRef] [PubMed]
56. Montes, C.; Cuadrillero, C.; Vilella, D. Maintenance of a laboratory colony of *Cimex lectularius* (Hemiptera: Cimicidae) using an artificial feeding technique. *J. Med. Entomol.* **2002**, *39*, 675–679. [CrossRef] [PubMed]
57. Campbell, B.; Miller, D. Insecticide Resistance in Eggs and First Instars of the Bed Bug, *Cimex lectularius* (Hemiptera: Cimicidae). *Insects* **2015**, *6*, 122–132. [CrossRef] [PubMed]
58. Linder, C.R.; Goertzen, L.R.; Heuvel, B.V.; Francisco-Ortega, J.; Jansen, R.K. The Complete External Transcribed Spacer of 18S26S-rDNA: Amplification and Phylogenetic Utility at Low Taxonomic Levels in Asteraceae and Closely Allied Families. *Mol. Phylogenet Evol.* **2000**, *14*, 285–303. [CrossRef] [PubMed]
59. White, T.J.; Bruns, T.; Lee, S.; Taylor, J. Amplification and direct sequencing of fungal ribosomal RNA genes for phylogenetics. In *PCR Protocols*; Innis, M.A., Gelfand, D.H., Sninsky, J.J., White, T.J., Eds.; Academic Press: San Diego, CA, USA, 1990; pp. 315–322.
60. Taberlet, P.; Gielly, L.; Pautou, G.; Bouvet, J. Universal primers for amplification of three non-coding regions of chloroplast DNA. *Plant. Mol. Biol.* **1991**, *17*, 1105–1109. [CrossRef] [PubMed]
61. Hamilton, M.B. Four primer pairs for the amplification of chloroplast intergenic regions with intraspecific variation. *Mol. Ecol.* **1999**, *8*, 521–523. [PubMed]

Sample Availability: Samples of the compounds are not available from the authors.

molecules

MDPI

Article

Antibacterial Activities of Metabolites from *Vitis rotundifolia* (Muscadine) Roots against Fish Pathogenic Bacteria

Kevin K. Schrader [1],*, Mohamed A. Ibrahim [2,3], Howaida I. Abd-Alla [2], Charles L. Cantrell [1] and David S. Pasco [3,4]

[1] United States Department of Agriculture, Agricultural Research Service, Natural Products Utilization Research Unit, National Center for Natural Products Research, Post Office Box 1848, University, MS 38677, USA; charles.cantrell@ars.usda.gov

[2] Chemistry of Natural Compounds Department, Pharmaceutical and Drug Industries Division, National Research Centre, Dokki, Giza 12622, Egypt; mmibrahi@olemiss.edu (M.A.I.); howaida_nrc@yahoo.com (H.I.A.-A.)

[3] National Center for Natural Products Research, School of Pharmacy, University of Mississippi, University, MS 38677, USA; dpasco@olemiss.edu

[4] Department of BioMolecular Sciences, School of Pharmacy, University of Mississippi, University, MS 38677, USA

* Correspondence: kevin.schrader@ars.usda.gov; Tel.: +1-662-915-1144; Fax: +1-662-915-1035

Received: 18 September 2018; Accepted: 22 October 2018; Published: 25 October 2018

Abstract: Enteric septicemia of catfish, columnaris disease and streptococcosis, caused by *Edwardsiella ictaluri*, *Flavobacterium columnare* and *Streptococcus iniae*, respectively, are the most common bacterial diseases of economic significance to the pond-raised channel catfish *Ictalurus punctatus* industry. Certain management practices are used by catfish farmers to prevent large financial losses from these diseases such as the use of commercial antibiotics. In order to discover environmentally benign alternatives, using a rapid bioassay, we evaluated a crude extract from the roots of muscadine *Vitis rotundifolia* against these fish pathogenic bacteria and determined that the extract was most active against *F. columnare*. Subsequently, several isolated compounds from the root extract were isolated. Among these isolated compounds, (+)-hopeaphenol (**2**) and (+)-vitisin A (**3**) were found to be the most active (bacteriostatic activity only) against *F. columnare*, with 24-h 50% inhibition concentrations of 4.0 ± 0.7 and 7.7 ± 0.6 mg/L, respectively, and minimum inhibitory concentrations of 9.1 ± 0 mg/L for each compound which were approximately 25X less active than the drug control florfenicol. Efficacy testing of **2** and **3** is necessary to further evaluate the potential for these compounds to be used as antibacterial agents for managing columnaris disease.

Keywords: antibacterial; channel catfish; columnaris disease; *Flavobacterium columnare*; stilbenes; muscadine; pyranoanthocyanin

1. Introduction

Two common diseases of channel catfish *Ictalurus punctatus* grown in ponds in the southeastern part of the United States of America (USA) are columnaris disease and enteric septicemia of catfish (ESC) [1,2]. The etiological agent for columnaris disease is the Gram-negative rod-shaped bacterium *Flavobacterium columnare* in the family Flavobacteriaceae [3]. The disease usually results in severe necrosis of gill tissue and skin ulceration from systemic infection. The Gram-negative bacterium *Edwardsiella ictaluri* (Enterobacteriaceae) is the etiological agent for ESC [2]. Gross lesions in channel catfish with ESC can include hemorrhaging at the base of the fins, on the belly, under the jaw and on the backs of infected fish, with small ulcers and/or depigmented lesions. Both diseases have high mortality rates and cost catfish producers millions of U.S. dollars annually [2].

Another common problem in fish species is the bacterial disease Streptococcosis. It can cause heavy economic losses of farmed freshwater fish including hybrid striped bass and tilapias [4]. The Gram-positive bacterium *Streptococcus iniae* is attributed as the cause of streptococcosis, which can result in very high mortality rates in freshwater fish. Catfish producers may manage columnaris disease and ESC by the application of medicated feed containing the antibiotic florfenicol (Aquaflor®; Intervet Inc., Millsboro, DE, USA), live attenuated vaccines [5] and nonantibiotic therapeutants such as 35% Perox-Aid® for external columnaris [2]. The potential treatments for columnaris disease with other inorganic agents such as potassium permanganate and copper sulfate pentahydrate have been cited [6]. The disadvantages of these therapeutants are their broad-spectrum toxicity towards non-target organisms (such as channel catfish) [7].

In the USA, only florfenicol (Aquaflor®) is approved for the treatment of streptococcal septicemia caused by *S. iniae* in freshwater-reared warm water finfish. Vaccinations may also be a good method for protection against this bacterial infection in Nile tilapia [8].

Because of the limitations of available management approaches for controlling the bacterial species responsible for ESC, columnaris disease and streptococcosis and due to public concerns about environmental impacts from the use of antibiotic-containing feed in agriculture, the discovery of environmentally safe, natural antibacterial compounds would benefit aquaculturists. Previous studies indicated that the *Vitis* species (muscadine) contain large amounts of bioactive phenolics, such as stilbenes, anthocyanins and flavonoids, with some of these compounds possessing antibacterial activities [9,10]. As part of our ongoing efforts to identify such active compounds against isolates of *F. columnare*, *E. ictaluri* and *S. iniae*, we evaluated crude extract and natural compounds from the roots of muscadine (*Vitis rotundifolia* Michx., family Vitaceae) using a rapid bioassay.

2. Results and Discussion

Among the three species of fish pathogenic bacteria tested, the crude extract from the roots of *V. rotundifolia* was found to be most active against *F. columnare*, with a 24-h 50% inhibition concentration (IC$_{50}$) of 16.5 ± 6.4 mg/L (clumping of cells can sometimes occur and result in larger variations of results between repeated bioassays) and a minimum inhibitory concentration (MIC) of 10.0 ± 0 mg/L (Table 1). Because the activity was an order of magnitude less active against *S. iniae* compared to *F. columnare* based on the MIC results (100.0 mg/L and 10.0 mg/L, respectively), the bioassay was not repeated for this test bacterial species. The crude extract was not toxic against *E. ictaluri* at the highest test concentration of 100.0 mg/L, therefore, the standard deviation was not calculated. While the relative-to-drug-control-florfenicol (RDCF) values for the 24-h IC$_{50}$ and MIC of the crude extract against *F. columnare* (60.8 mg/L and 100.0 mg/L, respectively) did not indicate strong activity compared to florfenicol, these results are typical for extracts that are compared to isolated pure active compounds because the active compounds are expected to be at lower concentrations in the initial crude extract. Because the crude extract was most active against *F. columnare*, isolated test compounds from the extract were only evaluated against *F. columnare* using the bioassay for the remainder of the study.

Table 1. Results of the bioassay evaluation of the crude extract from the roots of *Vitis rotundifolia* against fish pathogenic bacteria.

Bacteria Species	24-h IC$_{50}$ [a] (mg/L)	MIC [b] (mg/L)	24-h IC$_{50}$ RDCF [c]	MIC RDCF [c]
F. columnare	16.5 (6.4)	10.0 (0)	60.8 (32.9)	100.0 (0)
E. ictaluri	>100.0	>100.0	ND [d]	ND [d]
S. iniae	22.0 (0)	100.0 (0)	220.0 (0)	>1000.0

[a] 24-h IC$_{50}$ = 50% inhibition concentration, [b] MIC = Minimum inhibitory concentration, [c] RDCF = Relative-to-drug-control florfenicol; values below 1.0 indicate higher antibacterial activity compared to florfenicol. Mean 24-h IC$_{50}$ and MIC values ± standard deviation (SD) for florfenicol were 24-h IC$_{50}$ = 0.4 ± 0.1 mg/L and MIC = 0.1 ± 0 mg/L. [d] ND = not determined. Numbers in parentheses are the SD of the mean.

Four compounds were isolated from the root crude extract and identified as (+)-ampelopsin A (**1**), (+)-hopeaphenol (**2**), (+)-vitisin A (**3**) and the (+)-enantiomer of vitisin B (**4**) (Figure 1). Based on 24-h IC$_{50}$ results, compounds **2** and **3** were the most active against *F. columnare*, with 24-h IC$_{50}$ of 4.0 ± 0.7 and 7.7 ± 0.6 mg/L, respectively (Table 2). Based on the 24-h IC$_{50}$ results, compound **2** was found to be slightly more active against the pathogenic bacterium *F. columnare* than **3**. Subsequently, the 24-h IC$_{50}$ RDCF value of 6.8 for **2** also indicated strong activity and the 24-h IC$_{50}$ MTT of 8.9 ± 0.3 mg/L for **2** indicated less viable cells remaining compared to **3** (24-h IC$_{50}$ MTT = 16.3 ± 0 mg/L). Compound **1**, stilbene oligomer (viniferin), was not active against *F. columnare* at the highest test concentration of 47.0 mg/L. Compound **4** was less active than **2** and **3** against *F. columnare* based on 24-h IC$_{50}$ results and the MTT bioassay indicated no reduction in viable cells even at the highest test concentration of 90.7 mg/L.

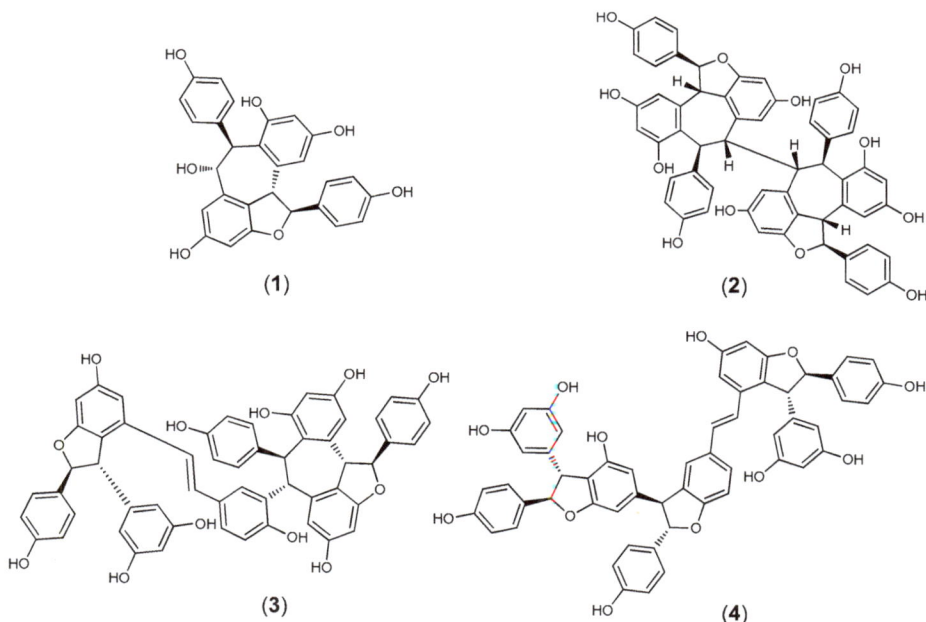

Figure 1. The chemical structures of (+)-ampelopsin (**1**), (+)-hopeaphenol (**2**), (+)-vitisin A (**3**) and (+)-vitisin B (**4**).

Table 2. The bioassay evaluation of compounds isolated from the crude extract of the roots of *V. rotundifolia* against *F. columnare*.

Test Compound	24-h IC$_{50}$ [a]	MIC [b]	24-h IC$_{50}$ RDCF [c]	MIC RDCF [c]	24-h IC$_{50}$ MTT [d]
Florfenicol	0.6 (0)	0.4 (0)			
1	>47.0	>47.0	ND [e]	ND [e]	ND [e]
2	4.0 (0.7)	9.1 (0)	6.8 (0)	25.3 (0)	8.9 (0.3)
3	7.7 (0.6)	9.1 (0)	13.1 (0)	25.3 (0)	16.3 (0)
4	41.3 (5.8)	9.1 (0)	70.0 (0)	25.3 (0)	>90.7

[a] 24-h IC$_{50}$ = 50% inhibition concentration in mg/L, [b] MIC = Minimum inhibitory concentration in mg/L, [c] RDCF = Relative-to-drug-control florfenicol; values below 1.0 indicate higher antibacterial activity compared to florfenicol, [d] MTT (cell viability) portion of the bioassay, [e] ND = not determined. Numbers in parentheses are the standard deviation of the mean.

The minimum bactericidal concentration (MBC) results for each test compound indicated no bactericidal activity against *F. columnare* at even the highest test concentrations (i.e., MBC > 100 µM). Therefore, the activity of compound 2 can be considered as bacteriostatic rather than bactericidal at the concentrations evaluated.

Stilbenes from the Vitaceae are thought to play a role in both animal and human health including their antimicrobial activity. Therefore, these compounds have been the subject of numerous studies during the past decade. They were reported to have activities against various pathogens, such as *Plasmopara viticola*, *Cladosporium cuccumerinum*, *Plasmopara viticola* and *Sphaeropsis sapinea* [11]. The hopeaphenol class of polyphenols are tetramers of resveratrol which is a *trans*-stilbene demonstrated to possess antibacterial activity against certain human pathogenic bacteria [12]. A previous study evaluated the antibacterial activity of (−)-hopeaphenol against ten animal and plant pathogenic bacteria (e.g., Gram-negative *Yersinia pseudotuberculosis* and *Pseudomonas aeruginosa*) but found no significant growth inhibition at test concentrations as high as 90.7 mg/L (100 µM) [13]. However, our current study demonstrated growth inhibition of *F. columnare* by (+)-hopeaphenol (2) at 9.1 mg/L (Table 2). The researchers in the previous study [13] suggested that low cell permeability due to the size and molecular weight of (−)-hopeaphenol and its subsequent interaction with bacterial secretion systems (e.g., toxin delivery system T3SS) at the cell surface rather than growth inhibition as the approach for targeting bacterial virulence. Our results with *F. columnare* indicate growth inhibition can occur at lower concentrations of (+)-hopeaphenol (2).

The pyranoanthocyanin vitisin A (3) has previously been isolated and identified from extracts of the grapevines *Vitis coignetiae* Pulliat. and *Vitis vinifera* L. (Vitaceae) [14] and from extracts of the roots of the grapevine *Vitis thunbergii* Siebold & Zucc. (Vitaceae) [15]. Although the antiplatelet and antioxidative activities of vitisin analogs were reported, specific antibacterial activities were not studied [15].

Efficacy studies of (+)-hopeaphenol (2) and (+)-vitisin A (3) as additives to fish feed and/or as therapeutants still needs to be performed to further evaluate their potential use in managing columnaris disease. Vitisin A (3) has been cited as a strong hepatoxic constituent of *Vitis coignetiae* [14], therefore, careful examination of the potential adverse health effects of vitisin A (3) on fish prior to any potential efficacy studies as an antibacterial compound against columnaris disease would need to be performed.

3. Materials and Methods

3.1. Plant Material

The crude root extract of *Vitis rotundifolia* in 95% EtOH (NPID 127513, 50 mg) was provided through the repository of the National Center for Natural Products Research, School of Pharmacy, University of Mississippi, University, MS, USA. The original specimen was collected in 2007 from a forest near Leon, FL, USA.

3.2. Extraction and Isolation

Samples of root from *V. rotundifolia* were placed in cells of an accelerated extraction system (ASE 300; Thermo Fisher Scientific, Waltham, MA, USA) and extracted with 95% EtOH three times and for 10 min per extraction. Approximately 250 mg of the 95% EtOH root extract was dissolved in 2000 µL of methanol and then exposed to HPLC separation which was conducted using Waters Prep 4000 HPLC system equipped with a UV-Diode detector (2996, Agilent Technologies, Inc., Santa Clara, CA, USA) controlled by Empower software (Rev. A. 10.02, Agilent Technologies, Inc., Santa Clara, CA, USA). The analysis of the extract was carried out on a RP-C18 column (250 × 21.2 mm; particle size 10 µm; Luna) at 25 °C and using the gradient system of eluent water, 0.1% AcOH (A) and acetonitrile, 0.1% AcOH (B) for the separation of target compounds. The gradient condition was as follows: 0–2 min (10% B), 2–45 min (10% B to 60% B) and 45–50 min (60% B to 100% B). The flow rate of the solvent was 10.0 mL/min and the injection volumes were 400 µL. All separations were carried out at wavelengths of 254, 280 and 325 nm with a run time of 50 min. Compounds **1–4** eluted at 28, 35, 38 and 42 min, respectively. NMR spectra were acquired on a Bruker 400 MHz NMR spectrometer (Bruker, Billerica, MA, USA) at 400 (^{1}H) and 100 MHz (^{13}C) in CD_3OD using the residual solvent as an internal standard (Supporting Information). Multiplicity determinations (DEPT) and 2D NMR spectra (HMQC, HMBC, NOESY) were obtained using standard Bruker pulse programs (Bruker III HD, Billerica, MA, USA). Acquisition of high-resolution mass data was acquired using AccuTOF (JMS-T100LC). Comparing the NMR data of the isolated metabolites with the previously reported has confirmed their identities as (+)-ampelopsin A (**1**), (+)-hopeaphenol (**2**) and (+)-vitisin A (**3**) [14,16,17]. The NMR data of compound (**4**) matched with the reported data for vitisin B [18], however it showed positive optical rotation indicating its identity as the (+)-enantiomer of vitisin B (**4**).

(+)-Ampelopsin A (**1**). For ^{1}H and ^{13}C-NMR data, see Supporting Information [17]. High-resolution ESI/MS: *m/z* 493.13312 [M + Na]$^+$; calculated for $C_{28}H_{22}NaO_7$, 493.12185. $[\alpha]^{25}_D$ + 183 (*c* 0.1, MeOH).

(+)-Hopeaphenol (**2**). For ^{1}H and ^{13}C-NMR data, see Supporting Information [16]. High-resolution ESI/MS: *m/z* 929.26922 [M + Na]$^+$; calculated for $C_{56}H_{42}NaO_{12}$, 929.25293. $[\alpha]^{25}_D$ + 201 (*c* 0.1, MeOH).

(+)-Vitisin A (**3**). For ^{1}H and ^{13}C-NMR data, see Supporting Information [14]. High-resolution ESI/MS: *m/z* 930.26681 [M + Na + H]$^{2+}$; calculated for $C_{56}H_{43}NaO_{12}$, 930.26522. $[\alpha]^{25}_D$ + 204 (*c* 0.1, MeOH).

(+)-Vitisin B (**4**). For ^{1}H and ^{13}C-NMR data, see Supporting Information [18]. High-resolution ESI/MS: *m/z* 930.26734 [M + Na + H]$^{2+}$; calculated for $C_{56}H_{43}NaO_{12}$, 930.26522. $[\alpha]^{25}_D$ + 55 (*c* 0.1, MeOH).

3.3. Microorganisms and Culture Material

The bacterial isolate of *F. columnare* [isolate ALM-00-173 (genomovar II)] was obtained from Dr. Covadonga Arias (Department of Fisheries and Allied Aquacultures, Auburn University, Auburn, AL, USA). In order to assure purity, cultures of *F. columnare* ALM-00-173 were maintained separately on modified Shieh (MS) agar plates (pH 7.2–7.4) at 29 ± 1 °C [19]. Prior to conducting the bioassay, individual colonies of *F. columnare* ALM-00-173 were used to prepare assay culture material by culturing in 75 mL of MS broth for at least 24 h at 29 ± 1 °C at 150 rpm on a rotary shaker (model C24KC; New Brunswick Scientific, Edison, NJ, USA). After overnight incubation, a 0.5 McFarland standard of *F. columnare* ALM-00-173 culture material was made by micropipetting cells from the broth culture to fresh MS broth [20].

The isolate of *E. ictaluri* (isolate S02-1039) was obtained from Mr. Tim Santucci (formerly with the College of Veterinary Medicine, Mississippi State University, Stoneville, MS, USA), and cultures of *E. ictaluri* were maintained at 29 ± 1 °C on 3.8% Mueller-Hinton (MH) agar plates (pH 7.3) (Becton, Dickinson and Company, Sparks, MD, USA) in order to assure purity. Prior to performing the bioassay, single colonies of *E. ictaluri* S02-1039 were used to prepare assay culture material by aseptically transferring bacterial cells from colonies on agar plates to 45 mL of 3.8% MH broth in order to produce a bacterial cell density of 0.5 McFarland standard.

A culture of *S. iniae* (isolate LA94-426) was provided by Dr. Ahmed Darwish (formerly with the U.S. Department of Agriculture, Agricultural Research Service, Harry K. Dupree Stuttgart National

Aquaculture Research Center, Stuttgart, AR, USA). In order to assure purity, cultures of *S. iniae* LA94-426 were maintained at 29 ± 1 °C on agar plates of Columbia CNA containing 5% sheep blood (Remel, Inc., Lenexa, KS, USA). The bioassay culture material of *S. iniae* LA94-426 was prepared in the same manner used for *F. columnare* ALM-00-173, except 3.8% MH broth was utilized and broth cultures were incubated for 18 h prior to preparing the 0.5 MacFarland standard.

3.4. Antibacterial Bioassay

The crude extract from the roots of *V. rotundifolia* and isolated test compounds were evaluated for antibacterial activity using a rapid 96-well microplate bioassay [20]. Florfenicol was utilized as a positive drug control and control wells were included in which no test material was added. The crude extract and test compounds were dissolved separately in technical grade 100% methanol while florfenicol was dissolved in technical grade 100% ethanol. The final test concentrations of the crude extract were 0.001, 0.01, 0.1, 1.0, 10.0, and 100.0 mg/L. Final concentrations of test compounds and florfenicol were 0.01, 0.1, 1.0, 10.0 and 100.0 µM. Three replications were used for each dilution of the crude extract, each test compound and florfenicol. Final results were converted to units of mg/L to allow comparisons with previous studies.

The 24-h 50% inhibition concentration (IC_{50}) and minimum inhibitory concentration (MIC) were determined using sterile 96-well polystyrene microplates (Corning Costar Corp., Acton, MA, USA) with flat-bottom wells. Crude extract, dissolved test compounds and florfenicol were initially micropippeted into separate microplate wells (10 µL/well), and the solvent was completely evaporated before 0.5 MacFarland bacterial culture was added to the microplate wells (200 µL/well). Microplates were incubated at 29 ± 1 °C (VWR model 2005 incubator; Sheldon Manufacturing Inc., Cornelius, OR, USA). A Packard model SpectraCount microplate photometer (Packard Instrument Company, Meriden, CT, USA) was used to measure the absorbance (630 nm) of the microplate wells at time 0 and after 24 h of incubation.

After 24 h of incubation, the cell viability of *F. columnare* was determined for the test compounds by using 3(4,5-dimethylthiazol-2-yl)-2,5-diphenyl tetrazolium bromide (MTT) (GenScript, Piscataway, NJ, USA) and previous procedures [20]. For the MTT bioassay, 40 µL of culture material from each growth-assay microplate well were aseptically micropipetted to a corresponding well in another sterile 96-well polystyrene microplate containing 10 µL of MTT (50 mg/10 mL phosphate buffered saline) per well. Each microplate was maintained for 4 h at 29 ± 1 °C and then 50 µL of lysing buffer [20% sodium dodecyl sulfate in 50% *N,N*-dimethylformamide (pH 4.7)] was added to each well. Microplates were then incubated for 20 h after which absorbance (570 nm) was measured (without mixing) using a Packard model SpectraCount microplate photometer. Microplate wells containing 3.8% MH broth, MTT and lysing buffer were used as blanks.

Means and standard deviations of absorbance measurements were calculated, graphed and compared to controls to determine the 24-h IC_{50} and MIC for the crude extract and each test compound [20]. The 24-h IC_{50} and MIC results for the crude extract and each test compound were divided by the respective 24-h IC_{50} and MIC results obtained for the drug control florfenicol to determine the relative-to-drug-control florfenicol (RDCF) values.

The minimum bactericidal concentration (MBC) of isolated compounds was determined as outlined previously [20]. Briefly, 5 µL of culture material were aseptically transferred from each growth bioassay microplate well to 3.8% MH agar plates and these plates were incubated at 29 ± 1 °C for 24 h. Plates were visually evaluated for growth and the MBC was determined to be the lowest concentration in which no growth was present on the agar plate.

4. Conclusions

We isolated four compounds from the root extract of *V. rotundifolia*. Among these compounds, (+)-hopeaphenol and (+)-vitisin A were found to possess the strongest activity against the fish pathogenic bacterium *F. columnare*. Although these two compounds possess bacteriostatic activity only,

Molecules **2018**, 23, 2761

efficacy testing will determine the potential of these two compounds for use in the management of columnaris disease.

Supplementary Materials: Supporting information is available online.

Author Contributions: K.K.S. and M.A.I. conceived the study. K.K.S., M.A.I., H.I.A.-A., C.L.C., and D.S.P. designed the study. K.K.S., M.A.I., and C.L.C. conducted the experiments. K.K.S., M.A.I., H.I.A.-A., C.L.C., and D.S.P. analyzed the data and wrote the manuscript.

Funding: This research received no external funding.

Acknowledgments: The technical assistance of Phaedra Page and Dewayne Harries, USDA-ARS-NPURU, is greatly appreciated. The authors wish to thank Amber Reichley, USDA-ARS-NPURU, for the HR-MS results. The mention of a trademark or proprietary product does not constitute a guarantee or warranty of the product by the U.S. Department of Agriculture, Agricultural Research Service and does not imply its approval to the exclusion of other products that may also be suitable.

Conflicts of Interest: The authors declare no conflict of interest.

References

1. Plumb, J.A.; Hanson, L.A. *Health Maintenance and Principal Microbial Diseases of Cultured Fishes*, 3rd ed.; Wiley-Blackwell: Ames, IA, USA, 2011; ISBN 9780813816937.
2. Wagner, B.A. The epidemiology of bacterial diseases in food-size channel catfish. *J. Aquat. Anim. Health* **2002**, *14*, 263–272. [CrossRef]
3. Durborow, R.M.; Thune, R.L.; Hawke, J.P.; Camus, A.C. *Columnaris Disease: A Bacterial Infection Caused by Flavobacterium columnare*; Southern Regional Aquaculture Center Publication No. 479; U.S. Department of Agriculture: Stoneville, MS, USA, 1998.
4. Shoemaker, C.A.; Klesius, P.H.; Evans, J.J. Prevalence of *Streptococcus iniae* in tilapia, hybrid striped bass, and channel catfish on commercial fish farms in the United States. *Am. J. Vet. Res.* **2001**, *62*, 174–177. [CrossRef] [PubMed]
5. Klesius, P.; Evans, J.; Shoemaker, C. A US perspective on advancements in fish vaccine development. *Aquac. Health Int.* **2006**, *4*, 20–21.
6. Plumb, J.A. *Health Maintenance and Principal Microbial Diseases of Cultured Fishes*; Iowa State University Press: Ames, IA, USA, 1999; ISBN 10: 081382298X.
7. Boyd, C.E.; Tucker, C.S. *Pond Aquaculture Water Quality Management*; Kluwer Academic Publishers: Boston, MA, USA, 1998; ISBN 10: 0412071819.
8. Shoemaker, C.A.; LaFrentz, B.R.; Klesius, P.H.; Evans, J.J. Protection against heterologous *Streptococcus iniae* isolates using a modified bacterin vaccine in Nile tilapia, *Oreochromis niloticus* (L.). *J. Fish. Dis.* **2010**, *33*, 537–544. [CrossRef] [PubMed]
9. Peng, S.C.; Cheng, C.Y.; Sheu, F.; Su, C.H. The antimicrobial activity of heyneanol A extracted from the root of taiwanese wild grape. *J. Appl. Microbiol.* **2008**, *105*, 485–491. [CrossRef] [PubMed]
10. Park, Y.J.; Biswas, R.; Phillips, R.D.; Chen, J. Antibacterial activities of blueberry and muscadine phenolic extracts. *J. Food Sci.* **2011**, *76*, M101–M105. [CrossRef] [PubMed]
11. Jeandet, P.; Douillet-Breuil, A.C.; Bessis, R.; Debord, S.; Sbaghi, M.; Adrian, M. Phytoalexins from the Vitaceae: Biosynthesis, phytoalexin gene expression in transgenic plants, antifungal activity, and metabolism. *J. Agric. Food. Chem.* **2002**, *50*, 2731–2741. [CrossRef] [PubMed]
12. Paulo, L.; Ferreira, S.; Gallardo, E.; Queiroz, J.A.; Domingues, F. Antimicrobial activity and effects of resveratrol on human pathogenic bacteria. *World J. Microbiol. Biotechnol.* **2010**, *26*, 1533–1538. [CrossRef]
13. Zetterström, C.E.; Hasselgren, J.; Salin, O.; Davis, R.A.; Quinn, R.J.; Sundin, C.; Elofsson, M. The resveratrol tetramer (-)-hopeaphenol inhibits type III secretion in the Gram-negative pathogens *Yersinia pseudotuberculosis* and *Pseudomonas aeruginosa*. *PLoS ONE* **2013**, *8*, e81969. [CrossRef] [PubMed]
14. Ito, J.; Gobaru, K.; Shimamura, T.; Niwa, M.; Takaya, Y.; Oshima, Y. Absolute configurations of some oligostilbenes from *Vitis coignetiae* and *Vitis vinifera* 'Kyohou'. *Tetrahedron* **1998**, *54*, 6651–6660. [CrossRef]
15. Huang, Y.-L.; Tsai, W.-J.; Shen, C.-C.; Chen, C.-C. Resveratrol derivatives from the roots of *Vitis thunbergii*. *J. Nat. Prod.* **2005**, *68*, 217–220. [CrossRef] [PubMed]
16. Ito, J.; Niwa, M.; Oshima, Y. A new hydroxystilbene tetramer named isohopeaphenol from *Vitis vinifera* 'Kyohou'. *Heterocycles* **1997**, *45*, 1809–1813. [CrossRef]

17. Chen, L.G.; Wang, C.C. Preparative separation of oligostilbenes from *Vitis thunbergii* var. *taiwaniana* by centrifugal partition chromatography followed by Sephadex LH-20 column chromatography. *Sep. Purif. Technol.* **2009**, *66*, 65–70. [CrossRef]
18. Oshima, Y.; Kamijou, A.; Ohizumi, Y.; Niwa, M.; Ito, J.; Hisamichi, K.; Takeshita, M. Novel oligostilbenes from *Vitis coignetiae*. *Tetrahedron* **1995**, *51*, 11979–11986. [CrossRef]
19. Decostere, A.; Henckaerts, K.; Haesebrouck, F. An alternative model to study the association of rainbow trout (*Oncorhynchus mykiss* L.) pathogens with the gill tissue. *Lab. Anim.* **2002**, *36*, 396–402. [CrossRef] [PubMed]
20. Schrader, K.K.; Harries, M.D. A rapid bioassay for bactericides against the catfish pathogens *Edwardsiella ictaluri* and *Flavobacterium columnare*. *Aquac. Res.* **2006**, *37*, 928–937. [CrossRef]

Sample Availability: Samples of the compounds **1**, **2**, **3**, and **4** are available from the authors.

![molecules logo] *molecules*

MDPI

Article

Anti-Leishmanial and Cytotoxic Activities of a Series of Maleimides: Synthesis, Biological Evaluation and Structure-Activity Relationship

Yongxian Fan [1], Yuele Lu [1], Xiaolong Chen [1,*], Babu Tekwani [2], Xing-Cong Li [2] and Yinchu Shen [1]

[1] Institute of Fermentation Engineering, College of Biotechnology and Bioengineering, Zhejiang University of Technology, 18# Chaowang Road, Hangzhou 310032, China; lilyfan@zjut.edu.cn (Y.F.); luyuele@zjut.edu.cn (Y.L.); syc@zjut.edu.cn (Y.S.)
[2] National Center for Natural Products Research, Research Institute of Pharmaceutical Sciences and Department of BioMolecular Sciences, School of Pharmacy, The University of Mississippi, University, MS 38677, USA; btekwani@olemiss.edu (B.T.); xcli7@olemiss.edu (X.-C.L.)
* Correspondence: richard_chen@zjut.edu.cn; Tel.: +86-571-88320571

Academic Editors: Muhammad Ilias and Charles L. Cantrell
Received: 25 September 2018; Accepted: 31 October 2018; Published: 5 November 2018

Abstract: In the present study, 45 maleimides have been synthesized and evaluated for anti-leishmanial activities against *L. donovani* in vitro and cytotoxicity toward THP1 cells. All compounds exhibited obvious anti-leishmanial activities. Among the tested compounds, there were 10 maleimides with superior anti-leishmanial activities to standard drug amphotericin B, and 32 maleimides with superior anti-leishmanial activities to standard drug pentamidine, especially compounds **16** ($IC_{50} < 0.0128$ µg/mL) and **42** ($IC_{50} < 0.0128$ µg/mL), which showed extraordinary efficacy in an in vitro test and low cytotoxicities ($CC_{50} > 10$ µg/mL). The anti-leishmanial activities of **16** and **42** were 10 times better than that of amphotericin B. The structure and activity relationship (SAR) studies revealed that 3,4-non-substituted maleimides displayed the strongest anti-leishmanial activities compared to those for 3-methyl-maleimides and 3,4-dichloro-maleimides. 3,4-dichloro-maleimides were the least cytotoxic compared to 3-methyl-maleimides and 3,4-non-substituted maleimides. The results show that several of the reported compounds are promising leads for potential anti-leishmanial drug development.

Keywords: anti-leishmanial activity; *Leishmania donovani*; maleimides; cytotoxicity; SAR

1. Introduction

Leishmaniasis, which is caused by several species of *Leishmania*, is one of the major tropical diseases defined by the World Health Organization (WHO), affecting about 12 million people [1,2]. A wide range of clinical manifestations are encompassed, including visceral leishmaniasis, cutaneous leishmaniasis, and mucocutaneous leishmaniasis. Among them, visceral leishmaniasis is the most severe form of the disease. Visceral leishmaniasis, also called black fever or Kala-azar, has a fatality rate as high as 100% within two years if untreated, and spontaneous cure is extremely rare [3,4]. Visceral leishmaniasis is found throughout the intertropical and temperate regions, and threatens around 350 million people in 88 countries. Up to now, few medicines have been available for leishmaniasis. Pentostam is the most widely used drug, which contains the heavy metal antimony. Other medicines, such as amphotericin B and its derivatives [5], liposomal amphotericin B [5], paromomycin [6], and miltefosine [7], have their individual problems, such as toxicity, poor efficacy, or high cost. Meanwhile, the emerging drug-resistant parasites have caused further problems for therapy of the disease. Therefore, the discovery of new types of medicines with novel chemical structures is highly desirable.

Recently, other candidate compounds, including buparvaquone [8], aminoquinolines [9], peptoids [10], (4-arylpiperazin-1-yl)(1-(thiophen-2-yl)-9H-pyrido[3,4-b]indol-3-yl) methanone derivatives [11], amino acid-triazole hybrids [12], biscoumarins [13], triazolyl quinoline derivatives [14], benzopiperidine, benzopyridine and phenyl piperazine based compounds [15], 2,5-diarylidene cyclohexanones [16], natamycin [17], piperazinyl-β-carboline derivatives [18], thiosemicarbazones [19], and so on, were evaluated and investigated. But for most of them, the IC_{50}s of anti-leishmanial activities still remained to be in micromolar ranges except thiosemicarbazones (two compounds had the IC_{50}s of 0.060 µg/mL and 0.068 µg/mL against promastigotes of *L. major*.) [19].

Maleimides, including natural products with maleimide core moiety, had excellent biological activities, including antimicrobial [20–28] and enzyme inhibition activities [29–31]. *N*-(4-Fluorophenyl)-dichloro-maleimide, significantly inhibiting fungal growth, was first developed and used to control the diseases of apple scab, rice blast and tomato late blight [32]. The MICs of *N*-butyl-maleimide and *N*-(4-phenylbutyl)-maleimide against 10 fungi were in the range of 0.48–15.63 µg/mL, similar to that of ampicillin, and had little toxicity to humans [33]. The antimicrobial mechanism was investigated and found that maleimides could interact preferably with the hydrophobic domains of target enzymes resulting in inactivation of sulfhydryl groups [29], which were essential for their catalytic activities. The activities were greatly affected by the structure of C=C double bond in the circle of diimide. For instance, *N*-ethyl-maleimide (NEM) and *N*-*tert*-butyl-maleimide inhibited β-(1,3)-glucan synthase with IC_{50} values 8.5 ± 1.1 µg/mL, 13.7 ± 2.3 µg/mL, respectively, and then influenced microbial growth [34]. In addition to great antimicrobial activities, maleimides had also been widely researched in medicine as antianxiety [35], anti-inflammatory [36], anticancer [37,38] and neuroprotective agents.

However, it should be noted that the reported studies were mainly focused on antimicrobial activity. No anti-leishmanial activity of maleimides has been reported before this. In the present work, a series of *N*-substituted maleimides, methyl-maleimides, and dichloro-maleimides were synthesized, and their anti-leishmanial effects in vitro and cytotoxicity toward THP1 cells were investigated. The results suggest that some of the synthesized maleimides might be developed as the anti-leishmanial drugs in the future.

2. Results and Discussion

2.1. Chemistry

All maleimides were synthesized employing two methods according to an improved procedure based on the reported methods [22,32,33], using amines and maleic anhydrides as starting materials (Scheme 1). Path A was a facile method by one-step reaction to prepare 3,4-dichloromaleimides and 3-methylmaleimides with shorter reaction time, especially for synthesis of the former. However, path B was used to prepare *N*-alkylmaleimides (1–4) by two-step reactions, dehydration and ring-closing reaction. Furthermore, path A had higher yields and an easier isolation method when compared to path B. The desired compound could be easily synthesized in good yield (more than 70%) in one or two steps, as either crystalline or oily compound.

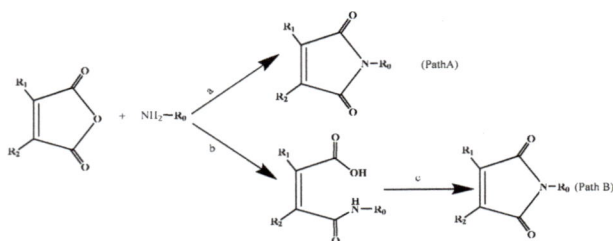

Scheme 1. Synthesis of *N*-substituted maleimide derivatives. Path A: a. CH_3COOH. Path B: b. toluene, 25–65 °C, 2–8 h; c. CH_3COONa, $(Et)_3N$, 101 °C, 10–24 h.

2.2. Biological Evaluations

2.2.1. Anti-leishmanial Activity

45 maleimides [20,29,32,33] (Figure 1) and two drugs, pentamidine (Figure 2) and amphotericin B (Figure 3), were evaluated against *L. donovani*, to identify the most active compounds that are worthy of further investigation (Table 1). Anti-leishmanial activity in vitro was described in terms of IC_{50}, which is the effective concentration ($\mu g/mL$) required to achieve 50% growth inhibition, with promastigotes in their exponential growth phase. Most of tested compounds had good anti-leishmanial activities (IC_{50}s for 41 compounds were less than 1 $\mu g/mL$). Among them, the IC_{50}s of **16** and **42** were less than 0.0128 $\mu g/mL$, much less than those of amphotericin B (IC_{50} = 0.12 $\mu g/mL$) and pentamidine (IC_{50} = 0.64 $\mu g/mL$), which reached nanogram grade. These activities were stronger than most compounds in literature [8–19]. There were another eight candidate compounds (**1, 3, 6, 7, 8, 13, 14,** and **41**) which had better activities than amphotericin B.

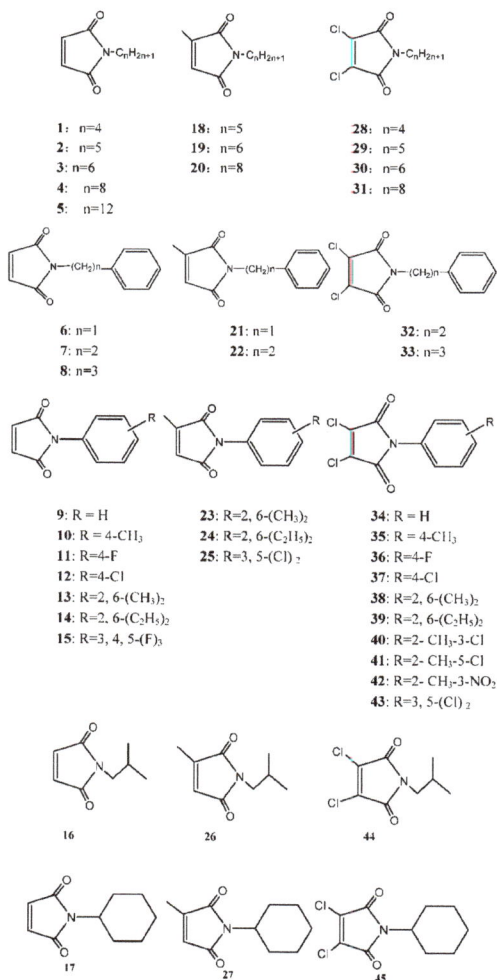

1: n=4
2: n=5
3: n=6
4: n=8
5: n=12

18: n=5
19: n=6
20: n=8

28: n=4
29: n=5
30: n=6
31: n=8

6: n=1
7: n=2
8: n=3

21: n=1
22: n=2

32: n=2
33: n=3

9: R = H
10: R = 4-CH₃
11: R=4-F
12: R=4-Cl
13: R=2, 6-(CH₃)₂
14: R=2, 6-(C₂H₅)₂
15: R=3, 4, 5-(F)₃

23: R=2, 6-(CH₃)₂
24: R=2, 6-(C₂H₅)₂
25: R=3, 5-(Cl)₂

34: R = H
35: R = 4-CH₃
36: R=4-F
37: R=4-Cl
38: R=2, 6-(CH₃)₂
39: R=2, 6-(C₂H₅)₂
40: R=2- CH₃-3-Cl
41: R=2- CH₃-5-Cl
42: R=2- CH₃-3-NO₂
43: R=3, 5-(Cl)₂

16

26

44

17

27

45

Figure 1. Chemical structures of the studied *N*-substituted maleimides.

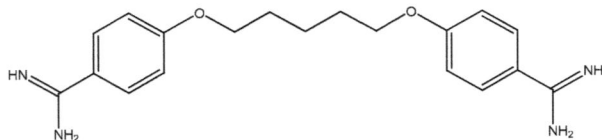

Figure 2. The chemical structure of pentamidine.

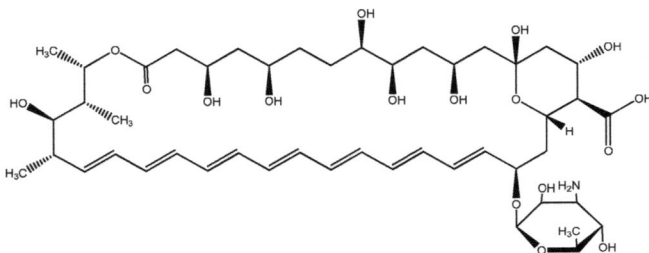

Figure 3. The chemical structure of amphotericin B.

Table 1. Anti-leishmanial activities and cytotoxicity of maleimides **1–45**, pentamidine and amphotericin B.

Compound	*L. donvani* IC$_{50}$ (µg/mL)	THP1 CC$_{50}$ (µg/mL)	SI	LogP
1	0.08	>10	>125.0	0.72
2	0.36	>10	>27.8	1.14
3	0.11	>10	>90.9	1.56
4	0.27	>10	>37.0	2.39
5	2.21	>10	>4.5	4.06
6	0.08	4.56	57.0	1.21
7	0.10	4.13	41.3	1.49
8	0.11	8.25	75.0	1.91
9	0.24	>10	>41.7	1.14
10	0.15	>10	>66.7	1.63
11	0.32	>10	>31.3	1.30
12	1.02	>10	>9.8	1.70
13	0.10	3.65	36.5	2.12
14	0.08	3.89	48.6	2.95
15	0.21	>10	>23.2	1.62
16	<0.0128	>10	>781.3	0.70
17	0.26	>10	>38.5	1.03
18	0.58	7.64	13.2	1.07
19	0.81	5.89	7.3	1.91
20	0.96	8.10	8.4	2.74
21	0.58	7.94	13.7	1.56
22	1.22	5.99	4.9	1.84
23	13.17	>10	>0.8	2.47
24	0.55	4.39	8.0	3.30
25	0.34	0.80	2.4	2.61
26	0.64	>10	>15.6	1.05
27	0.89	7.15	8.0	1.38
28	0.47	>10	>21.3	0.80
29	0.65	>10	>15.4	1.22
30	0.49	>10	>20.4	1.63
31	0.48	>10	>20.8	2.47
32	1.64	>10	>6.1	1.57
33	0.40	>10	>25.0	1.99
34	0.76	>10	>13.2	1.22

Table 1. *Cont.*

Compound	*L. donvani* IC$_{50}$ (µg/mL)	THP1 CC$_{50}$ (µg/mL)	SI	LogP
35	0.66	>10	>15.2	1.71
36	0.36	>10	>27.8	1.38
37	0.59	>10	>16.9	1.78
38	0.45	>10	>22.2	2.19
39	0.13	>10	>76.9	3.03
40	0.13	>10	>76.9	2.27
41	0.11	>10	>90.9	2.27
42	<0.0128	>10	>781.3	1.70
43	0.99	>10	>10.1	2.34
44	0.43	>10	>23.3	0.78
45	0.26	>10	>38.5	1.10
pentamidine	0.64	ND	-	2.84
amphotericin B	0.12	ND	-	2.30

2.2.2. Cytotoxicity

In order to verify the safety of maleimides, they were tested for cytotoxicity against human monocytic leukemia cells (THP1) 50% cytotoxic concentration values represent the concentration of compound required to kill 50% of the THP1 cells were calculated (CC$_{50}$). The selectivity indices were calculated using the formula SI = CC$_{50}$ /IC$_{50}$, against promastigotes. Interestingly, CC$_{50}$s of 31 out of 45 maleimides were greater than 10 µg/mL, which highlighted their safety on mammalian cells (Table 1). SIs of **16** and **42** were greater than 781.3.

2.3. *Structure-Activity Relationships*

2.3.1. Influences of Substituents at the 3- and 4-Positions of the Maleimide Ring

As shown in Table 1, the introduction of substituents at the 3- and 4-positions on the maleimide ring had different influences on the anti-leishmanial activities against *L. donovani*, depending on the type of introduced substituents. In general, 3,4-non-substituted maleimides (**1–17**) displayed very strong anti-leishmanial activities, with IC$_{50}$ values ranging from 0.0128 to 2.21 µg/mL, 0.336 on the average, comparing to 1.774 µg/mL for 3-methyl-maleimides and 0.501 µg/mL for 3,4-dichloro-maleimides. Especially, **1**, **3**, **6**, **7**, **8**, **13**, **14**, and **16** showed more interesting anti-leishmanial activities than the corresponding 3-methyl-maleimides and 3,4-dichloro-maleimides, which were superior or much superior to amphotericin B (IC$_{50}$ = 0.12 µg/mL). In them, 15 compounds had better activities than pentamidine (IC$_{50}$ = 0.64 µg/mL) except **5** (IC$_{50}$ = 2.21 µg/mL) and **12** (IC$_{50}$ = 1.02 µg/mL). Moreover, **16** and **42** displayed the strongest anti-leishmanial activity (IC$_{50}$ < 0.0128 µg/mL). However, there was no apparent regularity for the influences of variation in substituents at the 3- and 4-positions on the maleimide ring. As to cytotoxicity, 3,4-dichloro-maleimides were least cytotoxic, whose CC$_{50}$s were all higher than 10 µg/mL. Furthermore, 3-methyl-maleimides (**18–27**) were the most cytotoxic, whose CC$_{50}$s were all less than 10 µg/mL.

2.3.2. Influences of the *N*-Substituents

Influences of variation in the alkyl chain length. Results in Table 1 show that the *N*-alkyl substituents (**1** to **5**, **18** to **20** and **28** to **31**) had influences on anti-leishmanial activity. As the length of the *N*-alkyl chain increased, the anti-leishmanial activity decreased significantly. However, **1** exhibited the highest anti-leishmanial activity, with an IC$_{50}$ of 0.08 µg/mL, which was one of the compounds with the best activity. It could possibly be explained that with the change in the polarity and *N*-alkyl chain length, their ability to connect enzymes differs. Therefore, with variation in chain length, **1** to **5**, **18** to **20** and **28** to **31** had different influences on anti-leishmanial activities. It was observed that *N*-phenylalkyl substituents (**6** to **8**, **21** to **22** and **32** and **33**) showed a correlation with the anti-leishmanial activity against *L. donovani*. As the alkyl chain length of *N*-phenylalkyl substituents increased, the anti-leishmanial activities decreased gradually (except for **33**), which might be explained by the N-C distance between the two rings playing an important role in the anti-leishmanial activities of these compounds.

Influences of the substituted benzene ring. As for compounds with a mono-substituent on position 4 of phenyl ring, such as **10** to **12**, and **35** to **37**, it was obvious that the different groups had different influences on the inhibition. However, there was no apparent regularity for the effects of variation with a mono-substituent on position 4 of phenyl ring. Furthermore, compounds **13** to **14**, **24** to **25** and **38** to **43** with double substituents on benzene ring displayed much stronger anti-leishmanial activity against *L. donovani*. However, positions of the substituents had apparent influences on the inhibition behavior. For example, *N*-2-methyl-3-nitro-substituted compound (**42**) exhibited the highest anti-leishmanial activity (IC_{50} < 0.0128 µg/mL). And **13**, **14**, **39**, **40**, and **41** also showed very strong anti-leishmanial activity (IC_{50} ranged from 0.08–0.13 µg/mL). In spite of these factors, steric hindrances might affect anti-leishmanial activity. From the results in Table 1, it can be concluded that the logP value (predicted by Chemoffice 2014) is an important parameter influencing the anti-leishmanial activities against *L. donovani*. However, there was no apparent regularity between the IC_{50} values and log P values on the whole.

3. Experimental Section

3.1. Chemistry General Details

Starting materials were obtained from Aldrich and used as received. Melting points (MP) were measured with a WRS-1A melting point apparatus, and were uncorrected. ^{1}H-NMR spectra were recorded on a Bruker AVANCE III 500 spectrometer (Bruker, London, UK) at 500 MHz using tetramethylsilane (TMS) as an internal standard. Electrospray ionization-mass spectra (EIMS) were measured on a mass spectrometer (Thermo Fisher Scientific, LCQ/ADVANTAGE, 81 Wyman Street, Waltham, MA, USA). IR spectrum was recorded in KBr pellets on a Nicolet 6700 infrared spectrophotometer (Thermo Fisher Scientific, 81 Wyman Street, Waltham, MA, USA).

The structures of all the compounds were determined by IR, EI–MS, ^{1}H-NMR. Note that compounds **1–12**, **17**, **21**, **22**, **28–37** had been previously reported [27,33,39–43], which were coincident with the previous report by EI MS and ^{1}H-NMR. Physical and spectroscopic data of the other compounds were shown in the supplementary material.

3.2. Biology

In vitro anti-leishmanial assay. The antileishmanial activity of the compounds was tested in vitro against a culture of *L. donovani* promastigotes [44]. The promastigotes were grown in RPMI 1640 medium supplemented with 10% fetal calf serum (Gibco Chem. Co., 81 Wyman Street, Waltham, MA, USA) at 26 °C. A three day old culture was diluted to 5×10^5 promastigotes/mL. Drug dilutions were prepared directly in cell suspension in 96-well plates. Plates were incubated at 26 °C for 48 h and the growth of *Leishmania* promastigotes was determined by the Alamar blue assay as described earlier [45]. Standard fluorescence was measured on a Fluostar Galaxy plate reader (BMG Lab Technologies, Offenburg, Germany) at an excitation wavelength of 544 nm and an emission wavelength of 590 nm. Pentamidine and amphotericin B were used as the standard anti-leishmanial agents. IC_{50} values were computed from dose-response curves as above. The tested compounds were diluted with six concentrations (from 40–0.0128 µg/mL). Cytotoxicity assay. The in vitro cytotoxicity was determined against human monocytic leukemia cells (THP1) with a simple colorimetric method using the dye Alamar Blue [45]. THP1 suspensions were grown in RPMI-1640 medium supplemented with 10% FBS, 2 mM glutamine, 50 µg/mL gentamicin and 0.0025 mg/L of amphotericin B (Sigma) at 37 °C in a 5% CO_2 atmosphere. Cells were grown to a density between 0.2 and 1×10^6 cells/mL. Culture medium was replaced every 2–3 days with fresh growth medium. DMSO was used as the solvent, and the test compounds was with six concentrations from 10–0.0032 µg/mL.

4. Conclusions

In summary, a series of maleimides have been synthesized and evaluated for inhibitory activities against *L. donovani*. All compounds exhibited obvious anti-leishmanial activities, especially with compounds **16** and **42** showing extraordinary potency in an in vitro test and low cytotoxicity. The anti-leishmanial activities of the two compounds were 10 times better than that of amphotericin B. Therefore, further preclinical studies of **16** and **42** aimed at leishmaniasis are important for the therapy of this neglected disease. The cytotoxicity of these compounds was low, with nearly no toxicity (>10 µg/mL). Thus, compounds **16** and **42** are promising candidates for visceral leishmaniasis. Among the tested compounds, there were 10 maleimides with superior anti-leishmanial activities to amphotericin B, and 32 maleimides with superior anti-leishmanial activities to pentamidine. The SAR study showed that 3,4-non-substituted maleimides displayed the strongest anti-leishmanial activities compared to those 3-methyl-maleimides and 3,4-dichloro-maleimides. When the length of the alkyl side chain (*N*-alkyl and *N*-phenylalkyl) increased, the anti-leishmanial activities decreased significantly. There was no obvious regularity for the influences of variation with a mono-substituent on phenyl ring position 4. And the position of the substituent had an obvious influence on the inhibition behavior. As to cytotoxicity, 3, 4-dichloro-maleimides were least cytotoxic compared to 3-methyl-maleimides and 3,4-non-substituted maleimides.

Supplementary Materials: The following are available online, the structure data of all maleimides and the biological methods in detail.

Author Contributions: Conceptualization, Y.X. and X.C.; Methodology, Y.X. and B.T.; Software, Y.L.; Validation, X.C., Y.S., B.T. and X.-C.L.; Formal Analysis, Y.L.; Investigation, X.C.; Resources, X.C.; Data Curation, Y.X. and Y.F.; Writing-Original Draft Preparation, Y.F., X.C., Y.L. and X.-C.L.; Writing-Review & Editing, Y.F., Y.L. And X.-C.L.; Visualization, Y.L. and Y.X.; Supervision, Y.S. and X.C.; Project Administration, X.C.; Funding Acquisition, X.C.

Funding: This study received financial support from the National Natural Science Foundation of China (Grant No. 21572206 and No. 21172198).

Conflicts of Interest: The authors declare no conflict of interest.

References

1. World Health Organization. Available online: www.who.int/leishmaniasis/burden/en/ (accessed on 1 September 2017).
2. Akopyants, N.S.; Kimblin, N.; Secundino, N.; Patrick, R.; Peters, N.; Lawyer, P.; Dobson, D.E.; Beverley, S.M.; Sacks, D.L. Demonstration of genetic exchange during cyclical development of Leishmania in the sand fly vector. *Science* **2009**, *324*, 265–268. [CrossRef] [PubMed]
3. Chappuis, F.; Sundar, S.; Hailu, A.; Ghalib, H.; Rijal, S.; Peeling, R.W.; Alvar, J.; Boelaert, M. Visceral leishmaniasis: What are the needs for diagnosis, treatment and control. *Nat. Rev. Microbiol.* **2007**, *5*, 873–882. [CrossRef] [PubMed]
4. Sharma, U.; Singh, S. Immunobiology of leishmaniasis. *Indian J. Exp. Biol.* **2009**, *47*, 412–423. [PubMed]
5. Golenser, J.; Domb, B. New formulations and derivatives of amphotericin B for treatment of leishmaniasis. *Mini-Rev. Med. Chem.* **2006**, *6*, 153–162. [CrossRef] [PubMed]
6. Wiwanitkit, V. Interest in paromomycin for the treatment of visceral leishmaniasis (kala-azar). *Ther. Clin. Risk Manag.* **2012**, *8*, 323–328. [CrossRef] [PubMed]
7. Dorlo, T.P.C.; Balasegaram, M.; Beijnen, J.H.; de Vries, P.J. Miltefosine: A review of its pharmacology and therapeutic efficacy in the treatment of leishmaniasis. *J. Antimicrob. Chemother.* **2012**, *67*, 2576–2597. [CrossRef] [PubMed]
8. Garnier, T.; Mantyla, A.; Jarvinen, T.; Lawrence, J.; Brown, M.; Croft, S. In vivo studies on the antileishmanial activity of buparvaquone and its prodrugs. *J. Antimicrob. Chemother.* **2007**, *60*, 802–810. [CrossRef] [PubMed]

9. Kaur, K.; Jain, M.; Khan, S.; Jacob, M.R.; Tekwani, B.L.; Singh, S.; Singh, P.P.; Jain, R. Synthesis, antiprotozoal, antimicrobial, beta-hematin inhibition, cytotoxicity and methemoglobin (MetHb) formation activities of bis(8-aminoquinolines). *Bioorg. Med. Chem.* **2011**, *19*, 197–210. [CrossRef] [PubMed]

10. Bolt, H.L.; Denny, P.W.; Cobb, S.L. An Efficient Method for the Synthesis of Peptoids with Mixed Lysine-type/Arginine-type Monomers and Evaluation of Their Anti-leishmanial Activity. *J. Vis. Exp. Jove* **2016**, *117*, 1–12. [CrossRef] [PubMed]

11. Ashok, P.; Chander, S.; Chow, L.M.C.; Wong, I.L.K.; Singh, R.P.; Jha, P.N.; Sankaranarayanan, M. Synthesis and in-vitro anti-leishmanial activity of (4-arylpiperazin-1-yl)(1-(thiophen-2-yl)-9H-pyrido[3, 4-b]indol-3-yl)methanone derivatives. *Bioorg. Chem.* **2017**, *70*, 100–106. [CrossRef] [PubMed]

12. Masood, M.M.; Hasan, P.; Tabrez, S.; Ahmad, M.B.; Yadava, U.; Daniliuc, C.G.; Sonawane, Y.A.; Azam, A.; Rub, A.; Abid, M. Anti-leishmanial and cytotoxic activities of amino acid-triazole hybrids: Synthesis, biological evaluation, molecular docking and in silico physico-chemical properties. *Bioorg. Med. Chem. Lett.* **2017**, *27*, 1886–1891. [CrossRef] [PubMed]

13. Rahim, F.; Samreen, H.; Ullah, H.; Fakhri, M.I.; Salar, U.; Perveen, S.; Khan, K.M.; Iqbal Choudhary, M. Anti-leishmanial activities of synthetic biscoumarins. *J. Chem. Soc. Pak.* **2017**, *39*, 79–82.

14. Upadhyay, A.; Kushwaha, P.; Gupta, S.; Dodda, R.P.; Ramalingam, K.; Kant, R.; Goyal, N.; Sashidhara, K.V. Synthesis and evaluation of novel triazolyl quinoline derivatives as potential antileishmanial agents. *Eur. J. Med. Chem.* **2018**, *154*, 172–181. [CrossRef] [PubMed]

15. Chander, S.; Ashok, P.; Reguera, R.; Perez-Pertejo, M.; Carbajo-Andres, R.; Balana-Fouce, R.; Sankaranarayanan, M. Synthesis and activity of benzopiperidine, benzopyridine and phenyl piperazine based compounds against Leishmania infantum. *Exp. Parasitol.* **2018**, *189*, 49–60. [CrossRef] [PubMed]

16. Din, Z.U.; Trapp, M.A.; Soman de Medeiros, L.; Lazarin-Bidoia, D.; Garcia, F.P.; Peron, F.; Nakamura, C.V.; Rodriguez, I.C.; Wadood, A.; Rodrigues-Filho, E. Symmetrical and unsymmetrical substituted 2,5-diarylidene cyclohexanones as anti-parasitic compounds. *Eur. J. Med. Chem.* **2018**, *155*, 596–608. [CrossRef] [PubMed]

17. Awasthi, B.P.; Mitra, K. In vitro leishmanicidal effects of the anti-fungal drug natamycin are mediated through disruption of calcium homeostasis and mitochondrial dysfunction. *Apoptosis* **2018**, *23*, 420–435. [CrossRef] [PubMed]

18. Ashok, P.; Chander, S.; Smith, T.K.; Sankaranarayanan, M. Design, synthesis and biological evaluation of piperazinyl-β-carboline derivatives as anti-leishmanial agents. *Eur. J. Med. Chem.* **2018**, *150*, 559–566. [CrossRef] [PubMed]

19. Temraz, M.G.; Elzahhar, P.A.; El-Din, A.; Bekhit, A.; Bekhit, A.A.; Labib, H.F.; Belal, A.S.F. Anti-leishmanial click modifiable thiosemicarbazones: design, synthesis, biological evaluation and in silico studies. *Eur. J. Med. Chem.* **2018**, *151*, 585–600. [CrossRef] [PubMed]

20. Chen, X.L.; Zhang, L.J.; Li, F.G.; Fan, Y.X.; Wang, W.P.; Li, B.J.; Shen, Y.C. Synthesis and antifungal evaluation of a series of maleimides. *Pest Manag. Sci.* **2015**, *71*, 433–440. [CrossRef] [PubMed]

21. Li, W.; Fan, Y.X.; Shen, Z.Z.; Chen, X.L.; Shen, Y.C. Antifungal activity of simple compounds with maleic anhydride or dimethylmaleimide structure against Botrytis cinerea. *J. Pestic. Sci.* **2012**, *37*, 247–251. [CrossRef]

22. Shen, Z.Z.; Fan, Y.X.; Li, F.G.; Chen, X.L.; Shen, Y.C. Synthesis of a series of N-substituted dimethylmaleimides and their antifungal activities against *Sclerotinia sclerotiorum*. *J. Pest Sci.* **2013**, *86*, 353–360. [CrossRef]

23. Jens, R.A.; Irma, K.B.; Beata, A.C.; Anthony, C.W.A.; Edward, H.D.; Stephen, S.B.; Edward, T.C.; Vito, F.D. The synthesis and biological evaluation of two analogues of the C-Riboside showdomycin. *Aust. J. Chem.* **2005**, *58*, 86–93.

24. Thomas, B.; Stephan, A.S. Showdomycin as a versatile chemical tool for the detection of pathogenesis-associated enzymes in bacteria. *J. Am. Chem. Soc.* **2010**, *132*, 6964–6972.

25. Wu, M.D.; Cheng, M.J. Maleimide and maleic anhydride derivatives from the mycelia of *Antrodia cinnamomea* and their nitric oxide inhibitory activities in macrophages. *J. Nat. Prod.* **2008**, *71*, 1258–1261. [CrossRef] [PubMed]

26. Wael, A.Z.; Clarisse, B.F.; Fondja, Y.; Aqabamycins, A.-G. Novel nitro maleimides from a marine *Vibrio* species: I. taxonomy, fermentation, isolation and biological activities. *J. Antibiot.* **2010**, *63*, 297–301.

27. Frederic, Z.; Alain, V. Synthesis and antimicrobial activities of N-substituted imides. *IL Farmaco* **2002**, *57*, 421–426.

28. David, C.; Emmanuelle, S.S. Monohalogenated maleimides as potential agents for the inhibition of *Pseudomonas aeruginosa biofilm*. *Biofouling* **2010**, *26*, 379–385.

29. Silvia, N.L.; Maria, V.C. In vitro antifungal properties, structure-activity relationships and studies on the mode of action of *N*-phenyl, *N*-aryl, *N*-phenylalkyl maleimides and related compounds. *Arzneim-Forsch* **2005**, *55*, 123–132.

30. Slavica, A.; Dib, I.; Nidetzky, B. Selective modification of surface-exposed thiol groups in *Trigonopsis variabilis* D-amino acid oxidase using poly (ethylene glycol) maleimide and its effect on activity and stability of the enzyme. *Biotechnol. Bioeng.* **2007**, *96*, 9–17. [CrossRef] [PubMed]

31. Manas, K.S.; Debjani, D.; Dulal, P. Pyrene excimer fluorescence of yeast alcohol dehydrogenase: A sensitive probe to investigate ligand binding and unfolding pathway of the enzyme. *Photochem. Photobiol.* **2006**, *82*, 480–486.

32. Wu, P.; Hu, Y.Z. Synthesis of novel 1,4-benzoxazine-2,3-dicarboximides from maleic anhydride and substituted aromatic amines. *Synth. Commun.* **2009**, *39*, 70–84. [CrossRef]

33. Sortino, M.; Garibotto, F.; Cecheinel, F.V.; Gupta, M.; Enriz, R.; Zacchino, S. Antifungal, cytotoxic and SAR studies of a series of *N*-alkyl, *N*-aryl and *N*-alkylphenyl-1,4-pyrroledidiones and related compounds. *Bioorg. Med. Chem.* **2011**, *19*, 2823–2834. [CrossRef] [PubMed]

34. Natalia, S.; Joanna, B.M. Chemical reactivity and antimicrobial activity of *N*-substituted maleimides. *J. Enzyme Inhib. Med. Chem.* **2011**, *27*, 117–124.

35. Jerzy, K. Synthesis of new *N*-substituted cyclic imides with potential anxiolytic activity. XXV. derivatives of halogenodibenzo(e.h)bicyclo(2.2.2)otcane-2,3-dicarboximide. *Acta Pol. Pharm.* **2003**, *60*, 1183–1189.

36. Nara, L.M.; Gislaine, F.; Carla, S. *N*-antipyrine-3,4-dichloromaleimide, an effective cyclic imide for the treatment of chronic pain: the role of the glutamatergic system. *Anesth. Analg.* **2010**, *110*, 942–950.

37. Khan, M.I.; Baloch, M.K.; Ashfaq, M. Biological aspects of new organotin (IV) compounds of 3-maleimidopropionic acid. *J. Organomet. Chem.* **2004**, *689*, 3370–3378. [CrossRef]

38. Sosabowski, J.K.; Matzow, T.; Foster, J.M.; Finucane, C.; Ellison, D.; Watson, S.A.; Mather, S.J. Targeting of CCK-2 receptor-expressing tumors using a radiolabelled divalent gastrin peptide. *J. Nucl. Med.* **2009**, *50*, 2082–2089. [CrossRef] [PubMed]

39. Yu, X.Y.; Corten, C.; Gornerc, H.; Wolff, T.; Kuckling, D. Photodimers of *N*-alkyl-3, 4-dimethylmaleimides-product ratios and reaction mechanism. *J. Photochem. Photobiol.* **2008**, *198*, 34–44. [CrossRef]

40. Jiang, L.; Liu, F.; Zhang, D.K.; Wang, H.B. Synthesis and antifungal activity of 1-substitutedphenyl-3-(5-halobenzimidazol-2-yl) acylurea. *J. Pestic. Sci.* **2010**, *35*, 33–35. [CrossRef]

41. Takatori, K.; Hasegawa, T.; Nakano, S. Antifungal activities of *N*-substituted maleimide derivatives. *Microbiol. Immunol.* **1985**, *29*, 1237–1241. [CrossRef] [PubMed]

42. Trujillo-Ferrara, J.; Santillan, R.; Beltran, H.I. ^{1}H and ^{13}C-NMR spectra for a series of arylmaleamic acids, arylmaleimides, arylsuccinamic acids and arylsuccinimides. *Magn. Reson. Chem.* **1999**, *37*, 682–686. [CrossRef]

43. Sortino, M.; Zacchino, S.A. Efficient asymmetric hydrogenation of the C-C double bond of 2-methyl-*N*-phebylalkylmaleimides by *Aspergillus fumigatus*. *Tetrahedron Asym.* **2010**, *21*, 535–539. [CrossRef]

44. Rahman, A.A.; Samoylenko, V.; Jacob, M.R.; Sahu, R.; Jain, S.K.; Khan, S.I.; Tekwani, B.L.; Muhammad, I. Antiparasitic and antimicrobial indolizidines from the leaves of *Prosopis glandulosa* var. glandulosa. *Planta Medica* **2011**, *77*, 1639–1643. [CrossRef] [PubMed]

45. Mikus, J.; Steverding, D. A simple colorimetric method to screen drug cytotoxicity against Leishmania, using the dye Alamar Blue. *Parasitol. Int.* **2000**, *48*, 265–269. [CrossRef]

Sample Availability: Samples of the compounds are not available from the authors.

![molecules logo] *molecules*

MDPI

Article

Synthesis and Anti-Inflammatory Activities of Phloroglucinol-Based Derivatives

Ning Li [1,2], **Shabana I. Khan** [1,3], **Shi Qiu** [1] and **Xing-Cong Li** [1,3,*]

1   Nation Center for Natural Products Research, Research Institute of Pharmaceutical Sciences, School of Pharmacy, The University of Mississippi, University, MS 38677, USA; ahmulining@163.com (N.L.); skhan@olemiss.edu (S.I.K.); sqiu@olemiss.edu (S.Q.)
2   School of Pharmacy, Anhui Medical University, Hefei 230032, China
3   Department of Biomolecular Sciences, School of Pharmacy, The University of Mississippi, University, MS 38677, USA
*   Correspondence: xcli7@olemiss.edu; Tel.: +1-662-915-6742

Received: 7 November 2018; Accepted: 3 December 2018; Published: 7 December 2018

Abstract: The natural product phloroglucinol-based derivatives representing monoacyl-, diacyl-, dimeric acyl-, alkylated monoacyl-, and the nitrogen-containing alkylated monoacylphloro- glucinols were synthesized and evaluated for inhibitory activities against the inflammatory mediators such as inducible nitric oxide synthase (iNOS) and nuclear factor kappaB (NF-κB). The diacylphloroglucinol compound **2** and the alkylated acylphloroglucinol compound **4** inhibited iNOS with IC_{50} values of 19.0 and 19.5 μM, respectively, and NF-κB with IC_{50} values of 34.0 and 37.5 μM, respectively. These compounds may serve as leads for the synthesis of more potent anti-inflammatory compounds for future drug discovery.

Keywords: phloroglucniol; acylphloroglucinol; anti-inflammatory; iNOS; NF-κB

---

1. Introduction

The natural product phloroglucinol (**1**) and its derivatives have found a wide range of applications as pharmaceuticals, cosmetics, textiles, paints, and dyes due to their diverse biological activities [1,2]. More than 700 naturally occurring phloroglucinol derivatives have been reported, of which acylphloroglucinol derivatives comprise the largest group [1,3]. Structurally fascinating acylphloroglucinols possess antidepressant, antimicrobial, antiviral, antitumor, antioxidant, and anti-inflammatory activities [4–7]. The synthetic antispasmodic drug flopropione (3-propionyl-phloroglucinol) is a representative of this chemical class [8,9].

The anti-inflammatory activity of the naturally occurring phloroglucinol derivatives acting on diverse molecular targets has drawn our particular attention. For example, complex phlorotannins from edible brown algae showed strong inhibitory effects on nitric oxide (NO) production in lipopolysaccharide (LPS)-induced RAW 264.7 macrophage cells [10]. Purified oligomeric phlorotannins containing fucodiphlorethol A, tetraphlorethol B, tetrafucol A, tetraisofuhalol, and phlorofucofuroeckol A (Figure 1) also demonstrated the ability to inhibit NO production and suppressing iNOS and cyclooxygenase (COX)-2 [11]. The acylphloroglucinol hyperforin (Figure 1) from St. John's wort suppressed prostaglandin E_2 formation in vitro and in vivo through inhibition of microsomal PGE_2 synthase-1 [12]. 2-Acetyl-4-geranylphloroglucinol (Figure 1) isolated from the Rutaceaous plant *Melicope ptelefolia* is a strong dual inhibitor of 5-lipoxygenase (LOX) and COX-2 [13]. 2,4-Diacetylphloroglucinol (**2**) (Figure 1), a microbial metabolite of *Pseudomonas aeruginosa*, exhibited antimetastatic activity by mediating inhibition of reactive oxygen species (ROS), nuclear factor kappaB (NF-κB), B-cell lymphoma 2 (Bcl-2), matrix metalloproteinase-2 (MMP-2), vascular endothelial growth factor (VEGF) and primary inflammatory mediators such as tumor necrosis factor (TNF)-α, interleukin

(IL)-6, IL-1β and NO [14]. In an effort to identify synthetically available phloroglucinol-based anti-inflammatory compounds, we synthesized several acylphloroglucinols and evaluated their iNOS and NF-κB inhibitory activities.

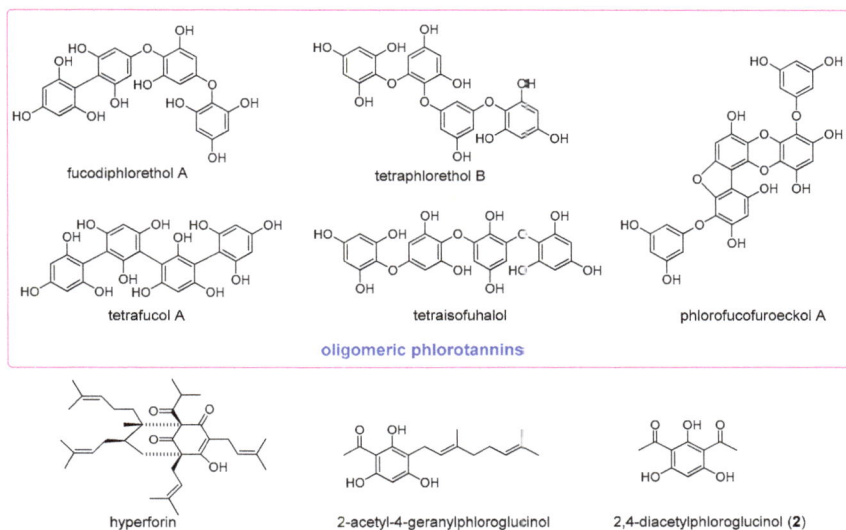

Figure 1. Representative phloroglucinol derivatives with anti-inflammatory activities.

2. Results and Discussion

Based on the concept that both natural and synthetic acylphloroglucinols possess potent biological activities [6,7,13–17], seven compounds representing monoacyl-, diacyl-, dimeric acyl-, alkylated monoacyl-, and the nitrogen-containing alkylated monoacylphloroglucinols were synthesized (Scheme 1). Diacylphloroglucinol (**2**) and monoacylphloroglucinol (**3**) were prepared first according to the procedures described previously [18–20]. Using **3** as a starting material, compounds **4** and **6** were synthesized, respectively, in a one-step reaction adapted from a previously reported procedure [17]. Blocking the two non-hydrogen bonding hydroxyl groups in **4** using MOMCl afforded compound **5**. This synthesis was intended to assess the role of the hydroxyl groups on the phloroglucinol ring for activity. The nitrogen-containing compound **7** was designed to alter chemical nature of the target molecule based on previous reports [21–23]. It was synthesized by reacting **3** with 1-(4-(bromomethyl)benzyl)-piperidine that was prepared from commercially available piperidine and *p*-xylylene dibromide in the presence of CH_2Cl_2 and K_2CO_3. As can be seen from Scheme 1, the reaction conditions for the synthesis of these compounds were mild as all the reactions were conducted at room temperature.

The identification of all the synthetic compounds was achieved by interpretation of NMR and MS data. Compounds **4**–**7** have not been reported prior to this study. It was noted that compound **7** produced an unusual ^{1}H-NMR spectrum in CD_3OD, displaying two sets of ^{1}H-NMR signals in a ratio of approximately 1.5:1 associated with the *para*-substituted phenyl ring and some protons of the piperidine ring (Supplementary Materials). However, when the compound was measured in $CDCl_3$ (with poor solubility), a single set of ^{1}H-NMR signals appeared. Use of the weakly basic NMR solvent C_5D_5N afforded a ^{1}H-NMR spectrum showing two sets of signals in a ratio of approximately 4:1 for the aromatic protons of the phenyl ring and the methylene attached to the piperidine ring. This interesting phenomenon can be explained by the fact that compound **7** is, to some extent, a zwitterionic molecule due to the presence of the acidic phenolic hydroxyl group on the phloroglucinol core and the basic amine functionality on the side chain. While CD_3OD can strongly induce the protonation/deuteriation of the nitrogen atom, C_5D_5N has the ability to facilitate the formulation of a zwitterion, thereby resulting in the presence of an additional set of NMR signals. The ^1H- and ^{13}C-NMR signals of **7** in C_5D_5N were unequivocally assigned using 2D NMR of ^1H-^1H correlation spectroscopy (COSY), heteronuclear single quantum coherence (HSQC) and hetero- nuclear multiple bond correlation (HMBC) experiments.

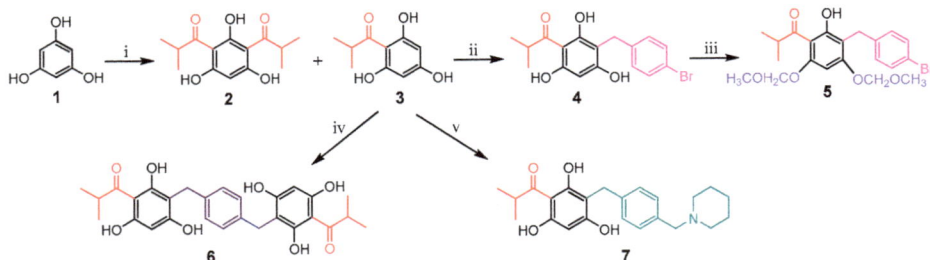

Scheme 1. Synthesis of compounds **2–7**. Reagents and conditions: (i) nitrobenzene, $AlCl_3$, rt for 0.5 h, then isobutyryl chloride, 65 °C for 21 h; (ii) NaOH, H_2O, TBAI, 1-bromo-4-(bromomethyl)benzene, hexane, rt for 48 h; (iii) acetone, K_2CO_3, CH_3OCH_2Cl, rt for 24 h; (iv) NaOH, H_2O, TBAI, *p*-xylylene dibromide, toluene, rt for 6 h; (v) piperidine, CH_2Cl_2, K_2CO_3, *p*-xylylene dibromide, NaOH, H_2O, MeOH, rt for 7 h.

In vitro anti-inflammatory testing of compounds **2–7** was conducted by assessing the inhibition of iNOS in LPS-induced macrophages (RAW 264.7) and inhibition of NF-κB in phorbol 12-myristate 13-acetate (PMA)-induced human chondrosarcoma cells (SW1353). As shown in Table 1, diacylphloroglucinol compound **2** showed the best activity, inhibiting iNOS and NF-κB with IC_{50} values of 19.0 and 34.0 μM, respectively, compared with the positive control parthenolide with IC_{50} values of 2.5 and 20.5 μM, respectively. The activity of alkylated acylphloroglucinol **4** against these two targets (IC_{50} 19.5 and 37.5 μM, respectively) was also very similar to compound **2**. When the hydroxyl groups in **4** were protected leading to compound **5**, iNOS inhibitory activity was decreased (IC_{50}, 50 μM) while NF-κB activity was slightly increased (IC_{50}, 29.0 μM). This suggests these two hydroxyl groups play a role for the activity and the differential activity between the two compounds may be associated with molecular lipophilicity. Dimeric acylphloroglucinol **6** inhibited iNOS to a similar extent (IC_{50}, 19.0 μM) but inhibition of NF-κB was weaker (IC_{50}, 85.0 μM). The nitrogen- containing compound **7** was only moderately inhibiting iNOS with an IC_{50} value of 49.0 μM, indicating the introduction of the amino group in this chemotype is not helpful. On the other hand, monoacylphloroglucinol **3** was only marginally inhibiting NF-κB with an IC_{50} value of 90.0 μM, indicating that additional acylation or alkylation on the phloroglucinol ring is necessary to enhance anti-inflammatory activity. Cytotoxicity testing using African monkey kidney fibroblast cell line (Vero) showed that compounds **2–7** were not cytotoxic up to a concentration of 50 μM

The anti-inflammatory activity observed for compound **2** against iNOS and NF-κB appears to be consistent with the previously reported data for the analogue 2,4-diacetylphloroglucinol [13]. Compound **2** was previously reported to possess potent in vitro antifungal activity against *Cryptococcus neoformans* [17]. An enhanced inflammatory cell response in *C. neoformans* infection has been observed via regulation of IL-23 and IL-12 [24]. Our findings suggest that the anti-inflammatory activity of compound **2** targeting iNOS and NF-κB would have a positive impact on its antifungal activity.

Table 1. Anti-inflammatory activity of synthetic acylphloroglucinols.

Compound	iNOS Inhibition (IC$_{50}$, µM)	NF-κB Inhibition (IC$_{50}$, µM)
2	19.0 ± 1.4	34.0 ± 8.5
3	NA	90.0 ± 14.1
4	19.5 ± 0.7	37.5 ± 3.5
5	50.0 ± 0	29.0 ± 15.5
6	19.0 ± 1.4	85.0 ± 21.2
7	49.0 ± 1.4	NA
Parthenolide	2.5 ± 0.2	20.5 ± 2.9

NA, not active at 50 µM for iNOS and 100 µM for NF-κB.

In conclusion, the natural product phloroglucinol-derived compounds were synthesized and evaluated for anti-inflammatory activity against iNOS and NF-κB as well as cytotoxicity against the mammalian Vero cells. This exploratory study has generated meaningful structure activity relationship information from the limited number of synthetic compounds. The diacylphloroglucinol compound **2** and the alkylated monoacylphloroglucinol compound **4** are dual inhibitors of iNOS and NF-κB. Further synthesis of these two chemotypes of compounds may afford compounds with improved anti-inflammatory activities which could be better candidates for future drug discovery.

3. Materials and Methods

3.1. General Experimental Procedures

The nuclear magnetic resonance (NMR) spectra using standard pulse programs were recorded at room temperature on a BrukerAvance DPX-400 spectrometer (Bruker, Billerica, MA, USA) operating at 400 (^{1}H) and 100 (^{13}C) MHz. The chemical shift (δ, ppm) values were calibrated using the residual NMR solvent and coupling constant (*J*) was reported in Herts (Hz). High-resolution time-of-flight mass spectrometry (TOF-MS) and electrospray ionisation mass spectrometry (ESI-MS) were measured on an Agilent series 1100 SL spectrometer (Agilent Technologies, Santa Clara, CA, USA) equipped with an ESI source. Thin layer chromatography (TLC) was performed on silica gel aluminum sheets (silica gel 60 F254, Merck, Darmstadt, Germany) and visualized by UV 254 nm and spraying 10% H$_2$SO$_4$ followed by heating. Column chromatography was done on normal-phase silica gel (230 × 400 mesh, J. T. Baker, Center Valley, PA, USA) and Sephadex LH-20 (75 µm, GE Healthcare Bio-Sciences AB, Uppsala, Sweden). Reactants and reagents including anhydrous phloroglucinol, nitrobenzene, aluminum chloride (AlCl$_3$), *n*-butyl ammonium iodide (TBAI), 1-bromo-4-bromomethyl-benzene, chloromethoxymethane (CH$_3$OCH$_2$Cl), toluene, and piperidine were purchased from Sigma-Aldrich (St. Louis, MO, USA) in appropriate grades and were used without further purification. The yield of each synthetic product after column chromatography is reported. The ^{1}H- and ^{13}C-NMR assignments for all synthetic products and key intermediates with appropriate numbering systems are shown in the Supplementary Materials. Compounds **2** and **3** were synthesized according to the protocols described previously [18–20] and their identification was made by comparison of their NMR spectroscopic data (see Supplementary Materials) with those reported in the literature [17].

3.2. Synthesis of Compound **4**

To a solution of compound **3** (3.0 g), NaOH (1.01 g), and TBAI (catalyzer, 500 mg) in H$_2$O (225 mL) was added 1-bromo-4-(bromomethyl)benzene (2.5 g) in hexane (20 mL) dropwise. The mixture was stirred at room temperature for 48 h. The precipitated crude product was collected by filtration. The residue was purified by chromatography on silica gel using acetone:hexane (1:2) to afford **4** (1.67 g, yield 30.0%).

1-[3-(4-Bromo-benzyl)-2,4,6-trihydroxy-phenyl]-2-methyl-propan-1-one (**4**).   ¹H-NMR (400 MHz, CD₃COCD₃) δ: 14.33 (s, OH), 7.35 (2H, d, *J* = 8.2 Hz, H-14,14′), 7.24 (2H, d, *J* = 8.2 Hz, H-13, 13′), 6.16 (1H, s, H-4), 3.99 (1H, m, H-8), 3.85 (2H, s, H-11), 1.13 (6H, d, *J* = 6.6 Hz, H-9, 10). ¹³C-NMR (100 MHz, CD₃COCD₃) δ: 211.0 (C-7), 165.7 (C-1), 162.8 (C-5), 160.7 (C-3), 142.4 (C-12), 131.6 (C-14, 14′), 131.5 (C-13, 13′), 119.4 (C-15), 107.3 (C-6), 104.2 (C-2), 95.2 (C-4), 39.5 (C-8), 28.0 (C-11), 19.6 (C-9, 10). MS ESI-(−) *m/z* = 363.0/365.2 [M − H]⁻ for C₁₇H₁₇BrO₄.

3.3. Synthesis of Compound 5

A suspension of compound **4** (346 mg) and K₂CO₃ (2.76 g) in acetone (35 mL) was stirred for 30 min, then CH₃OCH₂Cl (480 mg) was added. After stirring at room temperature for 48 h, the reaction mixture was filtered. The filtrate was concentrated, and the residue was purified by chromatography on silica gel with acetone:hexane (1:12) to give **5** (239 mg, yield 52.8%).

1-[3-(4-Bromo-benzyl)-2-hydroxy-4,6-bis-methoxymethoxy-phenyl]-2-methyl-propan-1-one (**5**). ¹H-NMR (400 MHz, CDCl₃) δ: 14.0 (s, OH), 7.35 (2H, d, *J* = 8.4 Hz, H-14, 14′), 7.21 (2H, d, *J* = 8.4 Hz, H-13, 13′), 6.46 (1H, s, H-4), 5.28 (2H, s, -OCH₂), 5.24 (1H, s, -OCH₂), 3.93 (2H, s, H-11), 3.82 (1H, m, H-8), 3.55 (3H, s, -OCH₃), 3.41 (3H, s, -OCH₃), 1.21 (6H, d, *J* = 6.7 Hz, H-9, 10). ¹³C-NMR (100 MHz, CDCl₃) δ: 210.7 (C-7), 164.0 (C-5), 160.5 (C-1), 158.9 (C-3), 140.6 (C-12), 131.0 (C-14, 14′), 130.4 (C-13, 13′), 119.2 (C-15), 110.7 (C-2), 106.0 (C-4), 94.9 (-OCH₂), 94.0 (-OCH₂), 91.4 (C-6), 56.8 (-OCH₃), 56.5 (-OCH₃), 39.8 (C-8), 27.7 (C-11), 19.4 (C-9, 10). MS ESI-(−) *m/z* = 451.0/453.2 [M − H]⁻ for C₂₁H₂₅BrO₆.

3.4. Synthesis of Compound 6

A suspension of compound **3** (98 mg) and NaOH (22 mg) in H₂O (7.5 mL) was stirred about 30–40 min until it became clear. To this solution was added TBAI (9.8 mg) and *p*-xylylene dibromide (144 mg) in toluene (4.5 mL) dropwise. After stirring at room temperature for 6 h, the reaction mixture was extracted with ethyl acetate (EtOAc) (3 × 20 mL). The combined EtOAc layer was concentrated under reduced pressure to afford a residue, which was purified by chromatography on silica gel using acetone:hexane (1:2) to give **6** (44 mg, yield 17.8%).

2-Methyl-1-{2,4,6-trihydroxy-3-[4-(2,4,6-trihydroxy-3-isobutyryl-benzyl)-benzyl]-phenyl}-propan-1-one (**6**). ¹H-NMR (400 MHz, CD₃COCD₃) δ: 14.19 (s, OH), 7.17 (4H, s, H-13, 13′, 14, 14′), 6.10 (2H, s, H-4, 4′), 4.02 (2H, m, H-8, 8′), 3.85 (4H, s, H-11, 11′), 1.14 (12H, d, *J* = 6.8 Hz, H-9, 9′, 10, 10′). ¹³C-NMR (100 MHz, CD₃COCD₃) δ: 210.1 (C-7, 7′), 164.8 (C-1, 1′), 162.0 (C-5, 5′), 159.7 (C-3, 3′), 138.9 (C-12, 12′), 128.1 (C-13, 13′, 14, 14′), 107.3 (C-6, 6′), 103.4 (C-2, 2′), 94.3 (C-4, 4′), 38.7 (C-8, 8′), 28.5 (C-11, 11′), 18.8 (C-9, 9′, 10, 10′). MS ESI-(−) *m/z* = 493.2 [M − H]⁻ for C₂₈H₃₀O₈.

3.5. Synthesis of Compound 7

To a solution of piperidine (2.55 g) in CH₂Cl₂ (100 mL) was added K₂CO₃ (5.46 g). After stirring at room temperature for 1 h, *p*-xylenedibromide (7.86 g) in CH₂Cl₂ (100 mL) was added dropwise, and the reaction continued for 3 h. The reaction mixture was filtered. The filtrate was concentrated under reduced pressure, and the residue was purified by chromatography on silica gel using acetone:hexane (1:7) to yield the intermediate 1-(4-(bromomethyl) benzyl)piperidine (2.0 g).

To a solution of compound **3** (196 mg) and NaOH (44 mg) in H₂O (60 mL) was added the intermediate (707 mg) in MeOH (40 mL) dropwise. After stirring at room temperature for 4 h, the reaction mixture was extracted with EtOAc (3 × 80 mL) and concentrated under reduced pressure. The residue was purified by chromatography on silica gel using CHCl₃:MeOH:acetone (10:1:1) to yield **7** (35 mg, yield 9.2%).

2-Methyl-1-[2,4,6-trihydroxy-3-(4-piperidin-1-ylmethyl-benzyl)-phenyl]-propan-1-one (**7**). ¹H-NMR (400 MHz, C₅D₅N) δ: 7.74 (2H, d, *J* = 6.0 Hz, H-13, 13′), 7.68 (2H, d, *J* = 6.0 Hz, H-14, 14′), 6.61 (1H, s, H-4), 4.37 (2H, s, H-11), 4.35 (1H, m, H-8), 4.23 (2H, s, H-16), 2.97 (4H, m, H-17, 17′), 1.77 (4H, m, H-18, 18′),

1.29 (6H, d, J = 6.7 Hz, H-9, 10), 1.28 (2H, m, H-19). ^{13}C-NMR (100 MHz, C$_5$D$_5$N) δ: 211.2 (C-7), 166.4 (C-1), 164.6 (C-3), 162.5 (C-5), 145.2 (C-12), 132.16 (C-14, 14′), 130.3 (C-13, 13′), 127.89 (C-15), 107.5 (C-6), 104.8 (C-2), 95.9 (C-4), 61.0 (C-16), 52.99 (C-17, 17′), 39.7 (C-8), 29.2 (C-11), 23.62 (C-18, 18′), 22.78 (C-19), 20.3 (C-9, 10). HRTOF-MS ESI-(+) m/z = 384.2170 [M + H]$^+$ for C$_{23}$H$_{30}$NO$_4$ (cal. 384.2176).

^{1}H-NMR for the isomer of **7** (400 MHz, C$_5$D$_5$N) δ: 7.78 (2H, d, J = 5.8 Hz, H-13, 13′), 7.63 (2H, d, J = 5.8 Hz, H-14, 14′), 6.61 (1H, s, H-4), 4.97 (2H, s, H-16), 4.37 (2H, s, H-11), 4.35 (1H, m, H-8), 2.97 (4H, m, H-17, 17′), 1.77 (4H, m, H-18, 18′), 1.36 (2H, m, H-19), 1.29 (6H, d, J = 6.7 Hz, H-9, 10). ^{13}C-NMR for the isomer of **7** (100 MHz, C$_5$D$_5$N) δ: 211.2 (C-7), 166.4 (C-1), 164.6 (C-3), 162.5 (C-5), 145.2 (C-12), 132.23 (C-14, 14′), 130.3 (C-13, 13′), 128.04 (C-15), 107.5 (C-6), 104.8 (C-2), 95.9 (C-4), 64.3 (C-16), 53.25 (C-17, 17′), 39.7 (C-8), 29.2 (C-11), 23.84 (C-18, 18′), 23.01 (C-19), 20.3 (C-9, 10).

3.6. Assay for iNOS Inhibition

The assay was performed in mouse macrophages (RAW264.7, obtained from ATCC)) cultured in phenol red free RPMI medium supplemented with 10% bovine calf serum and 100 U/mL penicillin G sodium, and 100 μg/mL streptomycin at 37 °C in an atmosphere of 5% CO$_2$ and 95% humidity. Cells were seeded in 96-well plates (50,000 cells/well) and incubated for 24 h for a confluency of 75% or more. Test samples diluted in serum free medium were added, and after 30 min of incubation, LPS (5 μg/mL) was added and cells were further incubated for 24 h. The concentration of nitric oxide (NO) was determined by measuring the level of nitrite in the cell culture supernatant by using Griess reagent. Percent inhibition of nitrite production by the test compound was calculated in comparison to the vehicle control. IC$_{50}$ values were obtained from concentration response curves. Parthenolide was used as a positive control [25].

3.7. Assay for NF-κB Inhibition

The assay was performed in human chondrosarcoma cells (SW1353, obtained from American Type Culture Collection (ATCC), Manassas, VA, USA). Cells were cultured in 1:1 mixture of DMEM/F12 supplemented with 10% fetal bovine serum (FBS), 100 U/mL penicillin G sodium, and 100 μg/mL streptomycin at 37 °C in an atmosphere of 5% CO$_2$ and 95% humidity. Cells (1.2 × 10^7) were washed once in an antibiotic and FBS-free DMEM/F12, and then resuspended in 500 μL of antibiotic-free DMEM/F12 containing 2.5% FBS. NF-κB luciferase plasmid construct was added to the cell suspension at a concentration of 50 μg/mL and incubated for 5 min at room temperature. The cells were electroporated at 160 V and one 70-ms pulse using BTX disposable cuvettes model 640 (4-mm gap) in a BTX Electro Square Porator T 820 (BTX I, San Diego, CA, USA). The transfected cells were plated to the wells of 96-well plates at a density of 1.25 × 10^5 cells per well. After 24 h, cells were treated with different concentrations of test compound for 30 min before the addition of PMA (70 ng/mL) and incubated for 8 h. Luciferase activity was measured using the Luciferase Assay kit (Promega). Light output was detected on a SpectraMax plate reader. Percent inhibition of luciferase activity was calculated as compared to vehicle control, and IC$_{50}$ values were obtained from concentration response curves. Parthenolide was used as positive control [25].

3.8. Cytotoxicity Assay

Cytotoxicity was determined against the mammalian cell line Vero (African green monkey kidney fibroblast) which was obtained from ATCC. The detailed assay procedure has been described previously [26]. In brief, cells were seeded in 96-well plates (10,000 cells /well), and after 24 h of incubation, they were treated with various dilutions of test samples for 48 h. The cell viability was determined by tetrazolium dye WST-8. Doxorubicin was included as drug control.

Supplementary Materials: The following are available online. Supplementary Materials, including MS, HRMS, 1D and 2D NMR spectra for the synthetic compounds, are available online.

Author Contributions: N.L.: chemical synthesis, data analysis and manuscript preparation; S.I.K.: bioassay and interpretation of data and manuscript editing; S.Q.: NMR analysis; X.-C.L.: project design and manuscript preparation.

Funding: This work was supported by the USDA Agricultural Research Service Specific Cooperative Agreement No. 58-6060-6-015.

Conflicts of Interest: The authors declare no conflict of interest.

References

1. Singh, I.P.; Sidana, J.; Bharate, S.B.; Foley, W.J. Phloroglucinol compounds of natural origin: Synthetic aspects. *Nat. Prod. Rep.* **2010**, *27*, 393–416. [CrossRef] [PubMed]

2. Schmidt, S.; Jurgenliemk, G.; Schmidt, T.J.; Skaltsa, H.; Heilmann, J. Bi-, tri-, and polycyclic acylphloroglucinols from *Hypericum empetrifolium*. *J. Nat. Prod.* **2012**, *75*, 1697–1705. [CrossRef] [PubMed]

3. Song, C.K.; Zhao, S.; Hong, X.T.; Liu, J.Y.; Schulenburg, K.; Schwab, W. A UDP-glucosyltransferase functions in both acylphloroglucinol glucoside and anthocyanin biosynthesis in strawberry (*Fragaria×ananassa*). *Plant J.* **2016**, *85*, 730–742. [CrossRef] [PubMed]

4. Zhao, J.; Liu, W.; Wang, J.C. Recent advances regarding constituents and bioactivities of plants from the genus *Hypericum*. *Chem. Biodivers.* **2015**, *12*, 309–349. [CrossRef] [PubMed]

5. Tanaka, N.; Kobayashi, J. Prenylated acylphloroglucinols and meroterpenoids from *Hypericum* plants. *Heterocycles* **2015**, *90*, 23–40.

6. Henry, G.E.; Campbell, M.S.; Zelinsky, A.A.; Liu, Y.B.; Bowen-Forbes, C.S.; Li, L.Y.; Nair, M.G.; Rowley, D.C.; Seeram, N.P. Bioactive Acylphloroglucinols from *Hypericum densiflorum*. *Phytother. Res.* **2009**, *23*, 1759–1762. [CrossRef] [PubMed]

7. Tanaka, N.; Tsuji, E.; Kashiwad, Y.; Kobayashi, J.I. Yezo'otogirins D-H, Acylphloroglucinols and meroterpenes from *Hypericum yezoense*. *Chem. Pharm. Bull.* **2016**, *64*, 991–995. [CrossRef]

8. Kondo, A.; Shimizu, K. Spanate (flopropione), its clinical evaluation for urolithiasis and effects on ureteral peristalsis and blood pressure in dogs. *Hinyokika Kiyo.* **1969**, *15*, 748–754.

9. Fukuyama, T.; Takahashi, Y.; Kuze, M.; Arai, E. Clinical experiences with spanate, a new antispasmodic. *Hinyokika Kiyo.* **1969**, *15*, 818–823.

10. Jung, H.A.; Jin, S.E.; Ahn, B.R.; Lee, C.M.; Choi, J.S. Anti-inflammatory activity of edible brown alga *Eisenia bicyclis* and its constituents fucosterol and phlorotannins in LPS-stimulated RAW264.7 macrophages. *Food Chem. Toxicol.* **2013**, *59*, 199–206. [CrossRef]

11. Lopes, G.; Sousa, C.; Silva, L.R.; Pinto, E.; Andrade, P.B.; Bernardo, J.; Mouga, T.; Valentão, P. Can phlorotannins purified extracts constitute a novel pharmacological alternative for microbial infections with associated inflammatory conditions? *PLoS ONE* **2012**, *7*. [CrossRef] [PubMed]

12. Koeberle, A.; Rossi, A.; Bauer, J.; Dehm, F.; Verotta, L.; Northoff, H.; Sautebin, L.; Werz, O. Hyperforin, an anti-Inflammatory constituent from St. John's Wort, inhibits microsomal prostaglandin E(2) synthase-1 and suppresses prostaglandin E(2) formation in vivo. *Front Pharmacol.* **2011**, *2*, 1–10. [CrossRef] [PubMed]

13. Shaari, K.; Suppaiah, V.; Wai, L.K.; Stanslas, J.; Tejo, B.A.; Israf, D.A.; Abas, F.; Ismail, I.S.; Shuaib, N.H.; Zareen, S.; Lajis, N.H. Bioassay-guided identification of an anti-inflammatory prenylated acylphloroglucinol from *Melicope ptelefolia* and molecular insights into its interaction with 5-lipoxygenase. *Bioorga. Med. Chem.* **2011**, *19*, 6340–6347. [CrossRef] [PubMed]

14. Veena, V.K.; Popavath, R.N.; Kennedy, K.; Sakthivel, N. In vitro antiproliferative, pro-apoptotic, antimetastatic and anti-inflammatory potential of 2,4-diacteylphloroglucinol (DAPG) by Pseudomonas aeruginosa strain FP10. *Apoptosis* **2015**, *20*, 1281–1295. [CrossRef] [PubMed]

15. Zhou, K.; Ludwig, L.; Li, S.M. Friedel-Crafts Alkylation of Acylphloroglucinols catalyzed by a fungal indole prenyltransferase. *J. Nat. Prod.* **2015**, *78*, 929–933. [CrossRef] [PubMed]

16. Zhu, H.C.; Chen, C.M.; Liu, J.J.; Sun, B.; Wei, G.Z.; Li, Y.; Zhang, J.W.; Yao, G.M.; Luo, Z.W.; Xue, Y.B.; Zhang, Y.H. Hyperascyrones A–H, polyprenylated spirocyclic acylphloroglucinol derivatives from *Hypericum ascyron* Linn. *Phytochemistry* **2015**, *115*, 222–230. [CrossRef] [PubMed]

17. Yu, Q.; Ravu, R.R.; Jacob, M.R.; Khan, S.I.; Agarwal, A.K.; Yu, B.Y.; Li, X.C. Synthesis of natural acylphloroglucinol-based antifungal compounds against *Cryptococcus* species. *J. Nat. Prod.* **2016**, *79*, 2195–2201. [CrossRef]

18. George, J.H.; Hesse, M.D.; Baldwin, E.; Adlington, R.M. Biomimetic synthesis of polycyclic polyprenylated acylphloroglucinol natural products isolated from *Hypericum papuanum*. *Org. Lett.* **2010**, *12*, 3532–3535. [CrossRef]

19. Pepper, H.P.; Lam, H.C.; Bloch, W.M.; George, J.H. Biomimetic total synthesis of (±)-Garcibracteatone. *Org. Lett.* **2012**, *14*, 5162–5164. [CrossRef]

20. Morkunas, M.; Dube, L.; Gotz, F.; Maier, M.E. Synthesis of the acylphloroglucinols rhodomyrtone and rhodomyrtosone B. *Tetrahedron* **2013**, *69*, 8559–8563. [CrossRef]

21. Liu, X.H.; Zhai, Z.W.; Xu, X.Y.; Yang, M.Y.; Sun, Z.H.; Weng, J.Q.; Tan, C.X.; Chen, J. Facile and efficient synthesis and herbicidal activity determination of novel 1,2,4-triazolo[4,3-*a*]pyridin-3(2*H*)-one derivatives via microwave irradiation. *Bioorg. Med. Chem. Lett.* **2015**, *25*, 5524–5528. [CrossRef]

22. Shen, Z.H.; Sun, Z.H.; Becnel, J.J.; Estep, A.; Wedge, D.E.; Tan, C.X.; Weng, J.Q.; Han, L.; Liu, X.H. Synthesis and mosquiticidal activity of novel hydrazone containing pyrimidine derivatives against *Aedes aegypti*. *Lett. Drug Des. Discov.* **2018**, *15*, 951–956. [CrossRef]

23. Zhang, L.J.; Yang, M.Y.; Sun, Z.H.; Tan, C.X.; Weng, J.Q.; Wu, H.K.; Liu, X.H. Synthesis and antifungal activity of 1,3,4-thiadiazole derivatives containing pyridine group. *Lett. Drug Des. Discov.* **2014**, *11*, 1107–1111. [CrossRef]

24. Kleinschek, M.A.; Muller, U.; Brodie, S.J.; Stenzel, W.; Kohler, G.; Blumenschein, W.M.; Straubinger, R.K.; McClanahan, T.; Kastelein, R.A.; Alber, G. IL-23 enhances the inflammatory cell response in *Cryptococcus neoformans* infection and induce a cytokine pattern distinct from IL-12. *J. Immunol.* **2006**, *176*, 1098–1106. [CrossRef]

25. Zhao, J.P.; Khan, S.I.; Wang, M.; Vasquez, Y.; Yang, M.H.; Avula, B.; Wang, Y.H.; Avonto, C.; Smillie, T.J.; Khan, I.A. Octulosonic acid derivatives from Roman chamomile (*Chamaemelum nobile*) with activities against inflammation and metabolic disorder. *J. Nat. Prod.* **2014**, *77*, 509–515. [CrossRef]

26. Zulfiqar, F.; Khan, S.I.; Rossa, S.A.; Ali, Z.; Khan, I.A. Prenylated flavonol glycosides from *Epimedium grandiflorum*: Cytotoxicity and evaluation against inflammation and metabolic disorder. *Phytochem. Lett.* **2017**, *20*, 160–167. [CrossRef]

Sample Availability: Samples of the compounds are not available.

molecules

Article

Fusaproliferin, a Fungal Mycotoxin, Shows Cytotoxicity against Pancreatic Cancer Cell Lines

Nazia Hoque [1,2,3], Choudhury Mahmood Hasan [4], Md. Sohel Rana [2], Amrit Varsha [5], Md. Hossain Sohrab [3,*] and Khondaker Miraz Rahman [5,*]

[1] Department of Pharmacy, East West University, Dhaka 1212, Bangladesh; nzh@ewubd.edu
[2] Department of Pharmacy, Jahangirnagar University, Savar, Dhaka 1342, Bangladesh; sohelrana.ju@gmail.com
[3] Pharmaceutical Sciences Research Division (PSRD), BCSIR Laboratories, Dhaka 1205, Bangladesh
[4] Department of Pharmaceutical Chemistry, University of Dhaka, Dhaka-1000, Bangladesh; cmhasan@gmail.com
[5] School of Cancer and Pharmaceutical Science, King's College London, 150 Stamford Street, London SE1 9NH, UK; amritvarsha7dec@gmail.com
* Correspondence: mhsohrab@bcsir.gov.bd (M.H.S.); k.miraz.rahman@kcl.ac.uk (K.M.R.); Tel.: +880-1720121525 (M.H.S.); +44-2078481891 (K.M.R.); Fax: +44-2078484295 (K.M.R.)

Received: 17 October 2018; Accepted: 4 December 2018; Published: 11 December 2018

Abstract: As a part of our ongoing research on endophytic fungi, we have isolated a sesterterpene mycotoxin, fusaproliferin (FUS), from a *Fusarium solani* strain, which is associated with the plant *Aglaonema hookerianum* Schott. FUS showed rapid and sub-micromolar IC_{50} against pancreatic cancer cell lines. Time-dependent survival analysis and microscopy imaging showed rapid morphological changes in cancer cell lines 4 h after incubation with FUS. This provides a new chemical scaffold that can be further developed to obtain more potent synthetic agents against pancreatic cancer.

Keywords: endophytic fungi; sesterterpene; cytotoxic activity; pancreatic cancer

1. Introduction

Pancreatic adenocarcinoma is a leading cause of adult cancer mortality. It is presently untreatable, with a 5-year survival rate of ~5% [1]. As early detection is difficult, most patients present with locally advanced or metastatic disease [2]. Therapeutic options are limited, and metastatic disease frequently develops after surgery [3]. Pancreatic cancer is the seventh leading cause of cancer-related deaths worldwide, and annually more than 200,000 deaths are attributed to pancreatic cancer every year [4,5]. Natural sources, particularly plants, represent an important source of new anticancer chemical scaffolds, and there is an increasing interest in searching for natural products with drug-like properties as potential leads for drug discovery projects [6,7].

Endophytic fungi are symbiotically associated with plants, capable of synthesizing bioactive compounds without causing any damage to the host [8]. Some of these compounds have proven useful for novel drug discoveries, and provide a defense against harmful pathogens for the plants [9,10]. As a part of our ongoing research on endophytic fungi [11–14], a *Fusarium solani* strain, isolated from the plant *Aglaonema hookerianum* Schott. (Family: Araceae), was investigated. *F. solani* has been proved as a potent source of structurally-diverse natural compounds with cytotoxic activity, such as karuquinone A and karuquinone B [15], 9-desmethylherbarine, 7-desmethylscorpinone and 7-desmethyl-6-methylbostrycoidin [13], camptothecin and 10-hydroxycamptothecin [16], as well as paclitaxel [17]. *F. solani* is considered a plant pathogen, and it accounts for more than 50% infections caused by *Fusarium* spp. Infections by other *Fusarium* spp. strains are relatively uncommon [18] Chemical investigation of the ethyl acetate extract of the *F. solani* led to the isolation of fusaproliferin (FUS), a mycotoxin which was first isolated from the Italian *F. proliferatum* strains, named "proliferin" [19] and later "fusaproliferin" [20]. The absolute stereochemistry of the compound was confirmed by Santini et al. in 1996 [21]. FUS is also produced by *Fusarium subglutinans* and fifteen other ex-type strains of *Fusarium* species [22,23]. FUS produced a toxic effect on *Artemia salina*, insect cells, and human B lymphocytes [24]. It was also reported to produce a teratogenic effect on chicken embryos [25]. In this study, we examined the anticancer activity of FUS against two pancreatic and two breast cancer cell lines, and compared the activity of the compound with that of gemcitabine and doxorubicin, the current drugs of choice for pancreatic and breast cancer, respectively.

2. Results and Discussion

FUS was obtained as a white gum. The structure of FUS (Figure 1) was confirmed by spectroscopic analysis (^{1}H, ^{13}C-NMR, DEPT-135, 2D-NMR and HR-ESIMS) and by comparison with the published spectral values [19]. Accurate mass measurement of FUS obtained by FT-ESI-MS yielded a parent mass at m/z 467.2778 in positive ionization mode, corresponding to the sodium adduct [M + Na]$^+$ with a molecular formula of $C_{27}H_{40}O_5$ (calcd. mass 467.2773, [$C_{27}H_{40}O_5$ + Na]$^+$), accounting for 8 degrees of unsaturation. The resonances at δ 170.9 and 207.9 ppm in the ^{13}C-NMR spectrum were characteristic of the presence of two carbonyl carbons of an ester and a ketone, respectively. The ^{1}H and ^{13}C-NMR data, in conjunction with the DEPT-135 spectrum (Figure S13), proved the presence of 27 carbon atom signals corresponding to six methyls (20-, 21-, 22-, 23-, 25-, and 27-), seven sp^3 methylenes (1-, 4-, 5-, 8-, 9-, 13- and 24-), three sp^3 methines (10-, 14- and 19-), three sp^2 methines (2-,6- and 12-), one sp^3 quaternary carbon (15-), five sp^2 quaternary carbons (3-, 7-, 11-, 17- and 18-), two carbonyl carbons of an ester (26-OCOCH$_3$) and a ketone (16-CO-). The presence of three sp^2 methines and five sp^2 quaternary carbons, along with one each of an ester and carbonyl moiety, proved the presence of six double bonds in this compound, and thus indicated that it was a bicyclic compound. After deducting the acetyl moiety 'OCOCH$_3$' (δ_H = 2.06, δ_C = 20.9 and 170.9 ppm), the compound consisted of 25 carbons, which indicated it as a sesterterpene.

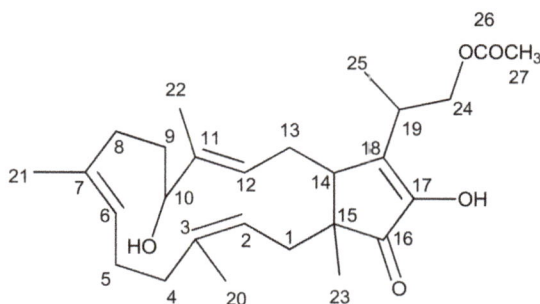

Figure 1. Structure of Fusaproliferin.

The cytotoxicity of FUS was determined against two pancreatic cancer cell lines, BxPc3 and MIA PaCa2, the ER-positive breast cancer cell line MCF7, and the triple negative breast cancer cell line MDA MB 231; gemcitabine was used as the positive control for the pancreatic cancer cell lines and doxorubicin for the breast cancer cell lines. FUS was active against all four cell lines tested (Figure 2 and Table 1) with sub to low micromolar IC$_{50}$, but the activity against the pancreatic cancer cell lines were notably better than the breast cancer cell lines. FUS was between 3 to 58 times more potent than gemcitabine in pancreatic cancer cell lines, but doxorubicin was superior against both breast cancer cell lines compared to FUS. The therapeutic utility of FUS was further investigated against WI38, a non-tumor lung fibroblast cell line. FUS was found to be cytotoxic against WI38 with a high micromolar IC$_{50}$ (Table 1), but was between 23 to 138 times more selective for the pancreatic cancer cell lines, and between 4.6 to 9.4 more selective for the breast cancer cell lines. This suggests a good therapeutic index against the pancreatic cancer cell lines which can be exploited if FUS is considered as a starting point for a medicinal chemistry program to develop a more potent analogue.

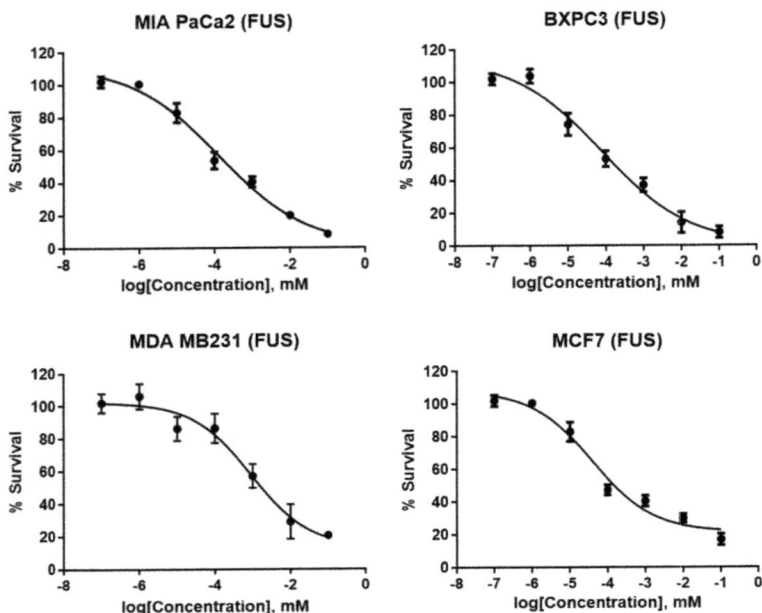

Figure 2. MTT cell-viability assay profile in pancreatic (MIA PaCa2 and BXPC3) and breast (MDA MB 231 and MCF7) cancer cell lines treated with FUS for 24 h.

Table 1. Cytotoxicity assay results of FUS against human tumor cells (IC$_{50}$ in μM).

Compound/Standard	MIA PaCa 2 (Pancreatic)	BXPC3 (Pancreatic)	MDA MB 231 (Breast)	MCF7 (Breast)	WI 38 (Lung Fibroblast)
FUS	0.13 ± 0.09	0.76 ± 0.24	1.9 ± 0.32	3.9 ± 0.75	18 ± 0.66
Gemcitabine	7.6 ± 0.66	2.2 ± 0.43	NT	NT	NT
Doxorubicin	NT	NT	0.06 ± 0.03	0.02 ± 0.018	NT

The relatively rapid cytotoxicity observed for FUS during the cell culture experiments led us to carry out a time-dependent cytotoxicity assay by monitoring percentage survival after 4- and 8-h post-incubation. FUS showed greater toxicity at both 4 and 8 h at 4 × IC$_{50}$ concentration in MIA PaCa2 cell line compared to gemcitabine. The differences were statistically significant ($p < 0.01$) (Figure 3a). Similarly, rapid toxicity was observed against MDA MB 231 cell line at 4 h ($p < 0.03$) and at 8 h ($p < 0.01$) compared to doxorubicin (Figures S5 and S6), although doxorubicin was notably more potent than FUS after 24 h incubation.

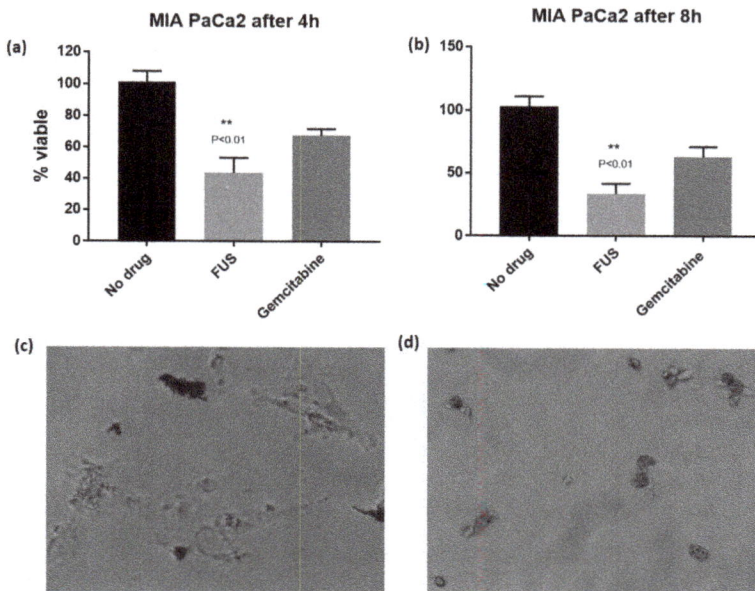

Figure 3. Effect of FUS on pancreatic cancer cell line. (**a**,**b**) FUS showed statistically significant rapid toxicity ($p < 0.01$) against MIA PaCa2 cell line after 4 h and 8 h incubation; (**c**,**d**) morphological changes observed in MIA PaCa2 cell lines after 4 h and 8 h incubation, respectively.

The morphological changes in the MIAPaCa2 cell line after incubating with FUS were monitored using a Nikon TS100 inverted microscope fitted with a camera. The cells appeared to show both apoptotic and necrotic damages within 4 h post-incubation, and the damages were fully evident at the 8-h time point (Figure 3c,d). These images, along with the survival analysis, point to the ability of the compound to induce severe stress resulting in rapid toxicity against the cell lines. This rapid cytotoxicity is intriguing and potentially a useful characteristic for an anticancer scaffold that can be developed against pancreatic cancer. Further studies are required to ascertain the mechanism of action of this compound, and will be reported in due course.

In summary, FUS is a known sesterterpene mycotoxin isolated from the endophytic fungus *F. solani*. This compound showed potent and rapid cytotoxicity against both pancreatic and breast cancer cell lines tested in this study. The complex structure and intriguing biological activity of FUS make it a good target for chemical synthesis and a lead structure for a medicinal chemistry project to develop a new anticancer drug against pancreatic cancer.

3. Experimental Section

3.1. Collection and Identification of the Plant Material

The aerial part of *A. hookerianum* was collected from Pablakhali, Rangamati, Chittagong Hill tracts, Bangladesh on 10 August 2014 and identified by the taxonomist of Bangladesh National Herbarium, Mirpur, Dhaka. A voucher specimen of the plant has been deposited (Accession no.: DACB 40633) in the herbarium for further reference (Figure S1).

3.2. General Experimental Procedures

The NMR spectra were recorded on a Bruker 400 MHz NMR spectrometer using $CDCl_3$. The HRMS spectrum was recorded on an Exactive Orbitrap by a Thermo Scientific mass spectrometer at King's College London, (London, UK), and the data were processed by Thermo XCalibur 2.2. Column chromatography was carried out on silica gel (70–230 mesh and 230–400 mesh, Merck, Darmstadt, Germany). Organic solvents, potato dextrose agar (PDA) medium, and TLC plates were purchased from Merck, Germany.

3.3. Isolation of Fungal Material

About 300 g of fresh and healthy parts of the plant (leaves, roots, and petioles) was cut with a sterile scalpel and stored at 4 °C in a sterile polyethene bag prior to use. Endophytic fungi were isolated from the fresh plant parts following the procedure, established at Pharmaceutical Sciences Research Division, BCSIR Laboratories, Dhaka, Bangladesh [11–14]. Total four endophytic fungi were isolated from different parts of *A. hookerianum* bearing the internal strain no. AHPE-3, AHPE-4 (Figure S2), AHLE-1 and AHLE-4. All the endophytic fungi were taxonomically identified up to genus level on the basis of macroscopic and microscopic morphological characters as *Fusarium* sp. (Figure S3). (AHPE-3), *Fusarium* sp. (AHPE-4), *Colletotrichum* sp. (AHLE-1) and *Colletotrichum* sp. (AHLE-4). The fungus AHPE-4 was selected for further investigation, based on the brine shrimp lethality bioassay data (Figure S4), and was cultured at a large scale to isolate bioactive secondary metabolites.

3.4. Molecular Identification of the Endophytic Fungus AHPE-4

For identification and differentiation, the Internal Transcript Spacer regions (ITS4 and ITS5) and the intervening 5.8S rRNA region was amplified and sequenced using electrophoretic sequencing on an ABI 3730 × 1 DNA analyzer (Applied Biosystems, Waltham, MA, USA) using Big Dye Terminator v 3.1 cycle sequencing kit (Thermo Fisher Scientific, Waltham, MA, USA). The ITS regions of the fungus were amplified using PCR (Hot Start Green Master Mix, Promega, Madison, WI, USA) and the universal ITS primers, ITS4 (5′-TCC GTA GGT GAA CCT GCG G-3′) and ITS5 (5′-GGA AGT AAA AGT CGT AAC AAG G-3′). The PCR products were purified and desalted using the Hot Start Green Master Mix (Cat: M7432, Promega, USA.) and sequenced on an ABI 3730 × 1 DNA analyzer (Applied Biosystems, USA). The sequences were aligned and prepared with the software Chromas (V 2.6.2, Technelysium, Brisbane, Australia) and matched against the nucleotide-nucleotide database (BLASTn) of the U.S. National Center for Biotechnology Information (NCBI) for final identification of the endophytic isolate. Finally, the sequence data (SI) were deposited in the Gen Bank database (accession number MG75792), which revealed 99% similarity other related fungal isolates of *F. solani* bearing accession numbers KX 497027, KJ863503, AB 190389, AY433805 etc. deposited in NCBI.

3.5. Extraction of the Fungal Material and Isolation of FUS

The fungus *F. solani* (AHPE-4), isolated from the petiole of the plant *A. hookerianum*, was cultivated at 28 ± 2 °C for 28 days on potato dextrose agar (PDA). The culture media were extracted with ethyl acetate for seven days in an air-tight, flat-bottom container with occasional shaking and stirring. This procedure was repeated three times to obtain the crude extract. The extract of endophytic fungi was then filtered using sterilized cotton filter followed by Whatman no. 1 filter papers. The solvent was evaporated with a rotary evaporator at low temperature (40 °C–50 °C) and reduced pressure.

The crude fungal extract (8 gm) was subjected to column chromatography for fractionation on silica gel (70–230 mesh) using gradients of petroleum ether/ethyl acetate, then ethyl acetate, followed by a gradient of ethyl acetate/methanol, and finally methanol, to afford a total of 15 fractions. These fractions were screened by TLC on silica gel under UV light and by spraying with vanillin-H_2SO_4 spray reagents. The column fraction of petroleum ether/15% ethyl acetate was subjected to preparative TLC on silica gel (toluene/20% ethyl acetate, 3 developments) to obtain FUS.

Fusaproliferin

18 mg, white, amorphous sticky mass; (^{1}H-NMR, CDCl$_3$): δ 2.40 (1H, dd, J = 10.8, 13.6 Hz, H-1′), 1.74 (1H, m, H-1″), 5.27 (1H, dd, J = 5.0, 10.2 Hz, H-2), 2.30 (1H, m, H-4′), 2.06 (1H, m, H-4″), 2.30 (1H, m, H-5′), 2.11 (1H, m, H-5″), 5.15 (1H, bs, H-6), 2.11 (1H, m, H-8′), 1.82 (1H, d, J = 9.2 Hz, H-8″), 1.82 (1H, d, J = 9.2 Hz, H-9′), 1.65 (1H, m, H-9″), 4.08 (1H, dd, J = 3.4, 9.8 Hz, H-10), 5.40 (1H, bt, H-12), 2.40 (1H, dd, J = 10.8, 13.6 Hz, H-13′), 1.95 (1H, m, H-13″), 2.69 (1H, dd, J = 2.0, 11.2 Hz, H-14), 2.80 (1H, sextet, H-19), 1.66 (3H, s, H-20), 1.66 (3H, s, H-21), 1.59 (3H, s, H-22), 1.02 (3H, s, H-23), 4.31 (1H, dd, J = 8.0, 10.4 Hz, H-24′), 4.27 (1H, dd, J = 7.2, 10.4 Hz, H-24″), 1.33 (3H, d, J = 6.8 Hz, H-25), 2.06 (3H, s, H-27), 5.56 (1H, s, 17- OH). ^{13}C-NMR: δ_C 39.1 (C-1), 121.4 (C-2), 138.2 (C-3), 40.3 (C-4), 23.8 (C-5), 124.3 (C-6), 132.9 (C-7), 34.9 (C-8), 29.7 (C-9), 76.5 (C-10), 136.5 (C-11), 128.9 (C-12), 28.7 (C-13), 49.6 (C-14), 49.0 (C-15), 207.9 (C-16), 147.3 (C-17), 146.7 (C-18), 33.7 (C-19), 15.5 (C-20), 15.3 (C-21), 10.4 (C-22), 16.2 (C-23), 66.4 (C-24), 14.5 (C-25), 170.9 (C-26), 20.9 (C-27). HRESIMS *m/z* 467.2778 [M + Na]$^+$ (calcd mass 467.2773, [C$_{27}$H$_{40}$O$_5$ + Na]$^+$).

3.6. Bioassays

3.6.1. Cell Culture

The MIA PaCa2 (pancreatic adenocarcinoma), BXPC3 (pancreatic adenocarcinoma), MDA-MB-231 (triple-negative breast cancer), MCF-7 (estrogen receptor positive breast cancer) cell lines were obtained from the American Type Culture Collection. The MIA PaCa2 cell line was maintained in Dulbecco's modified Eagle's medium (DMEM; Invitrogen, Carlsbad, CA, USA), supplemented with fetal bovine serum (10% *v/v*; Invitrogen), horse serum (2.5% *v/v*; Invitrogen) and penicillin-streptomycin (1% *v/v*, Invitrogen). The BXPC3 cell line was maintained in RPMI-1640 medium (DMEM; Invitrogen), supplemented with fetal bovine serum (10% *v/v*; Invitrogen), and penicillin-streptomycin (1% *v/v*, Invitrogen). The MDA MB 231 cell line was maintained in Dulbecco's modified Eagle's medium (DMEM; Invitrogen), supplemented with fetal bovine serum (10% *v/v*; Invitrogen), L-glutamine (2 mM; Invitrogen), non-essential amino acids (1×; Invitrogen) and penicillin-streptomycin (1% *v/v*, Invitrogen). The MCF7 cell line was maintained in Eagle's Minimum Essential medium supplemented with fetal bovine serum (10% *v/v*; Invitrogen), 0.01 mg/mL human recombinant insulin and penicillin-streptomycin (1% *v/v*, Invitrogen). During seeding, cells were counted using a Neubauer hemocytometer (Assistant, Hanover, Germany) by microscopy (Nikon, Melville, NY, USA) on a non-adherent suspension of cells that were washed in PBS, trypsinized, centrifuged at 8 °C at 8000 rpm for 5 min, and re-suspended in fresh medium.

3.6.2. MTT Assay

The cells were grown in normal cell culture conditions at 37 °C under a 5% CO$_2$ humidified atmosphere using an appropriate medium. The cell count was adjusted to 10^5 cells/mL and 2500 cells (MDA-MB-231) or 5000 cells (A4 and WI-38) were added per well. The cells were incubated for 24 h, and 1 μL of the appropriate inhibitor concentrations was added to the wells in triplicate. After 96 h of continuous exposure to each compound, the cytotoxicity was determined using the 3-(4,5-dimethylthiazol-2-yl)-2,5-diphenyltetrazolium bromide (MTT) (Lancaster Synthesis Ltd., Morecambe, Lancashire, UK) colorimetric assay. Absorbance was quantified by spectrophotometry at λ = 570 nm (Envision Plate Reader, PerkinElmer, Waltham, MA, USA). IC$_{50}$ values were calculated by a dose-response analysis using the Prism GraphPad Prism$^®$ software.

Supplementary Materials: The experimental procedures and ^{1}H, ^{13}C-NMR, DEPT-135 and HRMS spectra of FUS (PDF) are available online.

Author Contributions: Conceptualization, C.M.H. and M.H.S.; Formal analysis, C.M.H. and M.H.S.; Investigation, N.H., A.V., M.H.S. and K.M.R.; Methodology, N.H. and A.V.; Project administration, M.H.S.; Resources, K.M.R.; Supervision, C.M.H., M.S.R., M.H.S. and K.M.R.; Writing-original draft, N.H. and M.H.S.; Writing-review & editing, C.M.H., M.S.R. and K.M.R.

Funding: This research received no external funding. The APC was funded by King's College London ACAC11460.

Acknowledgments: The authors are thankful to the authority of BCSIR for supporting the work in this paper through providing the all lab facilities related to the work on endophytic fungi.

Conflicts of Interest: The authors declare no competing financial interest.

References

1. Kleeff, J.; Korc, M.; Apte, M.; La Vecchia, C.; Johnson, C.D.; Biankin, A.V.; Neale, R.E.; Tempero, M.; Tuveson, D.A.; Hruban, R.H.; et al. Pancreatic cancer. *Nat. Rev. Dis. Prim.* **2016**, *2*, 16022. [CrossRef] [PubMed]

2. Van Cutsem, E.; Vervenne, W.L.; Bennouna, J.; Humblet, Y.; Gill, S.; Van Laethem, J.-L.; Verslype, C.; Scheithauer, W.; Shang, A.; Cosaert, J. Phase III trial of bevacizumab in combination with gemcitabine and erlotinib in patients with metastatic pancreatic cancer. *J. Clin. Oncol.* **2009**, *27*, 2231–2237. [CrossRef]

3. Siriwardena, A.K.; Siriwardena, A.M. Pancreatic cancer. *BMJ Br. Med. J.* **2014**, *349*. [CrossRef] [PubMed]

4. Wong, M.C.S.; Jiang, J.Y.; Liang, M.; Fang, Y.; Yeung, M.S.; Sung, J.J.Y. Global temporal patterns of pancreatic cancer and association with socioeconomic development. *Sci. Rep.* **2017**, *7*, 3165. [CrossRef]

5. Vincent, A.; Herman, J.; Schulick, R.; Hruban, R.H.; Goggins, M. Pancreatic cancer. *Lancet* **2011**, *378*, 607–620. [CrossRef]

6. Ouyang, L.; Luo, Y.; Tian, M.; Zhang, S.Y.; Lu, R.; Wang, J.H.; Kasimu, R.; Li, X. Plant natural products: From traditional compounds to new emerging drugs in cancer therapy. *Cell Prolif.* **2014**, *47*, 506–515. [CrossRef] [PubMed]

7. Harvey, A.L.; Edrada-Ebel, R.; Quinn, R.J. The re-emergence of natural products for drug discovery in the genomics era. *Nat. Rev. Drug Disc.* **2015**, *14*, 111–129. [CrossRef]

8. Kumar, A.; Patil, D.; Rajamohanan, P.R.; Ahmad, A. Isolation, purification and characterization of vinblastine and vincristine from endophytic fungus *Fusarium oxysporum* isolated from *Catharanthus roseus*. *PLoS ONE* **2013**, *8*, e71805. [CrossRef]

9. Nascimento, A.M.d.; Conti, R.; Turatti, I.C.; Cavalcanti, B.C.; Costa-Lotufo, L.V.; Pessoa, C.; de Moraes, M.O.; Manfrim, V.; Toledo, J.S.; Cruz, A.K. Bioactive extracts and chemical constituents of two endophytic strains of *Fusarium oxysporum*. *Rev. Bras. Farmacogn.* **2012**, *22*, 1276–1281. [CrossRef]

10. Cui, Y.; Yi, D.; Bai, X.; Sun, B.; Zhao, Y.; Zhang, Y. Ginkgolide B produced endophytic fungus (*Fusarium oxysporum*) isolated from *Ginkgo biloba*. *Fitoterapia* **2012**, *83*, 913–920. [CrossRef]

11. Khan, M.I.H.; Sohrab, M.H.; Rony, S.R.; Tareq, F.S.; Hasan, C.M.; Mazid, M.A. Cytotoxic and antibacterial naphthoquinones from an endophytic fungus, *Cladosporium* sp. *Toxicol. Rep.* **2016**, *3*, 861–865. [CrossRef] [PubMed]

12. Chowdhury, N.S.; Sohrab, M.H.; Rony, S.R.; Sharmin, S.; Begum, M.N.; Rana, M.S.; Hasan, C.M. Identification and bioactive potential of endophytic fungi from *Monochoria hastata* (L.) Solms. *Bangladesh J. Bot.* **2016**, *45*, 187–193.

13. Chowdhury, N.S.; Sohrab, M.H.; Rana, M.S.; Hasan, C.M.; Jamshidi, S.; Rahman, K.M. Cytotoxic naphthoquinone and azaanthraquinone derivatives from an endophytic *Fusarium solani*. *J. Nat. Prod.* **2017**, *80*, 1173–1177. [CrossRef] [PubMed]

14. Khan, N.; Afroz, F.; Begum, M.N.; Rony, S.R.; Sharmin, S.; Moni, F.; Hasan, C.M.; Shaha, K.; Sohrab, M.H. Endophytic *Fusarium solani*: A rich source of cytotoxic and antimicrobial napthoquinone and aza-anthraquinone derivatives. *Toxicol. Rep.* **2018**, *5*, 970–976. [CrossRef] [PubMed]

15. Takemoto, K.; Kamisuki, S.; Chia, P.T.; Kuriyama, I.; Mizushina, Y.; Sugawara, F. Bioactive dihydronaphthoquinone derivatives from *Fusarium solani*. *J. Nat. Prod.* **2014**, *77*, 1992–1996. [CrossRef] [PubMed]

16. Shweta, S.; Zuehlke, S.; Ramesha, B.T.; Priti, V.; Kumar, M.P.; Ravikanth, G.; Spiteller, M.; Vasudeva, R.; Shaanker, U.R. Endophytic fungal strains of *Fusarium solani*, from *Apodytes dimidiata* E. Mey. ex Arn (Icacinaceae) produce camptothecin, 10-hydroxycamptothecin and 9-methoxycamptothecin. *Phytochemistry* **2010**, *71*, 117–122. [CrossRef] [PubMed]

17. Chakravarthi, B.; Das, P.; Surendranath, K.; Karande, A.A.; Jayabaskaran, C. Production of paclitaxel by *Fusarium solani* isolated from *Taxus celebica*. *J. Biosci.* **2008**, *33*, 259–267. [CrossRef]

18. Kosmidis, C.; Denning, D.W. Opportunistic and Systemic Fungi. In *Infectious Diseases*, 4th ed.; Elsevier: Amsterdam, The Netherlands, 2017; pp. 1681–1709.e3.

19. Randazzo, G.; Foglianoa, V.; Ritieni, A.; Rossin, L.M.E.; Scarallo, A.; Segred, A.L. Proliferin, a new sesterterpene from *Fusarium proliferatum. Tetrahedron* **1993**, *49*, 10883–10896. [CrossRef]

20. Ritieni, A.; Fogliano, V.; Randazzo, G.; Scarallo, A.; Logrieco, A.; Moretti, A.; Manndina, L.; Bottalico, A. Isolation and characterization of fusaproliferin, a new toxic metabolite from Fusarium proliferatum. *Nat. Toxins* **1995**, *3*, 17–20. [CrossRef]

21. Santini, A.; Ritieni, A.; Fogliano, V.; Randazzo, G.; Mannina, L.; Logrieco, A.; Benedetti, E. Structure and absolute stereochemistry of fusaproliferin, a toxic metabolite from Fusarium proliferatum. *J. Nat. Prod.* **1996**, *59*, 109–112. [CrossRef]

22. Fotso, J.; Leslie, J.F.; Smith, S.J. Production of beauvericin, moniliformin, fusaproliferin, and fumonisins B1, B2, and B3 by fifteen ex-type strains of *Fusarium* Species. *Appl. Environ. Microbiol.* **2002**, *68*, 5195–5197. [CrossRef] [PubMed]

23. Ritieni, A.; Monti, S.M.; Moretti, A.; Logrieco, A.; Gallo, A.; Ferracane, R.; Fogliano, V. Stability of fusaproliferin, a mycotoxin from *Fusarium* spp. *J. Sci. Food Agric.* **1999**, *79*, 1676–1680. [CrossRef]

24. Logrieco, A.; Moretti, A.; Fornelli, F.; Fogliano, V.; Ritieni, A.; Caiaffa, M.F.; Randazzo, G.; Bottalico, A.; Macchia, L. Fusaproliferin production by *Fusarium subglutinars* and its toxicity to *Artemia salina*, SF-9 insect cells, and IARC/LCL 171 human B lymphocytes. *Appl. Environ. Microbiol.* **1996**, *62*, 3378–3384. [PubMed]

25. Ritieni, A.; Monti, S.M.; Randazzo, G.; Logrieco, A.; Moretti, A.; Peluso, G.; Ferracane, R.; Fogliano, V. Teratogenic effects of fusaproliferin on chicken embryos. *J. Agric. Food Chem.* **1997**, *45*, 3039–3043. [CrossRef]

Sample Availability: Samples of the compounds are not available from the authors. However, the compounds can be extracted again from the natural source.

Article

Tetra-glucopyranosyl Diterpene *ent*-Kaur-16-en-19-oic Acid and *ent*-13(*S*)-Hydroxyatisenoic Acid Derivatives from a Commercial Extract of *Stevia rebaudiana* (Bertoni) Bertoni

Wilmer H. Perera [1,2], **Ion Ghiviriga** [3], **Douglas L. Rodenburg** [1], **Kamilla Alves** [1], **Frank T. Wiggers** [4], **Charles D. Hufford** [4,†], **Frank R. Fronczek** [5], **Mohamed A. Ibrahim** [4,6], **Ilias Muhammad** [4], **Bharathi Avula** [4], **Ikhlas A. Khan** [4] and **James D. McChesney** [1,*]

[1] Ironstone Separations, Inc., Etta, Oxford, MS 38627, USA; wilmer.perera@gmail.com (W.H.P.); douglasrodenburg@yahoo.com (D.L.R.); Kamilla_07@yahoo.com (K.A.)
[2] ORISE Fellow-Agricultural Research Service, Natural Product Utilization Research Unit, U.S. Department of Agriculture, University of Mississippi, Oxford, MS 38677, USA
[3] Department of Chemistry, University of Florida, Gainesville, FL 32611, USA; ion@chem.ufl.edu
[4] National Center for Natural Products Research, University of Mississippi, Oxford, MS 38677, USA; fwiggers@olemiss.edu (F.T.W.); mmibrahi@olemiss.edu (M.A.I.); milias@olemiss.edu (I.M.); bavula@olemiss.edu (B.A.); ikhan@olemiss.edu (I.A.K.)
[5] Department of Chemistry, Louisiana State University, Baton Rouge, LA 70803, USA; ffroncz@lsu.edu
[6] Chemistry of Natural Compounds Department, Pharmaceutical and Drug Industries Division, National Research Centre, Dokki, Giza 12622, Egypt
[*] Correspondence: jdmcchesney@yahoo.com; Tel.: +(303)808-4104
[†] Charles D. Hufford deceased on 15 May 2017.

Academic Editors: Muhammad Ilias and Charles L. Cantrell
Received: 29 November 2018; Accepted: 14 December 2018; Published: 15 December 2018

Abstract: *Stevia rebaudiana* and its diterpene glycosides are one of the main focuses of food companies interested in developing novel zero calorie sugar substitutes since the recognition of steviol glycosides as Generally Recognized as Safe (GRAS) by the United States Food and Drug Administration. Rebaudioside A, one of the major steviol glycosides of the leaves is more than 200 times sweeter than sucrose. However, its lingering aftertaste makes it less attractive as a table-top sweetener, despite its human health benefits. Herein, we report the purification of two novel tetra-glucopyranosyl diterpene glycosides **1** and **3** (rebaudioside A isomers) from a commercial *Stevia rebaudiana* leaf extract compounds, their saponification products compounds **2** and **4**, together with three known compounds isolated in gram quantities. Compound **1** was determined to be 13-[(2-*O*-β-D-glucopyranosyl-6-*O*-β-D-glucopyranosyl-β-D-glucopyranosyl) oxy]*ent*-kaur-16-en-19-oic acid-β-D-glucopyranosy ester (rebaudioside Z), whereas compound **3** was found to be 13-[(2-*O*-β-D-glucopyranosyl-3-*O*-β-D-glucopyranosyl-β-D-glucopyranosyl) oxy]*ent*-hydroxyatis-16-en-19-oic acid -β-D-glucopyranosy ester. Two new tetracyclic derivatives with no sugar at position C-19 were prepared from rebaudiosides **1** and **3** under mild alkaline hydrolysis to afford compounds **2** 13-[(2-*O*-β-D-glucopyranosyl-6-*O*-β-D-glucopyranosyl-β-D-glucopyranosyl) oxy]*ent*-kaur-16-en-19-oic acid (rebaudioside Z_1) and **4** 13-[(2-*O*-β-D-glucopyranosyl-3-*O*-β-D-glucopyranosyl-β-D-glucopyranosyl) oxy]*ent*-hydroxyatis-16-en-19-oic acid. Three known compounds were purified in gram quantities and identified as rebaudiosides A (**5**), H (**6**) and J (**7**). Chemical structures were unambiguously elucidated using different approaches, namely HRESIMS, HRESI-MS/MS, and 1D-and 2D-NMR spectroscopic data. Additionally, a high-quality crystal of iso-stevioside was grown in methanol and its structure confirmed by X-ray diffraction.

Molecules **2018**, *23*, 3328

Keywords: *Stevia rebaudiana*; diterpene glycosides; rebaudioside A isomers; 13(S)-hydroxyatisenoic acid derivative; iso-stevioside X-ray structure

1. Introduction

Diabetes is a chronic disease that affected 422 million of people worldwide in 2014 and is increasing considerably every year. Diabetes is characterized by insulin deficit or insulin insensitivity and consequently produces high blood sugar levels which are associated with other disorders [1,2]. Reducing or eliminating sugar consumption and replacing sucrose with healthier sweeteners is one approach to prevent and control diabetes.

Stevia rebaudiana (Bertoni) Bertoni and its steviol glycosides were conferred Generally Recognized as Safe status by the United States Food and Drug Administration in 2008 and in the European Union in 2011. Steviol glycosides from this Asteraceae plant have been gaining interest from companies and the general public since these natural compounds are calorie free sweeteners and have shown higher sweetener potency than sucrose [3,4]. Additionally, it was recently suggested that rebaudioside A potentiates the activity of a Ca^{2+} cation channel expressed in type II taste receptor cells and pancreatic β-cells (TRPM5) enhancing glucose-induced insulin secretion in a dependent manner. However, regardless of the potential of rebaudioside A as a sugar substitute to prevent and treat type 2 diabetes [5], the lingering aftertaste of this tetra-glucopyranosyl sugar substitute makes it unattractive to consumers.

In our continuing efforts to discover novel sugar substitutes with potential refined organoleptic properties, we describe herein the isolation and structure elucidation of two new tetra-glucopyranosyl diterpene glycosides, the preparation of their saponification products together with the isolation of three known compounds in gram quantities.

2. Results and Discussion

Structure Elucidation

Two new compounds were isolated from a commercial extract of *Stevia rebaudiana* using reversed-phase and high performance normal-phase chromatography [6]. Recently, two approaches have been described for a rapid detection of novel oligosaccharide arrangements linked at position C-13 in steviol glycosides and infer their C-19 linkage. One is based on tandem mass spectrometry dissociation patterns with ranging collision energies [7]. Thus, steviol glycosides with one monosaccharide unit or a less hindered disaccharide linked at C-19 (e.g., rebaudiosides I and U with 1-3 and 1-6 linkages, respectively) cleave the C-19 ester linkage with low collision energies (10 eV) while more hindered disaccharides (1-2 linkages) cleave with higher collision energies (40 eV).

Compound **1** was purified as an amorphous off-white solid with $[\alpha]^{25}_D$ −28.0 (c 0.1, MeOH). HR-ESIMS and HR-ESIMS/MS data of compound **1** showed a molecular ion at m/z 965.4208 $[M − H]^-$ (calculated m/z 965.4235 $[M − H]^-$), suggesting a molecular formula $C_{44}H_{70}O_{23}$. Both, the deprotonated molecular ion and an intense product ion m/z = 803.3703 Da resulted from the loss of one hexose at C-19 ($[M − H] − H_2O − 162$ Da)$^-$ were observed at 10 eV. Further sequential loss of three hexoses ($[M − H] − H_2O − 3 \times 162$ Da)$^-$ m/z = 641.3185, m/z = 479.2635 and m/z = 317.2146 from the C-13 moiety was observed at 70 eV. Acid hydrolysis of **1** furnished a mixture of aglycones and only D-glucose which was identified by comparison of the HPLC retention times of thiocarbamoyl thiazolidine derivatives prepared from sugar standards as previously described [8,9]. Hence compound **1** is a rebaudioside A isomer.

The second approach was based on comparing retention times of a pure steviol glycoside and its saponification product with those reported by RP-C18 HPLC [10]. Saponification condition was helpful to detect single monosaccharide, di and oligosaccharides linked at position C-19. Thus, compound **1** showed a retention time of 3.94 min that did not match with any previously reported steviol glycosides in a RP-C18 HPLC method [10]. The aglycone-C13 moiety was produced by mild alkaline hydrolysis corroborating the linkage of a single sugar at position C-19 (9.32 min) as deduced by MS. The aglycone with the C-13 portion showed an HPLC retention time that did not match with any of the steviol glycosides reported with a free carboxylic acid at C-19, suggesting the structural novelty of this compound, probably, in the C-13-oligosaccharide.

The assignments of the signals of the steviol aglycone started by identifying H-18 as the methyl protons coupling with C-19, which is the most deshielded carbon. H-18 also couples with a quaternary carbon, C-4, with methylene carbon, C-3, and with a methine carbon, C-5. The sequence H-3–H-2–H-1 could be followed in the COSY spectrum, and the protons on ring A assigned as axial or equatorial based on the number of large couplings. Both C-1 and C-5 couple with methyl protons at position 20, which also couples with a quaternary carbon, C-10 and with a methine carbon, C-9. The sequences H-6—H-7—H-9 and H-9—H-11—H-12 could be followed in the COSY spectrum. The C/D rings were supported by the HMBC correlations of H-9 with C-11, C-12, C-14 and C-15; H-14 with C-13, C-15 and C-16; and H-17 with C-13, C-15 and C-16.

We also observed four anomeric protons at δ 5.14, 5.19, 5.30, and 6.12 ppm supporting the MS information, all of them showing beta linkages.

The positions of attachment of the sugar moieties were established based on the 2D-NMR HMBC spectrum. ^3J HMBC correlations between anomeric proton H-1' (δ 6.12 ppm) and C-19 (δ 177.7 ppm) confirmed the attachment of one glucose unit at C-19. The position of H-2' was confirmed through the COSY correlation between H-2' (δ 4.17 ppm) and H-1' (δ 6.12 ppm). Additionally, the ^3J HMBC correlations between H-1'' (δ 5.19 ppm) and C-13 (δ 86.4 ppm) confirmed the attachment of the first glucose unit at C-13. The connections of other two glucoses were also established through ^3J HMBC correlations between H-1''' (δ 5.30 ppm), H-1'''' (δ 5.14 ppm) with C-2'' (δ 84.4 ppm) and C-6'' (δ 70.3 ppm) respectively. In the same way, the position of H-2'' was confirmed through the gDQCOSY correlation between H-2'' (δ 4.23 ppm) and H-1'' (δ 5.19 ppm). The positions of H-2 of the sugars linked at position C-13' were confirmed through their COSY correlations between H-2'' (δ 4.23 ppm) and H-1'' (δ 5.19 ppm), H-2''' (δ 4.11 ppm) and H-1''' (δ 5.30 ppm) and H-2'''' (δ 4.06 ppm) and H-1'''' (δ 5.14 ppm). Compound **1** was named as 13-[(2-*O*-β-D-glucopyranosyl-6-*O*-β-D-glucopyranosyl-β-D-glucopyranosyl)oxy]*ent*-kaur-16-en-19-oic acid-β-D-glucopyranosy ester (rebaudioside Z) and assigned the structure shown in Figure 1. ^{1}H- and ^{13}C-NMR chemical shifts are shown in Tables 1 and 2, respectively, and are typical signals of the *ent*-kaurene core [11].

Table 1. ^{1}H chemical shifts of diterpene glycosides **1–4**.

Moiety	Position	1 [a]	2 [b]	3 [b]	4 [b]
		δ (ppm)	δ (ppm)	δ (ppm) [c]	δ (ppm) [c]
Aglycone	1	0.73, 1.73	0.95, 2.13	0.92, 1.59	0.85, 1.56
	2	2.20, 1.43	1.96, 1.37	1.92, 1.40	2.01, 1.30
	3	2.28, 1.79	1.97, 1.54	2.20, 1.09	2.14, 0.88
	5	1.03	1.01	1.12	0.90
	6	1.91, 2.50	1.87, 1.87	1.78, 2.05	1.80, 1.98
	7	1.28, 1.28	1.52; 1.40	1.57, 1.16	1.53, 1.13
	9	0.87	0.96	1.09	1.04
	11	1.68, 1.68	1.79,1.63	1.63, 1.45	1.60, 1.44
	12	1.02, 2.35	0.83, 1.86	2.51	2.49
	13	-	-	4.04	4.05
	14	1.97, 2.74	1.51, 2.27	1.15, 2.51	1.14, 2.51
	15	2.06, 2.06	2.05, 2.10	1.89, 2.11	1.86, 2.08
	17	5.10, 5.73	4.85, 5.24	4.70, 4.84	4.68,4.82
	18	1.23	1.14	1.24	1.11
	20	1.31	1.03	0.89	0.95

Molecules **2018**, *23*, 3328

Table 1. *Cont.*

Moiety	Position	1 [a] δ (ppm)	2 [b] δ (ppm)	3 [b] δ (ppm) [c]	4 [b] δ (ppm) [c]
Glcβ-C$_{19}$	1′	6.12	-	5.43	-
	2′	4.17	-	3.39	-
	3′	3.98	-	3.43	-
	4′	4.33	-	3.39	-
	5′	4.22	-	3.39	-
	6′	4.43, 4.57	-	3.85, 3.71	-
Glc-C$_{13}$	1″	5.19	4.59	4.56	4.55
	2″	4.23	3.60	3.65	3.64
	3″	4.27	3.54	3.71	3.71
	4″	4.45	3.35	3.38	3.42
	5″	4.08	3.17	3.34	3.30
	6″	4.51, 4.78	3.78, 4.10	3.91, 3.68	3.85, 3.70
Glc(1–2)	1‴	5.30	4.62	4.81	4.81
	2‴	4.11	3.30	3.19	3.18
	3‴	4.23	3.46	3.35	3.34
	4‴	4.28	3.35	3.22	3.21
	5‴	3.91	3.36	3.33	3.30
	6‴	4.22, 4.57	3.68, 3.82	3.87, 3.66	3.85, 3.65
Glcβ(1-X) [d]	1⁗	5.14	4.61	4.65	4.65
	2⁗	4.06	3.21	3.28	3.27
	3⁗	4.27	3.46	3.39	3.38
	4⁗	4.12	3.35	3.31	3.33
	5⁗	4.03	3.36	3.37	3.36
	6⁗	4.22, 4.57	3.68, 3.82	3.90, 3.66	3.89, 3.65

(1) rebaudioside Z; (2) rebaudioside Z$_1$; (3) 13-[(2-*O*-β-D-glucopyranosyl-3-*O*-β-D-glucopyranosyl-β-D-glucopyranosyl) oxy]*ent*-hydroxyatis-16-en-19-oic acid-β-D-glucopyranosy ester; (4) 13-[(2-*O*-β-D-glucopyranosyl-3-*O*-β-D-glucopyranosyl-β-D-glucopyranosyl)oxy]*ent*-hydroxyatis-16-en-19-oic acid. All the sugars showed β linkage evidenced for the coupling constants of the anomeric protons (7.6 and 8 Hz), [a] in Pyr-d_5, [b] in MeOH-d_4, [c] diastereotopic protons in the aglycone are given in the order *pro-R, pro-S*. Protons in position 6 of glucose were not assigned sterically. Protons in position 17 are given in the order *pro-E, pro-Z*. [d] X = 6 for compound 1 and 2; X = 3 for compound 3 and 4.

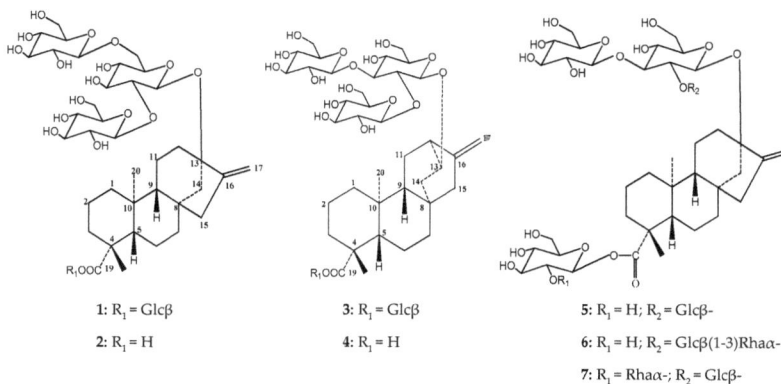

Figure 1. Chemical structures of purified compounds rebaudioside Z (**1**) and 13-[(2-*O*-β-D-glucopyranosyl-3-*O*-β-D-glucopyranosyl-β-D-glucopyranosyl)oxy]*ent*-hydroxyatis-16-en-19-oic acid-β-D-glucopyranosy ester (**3**) and chemically modified rebaudioside Z$_1$ (**2**) and 13-[(2-*O*-β-D-glucopyranosyl- 3-*O*-β-D-glucopyranosyl-β-D-glucopyranosyl) oxy]*ent*-hydroxyatis- 16-en-19-oic acid (**4**), rebaudioside A (**5**), rebaudioside H (**6**) and rebaudioside J (**7**).

Table 2. ^{13}C chemical shifts of diterpene glycosides **1–4**.

Moiety	Position	1 [a] δ (ppm)	2 [b] δ (ppm)	3 [b] δ (ppm)	4 [b] δ (ppm)
Aglycone	1	41.2	40.1	39.5	40.5
	2	19.9	20.9	18.4	19.3
	3	36.8	39.0	37.7	39.4
	4	44.5	45.6	43.7	44.7
	5	57.8	58.7	57.1	57.6
	6	22.7	23.7	19.9	20.6
	7	42.2	43.1	38.8	39.4
	8	43.3	43.1	33.9	33.8
	9	54.3	55.3	51.0	51.2
	10	40.3	40.9	38.1	38.0
	11	21.1	21.5	26.3	26.3
	12	38.9	42.6	41.0	41.0
	13	86.4	88.3	77.8	77.8
	14	45.1	45.8	37.7	37.7
	15	48.0	48.8	47.4	47.7
	16	155.0	154.2	146.7	147.0
	17	105.4	105.7	108.0	107.8
	18	28.8	30.1	27.5	29.1
	19	177.7	184.1	176.8	184.4
	20	16.0	17.3	11.9	12.2
Glcβ-C$_{19}$	1'	96.3	-	94.2	-
	2'	74.4	-	72.6	-
	3'	79.5	-	77.3	-
	4'	71.4	-	69.7	-
	5'	79.9	-	77.3	-
	6'	62.6	-	61.0	-
Glcβ-C$_{13}$	1''	98.3	97.7	100.5	100.3
	2''	84.4	81.8	79.0	78.8
	3''	78.6	78.5	86.0	85.9
	4''	71.8	71.8	68.8	68.6
	5''	77.8	77.8	76.1	76.1
	6''	70.3	70.0	61.4	61.2
Glcβ(1-2)	1'''	106.8	104.8	102.2	102.1
	2'''	75.7	76.3	74.6	74.5
	3'''	78.8	77.8	76.5	76.5
	4'''	72.1	71.4	70.6	70.6
	5'''	78.3	77.9	76.6	76.6
	6'''	63.3	62.8	61.9	61.8
Glcβ(1-X)	1''''	105.9	104.6	103.1	103.1
	2''''	77.2	75.3	73.9	73.9
	3''''	78.9	77.9	76.8	76.8
	4''''	72.6	71.7	70.1	70.1
	5''''	78.5	78.3	76.8	76.8
	6''''	63.3	62.8	61.2	61.1

(**1**) Rebaudioside Z; (**2**) rebaudioside Z$_1$; (**3**) 13-[(2-O-β-D-glucopyranosyl-3-O-β-D-glucopyranosyl-β-D-glucopyranosyl) oxy]*ent*-hydroxyatis-16-en-19-oic acid-β-D-glucopyranosy ester; (**4**) 13-[(2-O-β-D-glucopyranosyl-3-O-β-D-glucopyranosyl-β-D-glucopyranosyl) oxy]*ent*-hydroxyatis-16-en-19-oic acid. NMR spectra recorded in [a] Pyr-d_5 and [b] MeOH-d_4. X = 6 for compound **1** and **2**; X = 3 for compound **3** and **4**.

Compound **2** was prepared through mild alkaline hydrolysis of **1** to afford a new rebaudioside B isomer with a Glcβ(1-6)[Glcβ(1-2)]Glcβ1- arrangement at C-13 with retention of 9.32 min in the RP-C18 HPLC method that did not match with any of the previously reported steviol-C13 oligosaccharides [10]. HRESIMS/MS experiment of compound **2** showed a molecular ion at m/z 803.3721 [M − H]$^-$ (calculated m/z 803.3702 [M − H]$^-$), suggesting a molecular formula $C_{38}H_{60}O_{18}$.

A sequential loss of three glucoses units at C-13 portion were observed at 70 eV collision energy: m/z 641.3196 (−162 Da), m/z 479.2648 (−162 Da) and m/z 317.2115 Da. ^{1}H-NMR showed three anomeric protons corresponding to the oligosaccharide portion linked at position C-13 (δ_H 4.59; 4.61 and 4.62 ppm). The free carboxylic acid at position C-19 was evidenced by the signal at 184.1 ppm in the ^{13}C-NMR, less shielded than the observed for compound **1**. Full structural assignment was performed using 1D- and 2D-NMR (DQCOSY, HSQC and HMBC) experiments. Compound **2** was named as 13-[(2-*O*-β-D-glucopyranosyl-6-*O*-β-D-glucopyranosyl-β-D-glucopyranosyl) oxy]*ent*-kaur-16-en-19-oic acid (rebaudioside Z$_1$). ^{1}H and ^{13}C chemical shifts are presented in Table 1.

Compound **3** was purified as an amorphous off-white solid with $[\alpha]^{25}_D$ −20.0 (*c* 0.1, MeOH). HR-ESIMS/MS data of compound **3** showed a molecular ion at m/z 965.4297 [M − H]$^-$ (calculated m/z 965.4235 [M − H]$^-$), suggesting a molecular formula $C_{44}H_{70}O_{23}$. Both, the deprotonated molecular ion and an intense product ion m/z = 803.3735 Da resultant from the loss of one hexose at C-19 [M – H − H$_2$O − 162 Da]$^-$ were observed at 10 eV. Further sequential loss of three hexoses [M – H − H$_2$O − 3 × 162 Da]$^-$ m/z = 641.3195, m/z = 479.2710 and m/z = 317.2141 from C-13 moiety was observed at 70 eV. Compound **3** showed a HPLC retention time of 4.35 min that did not match with any diterpene glycosides in the RP-C18 method previously reported [10]. It was hydrolyzed under mild alkaline conditions corroborating the presence of a single monosaccharide attached at position C-19. Retention time of the saponification product showed a peak at 10.5 min that did not match with any aglycone-C13 previously described [10]. The rapid elution of compound **3**, <7 min, suggested that it is a highly substituted glycoside with a single sugar unit linked at C-19. Acid hydrolysis of **3** furnished a mixture of aglycones and only D-glucose which was identified by comparison of the HPLC retention times of thiocarbamoyl thiazolidine derivatives prepared from sugar standards, as previously described [8,9].

The aglycone signals were similar to those reported for steviol aglycone in compounds **1** and **2** except for δ_H 2.51 and δ_H 4.04 but signals were in good agreement with those reported for a similar aglycone previously found in *Stevia eupatoria* (Spreng.) Willd. and reported as 12-α-hydroxy-*ent*-kaur-16-en-19-oic acid based on the ^{1}H- and ^{13}C-NMR data together with chemical modifications [12]. Additionally, a triglucopyranosyl derivative isolated from *S. rebaudiana* (stevioside isomer) was also reported and was suggested a similar aglycone 12-α-hydroxy-*ent*-kaur-16-en-19-oic acid by comparison with previously reported data [13]. More recent NMR techniques allowed us to assign the aglycone as 13(*S*)-hydroxyastinoic acid. This type of aglycone was previously reported by converting microbiologically diterpene acids from *Helianthus* sp. with *Gibberella fujikuroi* [14]. The structure of the aglycone in **3** was inferred as described in Figure 2. The COSY spectrum reveals the sequence H-9–H-11ab–H-12. The coupling of H-17 with C-12 and C-15, seen in the HMBC spectrum, set fragment a. Coupling of H-15 with both C-8 and C-9 indicated that C-15 is bonded to C-8, as in fragment b. Couplings between the proton on the carbon carrying the oxygen (H-13, 4.04 ppm) with H-12 and H-14 were seen in the COSY spectrum, so they are vicinal, also corroborated for the HMBC correlation between H-13 and C-16. Protons in position 14 were identified by their coupling with H-13. They couple with C-8, C-9 and C-15, and therefore C-14 is bonded to C-8, as in fragment c. H-9 displayed a large coupling with C-14 and C-15, and the stereochemistry of fragment d was inferred. The stereochemistry of C-13 was assigned as *S* because of the large coupling of H-13 with C-16. The assignment of the diastereotopic protons was based on the large couplings between H-14 *pro-R* and C9, H-14 *pro-S* and C-15, H-15 *pro-S* and C-9, H-15 *pro-R* and C-14, H-11 *pro-S* and C-16, H-11 *pro-R* and C-13. A ROESY spectrum of compound **4**, in which the chemical shifts of the aglycone are very similar to those in **3**, confirmed this stereochemistry and these assignments, as it displayed nOes of H-20 with H-13 and H-14 *pro-R*, and of H-14 *pro-R* and H-11 *pro-R*.

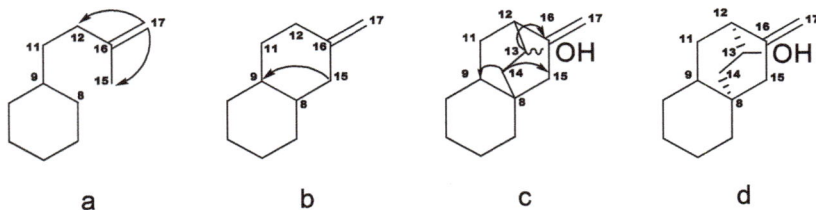

Figure 2. Structure elucidation for the 13(*S*)-hydroxyatisenoic acid. The arrows indicate cross-peaks in the HMBC spectrum between protons at the start position and carbons at the end position.

Four anomeric protons at δ_H 4.56, 4.65, 4.81, and 5.43 ppm were also observed supporting the MS information, all of them showing beta linkages. ^{1}H and ^{13}C chemical shifts are shown in Tables 1 and 2, respectively. The positions of attachment of the sugar moieties were established based on the HMBC spectrum. ^3J HMBC correlations between anomeric proton H-1' (δ 5.43 ppm) and C-19 (δ 176.8 ppm) confirmed the attachment of one glucose unit at C-19. The order of the protons in a sugar starting from the anomeric position was seen in the TOCSY spectra with increasing mixing times (Supporting Information). Additionally, the ^3J HMBC correlations between H-1'' (δ 4.56 ppm) and C-13 (δ 77.8 ppm) confirmed the attachment of the first glucose unit at C-13. The connections of the other two glucoses were also established through ^3J HMBC correlations between H-1''' (δ 4.81 ppm), H-1'''' (δ 4.65 ppm) with C-2'' (δ 79.0 ppm) and C-3'' (δ 86.0 ppm) respectively. In the same way, position of H-2'' was confirmed through the gDQCOSY correlation between H-2'' (δ 3.64 ppm) and H-1'' (δ 4.56 ppm). The order of the protons in a sugar starting from the anomeric position was seen in the TOCSY spectra with increasing mixing times, (Supporting Information). Compound **3** was named 13-[(2-*O*-β-D-glucopyranosyl-3-*O*-β-D-glucopyranosyl-β-D-glucopyranosyl)oxy]*ent*-hydroxyatis-16-en-19-oic acid-β-D-glucopyranosy ester and assigned structurally as shown in Figure 1.

Compound **4** was prepared through alkaline hydrolysis of **3** to afford a rebaudioside B isomer with a with a Glcβ(1-3)[Glcβ(1-2)]Glcβ$_1$- arrangement at C-13 with retention of 10.5 min that did not match with any of the previously reported diterpene glycosides with a free carboxylic acid previously reported in the RP-C18 HPLC method [10]. HRESIMS/MS experiment of compound **4** showed a molecular ion at *m/z* 803.3682 [M − H]$^-$ (calculated *m/z* 803.3702 [M − H]$^-$), suggesting a molecular formula C$_{38}$H$_{60}$O$_{18}$. A sequential loss of three glucoses units at C-13 portion at collision energy of 70 eV was observed: *m/z* 641.3162 (−162 Da), *m/z* 479.2703 (−162 Da) and *m/z* 317.2111 Da. ^{1}H-NMR showed three anomeric protons corresponding to the oligosaccharide portion linked at position C-13 (δ_H 4.55; 4.65 and 4.81 ppm). The free carboxylic acid at position C-19 was evidenced by the signal at 184.4 ppm in the ^{13}C-NMR, less shielded than that observed for compound **3**. Full structural assignment was performed using 1D- and 2D-NMR experiments. Compound **4** was named as 13-[(2-*O*-β-D-glucopyranosyl-3-*O*-β-D-glucopyranosyl-β-D-glucopyranosyl)oxy]*ent*-hydroxyatis-16-en-19-oic acid. ^{1}H- and ^{13}C-NMR chemical shifts are presented in Table 1. Additionally, glycosylation sites, sugar arrangements, retention times and [M − H]$^-$ of compounds **1-5** including other related isomers were presented in Table 3.

Table 3. Glycosylation sites, sugar arrangements, HPLC retention times, molecular weights and type of aglycones of rebaudioside A and B isomers.

DG	Glycosylation Sites			RT (min) [a]		$[M - H]^-$ (m/z)
	C-12 (R_2)	C-13 (R_2)	C-19 (R_1)	DG	Agly [b]	Experimental
Rebaudioside Z (**1**)	-CH$_2$	Glcβ(1-2)[Glcβ(1-6)]-Glcβ1-	Glcβ1-	3.94	I	965.4222
Compound **3**	-CH	Glcβ(1-2)[Glcβ(1-3)]-Glcβ$_1$- and -H	Glcβ1-	4.35	II	965.4297
Rebaudioside A	-CH$_2$	Glcβ(1-2)[Glcβ(1-3)]-Glcβ1-	Glcβ1-	7.11	I	965.4178
Iso-rebaudioside A	-CH$_2$	Glcβ(1-2)[Glcβ(1-3)]-Glcβ1-	Glcβ1-	7.91	III	965.4204
Rebaudioside Z$_1$ (**2**)	-CH$_2$	Glcβ(1-2)[Glcβ(1-6)]-Glcβ1-	H	9.32	I	803.3721
Compound **4**	-CH	Glcβ(1-2)[Glcβ(1-3)]-Glcβ$_1$- and -H	H	10.5	II	803.3682
Rebaudioside B	-CH$_2$	Glcβ(1-2)[Glcβ(1-6)]-Glcβ1-	H	14.32	I	803.3692
Iso-rebaudioside B	-CH$_2$	Glcβ(1-2)[Glcβ(1-3)]-Glcβ1-	H	14.44	III	803.3682

[a] Reversed-phase C18 high performance liquid chromatography method. [b] Aglycone structures are presented in Figure 3. Compound **3**: 13-[(2-*O*-β-D-glucopyranosyl-3-*O*-β-D-glucopyranosyl-β-D-glucopyranosyl) oxy]*ent*-hydroxyatis-16-en-19-oic acid-β-D-glucopyranosy ester; compound **4**: 13-[(2-*O*-β-D-glucopyranosyl-3-*O*-β-D-glucopyranosyl-β-D-glucopyranosyl) oxy]*ent*-hydroxyatis-16-en-19-oic acid.

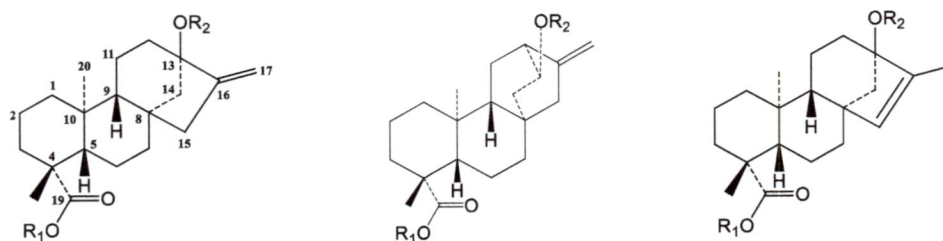

Figure 3. Chemical structures of steviol (I), hydroxyatisenoic acid (II) and *endo*-steviol aglycone (III).

Additionally, three known compounds were purified in gram quantities and identified as rebaudiosides A (**5**), H (**6**) and J (**7**). The purification of the degradation products formed under acidic condition from rebaudioside A and stevioside was recently reported [8]. Preparation of a high-quality crystal of iso-stevioside, one of the by-products, allowed the confirmation of its structure by X-ray diffraction (Figure 4).

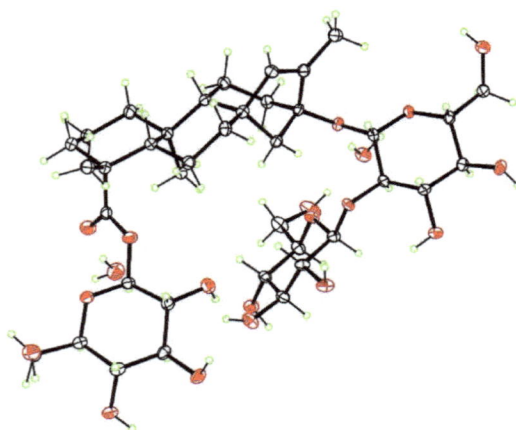

Figure 4. X-ray structure of iso-stevioside dihydrate, with ellipsoids at the 50% level (ORTEP).

Rebaudioside A, one of the major compounds from *S. rebaudiana*, has found use as the main table-top and additive in beverages due to it being more than 200 times sweeter than sucrose and its potential human health benefits [5]. However, this compound certainly interacts with bitter taste receptors through several possible mechanisms by which the bitter aftertaste of rebaudioside A may suppress the sweet gustatory receptors activity. As far as we know, there is no complete study showing the relationship between structure-organoleptic properties of steviol glycosides. Only a couple of rebaudioside A isomers with sugar arrangement at position C-13 as follows Glcβ(1-6)[Glcβ(1-3)]-Glcβ$_1$- and Glcβ(1-2)[Fruβ(1-3)]-Glcβ$_1$- were reported [14]. The novel rebaudioside A isomers herein described, rebaudioside Z Glcβ(1-6)[Glcβ(1-2)]-Glcβ$_1$- at C-13 and 13-[(2-O-β-D-glucopyranosyl-3-O-β-D-glucopyranosyl-β-D-glucopyranosyl) oxy]*ent*-hydroxyatis-16-en-19-oic acid-β-D-glucopyranosy ester Glcβ(1-3)[Glcβ(1-2)]-Glcβ$_1$- at C-13 with a different aglycone may serve as models to provide important findings to better understand the relationship between sugar arrangement and positions with sweet/bitter flavors. To date, several steviol glycoside isomers have been isolated from *S. rebaudiana* mainly with different sugar arrangements, position of the attachment to the aglycone and also with different aglycones as is the case of stevioside (with Glcβ(1-2)Glcβ$_1$- at C-13)/12-α-[(2-*O*-β-D-glucopyranosyl-β-D-glucopyranosyl) oxy]ent-kaur-16-en-19-oic acid-β-D-glucopyranosyl ester (Glcβ(1-2)Glcβ$_1$- at C-12) [13]; rebaudioside E (with Glcβ(1-2)Glcβ$_1$- at C-19) /rebaudioside Y (Glcβ(1-6)Glcβ$_1$- at C-19) [15]; rebaudioside F (with Xylβ(1-2)[Glcβ(1-3)]-Glcβ$_1$- at C-13)/rebaudioside R (Glcβ(1-2)[Glcβ(1-3)]-Xylβ$_1$- at C-13)/rebaudioside F isomer (with Glcβ(1-2)[Xylβ(1-3)]-Glcβ$_1$- at C-13) [14,16,17]; rebaudioside D (with Glcβ(1-2)Glcβ$_1$- at C-19)/rebaudioside I (Glcβ(1-3)Glcβ$_1$- at C-19) [17,18]. Additionally, several compounds differing in the type of sugar in a specific position could also be compared for better understanding of the organoleptic properties e.g., rebaudioside C (Rhaα(1-2)[Glcβ(1-3)]-Glcβ$_1$- at C-13)/6-deoxyGlcβ(1-2)[Glcβ(1-3)]Glcβ$_1$-/rebaudioside F [18,19]; rebaudioside A/Glcβ(1-2)[Fruβ(1-3)]-Glcβ$_1$- at C-13 among others. Additionally, diterpene glycosides with an endocyclic double bond (C 15) could be compared with their exocyclic double bond isomers as in the case of iso-rebaudioside A/rebaudioside A; iso-stevioside/stevioside among other pairs of compounds.

3. Materials and Methods

3.1. Chemicals

Acetonitrile and water for HPLC and Silica gel 60 F254 HPTLC plates were purchased from EMD Millipore (Cincinnati, OH, USA). The bulk acetonitrile, methanol, methyl *tert*-butyl ether (MTBE) acetic acid, ethyl acetate, and isopropyl alcohol (IPA) were purchased from Reagents (Nashville, TN, USA). Flash silica was purchased from Sorbent Technologies (Atlanta, GA, USA).

3.2. General Experimental Procedures

The mass detector was a quadrupole time of flight (Model G6530A, Agilent, Palo Alto, CA, USA) equipped with an electrospray ionization interface and was controlled by Agilent software (A.05.00, Agilent MassHunter Work Station, Palo Alto, CA, USA). All acquisitions were performed under negative ionization mode with a capillary voltage of 3500 V. Nitrogen was used as nebulizer gas (30 psig) as well as drying gas at 10 L/min at drying gas temperature of 300 °C. The voltage of PMT, fragmentor and skimmer was set at 750 V, 100 V and 65 V respectively. Full scan mass spectra were acquired from *m/z* 100–1700. Data acquisition and processing was done using the MassHunter Workstation software (Qualitative Analysis Version B.07.00).

NMR spectra were acquired either at the University of Mississippi (Oxford, MS, USA) on an Avance NMR spectrometer (Bruker, Billerica, MA, USA) equipped with a Bruker 5 mm C13/H1-F19 cryoprobe or at the University of Florida (Gainesville, FL, USA) on an Inova spectrometer (Varian, Palo Alto, CA, USA) equipped with a Varian 5 mm H1/C13/P31-N15 indirect detection probe both operating at 500 MHz for proton and 125 MHz for carbon and using z-axis pulsed-field gradients. The temperature was set at 25 °C and chemical shifts (δ) were reported in ppm and referenced to tetramethylsilane or solvent signals using similar experimental conditions as previously reported [20].

3.3. Plant Material

The starting material was a partially processed commercially available extract of *Stevia rebaudiana* with Lot # SRE50-14091 purchased from American Mercantile (Memphis, TN, USA). HPLC comparison of that extract with several other *S. rebaudiana* extracts purchased from various sources showed high similarities, differing only in the relative concentrations of specific glycosides but not their presence or absence.

3.4. Isolation Procedure

Commercially available *S. rebaudiana* leaf extract (1.5 kg) was dissolved in methanol or 10% aqueous methanol at about 200 mg/mL and allowed to crystallize. The crystalline products were rebaudioside A and stevioside, which accounted for approximately 50% of the starting mass.

Pools rich in rebaudioside N from several large-scale chromatographies for isolation of rebaudioside C in quantity [21] were combined to obtain 140 g of extract. All this material was fractionated on a high efficiency reverse-phase chromatography column (7.5 i.d. ×50 cm, 10 μm spherical C18 gel) [20]. The column was loaded with the 140 grams dissolved in distilled water, the column eluted with 3 liters of 0.5% acetic acid in water and then switched to 15:85 acetonitrile: 0.1% acetic acid in water (3 liters); 25:75 acetonitrile: 0.1% acetic acid in water (3 liters); 40:60 acetonitrile: 0.1% acetic acid in water (3 liters); and column washed with 90:10 acetonitrile: 0.1% acetic acid in water and 100% methanol (1 liter each, washes combined and passed to waste.). 500 mL fractions collected and analyzed by HPLC. Similar fractions combined and fractions rich in rebaudioside N yielded ~22 grams. This pool was adsorbed onto celite (120 g), divided into three portions, each portion packed into a load column and chromatographed on a high efficiency normal-phase chromatography column (7.5 i.d. ×50 cm, 10 μm spherical silica gel) using Reb N mobile phase (Reb C mobile phase [100:18:14; EtOAc:MeOH:H_2O with 0.1% AcOH] with additional 10 parts methanol and 10 parts water and 5 parts acetic acid). Column analysis [21] allowed combination of fractions rich in rebaudioside N. This pool spontaneously crystallized, the crystals filtered and the MLs dried, redissolved and a second crop obtained.

The supernatant second crop (7.02 g) was absorbed onto 70 g of Celite and subjected to a high efficiency normal-phase chromatography (7.5 i.d. ×50 cm, 10 μm spherical silica gel) with acetonitrile: H_2O: AcOH (88:12:0.01 v/v/%). 2 × 1 L forerun were initially collected, followed by 48 fractions of 120 mL. All fractions were analyzed by HPLC methods and five main fractions were pooled based on results from column analysis [21]. *Chromatography 1*, fraction 1.1 (0.093 g); fraction 2.1 (5.858 g); fraction 3.1 (0.64 g); fraction 4.1 (0.328 g) and fraction 5.1 (0.109 g). Fraction 2.1 was re-chromatographed in high efficiency normal-phase chromatography (7.5 i.d. ×50 cm) using MTBE: MeOH: H_2O: AcOH (100:30:12.5:0.01). 2 × 1 L forerun and 40 × 120 mL fractions were collected and analyzed by HPLC. Five main fractions were selected by column analysis [21]. *Chromatography 2*, fraction 1.2 (1.3 g); fraction 2.2 (2.09 g); fraction 3.2 (1.49 g); fraction 4.2 (0.088 g) and fraction 5.2 (0.065 g). Fraction 1.2 (1.3 g) was chromatographed in a high efficiency reversed-phase column (7.5 i.d. ×50 cm, 10 μm spherical C18 gel) using ACN:H_2O:AcOH (25:75:0.01). 2 × 1 L forerun and 25 × 120 mL fractions were collected and analyzed by HPLC. Column analysis [21] allowed us to select two main fractions. *Chromatography 3*, fraction 1.3 (0.208 g); fraction 2.3 (0.922 g). Fraction 2.2 (2.09 g) and fraction 2.3 (0.922 g) were combined and digested with MeOH to afford 2.2 g of solids which were chromatographed in a high efficiency reversed-phase column (7.5 i.d. ×50 cm) using ACN:H_2O:AcOH (23:77:0.01). 6 × 1 L + 1 × 0.7 L forerun, and 37 × 120 mL fractions were collected and analyzed by HPLC. Column analysis allowed us to select three main fractions. *Chromatography 4*, fraction 1.4 (0.546 g); fraction 2.4 (0.713 g) and fraction 3.4 (1.165 g). Fraction 3.4 (1.165 g) was absorbed onto 10 g of celite and chromatographed in high efficiency normal-phase chromatography (7.5 i.d. ×50 cm) using "Reb C" mobile phase 2% MeOH [Reb C mobile phase = EtOAc:MeOH:H_2O:AcOH (100:18:14:0.1; $v/v/v$/%]. 1 × 1 L forerun and 48 × 120 mL fractions were collected and analyzed by HPLC. Column analysis allowed us to select five main pools. *Chromatography 5*, fraction 1.5 (67.1 mg); fraction 2.5 (135 mg); fraction 3.5 (226 mg); fraction 4.5 (179 mg) and fraction 5.5 (179 mg). Compound **1** (226 mg; 0.02% yield) was obtained from fraction 3.5.

Fractions rich in rebaudioside H were pooled (3.7 g) and chromatographed over a RP-C18 (7.5 i.d. ×50 cm) column with H_2O:AcOH (100:0.1 v/v/%) and ACN:H_2O:AcOH (5:95:0.1 v/v/%) to afford seven main fractions. *Chromatography 6*, fraction 1.6 (175 mg); fraction 2.6 (120 mg); fraction 3.6 (108 mg); fraction 4.6 (169 mg); fraction 5.6 (1.8 g, rebaudioside H); fraction 6.6 (156 mg) and fraction 7.6 (13 mg). *Chromatography 7*, fraction 1.6 (175 mg) was submitted to a RP-C18 (250 × 10 mm, 5 μm) chromatography using H_2O: AcOH (100:0.1) and ACN: H_2O: AcOH (10:90:0.1) to afford two main fractions., fraction 1.7 (110 mg; 0.007% yield) and fraction 2.7 (48.3 mg). Compound **3** was obtained from fraction 1.7. Additionally, several hundred grams from the initial crystallization and subsequent chromatographies of rebaudioside A after processing 1.5 kg of commercial extract. Rebaudiosides J (1 g) and H (1.8 g) were also isolated.

3.5. Alkaline Hydrolysis of Compounds 1 and 3

Compounds **1** (40 mg) and **3** (30 mg) were heated individually with NaOH (1 N) at 80 °C over 1 h. Each reaction mixture was cooled over 5 min and neutralized with two or three drops of acetic acid glacial [18] with further cleanup through a Strata RP-C18-E cartridge (500 mg/6 mL) (Phenomenex, Torrance, CA, USA). Elution with a stepwise gradient of 1.5 mL volume each, water, acetonitrile: water (2:8) and methanol to produce clean compounds **2** (27 mg; 68% yield) and **4** (15 mg; 50% yield).

3.6. Physicochemical Parameters of Compounds

Rebaudioside Z (**1**): Amorphous off-white solid; $[\alpha]^{25}_D$ −23.0 (*c* 0.1, MeOH). HR-ESIMS/MS *m/z* 965.4222 [M − H]⁻ (calculated for $C_{44}H_{70}O_{23}$, 966.4309), *m/z* 803.3703 at 10 eV collision energy, loss of one hexose (−162 Da) from C-19 moiety, *m/z* 641.3185 (−162 Da), 479.2635 (−162 Da) and 317.2146 (−162 Da), loss of three hexoses from C-13 moiety at 70 eV collision energy. ^{1}H- and ^{13}C-NMR spectroscopic data are shown in Tables 1 and 2.

Rebaudioside Z_1 (**2**): Amorphous off-white solid; $[\alpha]^{25}_D$ −46.0 (*c* 0.1, MeOH). HR-ESIMS/MS *m/z* 803.3721 [M − H]⁻ (calculated for $C_{38}H_{60}O_{18}$, 804.3781), *m/z* 641.3196 (−162 Da), 479.2648 (−162 Da) and 317.2115 (−162 Da), loss of three hexoses from C-13 moiety at 70 eV collision energy. ^{1}H- and ^{13}C-NMR spectroscopic data are shown in Tables 1 and 2.

13-[(2-O-β-D-Glucopyranosyl-3-O-β-D-glucopyranosyl-β-D-glucopyranosyl) oxy]ent-hydroxyatis-16-en-19-oic acid-β-D-glucopyranosy ester (**3**): Amorphous off-white solid; $[\alpha]^{25}_D$ −22.0 (*c* 0.1, MeOH), HR-ESIMS/MS *m/z* 965.4297 [M − H]⁻ (calculated for $C_{44}H_{70}O_{23}$, 966.4309), *m/z* 803.3735 at 10 eV collision energy (−162 Da), loss of one hexoses from C-19 moiety, *m/z* 641.3195 (−162 Da), 479.2703 (−162 Da) and 317.2141 (−162 Da), loss of three hexoses from C-13 moiety at 70 eV collision energy. ^{1}H- and ^{13}C-NMR spectroscopic data are shown in Tables 1 and 2.

13-[(2-O-β-D-Glucopyranosyl-3-O-β-D-glucopyranosyl-β-D-glucopyranosyl) oxy]ent-hydroxyatis-16-en-19-oic acid (**4**): Amorphous off-white solid; $[\alpha]^{25}_D$ −20.0 (*c* 0.1, MeOH). HRESIMS/MS *m/z* 803.3682 [M − H]⁻ (calculated for $C_{38}H_{60}O_{18}$, 803.3702), *m/z* 641.3162 (−162 Da), *m/z* 479.2703 (−162 Da) and *m/z* 317.2111 Da at 70 eV collision energy. ^{1}H- and ^{13}C-NMR spectroscopic data are shown in Tables 1 and 2.

3.7. RP-C18 HPLC Analysis

Analyses were performed with a Hewlett Packard Agilent 1100 Series system equipped with a G1311A quaternary pump, a G1322 degasser, a G1316A oven, G1313A autosampler and a G1315A diode array detector. Acetonitrile and water for HPLC were purchased from EMD Millipore (Cincinnati, OH, USA). The elution was performed with 0.01 M phosphoric acid (A) and acetonitrile (B) with a flow rate set at 1 mL/min. All the analyses were performed with Phenomenex columns. After each analysis, the column was washed and equilibrated appropriately. 10 μL of compound **1**–**4** were injected in the RP C-18 Luna (2), Phenomenex (250 × 4.6 mm, 5 μm) column at 30 °C. The elution was performed using gradient of elution as follows: 0–5 min, 32% B; 5–13 min, 32–41% B; 13–16 min, 41–43% B; 16–17 min: 43–50% B; 17–23 min, 50% B. The chromatogram was recorded at 205 nm and the flow rate set at 1 mL/min [10].

3.8. Determination of the Sugar Unit Absolute Configuration

Compounds **1** and **3** (1 mg) were hydrolyzed with HCl (1 N) at 80 °C over 2 h followed by liquid-liquid partition with ethyl acetate (2 × 1 mL). The aqueous layer was neutralized with silver carbonate and the supernatant was recovered and heated with L-cysteine methyl ester in pyridine for 1 h at 60–70 °C. The mixture was dried in a vacuum oven at 40 °C. After dryness, 400 μL of pyridine and 100 μL of phenylisothiocyanate were added and heated for an additional hour at 60–70 °C to form the thiocarbamoyl thiazoline derivatives. Reaction mixtures were analyzed by the HPLC method previously reported [9]. The absolute configuration of the sugars was determined by comparison of the HPLC retention times of the prepared thiocarbamoyl thiazolidine derivatives with appropriate standards.

3.9. X-ray Crystallography of Iso-Stevioside

The crystal structure and absolute configuration of iso-stevioside dihydrate were determined from a colorless crystal of dimensions 0.45 × 0.14 × 0.02 mm, using data collected at T = 90 K with Cu Kα radiation on an APEX-II DUO CCD diffractometer (Bruker, Madison, WI, USA) equipped with a microfocus source and a Cryostream cooler (Oxford, Cryosystems, Oxford, UK). The structure was solved using the program SHELXS-97 (University of Göttingen, Germany) and refined anisotropically by full-matrix least-squares on F^2 using SHELXL-2014/7 (University of Göttingen, Germany) [22]. All H atoms were visible in difference maps, but were placed in idealized positions for the refinement, except for those of OH groups and water molecules, which were refined. The absolute configuration was determined from the Flack [23] parameter of 0.01(5) based on resonant scattering of the light atoms and 2804 quotients. The reported configuration has C4(*R*), C5(*S*), C8(*R*), C9(*R*), C10(*S*), C13(*S*) and is in agreement with the known configurations of the β-D-glucopyranose moieties. Crystal data: $C_{38}H_{60}O_{18} \cdot 2H_2O$, Mr = 840.89, monoclinic space group P2₁, a = 13.2330(6) Å, b = 8.1081(4) Å, c = 19.1556(8) Å, β = 105.655(2)°, V = 1979.05(16) Å³, Z = 2, Dx = 1.411 g cm⁻³, θ_{max} = 68.3°, R = 0.031 for all 6901 unique data and 571 refined parameters. Supplementary crystallographic data for iso-stevioside dihydrate are contained in Cambridge Structural Database deposition CCDC-1879103; this data can be obtained free of charge via www.ccdc.cam.ac.uk/conts/retrieving.html (or from the Cambridge Crystallographic Data Centre, 12 Union Road, Cambridge CB2 1EZ, UK; fax: (+44) 1223-336-033 or e-mail: deposit@ccdc.cam.ac.uk)

4. Conclusions

Two new rebaudioside A isomers, rebaudiosides Z (**1**) and 13-[(2-*O*-β-D-glucopyranosyl-3-*O*-β-D-glucopyranosyl-β-D-glucopyranosyl) oxy]*ent*-hydroxyatis-16-en-19-oic acid-β-D-glucopyranosy ester (**3**) were isolated from a partially processed commercial extract of *S. rebaudiana* and two new rebaudioside B isomers, rebaudiosides Z₁ and 13-[(2-*O*-β-D-glucopyranosyl-3-*O*-β-D-glucopyranosyl-β-D-glucopyranosyl) oxy]*ent*-hydroxyatis-16-en-19-oic acid were prepared and purified for the first time, respectively. An additional three known compounds, rebaudiosides A, H and J, were purified in gram quantities. Scarce rebaudioside A isomers have been reported from *Stevia rebaudiana* differing in sugar arrangement and type of sugar at position C-13 (Glcβ(1-6)[Glcβ(1-3)]-Glcβ₁-) and Glcβ(1-2)[Fruβ(1-3)]-Glcβ₁-). However, herein we describe the occurrence of two rebaudioside A isomers with (Glcβ(1-6)[Glcβ(1-2)]-Glcβ₁-) and (Glcβ(1-3)[Glcβ(1-2)]-Glcβ₁-) at position C-13. However, compound **3** showed a different aglycone and was found to be a 13(*S*)-hydroxyatisenoic acid type. This finding may contribute to better understanding of the relationship between sweet/bitter taste of Stevia glycosides with their sugar arrangements. Several new compounds have been reported in recent years, however, a systematic study comparing organoleptic properties with structure of diterpene glycosides is still not available in literature.

Supplementary Materials: The supporting information are available online.

Author Contributions: W.H.P., C.D.H., I.M., I.A.K., and J.D.M. conceived and designed the study. W.H.P., K.A., D.L.R. purified compounds; W.H.P., I.G., D.L.R., K.A., F.T.W., F.R.F., M.A.I., and B.A. conducted the experiments. All authors shared analyzing the data and writing the manuscript.

Funding: This research received no external funding.

Acknowledgments: The project was supported by National Center for Natural Product Research, School of Pharmacy, University of Mississippi, University, MS 38677, USA. This work was supported in part by the USDA Agricultural Research Service Specific Cooperative Agreement No. 58-6060-6-015.

Conflicts of Interest: The authors declare no conflict of interest.

References

1. Whiting, D.R.; Guariguata, L.; Weil, C.; Shaw, J. IDF diabetes atlas: Global estimates of the prevalence of diabetes for 2011 and 2030. *Diabetes Res. Clin. Pract.* **2011**, *94*, 311–321. [CrossRef] [PubMed]
2. DeFronzo, R.A.; Ferrannini, E. Insulin resistance: A multifaceted syndrome responsible for NIDDM, obesity, hypertension, dyslipidemia, and atherosclerotic cardiovascular disease. *Diabetes Care* **1991**, *14*, 173–194. [CrossRef] [PubMed]
3. Midmore, D.; Rank, A. *A Report for the Rural Industries Research and Development Corporation, RIRDC Project No UCQ-16A*; Rural Industries Research and Development Corporation: ACT, Barton, Australia, 2002.
4. Ohtani, K.; Yamasaki, K. Methods to improve the taste of the sweet principles of *Stevia rebaudiana*. In *Stevia: The Genus Stevia*; Kinghorn, A.D., Ed.; Taylor & Francis: London, UK, 2002; pp. 138–159.
5. Philippaert, K.; Pironet, A.; Mesuere, M.; Sones, W.; Vermeiren, L.; Kerselaers, S.; Pinto, S.; Segal, A.; Antoine, N.; Gysemans, C. Steviol glycosides enhance pancreatic beta-cell function and taste sensation by potentiation of TRPM5 channel activity. *Nat. Commun.* **2017**, *8*, 1–16. [CrossRef] [PubMed]
6. McChesney, J.; Rodenburg, D. Chromatography Methods. U.S patent 8,801,924B2, 3 January 2014.
7. Perera, W.H.; Avula, B.; Khan, I.A.; McChesney, J.D. Assignment of sugar arrangement in branched steviol glycosides using electrospray ionization quadrupole time-of-flight tandem mass spectrometry. *Rapid Commun. Mass Spectrom.* **2017**, *31*, 315–324. [CrossRef] [PubMed]
8. Perera, W.H.; Docampo, M.L.; Wiggers, F.T.; Hufford, C.D.; Fronczek, F.R.; Avula, B.; Khan, I.A.; McChesney, J.D. Endocyclic double bond isomers and by-products from rebaudioside A and stevioside formed under acid conditions. *Phytochem Lett.* **2018**, *25*, 163–170. [CrossRef]
9. Tanaka, T.; Nakashima, T.; Ueda, T.; Tomii, K.; Kouno, I. Facile Discrimination of Aldose Enantiomers by Reversed-Phase HPLC. *Chem. Pharm. Bull.* **2007**, *55*, 899–901. [CrossRef] [PubMed]
10. Perera, W.H.; Ghiviriga, I.; Rodenburg, D.L.; Carvalho, R.; Alves, K.; McChesney, J.D. Development of a high-performance liquid chromatography procedure to identify known and detect novel C-13 oligosaccharide moieties in diterpene glycosides from *Stevia rebaudiana* (Bertoni) Bertoni (Asteraceae): Structure elucidation of rebaudiosides V and W. *J. Sep. Sci.* **2017**, *40*, 3771–3781. [CrossRef] [PubMed]
11. Kohda, H.; Kasai, R.; Yamasaki, K.; Murakami, K.; Tanaka, O. New sweet diterpene glucosides from *Stevia rebaudiana*. *Phytochemistry* **1976**, *15*, 981–983. [CrossRef]
12. Ortega, A.; Morales, F.; Salmon, M. Kaurenic acid derivatives from *Stevia eupatoria*. *Phytochemistry* **1985**, *24*, 1850–1852. [CrossRef]
13. Ibrahim, M.A.; Rodenburg, D.L.; Alves, K.; Fronczek, F.R.; McChesney, J.D.; Wu, C.; Nettles, B.J.; Venkataraman, S.K.; Jaksch, F.J. Minor diterpene glycosides from the leaves of *Stevia rebaudiana*. *J. Nat. Prod.* **2014**, *77*, 1231–1235. [CrossRef] [PubMed]
14. Chaturvedula, V.; Prakash, I.J. Additional minor diterpene glycosides from *Stevia rebaudiana*. *Nat. Prod. Commun.* **2011**, *6*, 1059–1062. [PubMed]
15. Perera, W.H.; Ramsaroop, T.; Carvalho, R.; Rodenburg, D.L.; McChesney, J.D. A silica gel orthogonal high-performance liquid chromatography method for the analyses of steviol glycosides: Novel tetra-glucopyranosyl steviol. *Nat. Prod. Res.* **2018**, 1–9. [CrossRef] [PubMed]
16. Starratt, A.N.; Kirby, C.W.; Pocs, R.; Brandle, J.E. Rebaudioside F, a diterpene glycoside from *Stevia rebaudiana*. *Phytochemistry* **2002**, *59*, 367–370. [CrossRef]
17. Ibrahim, M.A.; Rodenburg, D.L.; Alves, K.; Perera, W.H.; Fronczek, F.R.; Bowling, J.; McChesney, J.D. Rebaudiosides R and S, minor diterpene glycosides from the leaves of *Stevia rebaudiana*. *J. Nat. Prod.* **2016**, *79*, 1468–1472. [CrossRef] [PubMed]

18. Ohta, M.; Sasa, S.; Inoue, A.; Tamai, T.; Fujita, I.; Morita, K.; Matsuura, F. Characterization of Novel Steviol Glycosides from Leaves of *Stevia rebaudiana* Morita. *J. Appl. Glycosci.* **2010**, *57*, 199–209. [CrossRef]

19. Ceunen, S.; Geuns, J.M. Steviol glycosides: Chemical diversity, metabolism, and function. *J. Nat. Prod.* **2013**, *76*, 1201–1228. [CrossRef] [PubMed]

20. Perera, W.H.; Ghiviriga, I.; Rodenburg, D.L.; Alves, K.; Bowling, J.J.; Avula, B.; Khan, I.A.; McChesney, J.D. Rebaudiosides T and U, minor C-19 xylopyranosyl and arabinopyranosyl steviol glycoside derivatives from *Stevia rebaudiana* (Bertoni) Bertoni. *Phytochemistry* **2017**, *135*, 106–114. [CrossRef] [PubMed]

21. Rodenburg, D.L.; Alves, K.; Perera, W.H.; Ramsaroop, T.; Carvalho, R.; McChesney, J.D. Development of HPLC analytical techniques for diterpene glycosides from *Stevia rebaudiana* (Bertoni) Bertoni: Strategies to scale-up. *J. Braz. Chem. Soc.* **2016**, *27*, 1406–1412.

22. Sheldrick, G.M. A short history of SHELX. *Acta Crystallogr.* **2008**, *A64*, 112–122. [CrossRef] [PubMed]

23. Flack, H.D. Crystal Structure and Synthesis of 3β-(p-Iodobenzoyloxy)-16α,17α-Epoxypregn-4-En-6,20-Dione. *Acta Crystallogr.* **1983**, *A39*, 876–881. [CrossRef]

Sample Availability: Samples of the compounds **1–7** are available from the authors.

![molecules](molecules logo)

molecules

MDPI

Article

Lignans from the Twigs of *Litsea cubeba* and Their Bioactivities

Xiuting Li [1,†], Huan Xia [2,†], Lingyan Wang [2], Guiyang Xia [2], Yuhong Qu [2], Xiaoya Shang [3,*] and Sheng Lin [2,*]

1   Beijing Advanced Innovation Center for Food Nutrition and Human Health,
    Beijing Technology and Business University, Beijing 100048, China; lixt@btbu.edu.cn
2   State Key Laboratory of Bioactive Substance and Function of Natural Medicines, Institute of Materia Medica,
    Chinese Academy of Medical Sciences and Peking Union Medical College, Beijing 100050, China;
    xiahuan@imm.ac.cn (H.X.); wanglingyan@imm.ac.cn (L.W.); xiaguiyang@imm.ac.cn (G.X.);
    qyhcxl28@126.com (Y.Q.)
3   Beijing Key Laboratory of Bioactive Substances and Functional Foods, Beijing Union University,
    Beijing 100023, China
*   Correspondence: shangxiaoya@buu.edu.cn (X.S.); lsznn@imm.ac.cn (S.L.);
    Tel.: +86-10-62004533 (X.S.); +86-10-60212110 (S.L.)
†   These authors contributed equally to this work.

Academic Editor: Muhammad Ilias
Received: 18 December 2018; Accepted: 4 January 2019; Published: 16 January 2019

Abstract: *Litsea cubeba*, an important medicinal plant, is widely used as a traditional Chinese medicine and spice. Using cytotoxicity-guided fractionation, nine new lignans **1–9** and ten known analogues **10–19** were obtained from the EtOH extract of the twigs of *L. cubeba*. Their structures were assigned by extensive 1D- and 2D-NMR experiments, and the absolute configurations were resolved by specific rotation and a combination of experimental and theoretically calculated electronic circular dichroism (ECD) spectra. In the cytotoxicity assay, 7′,9-epoxylignans with feruloyl or cinnamoyl groups (compounds **7–9**, **13** and **14**) were selectively cytotoxic against NCI-H1650 cell line, while the dibenzylbutyrolactone lignans **17–19** exerted cytotoxicities against HCT-116 and A2780 cell lines. The results highlighted the structure-activity relationship importance of a feruloyl or a cinnamoyl moiety at C-9′ or/and C-7 ketone in 7′,9-epoxylignans. Furthermore, compound **11** was moderate active toward protein tyrosine phosphatase 1B (PTP1B) with an IC$_{50}$ value of 13.5 μM, and compounds **4–6**, **11** and **12** displayed inhibitory activity against LPS-induced NO production in RAW264.7 macrophages, with IC$_{50}$ values of 46.8, 50.1, 58.6, 47.5, and 66.5 μM, respectively.

Keywords: *Litsea cubeba*; cytotoxicity; isolation and elucidation; lignans

1. Introduction

Plants from the *Litsea* species (Lauraceae) are widely distributed in tropical or subtropical areas. *Litsea cubeba*, mainly grown in the east and south of China, is broadly used as a traditional Chinese medicine and spice. "Bi-cheng-qie" and "dou-chi-jiang", the dried fruits and roots of *L. cubeba*, respectively, have been documented in the Chinese Pharmacopoeia and *Chinese Materia Medica* as two important traditional Chinese medicines for the treatment of various ailments, including coronary disease, cerebral apoplexy, asthma, and rheumatic arthritis [1–3]. Moreover, *Litsea cubeba* fruits are also important spices and great sources of essential oils which are often used as flavor enhancers in foods, cigarettes, and cosmetics [4]. Previous phytochemical investigation of the fruits and roots of *L. cubeba* have reported the discovery of aporphine-type alkaloids, lignans, and phenolic constituents [5–11]. Among them, aporphine-type alkaloids and lignans were considered as the major active principles of this plant due to their antithrombotic, anti-inflammatory, and antinociceptive properties [8,9,12–15]. Since there are few reports on the phytochemicals of twigs of *L. cubeba*, a recent study on *L. cubeba* twigs by our group led to the characterization of 36 aromatic glycosides from the the water-soluble fraction of an ethanolic extract. Interestingly, some lignan glycosides showed potent hepatoprotective and HDAC1 inhibitory activity [16,17]. In the present study, we have investigated the constituents of the EtOAc-soluble fraction of the ethanolic extract of *L. cubeba* twigs. Bioassay-guided isolation of a fraction with cytotoxicity against HCT-116, NCI-H1650, and A2780 cell lines (IC$_{50}$ = 28.3, 11.5, and 16.8 µg/mL, respectively) led to the discovery of nine new lignans 1–9 and ten analogues 10–19 (Figure 1). The structures of 1–9 were elucidated by spectroscopic methods, and their absolute configurations were determined by optical rotations and a combination of experimental and theoretically calculated electronic circular dichroism (ECD) spectra. Detailed herein are the isolation, structural elucidation, and bioactivity assay of compounds 1–19.

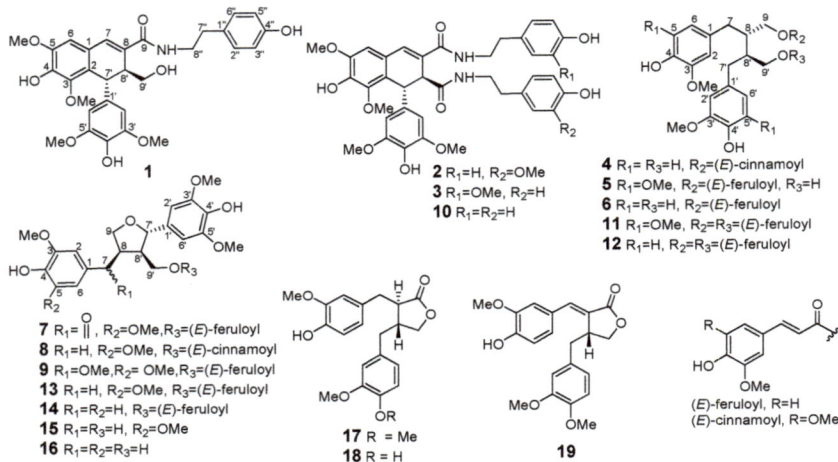

Figure 1. The structures of compounds 1–19.

2. Results and Discussions

2.1. Structure Elucidation

The EtOAc extract of the twigs of *L. cubeba* was subjected to column chromatography on silica gel to give 13 fractions (F$_1$–F$_{13}$). Cytotoxicity assays found that F$_9$ displayed potent activities against HCT-116, NCI-H1650, and A270 cell lines. Fractionation of F$_9$ by Sephadex LH-20, RP-18, preparative TLC, and preparative HPLC led to the discovery of nine new lignans 1–9 and the ten known ones 10–19.

Compound 1 was obtained as a white amorphous powder. The presence of amide (1643 cm^{-1}), aromatic ring (1611, 1516, and 1459 cm^{-1}), and hydroxy (3372 cm^{-1}) functionalities were evident in its IR spectrum. Its molecular formula of C$_{30}$H$_{33}$NO$_9$ with fifteen degrees of unsaturation was established by HREIMS based on the [M + H]$^+$ ion at *m/z* 552.2234 (calcd. 552.2228) and ^{13}C-NMR spectrum. In the ^{1}H-NMR spectrum recorded in acetone-*d$_6$*, the signals for an aromatic singlet integrated for two protons at δ 6.39 (2H, s, H-2' and H-6'), a methoxy singlet integrated for six protons at δ 3.67 (6H, s, OMe×2), suggested a 1-substituted-3,5-dimethoxy-4-hydroxybenzene ring in 1. Signals of a singlet proton at δ 6.74 and two methoxy protons at δ 3.86 and 3.58 revealed a pentasubstituted aromatic ring attached two methoxy groups. These ^{1}H-NMR signals, together with another two singlet protons at δ 7.19 and 4.62, were indicative of a typical skeleton of 2,7'-cyclolignan-7-en such as thomasic acid [18]. Additionally, the ^{1}H-NMR spectrum of **1** displayed characteristic signals for a tyramine group with resonances at δ_H 6.98 (2H, d, J = 8.5 Hz, H-2″ and H-6″), 6.71 (2H, d, J = 8.5 Hz, H-3″ and H-5″), 2.69 (2H, t, J = 7.5 Hz, H$_2$-7″), and 3.39 (2H, dt, J = 7.5, 4.5 Hz, H$_2$-8″). The ^{13}C-NMR spectrum of **1** displayed 30 carbon signals, of which twelve could be assigned to be a tyramine moiety (δ_C 131.2, 130.5 × 2, 116.0 × 2, 156.6, 35.6, 42.2) and four methoxy groups (δ_C 56.6 × 2, 56.5, 60.4), and the remaining eighteen carbons were consistent with the 2,7′-cyclolignan-7-en skeleton. The complete ^{1}H- and ^{13}C-NMR assignments of **1** were made by a combination of 1D- and 2D-NMR experiments. In the HMBC spectrum of **1**, the two or three bonds long range correlations from H-6 to C-2, C-4, and C-7, from H-7 to C-2, C-6, C-9, and C-8′, from H-7′ to C-3, C-8, C-2′ (C-6′), and C-9′, from H-8′ to C-2, C-7, C-9, and C-1′, from H$_2$-9′ to C-8 and C-7′, and from the methoxy protons at δ_H 3.58 to C-3′ (C-5′) (Figure 2) confirmed the 2,7′-cyclolignan-7-en type lignan containing a 3,5-dimethoxy-4-hydroxy-benzene moiety. The NOESY correlation observed between H-6 and the methoxy protons at δ_H 3.86 together with the HMBC correlation observed for these methoxy protons and C-5 gave the evidence for the location of one methoxy group at C-5. Key HMBC cross-peaks, such as between methoxy protons at δ_H 3.58 and C-3, as well as between OH proton at δ_H 7.76 and C-4, served to locate this methoxy and OH group at C-3 and C-4, respectively. Furthermore, the tyramine was linked to C-9 to form an amine bond, according to the HMBC correlations from both H$_2$-8″ and NH proton to C-9. Therefore, these data completed the planar structure of **1** as *N*-[2-(4-hydroxyphenyl)-ethyl]-4,4′,9′-trihydroxy-3,5,3′,5′-tetramethoxy-2,7′-cyclolignan-7-en-9-amide. H-7′ appearing as a singlet suggested the dihedral angle for the vicinal protons of H-7′ and H-8′ was nearly 90°, requiring a *trans* relationship of H-7′ and H-8′. This assignment was also supported by the NOESY correlations of H-7′ with H$_2$-9′, and H-8′ with H-2′ (H-6)′. Finally, the negative optical rotation of **1** demonstrated the 7′*R*,8′*S* absolute configuration of **1** [18,19]. Hence, compound **1** was defined as (−)-(7′*R*,8′*S*)-*N*-[2-(4-hydroxyphenyl)-ethyl]-4,4′,9′-trihydroxy-3,5,3′,5′-tetramethoxy-2,7′-cyclolignan-7-en-9-amide.

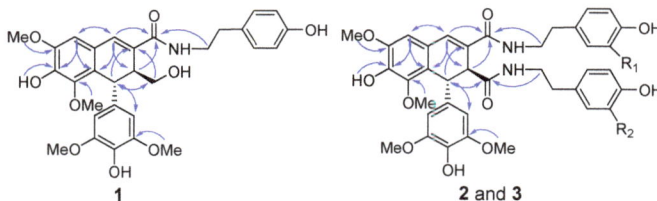

Figure 2. The key HMBC correlations of **1–3**.

Compound **2** was isolated as a white amorphous powder. The IR spectrum exhibited absorptions of hydroxy (3362 cm^{-1}), amide (1649 cm^{-1}), and aromatic (1612 and 1516 cm^{-1}) moieties. Its molecular formula was deduced as $C_{39}H_{42}N_2O_{11}$ from the negative HRESIMS at *m/z* 713.2719 [M − H]$^-$ (calcd. 713.2716) and the ^{13}C-NMR spectrum. This indicated twenty degrees of unsaturation. The NMR spectra of **2** were very similar to those of compound **10**, a known lignan diamide that was also isolated from this plant [20], with the only difference being the replacement of one of a tyramine group by a 3-methoxytyramine moiety (Table 1; Table 2). In the HMBC spectrum of **2**, H$_2$-7‴ showed HMBC correlations with the amide carbon at δ_C 171.4, which indicated that the 3-methoxytyramine moiety was connected to C-9′ via an amide bond (Figure 2). In the 1D NOE difference spectrum of **2**, H-8′ was enhanced upon irradiation of H-2′ (H-6′). This enhancement, together with H-7′ presented in a singlet, revealed a *trans*-vicinal orientation of H-7′ and H-8′. Finally, on the basis of the negative optical rotation of **2** and biosynthetic considerations, the structure of compound **2** was defined as (−)-(7′*R*,8′*S*)-*N*^{1}-[2-(4-hydroxyphenyl)-ethyl]-*N*^{2}-[2-(4-hydroxy-3-methoxyphenyl)-ethyl]-4,4′-dihydroxy-3,5,3′,5′-tetramethoxy-2,7′-cyclolignan-7-en-9,9′-diamide.

Table 1. ^{1}H-NMR Data (δ_H (mult, J, Hz)) of Compounds **1–9** in Acetone-d_6 [a].

No.	1	2	3	4	5	6	7	8	9
2				6.71 d (1.5)	6.42 s	6.70 d (1.8)	7.39 s	6.57 s	6.67 s
5				6.71 d (7.5)		6.71 d (7.8)			
6	6.74 s	6.69 s	6.60 s	6.61 dd (7.5, 1.5)	6.42 s	6.61 dd (7.8, 1.8)	7.39 s	6.57 s	6.67 s
7	7.19 s	7.18 s	7.21 s	2.80 dd (13.5, 7.0); 2.62 dd (13.5, 8.0)	2.79 dd (14.2, 7.2); 2.62 dd (14.2, 8.4)	2.80 dd (13.8, 6.6); 2.62 dd (13.8, 8.4)	4.57 m	2.91 dd (13.2, 5.4); 2.59 dd (13.2, 10.2)	4.35 d (6.5)
8				2.32 m	2.31 m	2.31 m		2.82 m	2.84 m
9				4.36 dd (11.5, 6.5); 4.11 dd (11.5, 6.0)	4.42 dd (10.8, 6.0); 4.10 dd (10.8, 6.0)	4.36 dd (11.4, 6.6); 4.11 dd (11.4, 6.0)	4.35 t (8.0); 4.21 t (8.0)	4.04 dd (8.4, 6.6); 3.74 dd (8.4, 6.6)	4.14 t (8.5); 4.04 t (8.5)
2'	6.39 s	6.38 s	6.38 s	6.73 d (1.5)	6.44 s	6.73 d (1.8)	6.78 s	6.68 s	6.63 s
5'				6.69 d (7.5)		6.69 d (7.8)			
6'	6.39 s	6.38 s	6.38 s	6.61 dd (7.5, 1.5)	6.44 s	6.61 dd (7.8, 1.8)	6.78 s	6.68 s	6.63 s
7'	4.62 s	5.03 s	5.03 s	2.70 dd (13.5, 7.0); 2.63 dd (13.5, 8.0)	2.70 dd (14.2, 7.2); 2.63 dd (14.2, 8.4)	2.70 dd (13.8, 6.6); 2.63 dd (13.8, 8.4)	4.74 d (7.5)		4.82 d (5.5)
8'	3.14 dd (7.5, 7.5)	3.66 s	3.67 s	1.99 m	1.99 m	1.99 m	3.01 m	2.61 m	
9'	3.59 m 3.28 m			3.67 m; 3.59 m	3.67 m; 3.61 m	3.69 m; 3.59 m	4.16 d (6.5)	4.53 dd (11.4, 6.6); 4.30 dd (11.4, 7.8)	
2''	6.98 d (8.5)	6.98 d (8.5)	6.79 d (1.8)	7.00 s	7.32 s	7.32 d (1.8)	7.06 d (1.5)	6.98 s	7.27 d (2.0)
3''	6.71 d (8.5)	6.72 d (8.5)							
5''	6.71 d (8.5)				6.86 d (8.4)	6.81 d (8.4)	6.82 d (8.5)		6.85 d (8.0)
6''	6.98 d (8.5)	6.98 d (8.5)	6.61 dd (7.8, 1.8)	7.00 s	7.13 dd (8.4, 1.8)	7.13 dd (8.4, 1.8)	6.96 dd (8.5, 1.5)	6.98 s	7.11 dd (8.0, 2.0)
7''	2.69 t (7.5)	2.70 t (7.0)	2.72 t (7.2)		7.58 d (15.6)	7.57 d (15.6)	7.16 d (16.0)	7.47 d (16.2)	7.49 d (15.5)
8''	3.39 dt (7.4, 4.5)	3.41 t (6.0)	3.47 m, 3.39 m		6.42 d (15.6)	6.41 d (15.6)	5.89 d (16.0)	6.39 d (16.2)	6.34 d (15.5)
2'''		6.79 d (1.5)	6.93 d (8.4)						
3'''			6.70 d (8.4)						
5'''		6.69 d (8.0)	6.70 d (8.4)						
6'''		6.55 dd (8.0, 1.5)	6.93 d (8.4)						
7'''		2.58 t (7.0)	2.56 t (7.2)						
8'''		3.28 t (7.0)	3.29 m, 3.21 m						
OMe-3	3.58 s	3.69 s	3.69 s	3.75 s	3.73 s	3.75 s	3.84 s	3.79 s	3.80 s
OMe-5	3.86 s	3.85 s	3.85 s		3.73 s		3.84 s	3.79 s	3.80 s
OMe-7									3.17 s
OMe-3'	3.67 s	3.67 s	3.67 s	3.75 s	3.73 s	3.75 s	3.83 s	3.88 s	3.77 s
OMe-5'	3.67 s	3.67 s	3.67 s		3.73 s		3.83 s	3.88 s	3.77 s
OMe-3''		3.80 s	3.78 s	3.88 s	3.91 s	3.90 s	3.91 s	3.79 s	3.91 s
OMe-5''				3.88 s				3.79 s	
OH-4	7.76 s	7.78 s	7.79 s	7.29 s	6.91 s	7.27 s		7.09 s	
OH-4'	6.90 s	6.91 s	6.91 s	7.26 s	6.89 s	7.24 s		6.98 s	
OH-4''		8.08 s	7.21 s	7.75 s	8.15 s	8.12 s		7.77 s	
OH-4'''		7.26 s	8.07 s						
NH	7.45 t (4.5)	7.81 t (4.5), 7.59 t (4.5)	7.72 t (4.5), 7.59 t (4.5)						

[a] ^{1}H-NMR data (δ) were measured at 600 MHz or 500 MHz. The assignments were based on ^{1}H-^{1}H COSY, HSQC, and HMBC experiments.

Table 2. ^{13}C-NMR Data (δ_C) for Compounds **1–9** in Acetone-d_6 ^{a}.

No.	1	2	3	4	5	6	7	8	9
1	132.0	123.8	123.8	132.9	132.4	132.9	129.6	131.8	134.5
2	124.6	126.5	126.5	113.2	107.2	113.2	107.2	106.9	106.9
3	147.0	146.4	146.4	148.1	148.5	148.1	148.4	148.9	148.5
4	141.8	142.4	142.4	145.5	134.9	145.5	142.2	135.2	136.0
5	148.2	148.1	148.1	115.5	148.5	115.5	148.4	148.9	148.5
6	108.0	108.3	108.2	122.3	107.2	122.3	107.2	106.9	106.9
7	131.5	132.5	133.6	35.4	35.9	35.4	198.2	34.2	82.6
8	124.4	128.3	128.4	40.7	40.6	40.8	47.6	43.6	48.1
9	169.1	169.8	169.6	65.2	65.2	65.2	71.1	73.3	70.3
1′	135.9	135.1	135.1	133.4	131.8	133.4	132.9	134.6	131.7
2′	106.4	106.4	106.4	113.2	107.3	113.2	104.7	104.3	104.4
3′	148.3	148.4	148.4	148.1	148.5	148.1	148.6	148.7	148.8
4′	135.3	135.5	135.5	145.6	135.0	145.5	136.3	136.0	136.5
5′	148.3	148.4	148.4	115.4	148.5	115.4	148.6	148.7	148.8
6′	106.4	106.4	106.4	122.3	107.3	122.3	104.7	104.3	104.4
7′	39.0	39.5	39.6	34.9	35.4	35.0	84.9	84.5	85.1
8′	46.1	49.1	49.1	44.1	44.1	44.2	51.5	50.3	49.4
9′	64.6	171.4	171.4	62.1	62.1	62.1	62.8	63.4	63.6
1″	131.2	131.1	131.7	126.1	127.4	127.5	127.2	126.0	127.3
2″	130.5	130.6	113.1	106.8	111.3	11.3	111.0	106.7	111.3
3″	116.0	116.1	148.2	148.9	148.7	148.8	148.6	148.6	148.7
4″	156.6	156.7	145.9	139.4	150.1	150.1	149.9	139.5	150.1
5″	116.0	116.1	115.7	148.9	116.1	116.0	115.9	148.6	116.1
6″	130.5	130.6	122.0	106.8	123.9	124.0	123.7	106.7	123.8
7″	35.6	35.5	36.0	145.9	145.6	145.6	145.6	146.2	145.8
8″	42.2	42.4	42.3	116.2	116.0	116.0	115.1	115.9	114.8
9″				167.5	167.6	167.5	166.7	167.3	167.3
1‴		131.8	131.2						
2‴		113.0	130.6						
3‴		148.2	116.0						
4‴		145.8	156.6						
5‴		115.6	116.0						
6‴		122.0	130.6						
7‴		36.1	35.7						
8‴		41.9	42.1						
OMe-3	60.4	60.3	60.3	56.5	56.5	56.1	56.7	56.6	56.6
OMe-5	56.5	56.2	56.2		56.5		56.7	56.6	56.6
OMe-7									56.1
OMe-3′	56.6	56.7	56.7	56.1	56.4	56.1	56.6	56.7	56.6
OMe-5′	56.6	56.7	56.7		56.4		56.6	56.7	56.6
OMe-3″			56.5	56.7	56.3	56.3	56.3	56.6	
OMe-5″				56.7				56.6	
OMe-3‴		56.6							

^{a} ^{13}C-NMR data (δ) were measured at 150 MHz or 125 MHz. The assignments were based on ^{1}H-^{1}H COSY, HSQC, and HMBC experiments.

Compound **3** gave the same molecular formula, $C_{39}H_{42}N_2O_{11}$, as that of **2** by analysis of the HRESIMS. Compound **3** shared almost identical UV, IR, and ^{1}H- and ^{13}C-NMR features to those of **2**, which suggested that they both contained the 4,4′-dihydroxy-3,5,3′,5′-tetramethoxy-2,7′-cyclolignan-7-en-9,9′-diamide core, a tyramine, and a 3-methoxytyramine moieties.

Further analysis of 2D-NMR data permitted the tyramine and 3-methoxytyramine moieties to be located at C-9′ and C-9 in **3**, the reverse of **2**, via the amide bonds (Figure 2), respectively. Analysis of the 1D NOE difference spectrum of **3** and its optical rotation indicated that **3** had the same absolute configuration as **2**. Therefore, the structure of **3** was confirmed as (−)-(7′R,8′S)-

N^1-[2-(4-hydroxy-3-methoxyphenyl)-ethyl]-N^2-[2-(4-hydroxyphenyl)-ethyl]-4,4′-dihydroxy-3,5,3′,5′-tetramethoxy-2,7′-cyclolignan-7-en-9,9′-diamide.

Compound **4** was obtained as a yellow solid and its molecular formula was deduced as $C_{31}H_{36}O_{10}$ from HRESIMS. The IR spectrum exhibited absorption bands at 3391, 1608, and 1516 cm^{-1} due to the aromatic and hydroxy groups. The NMR data of **4** showed signals similar with secoisolariciresinol (Table 1; Table 2) [21,22]. However, both the H_2-9 and C-9 were shifted downfield when compared with secoisolariciresinol. Besides, the ^{1}H- and ^{13}C-NMR signals attributed to a *trans*-cinnamyloxy unit were present (Table 1; Table 2). These were consistent with the substitution of the *trans*-cinnamyloxy at C-9, which was verified by the key HMBC correlation from H_2-9 to C-9″. The positive optical rotation of **4** supported the same (8*S*,8′*S*) configuration as that of the known compound (+)-(8*S*,8′*S*)-9,9′-di-*O*-(*E*)-feruloylsecoisolariciresinol (**11**), which has been also isolated from this plant [12]. The (8*S*,8′*S*) configuration was confirmed by the evidence that compound **4** showed optical rotation opposite to that of (−)-1-*O*-feruloylsecoisolariciresinol [21]. Thus, the structure of **4** was defined as (+)-(8*S*,8′*S*)-9-*O*-(*E*)-cinnamoylsecoisolariciresinol.

The molecular formula of compound **5** was $C_{32}H_{38}O_{11}$ from the HRESIMS data. Analysis of the 1D- and 2D-NMR data revealed that its planar structure was completely identical to the known lignan, (−)-(8*R*,8′*R*)-9-*O*-(*E*)-feruloyl-5,5′-dimethoxysecoisolariciresinol, but their specific rotation was inverse [23]. Taking into account that **4** was the 5-methoxy analogue of **5** and they displayed similar specific rotation, it is proposed that they both have the (8*S*,8′*S*) configuration. Thus, the structure of **5** was defined as (+)-(8*S*,8′*S*)-9-*O*-(*E*)-feruloyl-5,5′-dimethoxysecoisolariciresinol.

The planar structure of **6** was proved to be identical to (−)-(8*R*,8′*R*)-9-*O*-(*E*)-feruloyl-secoisolariciresinol (different nomenclature was used in literature [21]) after analysis of the HRMS, and 1D- and 2D-NMR data of **6**. However, the optical rotation of **6** was opposite for (−)-(8*R*,8′*R*)-9-*O*-(*E*)-feruloyl-secoisolariciresinol [21]. Thus, the structure of **6** was determined as (+)-(8*S*,8′*S*)-9-*O*-(*E*)-feruloyl-secoisolariciresinol.

Compound **7**, an amorphous powder, was determined to have the molecular formula of $C_{32}H_{34}O_{12}$ by HRESIMS. The NMR spectra of **7** were similar to the co-occurring (+)-9′-*O*-*trans*-feruloyl-5,5′-dimethoxylariciresinol (**13**) [24], with the only difference being the replacement of the CH_2 group by a ketone. These data demonstrated the presence of a ketone moiety at C-7 in **7**. This inference was confirmed by the HMBC cross-peak of H-2(6)/C-7, H_2-9/C-7, and H-8′/C-7. The coupling constant of H-7′ (*J* = 7.5 Hz) indicated a *trans* relationship of H-7′/H-8′. The presence of correlations of H-7′/H_2-9′ and H-2(6)/H-8′ and the absence of H-8/H_2-9′ were observed in the NOESY spectrum of **7**, which confirmed that H-7′ was oriented opposite to H-8 and H-8′. The absolute configuration of **7** was established by quantum chemical ECD calculation (Supplementary Materials). The calculated ECD curve for 8*R*,7′*S*,8′*R*-isomer matched well with the experimental ECD spectrum of **7** (Figure 3), which suggested compound **7** had the (8*R*,7′*S*,8′*R*) absolute configurations. Based on these observations, the structures of **7** was assigned as (+)-(8*R*,7′*S*,8*R*′)-9′-*O*-(*E*)-feruloyl-5,5′-dimethoxylariciresinol-7-one.

The molecular formula of compound **8** was $C_{33}H_{38}O_{12}$ as indicated by the HRESIMS. The NMR spectra of **8** and (+)-9′-*O*-*trans*-feruloyl-5,5′-dimethoxylariciresinol were closely comparable [24], except for the replacement of (*E*)-feruloyl group by the (*E*)-cinnamoyl group. The structure of **8** was confirmed by the 2D-NMR HSQC, COSY, HMBC, and NOESY data. Also, the NOESY correlations of H-7′/H_2-9′ and H_2-7/H_2-9′ revealed that compounds **7** and **8** have the same relative configuration. Therefore, on the basis of the positive optical rotation of **8** and biosynthetic considerations, the structure of **8** was deduced as (+)-(8*R*,7′*S*,8′*R*)-9′-*O*-(*E*)-cinnamoyl-5,5′-dimethoxylariciresinol.

Compound **9** was shown to have the molecular formula of $C_{33}H_{38}O_{12}$, as established by the HRESIMS. The ^{1}H- and ^{13}C-NMR spectra of **9** closely resembled those of **7**, the only discernable difference being the presence of a new methoxy moiety and lack of a ketone moiety in **9**, suggesting that compound **9** contains a methoxy moiety rather than a ketone moiety at C-7. This was confirmed from the COSY correlation of H-7/H-8 and HMBC correlation of OMe/C-7. In the NOESY spectrum of

9, the NOE correlations of H-7/H$_2$-9' and H-7'/H$_2$-9' also verified that H-7' was oriented opposite to H-8 and H-8'. Thus, the structure of **9** was defined as 9'-*O*-(*E*)-feruloyl-5,7,5'-trimethoxy-lariciresinol.

The known compounds were identified as 1,2-dihydro-6,8-dimethoxy-7-hydroxy-1-(3,5-dimethoxy-4-hydroxyphenyl)-N^1,N^2-bis-[2-(4-hydroxypeenyl)ethyl]-2,3-naphthalene dicarboxamide (**10**) [20], (+)-9,9'-*O*-di-(*E*)-feruloyl-5,5'-dimethoxy secoisolariciresinol (**11**) [25], (+)-9,9'-*O*-di-(*E*)-feruloyl-secoisolariciresinol (**12**) [12], (+)-9'-*O*-(*E*)-feruloyl-5,5'-dimethoxylariciresinol (**13**) [24], (+)-9'-*O*-(*E*)-feruloyl-5'-methoxylariciresinol (**14**) [26], (+)-5,5'-dimethoxylariciresinol (**15**) [27], (+)-5'-methoxylariciresinol (**16**) [28], arctigenin (**17**), matairesinol (**18**) [29], and (7*E*,8*R*')- didehydroarctigenin (**19**) [30], respectively, by spectroscopic analysis and comparison of the data obtained with literature values.

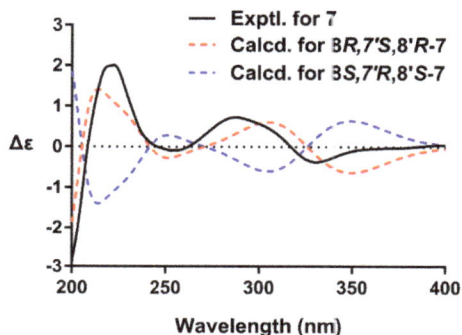

Figure 3. The experimental ECD spectrum of **7** (black), and the calculated ECD spectra of (8*R*,7'*S*,8'*R*)-**7** (red) and (8*S*,7'*R*,8'*S*)-**7** (blue).

2.2. Biological Activities of Compounds **1–19**

2.2.1. Cytotoxic Activity

The task of IC$_{50}$ assessment for all isolates against human colon cancer (HCT-116), human non-small-cell lung carcinoma (NCI-H1650), and human ovarian cancer (A2780) cell lines began immediately following the purification and characterization of each lignan.

Of the compounds, only 7',9-epoxylignans with feruloyl or cinnamoyl group (compounds **7–9**, **13** and **14**) were selectively cytotoxic against NCI-H1650 cell line, with IC$_{50}$ values of less than 20 μM. These results suggested the presence of a feruloyl or a cinnamoyl moiety at C-9' in 7',9-epoxylignans is essential for cytotoxicity against NCI-H1650 cell line. It is noteworthy that compound **7** displayed 4-6 folds more active than **8**, **9**, **13**, and **14**, indicating that the presence of the C-7 ketone could enhance the bioactivity. In addition, the dibenzylbutyrolactone lignans (**17–19**) exerted cytotoxicities against HCT-116 and A2780 cell lines, with IC$_{50}$ values ranging from 0.28 to 18.47 μM (Table 3), but less potent than the positive control taxol (IC$_{50}$ = 0.005 and 0.02 μM, respectively). Interestingly, the addition of the double bond at C-7–C-8 on **19** resulted in 4–40 folds less active than **17** and **18**. This implied that the C-7–C-8 double bond could reduce the cytotoxicity, especially against the A2780 cell line.

2.2.2. Inhibitory Activity of Protein Tyrosine Phosphatase 1B

The isolates were also evaluated for inhibitory activities against protein tyrosine phosphatase 1B (PTP1B). Only compound **11** was moderate active toward PTP1B with an IC$_{50}$ value of 13.5 μM. The positive control oleanolic acid gave an IC$_{50}$ value of 3.82 μM.

2.2.3. Anti-Inflammatory Activity

The inhibitory activity of compounds **1–19** against LPS-induced NO production in RAW264.7 macrophages was examined in this study. As a result, compounds **4–6**, **11** and **12** displayed inhibitions against LPS-induced NO production in RAW264.7 macrophages, with IC_{50} values of 46.8, 50.1, 58.6, 47.5, and 66.5 μM, respectively. Dexamethasone was used as positive control with an IC_{50} value of 9.5 μM.

Table 3. Cytotoxicity of Compounds **1–19** to HCT-116, NCI-H1650, and A2780 Cell Lines.

Compound	IC_{50} (μM)		
	HCT-116	NCI-H1650	A2780
1	>20	>20	>20
2	>20	>20	>20
3	>20	>20	>20
4	>20	>20	>20
5	>20	>20	>20
6	>20	>20	>20
7	>20	2.47	>20
8	>20	11.25	>20
9	>20	13.16	>20
10	>20	>20	>20
11	>20	>20	>20
12	>20	>20	>20
13	>20	9.68	>20
14	>20	10.52	>20
15	>20	>20	>20
16	>20	>20	>20
17	3.25	>20	0.28
18	13.95	>20	1.53
19	18.47	>20	12.8
Taxol [a]	0.005	1.28	0.02

[a] Taxol was used as a positive control.

3. Materials and Methods

3.1. General Experimental Procedures

Optical rotations were measured on an Autopol III automatic polarimeter (Rudolph Research, Hackettstown, NJ, USA). UV spectra were measured on a Cary 300 spectrometer (Agilent, Melbourne, Australia). ECD spectra were recorded on a J-815 spectrometer (JASCO, Tokyo, Japan). IR spectra were acquired on an Impact 400 FT-IR Spectrophotometer (Nicolet, Madison, WI, USA). Standard pulse sequences were used for all NMR experiments, which were run on either a Bruker spectrometer (600 MHz for ^{1}H or 150 MHz for ^{13}C, Karlsruhe, Germany) or a Varian INOVA spectrometer (500 MHz for ^{1}H or 125 MHz for ^{13}C, Palo Alto, CA, USA) equipped with an inverse detection probe. Residual solvent shifts for acetone-d_6 were referenced to δ_H 2.05, δ_C 206.7 and 29.9, respectively. Accurate mass measurements were obtained on a Q-Trap LC/MS/MS (Turbo ionspray source) spectrometer (Sciex, Toronto, ON, Canada). Column chromatography (CC) was run using silica gel (200–300 mesh, Qingdao Marine Chemical Inc., Qingdao, China), and Sephadex LH-20 (Pharmacia Biotech AB, Uppsala, Sweden). HPLC separation was done on Waters HPLC components (Milford, MA, USA) comprising of a Waters 600 pump, a Waters 600 controller, a Waters 2487 dual λ absorbance, with GRACE preparative (250 × 19 mm) Rp C_{18} (5 μm) columns.

3.2. Plant Material

The twigs of *Litsea cubeba* were collected in Zhaotong, Yunnan Province, People's Republic of China, in May 2013, and identified by Prof. Gan-Peng Li at Yunnan Minzu University. A herbarium specimen was deposited in at the Herbarium of the Department of Medicinal Plants, Institute of Materia Medica, Beijing 100050, People's Republic of China (herbarium No. 2013-05-10).

3.3. Extraction and Isolation

The air-dried twigs of *L. cubeba* (12 kg) were ground and extracted using 30.0 L of 95% EtOH under ambient temperature for 3 × 48 h. The EtOH extract was concentrated in vacuo and the residue was suspended in H_2O, then partitioned with EtOAc, to afford EtOAc and H_2O soluble extracts.

The EtOAc fraction (300 g) was chromatographed over silica gel (1500 g), eluting with a gradient of acetone (0–100%) in petroleum ether, and 13 fractions (F_1–F_{13}) was obtained based on the TLC analysis. The F_9 (12.0 g), which showed potent cytotoxicity against HCT-116, NCI-H1650, and A270 cell lines, was subjected to the reversed-phase flash chromatography over C-18 silica gel, eluting with a step gradient from 20 to 95% MeOH in H_2O, to give 15 fractions (F_{9-1}–F_{9-15}). F_{9-8} (1.5 g) was separated on Sephadex LH-20 eluting with petroleum $CHCl_3$-MeOH (1:1) to give three subfractions, and the first subfraction was purified by reversed-phase preparative HPLC (RP$_{18}$, 5 μm, 254 nm, MeOH-H_2O, 75:25) to yield **1** (9.2 mg). The second and third subfractions were further purified by preparative TLC developed with $CHCl_3$-MeOH (15:1) to afford **15** (52 mg), **16** (35mg), and **18** (29 mg). F_{9-9} (1.0 g) was fractionated on a Sephadex LH-20 column using $CHCl_3$-MeOH (1:1) as the eluent to yield five corresponding subfractions. Compound **10** (55 mg) was crystallized from a Me_2CO solution of the second subfraction. The third subfraction was further purified by preparative TLC with $CHCl_3$-MeOH (20:1) to give **17** (17 mg) and **19** (8 mg). The fourth subfraction was purified by reversed-phase preparative HPLC (RP$_{18}$, 5 μm, 254 nm, MeOH-H_2O, 85:15) to give **2** (56 mg), **3** (21 mg), and **14** (23 mg). Using the same HPLC system, the fifth subfraction afforded **7** (27 mg), **8** (12 mg) and **9** (8 mg), and **13** (17 mg). F_{9-10} (1.2 g) was chromatographed over Sephadex LH-20 eluting with $CHCl_3$-MeOH (1:1), and then further separated by reversed-phase preparative HPLC (RP$_{18}$, 5 μm, 254 nm, MeOH-H_2O, 90:10), to afford **4** (8 mg) and **5** (5 mg). F_{9-11} (0.8 g) was fractionated on a Sephadex LH-20 column with $CHCl_3$-MeOH (1:1) as the eluent to give three subfractions. The second and third subfractions were further purified by reversed-phase preparative HPLC (RP$_{18}$, 5 μm, 254 nm, MeOH-H_2O, 90:10) to afford **6** (12 mg), **11** (23 mg), and **12** (15 mg).

3.4. (−)-(7′R,8′S)-N-[2-(4-Hydroxyphenyl)-ethyl]-4,4′,9′-trihydroxy-3,5,3′,5′-tetramethoxy-2,7′-cyclo-lignan-7-en-9-amide (1)

White, amorphous powder. $[\alpha]_D^{20}$ −35.0 (c 0.1, MeOH); UV (MeOH) λ_{max} (log ε) 204 (4.04), 200 (2.32), 245 2.12), 324 (1.13) nm; IR (KBr) ν_{max} 3372, 2935, 2849, 1643, 1611, 1516, 1459, 1427, 1329, 1286, 1218, 1115, 1030, 961, 912, 834, 646 cm^{-1}; ^{1}H-NMR (acetone-d_6, 500 MHz) and ^{13}C-NMR (acetone-d_6, 125 MHz) data, see Table 1; Table 2; ESIMS *m/z* 574 [M + Na]$^+$ and 550 [M − H]$^-$; HRESIMS *m/z* 552.2234 [M + H]$^+$ (calcd. for $C_{30}H_{34}NO_9$, 552.2228) and 574.2048 [M + Na]$^+$ (calcd. for $C_{30}H_{33}NO_9Na$, 574.2048).

3.5. (−)-(7′R,8′S)-N^1-[2-(4-Hydroxyphenyl)-ethyl]-N^2-[2-(4-hydroxy-3-methoxyphenyl)-ethyl]-4,4′-dihydro-xy-3,5,3′,5′-tetramethoxy-2,7′-cyclolignan-7-en-9,9′-diamide (2)

White, amorphous power. $[\alpha]_D^{20}$ −23.0 (c 0.1, MeOH); UV (MeOH) λ_{max} (log ε) 204 (4.11), 250 (0.86), 281 (0.30), 328 (0.42) nm; IR (KBr) ν_{max} 3362, 2919, 2851, 1736, 1649, 1612, 1516, 1464, 1424, 1372, 1328, 1274, 1217, 1115, 1035, 890, 834, 802, 721, 640 cm^{-1}; ^{1}H-NMR (acetone-d_6, 600 MHz) and ^{13}C-NMR (acetone-d_6, 150 MHz) data, see Table 1; Table 2; ESIMS *m/z* 713 [M − H]$^-$; HRESIMS *m/z* 713.2719 [M − H]$^-$ (calcd. for $C_{39}H_{41}N_2O_{11}$, 713.2716).

3.6. (−)-(7′R,8′S)-N^1-[2-(4-Hydroxy-3-methoxyphenyl)-ethyl]-N^2-[2-(4-hydroxyphenyl)-ethyl]-4,4′-dihydro-xy-3,5,3′,5′-tetramethoxy-2,7′-cyclolignan-7-en-9,9′-diamide (**3**)

White, amorphous power. $[\alpha]_D^{20}$ −25.0 (c 0.1, MeOH); UV (MeOH) λ_{max} (log ε) 204 (4.12), 248 (0.82), 285 (0.27), 333 (0.45) nm; IR (KBr) ν_{max} 3391, 2920, 2851, 1647, 1611, 1541, 1517, 1465, 1425, 1367, 1278, 1203, 1116, 1035, 932, 888, 829, 801, 722, 650, 599 cm^{-1}; ^{1}H-NMR (acetone-d_6, 600 MHz) and ^{13}C-NMR (acetone-d_6, 150 MHz) data, see Table 1; Table 2; ESIMS *m/z* ESIMS *m/z* 713 [M − H]$^-$; HRESIMS *m/z* 713.2715 [M − H]$^-$ (calcd. for $C_{39}H_{41}N_2O_{11}$, 713.2716).

3.7. (+)-(8S,8′S)-9-O-(E)-Cinnamoyl-secoisolariciresinol (**4**)

Yellow solid. $[\alpha]_D^{20}$ +18.2 (c 0.05, MeOH); UV (MeOH) λ_{max} (log ε) 204 (4.12), 230 (0.82), 287 (0.39), 329 (0.78) nm; IR (KBr) νmax 3391, 2920, 2850, 1683, 1645, 1608, 1516, 1463, 1428, 1375, 1341, 1272, 1237, 1155, 1119, 1033, 875, 820, 799, 721, 631 cm^{-1}; ^{1}H-NMR (acetone-d_6, 500 MHz) and ^{13}C-NMR (acetone-d_6, 125 MHz) data, see Table 1; Table 2; ESIMS *m/z* 567 [M − H]$^-$; HRESIMS *m/z* 569.2387 [M + H]$^+$ (calcd. for $C_{31}H_{37}NO_{10}$, 569.2381) and 591.2204 [M + Na]$^+$ (calcd. for $C_{31}H_{36}O_{10}Na$, 591.2201).

3.8. (+)-(8S,8′S)-9-O-(E)-Feruloyl-5,5′-dimethoxysecoisolariciresinol (**5**)

Yellow solid. $[\alpha]_D^{20}$ +22.2 (c 0.05, MeOH); UV (MeOH) λ_{max} (log ε) 206 (4.22), 234 (0.84), 284 (0.36), 326 (0.82) nm; IR (KBr) ν_{max} 3394, 2921, 2850, 1696, 1604, 1517, 1461, 1428, 1370, 1328, 1273, 1218, 1161, 1117, 1033, 984, 915, 825, 721, 645, 604 cm^{-1}; ^{1}H-NMR (acetone-d_6, 600 MHz) and ^{13}C-NMR (acetone-d_6, 150 MHz) data, see Table 1; Table 2; HRESIMS *m/z* 621.2299 [M + Na]$^+$ (calcd. for $C_{32}H_{38}O_{11}Na$, 621.2306).

3.9. (+)-(8S,8′S)-9-O-(E)-Feruloyl-secoisolariciresinol (**6**)

Yellow solid. $[\alpha]_D^{20}$ +25.2 (c 0.1, MeOH); IR (KBr) ν_{max} 3367, 2928, 2855, 1683, 1601, 1516, 1454, 1431, 1375, 1271, 1207, 1154, 1033, 935, 846, 801, 724 cm^{-1}; ^{1}H-NMR (acetone-d_6, 600 MHz) and ^{13}C-NMR (acetone-d_6, 150 MHz) data, see Table 1; Table 2; HRESIMS *m/z* 537.2134 [M − H]$^-$ (calcd. for $C_{30}H_{33}O_9$, 537.2130).

3.10. (+)-(8R,7′S,8′R)-9′-O-(E)-Feruloyl-5,5′-dimethoxylariciresinol-7-one (**7**)

Amorphous powder. $[\alpha]_D^{20}$ +19.5 (c 0.1, MeOH); UV (MeOH) λ_{max} (log ε) 211 (4.01), 234 (2.12), 318 (1.96) nm; ECD (MeOH) 331 ($\Delta\varepsilon$ − 0.37), 288 ($\Delta\varepsilon$ + 0.73), 222 ($\Delta\varepsilon$ + 2.01); IR (KBr) ν_{max} 3409, 2940, 2843, 1701, 1665, 1604, 1516, 1461, 1425, 1371, 1323, 1271, 1215, 1169, 1116, 1032, 983, 912, 845, 827, 765, 712, 662 cm^{-1}; ^{1}H-NMR (acetone-d_6, 500 MHz) and ^{13}C-NMR (acetone-d_6, 125 MHz) data, see Table 1; Table 2; ESIMS *m/z* 609 [M − H]$^-$; HRESIMS *m/z* 609.1980 [M − H]$^-$ (calcd. for $C_{32}H_{33}O_{12}$, 609.1978).

3.11. (+)-(8R,7′S,8′R)-9′-O-(E)-Cinnamoyl-5,5′-dimethoxylariciresinol (**8**)

Amorphous powder. $[\alpha]_D^{20}$ +23.0 (c 0.1, MeOH); IR (KBr) ν_{max} 3425, 2937, 2845, 1703, 1612, 1516, 1461, 1427, 1331, 1282, 1218, 1154, 1117, 1041, 980, 913, 832, 719 cm^{-1}; ^{1}H-NMR (acetone-d_6, 600 MHz) and ^{13}C-NMR (acetone-d_6, 150 MHz) data, see Table 1; Table 2; HRESIMS *m/z* 625.2297 [M − H]$^-$ (calcd. for $C_{33}H_{37}O_{12}$, 625.2291).

3.12. 9′-O-(E)-Feruloyl-5,7,5′-trimethoxylariciresinol (**9**)

Amorphous powder. $[\alpha]_D^{20}$ +21.0 (c 0.1, MeOH); IR (KBr) ν_{max} 3395, 2933, 2849, 1701, 1610, 1517, 1462, 1428, 1372, 1324, 1270, 1214, 1159, 1116, 1033, 983, 909, 831, 703 cm^{-1}; ^{1}H-NMR (acetone-d_6, 500 MHz) and ^{13}C-NMR (acetone-d_6, 125 MHz) data, Table 1; Table 2; HRESIMS *m/z* 625.2297 [M − H]$^-$ (calcd. for $C_{33}H_{37}O_{12}$, 625.2291).

3.13. Cytotoxicity Assay

The cytotoxic activity was determined against human colon cancer (HCT-116), human non-small-cell lung carcinoma (NCI-H1650), and human ovarian cancer (A2780) cell lines which were bought from the Cell Bank of Shanghai Institute of Cell Biology (Chinese Academy of Sciences) and originally obtained from the American Type Culture Collection (ATCC, Rockville, MD, USA). Cells were grown in RPMI 1640 (GIBCO, New York, NY, USA) supplemented with 10% fetal calf serum (Life Technologies, Carlsbad, CA, USA), penicillin G (100 U/mL), and streptomycin (100 µg/mL) at 37 °C in a 5% CO_2 and seeded in 96-well plates (CLS3635, Corning®, Sigma, Santa Clara, CA, USA) at a cell density of 3000 per well over night, and then were treated with various diluted concentrations (each concentration was arranged triple) of compounds **1–19**, which were prepared with DMSO (Sigma) to 100 µM stock solution and stored in −20 °C in advance. After 24 h of treatment, 10 µL of MTT (5 mg/mL in PBS) was then added directly to all wells and the plates were placed in the dark at 37 °C for 3 h incubation. Cell viability was measured by observing absorbance at 570 nm on a SpectraMax[190] microplate reader (Molecular Devices, Silicon Valley, CA, USA). IC_{50} values were calculated using Microsoft Excel software (version 2010, Redmond, WA, USA). Taxol was used as a positive control.

3.14. PTP1B Inhibition Assay

The recombinant GST-hPTP1B (gluthathione *S*-transferase-human protein tyrosine phosphatase 1B) bacteria pellets were purified by a GST bead column. The dephosphorylation of *para*-nitrophenyl phosphate (*p*-NPP) was catalyzed to *para*-nitrophenol by PTP1B. Enzyme activity involving an end-point assay, which intensified the yellow color, was measured at a wavelength of 405 nm. All compounds were dissolved in 100% dimethyl sulfoxide (DMSO), and reactions, including controls, were performed at a final concentration of 10% DMSO. Selected compounds were first evaluated for their ability to inhibit the PTPase reaction at a 10 µM concentration at 30 °C for 10 min, in a reaction system with 3 mM *p*-NPP in HEPES assay buffer (pH 7.0). The reaction was initiated by addition of the enzyme and quenched by addition of 1 M NaOH. The amount of the produced *p*-nitrophenol was determined at 405 nm using a microplate spectrophotometer (uQuant, Bio-Tek, Winooski, VT, USA). IC_{50} values were evaluated using a sigmoidal dose-response (variable slope) curve-fitting program of GraphPad Prism 4.0 software (La Jolla, CA, USA). Oleanolic acid was used as a positive control.

3.15. Nitric Oxide (NO) Production in RAW264.7 Macrophages

The RAW 264.7 macrophages were cultured in The RPMI 1640 medium (Hyclone, Logan, UT, USA) containing 10% FBS. The compounds were dissolved in DMSO and further diluted in medium to produce different concentrations. The cell mixture and culture medium were dispensed into 96-well plates (2 × 105 cells/well) and maintained at 37 °C under 5% CO2. After preincubation for 24 h, serial dilutions of the test compounds were added into the cells, up to the maximum concentration 25 µM, then added with LPS to a concentration 1 µg/mL and continued to incubate for 18 h. The amount of NO was assessed by determined the nitrite concentration in the cultured RAW264.7 macrophage supernatants with Griess reagent. Aliquots of supernatants (100 µL) were incubated, in sequence, with 50 µL 1% sulphanilamide and 50 µL 1% naphthylethylenediamine in 2.5% phosphoric acid solution. The sample absorbance was measured at 570 nm by a 2104 Envision Multilabel Plate Reader (PerkinElmer, Inc., Waltham, MA, USA). Dexamethasone was used as a positive control.

4. Conclusions

In summary, bioassay-guided isolation of cytotoxic fractionsof the twigs of *L. cubeba*revealed the presence of nine new lignans **1–9** and ten analogues **10–19**. Initially, all of the isolated compounds were evaluated against HCT-116, NCI-H1650, and A2780 tumor cell lines. Of the compounds, only 7′,9-epoxylignans with feruloyl or cinnamoyl group (**7–9**, **13** and **14**) were selectively cytotoxic against NCI-H1650 cell line, with IC_{50} values of less than 20 µM, whereas, the dibenzylbutyrolactone lignans

17–19 exerted cytotoxicity against HCT-116 and A2780 cell lines, with IC$_{50}$ values ranging from 0.28 to 18.47 μM. The results highlighted the structure-activity relationship importance of a feruloyl or a cinnamoyl moiety at C-9′ or/and C-7 ketone in 7′,9-epoxylignans. The isolates were also examined for inhibitory activities against PTP1B and LPS-induced NO production in RAW264.7 macrophages. As a result, compound **11** was moderate active toward PTP1B with an IC$_{50}$ value of 13.5 μM and compounds **4–6**, **11** and **12** displayed inhibitions against LPS-induced NO production in RAW264.7 macrophages, with IC$_{50}$ values of 46.8, 50.1, 58.6, 47.5, and 66.5 μM, respectively. The present results provide additional phytochemical and bioactive information of this medicinal and spiced plant.

Supplementary Materials: The following are available online, IR, UV, HRMS, NMR and ECD spectra of compounds **1–9** as well as other supporting data.

Author Contributions: X.S. conceived and designed the experiments; X.L. and Y.Q. realized the evaluation of bioactivities; H.X., L.W. and G.X. performed the isolation, structural elucidation and wrote the paper; S.L. analyzed the results and revised the paper.

Funding: Financial supports from the Beijing Advanced Innovation Center for Food Nutrition and Human Health, Beijing Technology and Business University (No. 20171040) and the National Natural Science Foundation of China (NNSFC; Nos. 81522050 and 81773589), and the Key projects of the Beijing Natural Sciences Foundation and Beijing Municipal Education Committee (No. KZ201811417049).

Conflicts of Interest: The authors declare no conflict of interest.

References

1. National Pharmacopoeia Commission. *Chinese Pharmacopoeia*; China Medical Science and Technology Press: Beijing, China, 2015.
2. Editorial Committee of Chinese Materia Medica, State Administration Bureau of Traditional Chinese Medicine. *Chinese Materia Medica (Zhonghua Bencao)*; Shanghai Science & Technology Press: Shanghai, China, 1999.
3. Zhang, S.Y.; Guo, Q.; Gao, X.L.; Guo, Z.Q.; Zhao, Y.F.; Chai, X.Y.; Tu, P.F. A phytochemical and pharmacological advance on medicinal plant *Litsea cubeba* (Lauraceae). *Chin. J. Chin. Mater. Med.* **2014**, *39*, 769–776.
4. Li, W.R.; Shi, Q.S.; Liang, Q.; Xie, X.B.; Huang, X.M.; Chen, Y.B. Antibacterial activity and kinetics of *Litsea cubeba* oil on *Escherichia coli*. *PLoS ONE* **2014**, *9*, e110983. [CrossRef] [PubMed]
5. Lee, S.S.; Chen, C.K.; Huang, F.M.; Chen, C.H. Two dibenzopyrrocoline alkaloids from *Litsea cubeba*. *J. Nat. Prod.* **1996**, *59*, 80–82. [CrossRef]
6. Lee, S.S.; Lin, Y.J.; Chen, C.K.; Liu, K.C.S.; Chen, C.H. Quaternary alkaloids from *Litsea cubeba* and *Cryptocarya konishii*. *J. Nat. Prod.* **1993**, *56*, 1971–1976. [CrossRef]
7. Wu, Y.C.; Liou, J.Y.; Duh, C.Y.; Lee, S.S.; Lu, S.T. Litebamine, a novel phenanthrene alkalord from Quaternary alkaloids from *Litsea cubeba*. *Tetrahedron Lett.* **1991**, *32*, 4169–4170. [CrossRef]
8. Feng, T.; Xu, Y.; Cai, X.H.; Du, Z.Z.; Luo, X.D. Antimicrobially active isoquinoline alkaloids from *Litsea cubeba*. *Planta Med.* **2009**, *75*, 76–79. [CrossRef]
9. Zhang, S.Y.; Guo, Q.; Cao, Y.; Zhang, Y.; Gao, X.L.; Tu, P.F.; Chai, X.Y. Alkaloids from roots and stems of *Litsea cubeba*. *Chin. J. Chin. Mater. Med.* **2014**, *39*, 3964–3968.
10. Guo, Q.; Bai, R.F.; Su, G.Z.; Zhu, Z.X.; Tu, P.F.; Chai, X.Y. Chemical constituents from the roots and stems of *Litsea cubeba*. *J. Asian Nat. Prod. Res.* **2015**, *1*, 51–58.
11. Zhang, S.Y.; Zhang, Q.; Guo, Q.; Zhao, Y.F.; Gao, X.L.; Chai, X.Y.; Tu, P.F. Characterization and simultaneous quantification of biological aporphine alkaloids in *Litsea cubeba* by HPLC with hybrid ion trap time-of-flight mass spectrometry and HPLC with diode array detection. *J. Sep. Sci.* **2015**, *38*, 2614–2624. [CrossRef]
12. Guo, Q.; Zeng, K.W.; Gao, X.L.; Zhu, Z.X.; Zhang, S.Y.; Chai, X.Y.; Tu, P.F. Chemical constituents with NO production inhibitory and cytotoxic activities from *Litsea cubeba*. *J. Nat. Med.* **2015**, *69*, 94–99. [CrossRef]
13. Lin, B.; Zhang, H.; Zhao, X.X.; Rahman, K.; Wang, Y.; Ma, X.Q.; Zheng, C.J.; Zhang, Q.Y.; Han, T.; Qin, L. Inhibitory effects of the root extract of *Litsea cubeba* (lour.) pers. on adjuvant arthritis in rats. *J. Ethnopharmacol.* **2013**, *147*, 327–334. [CrossRef] [PubMed]
14. Xie, H.H.; Zhang, F.X.; Wei, X.Y.; Liu, M.F. A review of the studies on *Litsea* alkaloids. *J. Trop. Subtrop. Bot.* **1999**, *7*, 87–92.

15. Zhang, W.; Hu, J.F.; Lv, W.W.; Zhao, Q.C.; Shi, G.B. Antibacterial, antifungal and cytotoxic isoquinoline alkaloidsfrom *Litsea cubeba*. *Molecules* **2012**, *17*, 12950–12960. [CrossRef]

16. Wang, L.Y.; Chen, M.H.; Wu, J.; Sun, H.; Liu, W.; Qu, Y.H.; Li, Y.C.; Wu, Y.Z.; Li, R.; Zhang, D.; et al. Bioactive glycosides form the twigs of *Litsea cubeba*. *J. Nat. Prod.* **2017**, *80*, 1808–1818. [CrossRef]

17. Wang, L.Y.; Qu, Y.H.; Li, Y.C.; Wu, Y.Z.; Li, R.; Guo, Q.L.; Wang, S.J.; Wang, Y.N.; Yang, Y.C.; Lin, S. Water soluble constituents from the twigs of *Litsea cubeba*. *Chin. J. Chin. Mater. Med.* **2017**, *42*, 2704–2713.

18. Wallis, A.F.A. Stereochemistry of cyclolignan—A revised structure of thomasic acid. *Tetrahedron Lett.* **1968**, *9*, 5287–5288. [CrossRef]

19. Assoumatine, T.; Datta, P.K.; Hooper, T.S.; Yvon, B.L.; Charlton, J.L. A short asymmetric synthesis of (+)-lyoniresinol dimethyl ether. *J. Org. Chem.* **2004**, *69*, 4140–4144. [CrossRef]

20. Chaves, M.H.; Roque, N.F. Amides and lignanamides from *Porcelia macrocarpa*. *Phytochemistry* **1997**, *46*, 879–881. [CrossRef]

21. Moon, S.S.; Rahman, A.A.; Kim, J.Y.; Kee, S.H. Hanultarin, a cytotoxic lignan as an inhibitor of actin cytoskeleton polymerization from the seeds of *Trichosanthes kirilowii*. *Bioorg. Med. Chem.* **2008**, *16*, 7264–7269. [CrossRef]

22. Park, H.B.; Lee, K.H.; Kim, K.H.; Lee, I.K.; Noh, H.J.; Choi, S.U.; Lee, K.R. Lignans from the roots of *Berberis amurensis*. *Nat. Prod. Sci.* **2009**, *15*, 17–21.

23. Zhao, Q.; Liu, J.; Wang, F.N.; Liu, G.F.; Wang, G.Z.; Zhang, K. Lignans from branch of *Hypericum petiolulatum*. *Chin. J. Chin. Mater. Med.* **2009**, *34*, 1373–1376.

24. Chen, J.J.; Wang, T.Y.; Hwang, T.L. Neolignans, a coumarinolignan, lignan derivatives, and a chromene: Anti-inflammatory constituents from *Zanthoxylum avicennae*. *J. Nat. Prod.* **2008**, *71*, 212–217. [CrossRef] [PubMed]

25. Chen, T.H.; Huang, Y.H.; Lin, J.J.; Liau, B.C.; Wang, S.Y.; Wu, Y.C.; Jong, T.T. Cytotoxic lignans esters from *Cinnamomum osmophloeum*. *Planta Med.* **2010**, *76*, 613–619. [CrossRef]

26. Hsiao, J.J.; Chiang, H.C. Lignans from the wood of *Aralia bipinnata*. *Phytochemistry* **1995**, *39*, 899–902. [CrossRef]

27. Achenbach, H.; Stöcker, M.; Constenla, M.A. Flavonoid and other constituents of *Bauhinia manca*. *Phytochemistry* **1988**, *27*, 1835–1841. [CrossRef]

28. Duh, C.Y.; Phoebe, C.H., Jr.; Pezzuto, J.M.; Kinghorn, A.D.; Farnsworth, N.R. Plant anticancer agents, XLII. Cytotoxic constituents from *Wikstroemia elliptica*. *J. Nat. Prod.* **1986**, *49*, 706–709. [CrossRef]

29. Umehara, K.; Sugawa, A.; Kuroyanagi, M.; Ueno, A.; Taki, T. Studies on differentiation-inducers from Arctium Fructus. *Chem. Pharm. Bull.* **1993**, *41*, 1774–1779. [CrossRef]

30. Wang, H.Y.; Yang, J.S. Chemical components from *Arctimu lappa*. *Acta Pharm. Sin.* **1993**, *28*, 911–917.

Sample Availability: Samples of the compounds **1–19** are available from the authors.

molecules

Article

Terpenes from *Zingiber montanum* and Their Screening against Multi-Drug Resistant and Methicillin Resistant *Staphylococcus aureus*

Holly Siddique, Barbara Pendry and M. Mukhlesur Rahman *

School of Health, Sports and Bioscience, University of East London, Stratford Campus, Water Lane, London E15 4LZ, UK; h.siddique@uel.ac.uk (H.S.); b.pendry@uel.ac.uk (B.P.)
* Correspondence: m.rahman@uel.ac.uk; Tel.: +44-2082234299

Academic Editors: Muhammad Ilias and Charles L. Cantrell
Received: 31 December 2018; Accepted: 20 January 2019; Published: 22 January 2019

Abstract: Bioassay directed isolation of secondary metabolites from the rhizomes of *Zingiber montanum* (Fam. Zingiberaceae) led to the isolation of mono-, sesqui-, and di-terpenes. The compounds were characterized as (*E*)-8(17),12-labdadiene-15,16-dial (**1**), zerumbol (**2**), zerumbone (**3**), buddledone A (**4**), furanodienone (**5**), germacrone (**6**), borneol (**7**), and camphor (**8**) by analysing one-dimensional (1D) (^{1}H and ^{13}C) and two-dimensional (2D) (COSY, HSQC, HMBC, and NOESY) NMR data and mass spectra. Among these terpenes, compounds **1** and **2** revealed potential antibacterial activity (minimum inhibitory concentrations (MIC) values 32–128 µg/mL; 0.145–0.291 mM)) against a series of clinical isolates of multi-drug resistant (MDR) and Methicillin resistant *Staphylococcus aureus* (MRSA).

Keywords: antimicrobial resistance; multi-drug resistant (MDR); methicillin resistant *Staphylococcus aureus* (MRSA); *Zingiber monatnum*; terpenes; (*E*)-8(17),12-labdadiene-15,16-dial; zerumbol

1. Introduction

Antimicrobial resistance has increasingly become a major public health issue that currently claims 700,000 lives every year. It is predicted that if no action is taken, there will be approximately 10 million deaths each year globally by 2050, which will be more than the predicted number of deaths by cancer [1] and will cause a cumulative 100 trillion USD of economic output due to the rise of drug-resistant infections [1]. The morbidity due to resistant infection has doubled since 2007, which equals the burden of HIV, influenza, and tuberculosis [1]. As antibiotic resistance occurs naturally, its mishandling and misuse in humans and animals accelerate the process of development of resistant infection [2]. Infectious diseases, like pneumonia, tuberculosis, gonorrhoea, and salmonellosis are becoming difficult to treat because the antibiotics that are used to treat them are becoming less effective [2]. The gram-positive bacterium *Staphylococcus aureus* relates to an extensive range of infection of skin and soft tissue, pneumonia, endocarditis, sepsis and bacteremia [3] that causes nosocomial infection (resistant to methicillin and vancomycin). Therefore, it is no doubt important to discover new antibiotics to act against multi-drug resistant (MDR) and Methicillin resistant *Staphylococcus aureus* (MRSA).

Since the accidental discovery of penicillin from *Penicillium notatum*, a huge number of antibiotics have been developed from microbes. However, the development of resistance of existing antibiotics to pathogenic microorganisms necessitates the development of new antibiotics from natural sources, including plants, microbes, and marine resources. Medicinal plants have been under-exploited for antimicrobial drug discovery, although plants are considered as leads for the development of several medicines. Some key examples of plant derived medicines include cardioactive digoxin from *Digitalis lanata* [4], anticancer vincristine and vinblastine from *Catharanthus roseus* [5], analgesic morphine from *Papaver somniferum* [6], antimalarial artemisinin from *Artemisia annua* [7], and antiinflamatory salicin from the bark of the willow tree *Salix alba* L. [8]. There are also significant reports of medicinal plants being used as systemic and topical antimicrobial agents in Ayurvedic [9] and Traditional Chinese Medicine [10], as well as in western herbal medicine [11] due to their self-protection strategy to counter bacteria and fungi in their own environment. Hyperforin isolated from the medicinal plant *Hypericum perforatum* exhibited antimicrobial activity with minimum inhibitory concentrations (MIC) value of 0.1 mg/L against methicillin-resistant *Staphylococcus aureus* (MRSA) and penicillin-resistant variants [12]. A series of new acylphloroglucinol isolated from *Hypericum olympicum* showed highly promising antibacterial activity (MICs 0.50–1 µg/mL) against a series of clinical isolates of MRSA strains [13].

Zingiber montanum (Fam. Zingiberaceae), an herbaceous plant that produces a clump of leaves from large rhizomes, is indigenous to Bangladesh, India, Malaysia, Thailand, Indonesia, and Sri Lanka [14]. Traditionally, it has been used for the treatment of asthma, sprains, muscular pain, inflammation, wounds, and as a mosquito repellent, a carminative, and an antidysentery agent [15,16]. *Z. montanum* has been reported to exhibit antioxidant [17], radioprotective [17,18], antiulcer [19], and anti-inflammatory [20] properties. In regard to the phytochemical investigation on *Z. montanum*, a number of monoterpene and sesquiterpene hydrocarbons have recently been reported using gas chromatography-flame ionization detection and gas chromatography-mass spectrometry [21]. The antibacterial, antifungal, allelopathic, and acetylcholinesterase inhibitory activities of these terpenes have also been reported [21]. As part of our research into anti-infective secondary metabolites from Bangladesh medicinal plants, the authors report the bioassay directed isolation and identification of a total of eight terpenes from *Z. montanum* and also their antibacterial activity against a panel of clinical isolates of multi-drug resitant (MDR) and methicillin resistance *Staphycococcus aureus* (MRSA) strains.

2. Results and Discussion

The *n*-hexane, $CHCl_3$, and MeOH extracts from the rhizomes of *Z. montanum* were initially screened for antibacterial activity (Table 1) against clinical isolates of MRSA strains. Whilst MeOH extract did not exhibit any activity at a concentration of 512 µg/mL, both n-hexane and $CHCl_3$ extracts showed activity against the MRSA strains tested with MICs of 64–256 µg/mL. Vacuum liquid chromatography (VLC) fractionation on active crude extracts, followed by further purification using column chromatography over Sephadex LH20, solid phase extraction (SPE), and/or preparative TLC led to the isolation of seven terpenes (**2–8**) from n-hexane extract and two terpenes (**1 and 8**) from the $CHCl_3$ extract. Among these compounds, **1** and **2** exhibited promising antibacterial activity against MRSA strains with MICs of 64–128 µg/mL (0.145–0.291 mM).

Table 1. Antibacterial activity of crude extracts against standard, multi-drug resistant (MDR) and methicillin-resistant strains of *Staphylococcus aureus* in µg/mL.

Extracts/Antibiotic	MICs in µg/mL				
	SA1199B	XU212	EMRSA15	RN4229	ATCC25941
n-Hexane	128	128	256	256	128
Chloroform	64	128	128	128	256
Methanol	>512	>512	>512	>512	>512
Norfloxacin	32	64	16	8	16

Compound **1** was isolated as colourless amorphous powder from the $CHCl_3$ extract of *Z. montanum*. The molecular formula of **1** was established as $C_{20}H_{30}O_2$ from the $[M + H]^+$ at *m/z* 303.23122 (calculated for $C_{20}H_{31}O_2$ at 303.23240) in the high resolution of mass spectrometry. The ^{1}H-NMR (600 MHz, $CDCl_3$, Table 2) spectrum showed the presence of two sets of aldehyde protons (δ_H 9.40 and 9.63), one olefinic proton resonating at δ_H 6.76 (*J* = 6.6 Hz), exomethylene protons at 4.36 and 4.86, three sets of methyl protons, and a number of peaks for methine and methylene protons. The ^{13}C-NMR spectrum showed the presence of a total of 20 carbons including two aldehyde carbons (193.8 and 197.5), an exomethylene (108.1), an olefinic methine (135.0), two aliphatic methines, four quaternary carbons, three methyl carbons, and seven methylene carbons. In HMBC, two sets of methyl protons at δ_H 0.82 (δ_C 22.1 from HSQC) and δ_H 0.89 (δ_C 33.7 from HSQC) showed a common ^2J connectivity to a carbon at 33.7 (C-4) and ^3J connection with methylene carbon at δ_C 42.0 (C-3; δ_H 1.41 and 1.18 from HSQC) and methine carbon at δ_C 55.8 (C-5; δ_H 1.13 from HSQC). H-5 revealed ^3J HMBC correation to methylene carbons at 42.0 (C-3), 38.1 (C-7; δ_H 2.02 and 2.42 from HSQC), methine carbon at 56.6 (C-9; δ_H 1.90 from HMQC), and methyl carbon at 14.6 (C-20; δ_H 0.72 from HSQC). H-9 exhibited HMBC interactions to C-10 (by ^2J), C-11 (by ^2J), C-12 (160.4 by ^3J; δ_H 6.76 from HSQC), C-17 (108.1 by ^3J; δ_H 4.36 and 4.86 from HSQC), and C-20 (by ^3J). H-12 showed ^3J HMBC connectivity to C-9, methylene carbon at 39.6 (C-14; δ_H 3.41 and 3.46 from HSQC), and aldehydic carbon at 193.8 (C-16; δ_H 9.40 from HSQC). The other aldehydic proton at H 9.63 (H-15; C 197.5 from HSQC) showed ^3J HMBC correation to a quaternary carbon at 135.0 (C-13). Accordingly, structure of 1 was confirmed as (*E*)-8(17),12-labdadiene-15,16-dial [22]. The NMR spectra of compound 1 are available in Supplementary Material. This compound has previously reported from *Alpinia chinensis* [22] and *Curcuma heyneana* [23]. This is the first report of its isolation from the genus *Zingiber*.

Table 2. ^{1}H- (600 MHz), ^{13}C- (150 MHz) NMR and HMBC (600 MHz) data of 1 in CDCl$_3$.

Position	1H	13C	HMBC	
			^2J	^3J
1	1.06, m, 1H; 1.68, m, 1H	39.4	-	C-9
2	1.50, m, 1H; 1.57, m, 1H	19.5	C-3	C-4
3	1.18, m, 1H; 1.41, m, 1H	42	-	C-18, C19
4	-	33.6	-	-
5	1.13, m, 1H	55.8	C6	C3, C7, C9, C19, C20
6	1.34, m, 1H; 1.75, m, 1H	24.4	C5, C7	C10
7	2.02, m, 1H; 2.42, m, 1H	38.1	C8	C5, C17
8	-	148.4	-	-
9	1.90, m, 1H	56.6	C10, C11	C12, C17, C20
10	-	39.8	-	-
11	2.31, m, 1H; 2.49, m, 1H	24.8	C9	C8, C13
12	6.76, t, *J* = 6.6 Hz, 1H	160.4	-	C9, C14, C16
13	-	135	-	-
14	3.41, d, *J* = 16.8 Hz, 1H	39.6	C13	C12, C16
	3.46, d, *J* = 16.7 Hz, 1H			
15	9.63, t, *J* = 14.4 Hz, 1H	197.5	C14	C13
16	9.40, s, 1H	193.8	C13	C12, C14
17	4.36, s, 1H; 4.86, s, 1H	108.1	C8	C7, C9
18	0.88, s, 3H	33.7	-	C3, C5, C19
19	0.82, s, 3H	22.1	-	C3, C5, C18
20	0.72, s, 3H	14.6	C10	C5, C9

Compound 2 was isolated colourless oil from the *n*-hexane extract of *Z. montanum*. The IR spectrum revealed the presence of a hydroxyl group (3300 cm^{-1}) and double bond (1610 cm^{-1}). The high resolution of mass spectroscopy showed the [M + H]$^+$ at *m/z* 221.18956 (calculated for C$_{15}$H$_{25}$O, at 221.19054), which confirmed the molecular formula of 2 as C$_{15}$H$_{24}$O. The ^{1}H-NMR spectrum (CDCl$_3$, 600 MHz, Table 3) of 2 showed the presence of four methyl singlets resonating at δ_H 1.04, 1.06, 1.43, 1.65, four sets of olefinic protons at 4.82 (dd, *J* = 10.2, 4.4 Hz), 5.20 (d, *J* = 7.5 Hz), 5.23 (d, *J* = 16.2 Hz), and 5.56 (dd, *J* = 16.2, 7.5 Hz), an oxymethine proton at 4.63 as doublet (*J* = 7.5 Hz), and also couple of methylene protons peaks between 1.87–2.35 Hz. The ^{13}C-NMR spectrum (150 MHz, CDCl$_3$) revealed the presence of a total of 15 carbons, including an oxymethine carbon at 78.8. The DEPT135 identified four methyl, three methylene, one oxymethine, four olefinic methine, and the remaining three as quaternary carbons. Among the later three quaternary carbons, one at 37.3 is aliphatic and remaining two are connected double bonds. The complete structure of this compound was established by two-dimensional (2D) NMR spectra, predominantly by HSQC and HMBC. In the ^{1}H-^{1}H COSY spectrum, the *trans* double bonded protons showed expected interaction between them. In the HMBC experiment, the *trans* double bonded proton at 5.23 (δ_C 139.5 from HSQC) and methyl protons at 1.65 (H-12; δ_C 12.8 from HSQC) revealed a common ^3J interaction to an oxymethine carbon at 78.8 (C-1). Olefinic protons at δ_H 4.82 (H-7; δ_C 125.0 ppm from HSQC) and 5.20 (H-3; δ_C 124.8 ppm from HSQC) and methyl protons at 1.43 (H-13; δ_C 15.3 ppm from HSQC) showed ^3J correlation to a methylene carbon at 39.5 (C-5; δ_H 2.35 ppm from HSQC). Protons at 2.20 and 2.24 (H-4; δ_C 24.4 from HSQC) exhibited ^3J correlations to quaternary carbons at 142.2 (C-2) and 133.2 (C-6). Two sets of methyl protons at 1.04 (H-14; δ_C 24.9) and 1.06 (H-15; δ_C 29.9) revealed common ^2J correlation to a quaternary carbon at 37.3 (C-9) and ^3J interaction to methine carbon at 139.5 (C-10, δ_H 5.23) and methylene at 42.4 (C-8, δ_H 1.87, 2.32). H-10 also revealed ^3J interaction to both methyl group carbons at 24.9 (C-14) and 29.9 (C-15). The COSY experiment exhibited usual interaction (H-10 to H-11; H-4 to both H-3 and H-5; H-7 to H-8). Accordingly, compound **2** was identified as (2Z,6Z,10E)-2,6,9,9-tetramethylcycloundeca-2,6,10-trien-1-ol, commonly known as zerumbol (**2**) [24]. The NMR spectra of compound **2** are available in Supplementary Material. Compounds **3–8** were identified as zerumbone (**3**) [23,25], buddledone A (**4**) [26], germacrone (**5**) [23,27], furanodienone (**6**) [28], borneol (**7**) [29], and camphor (**8**) [29]. Among these compounds, zerumbol (**2**), buddledone A (**4**), germacrone (**5**), and furanodienone (**6**) have been reported first time from the genus *Zingiber*, while zerumbone (**3**) was reported from *Z. zerumbet* [30]. Chemical structures of compounds **1–8** are incorporated in Figure 1.

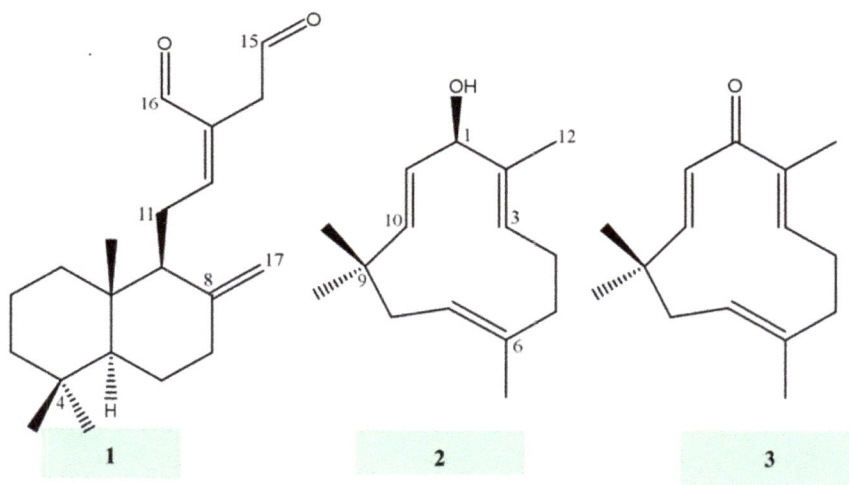

Figure 1. *Cont.*

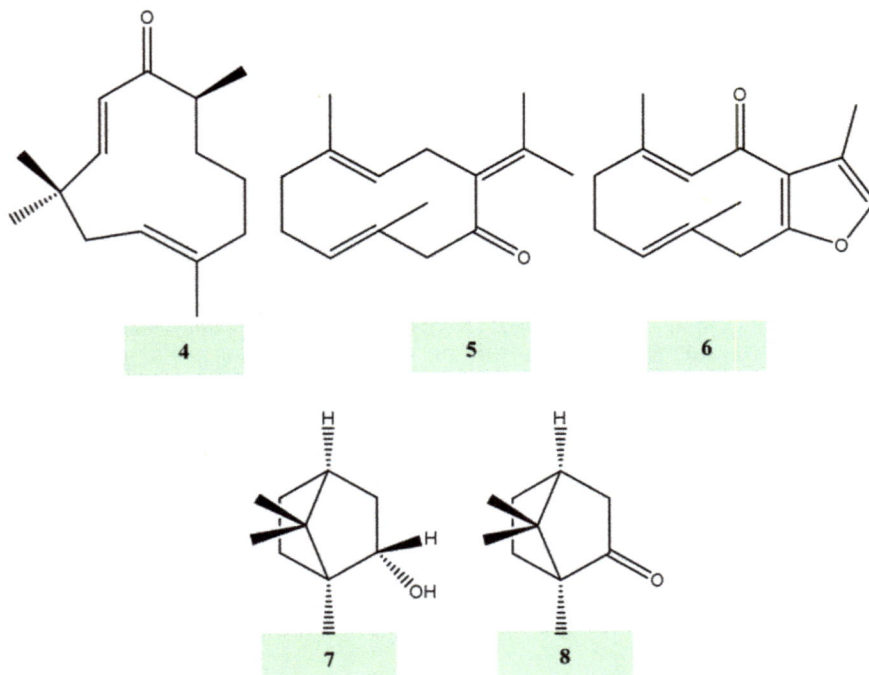

Figure 1. Chemical structures of terpenes isolated from *Z. montanum.*

Table 3. ^{1}H- (600 MHz) and ^{13}C- (150 MHz) NMR and HMBC (500 MHz) data of **2** in CDCl$_3$.

Position	1H	13C	HMBC	
			^2J	^3J
1	4.63, d, *J* = 7.5 Hz, 1H	78.8	-	C3, C10, C12
2	-	142.2	-	-
3	5.20, d, *J* = 7.5 Hz, 1H	124.8	-	C1, C12
4	2.20, m, 1H; 2.24, m, 1H	24.4	C3	C6
5	2.35, m, 2H	39.5	-	C3, C13
6	-	133.2	-	-
7	4.82, dd, *J* = 10.2, 4.4 Hz, 1H	125	-	C5, C9
8	1.87, m, 1H; 2.32, m, 1H	42.4	C7	C6, C10
9	-	37.3	-	-
10	5.23, d, *J* = 16.2 Hz, 1H	139.5	-	C1, C14, C15
11	5.56, dd, *J* = 16.2, 7.5 Hz, 1H	131.2	C1	C9, C13
12	1.65, s, 3H	12.8	C2	C1, C3
13	1.43, s, 3H	15.3	C6	C5, C7
14	1.04, s, 3H	24.9	C9	C8, C10, C15
15	1.06, s, 3H	29.9	C9	C8, C10, C14

Compounds **1–8** were assessed for their antibacterial activities against multi-drug resistant and methicillin resistant *Staphylococcus aureus*, notably SA1199B, XU212, RN4229, EMRSA15, MRSA27819, and MRSA340702. The minimum inhibitory concentrations (MICs) of these compounds are presented in Table 4. Norfloxacin was used as positive control for the comparison of antibacterial potencial of these compounds. Among these compounds, **1** and **2** displayed the highest activities with MICs in the range of 32–128 µg/mL (0.145–0.291 mM) against the test organisms. Compound **1** is a labdane diterpene with exomethylene at C-8, an olefine at C-12, and two aldehyde groups at C-16 and 17. The presence of these groups and unsaturations could account for significant antibacterial activity against MRSA strains. Although compounds **2** and **3** are structurally very similar, they differ in activity. We suggest that the presence of a hydroxyl group instead of carbonyl group at C-1 might make compound **2** more active than compound **3**. The antibacterial activity of compounds **3–8** were above 128 µg/mL (0.557–0.842 mM), the highest concentrations at which the compounds were tested. Monoterpene and sesquiterpenes from *Zingiber* were reported with antibacterial activity against *Staphylococcus aureus* (MTCC 96) and *Staphylococcus epidermidis* (MTCC 435) [21].

Table 4. Minimum inhibitory concentrations (MICs) (in mM) of compounds (**1–8**) against standard, multi-drug resistant (MDR) and methicillin-resistant strains of *Staphylococcus aureus*.

Compound	SA1199B	XU212	ATCC25941	RN4220	EMRSA15	MRSA27819	MRSA340702
1	0.212	0.424	0.212	0.212	0.212	0.424	0.424
2	0.291	0.582	0.582	0.582	0.145-0.291	0.582	>0.582
3	>0.587	>0.587	>0.587	>0.587	>0.587	>0.587	>0.587
4	>0.582	>0.582	>0.582	>0.582	>0.582	>0.582	>0.582
5	>0.587	>0.587	>0.587	>0.587	>0.587	>0.587	>0.587
6	>0.557	>0.557	>0.557	>0.557	>0.557	>0.557	>0.557
7	>0.831	>0.831	>0.831	>0.831	>0.831	>0.831	>0.831
8	>0.842	>0.842	>0.842	>0.842	>0.842	>0.842	>0.842
Norfloxacin	0.100	0.200	0.050	0.025	0.050	0.100	0.401

3. Material and Methods

3.1. General

Analytical solvents, such as *n*-hexane, chloroform, ethyl acetate, acetone, and methanol were purchased from Fisher Scientific, Loughborough, UK. Dimethyl sulfoxide, sodium chloride, norfloxacin and 3-[4,5-dimethylthiazol-2-yl]-2,5-iphenyltetrazolium bromide (MTT) used during antibacterial assay were purchased from Sigma Aldrich (Dorset, UK). Silica gel 60H used for vacuum liquid chromatography was purchased from Merck Millipore, UK. Sephadex LH 20 used for gel filtration chromatography was purchased from GE healthcare, Uppasala, Sweden. Prepacked silica column (normal phase) used for solid phase extraction (SPE) was purchased from Phenomenex, Cheshire, UK. Analytical and preparative TLC carried out on 0.2 mm silica gel 60 F_{254} was purchased from Merck, Darmstadt, Germany. Spots on the TLC plates were visualized under short UV (254 nm) and long UV (366 nm), and also by spraying them with 1% vanillin in concentrated H_2SO_4 followed by heating at 100 °C for 3–6 min. The NMR spectroscopy was performed with Bruker AMX 600 NMR spectrometer, Coventry, UK (600 MHz for ^{1}H, and 150 MHz for ^{13}C) in the Department of Chemistry at University College London (London, UK). High Resolution Mass Spectrometry (HRMS) was performed in Liverpool John Moores University (Liverpool, UK). IR spectroscopy was recorded on Agilent FT-IR (Cary 630, Stockport, UK).

3.2. Plant Material

The rhizomes of *Zingiber montanum* were collected from Bangladesh National Botanical Garden, Dhaka, Bangladesh in September 2016. The plant was identified by the Bangladesh National Herbarium, Mirpur, Dhaka, Bangladesh, where a voucher specimen (DACB 43550) of this collection was deposited.

3.3. Extraction and Isolation of Compounds

The rhizomes of *Z. montanum* were sun dried for 2–3 days, followed by drying in the oven at a temperature of 30–35 °C for 30 min prior to grinding. Subsequently, the plant materials were ground into fine powders using a grinder. The ground plant material (242 g) was Soxhlet extracted with solvents of increasing polarity: *n*-hexane, chloroform, and methanol (approximately 700 mL, 10–15 cycles each). Each of the extracts was concentrated using rotary evaporator under reduced pressure at a maximum temperature of 40 °C to yield 9.38 g, 10.22 g, and 23.0 g of *n*-hexane, chloroform, and methanol extracts, respectively. The antibacterial screening was performed on these crude extracts against clinical isolates of MRSA strains. Hexane (MICs 128–256 µg/mL) and chloroform (MICs 64–256 µg/mL) extracts appeared to be active and they were further fractionated by vacuum liquid chromatography (VLC). A portion of *n*-hexane (6.5 g) or chloroform (7.6 g) were adsorbed into silica gel (70–230 mesh) and loaded into VLC column, which was uniformly packed with VLC grade silica gel (60H), followed by eluting with stepwise gradient of mobile phase initially with mixture of *n*-hexane and ethyl acetate and then with EtOAc and MeOH mixtures. The eluted fractions (200 mL each) were evaporated using rotary evaporator and analysed by TLC. Based on TLC results, the similar fractions were bulked together for further purifications by solid phase extraction (SPE), column chromatography over Sephadex LH20, and/or preparative TLC. The basic principle of SPE is similar to VLC, but SPE was used in smaller scale fractionation or further purification of compounds from the VLC fractions or pooled fractions from Sephadex LH20 column chromatography. For column chromatography over Sephadex LH20, the glass column was packed with the slurry of Sephadex LH-20, which was soaked in the solvent (50% chloroform in *n*-hexane or 100% chloroform) half an hour prior to the packing of the column. The sample was dissolved in a small amount of appropriate solvent and then applied onto the top of the adsorbent. The column was eluted with 50–75% chloroform in *n*-hexane, followed by 100% chloroform and then CHCl$_3$ + MeOH mixtures of increasing polarity. During preparative TLC, the sample was applied uniformly as band in the sample application zone (2 cm above from the bottom edge of TLC plate) on commercially available TLC aluminium plates (pre-coated silica gel 60 PF$_{254}$). The TLC plates were developed with appropriate mobile phase up to the upper edge of plates. In addition, the multiple development technique was also adapted for a better accomplishment of separation of compounds of very similar polarity.

VLC fraction eluted with 5–10% of EtOAc in *n*-hexane of *n*-hexane extract was subjected to column chromatography over Sephadex LH20. PTLC (mobile phase 15% EtOAc in hexane) on Sephadex column eluted with 50% CHCl$_3$ in *n*-hexane yielded compounds **2** (3 mg) and **3** (10 mg), whereas compound **4** (9 mg) was obtained from sephadex column eluted with 100% chloroform. SPE on the VLC fraction eluted with 15% of EtOAc in *n*-hexane of *n*-hexane extract provided six sub-fractions. Preparative TLC (mobile phase 4% EtOAc in hexane plus two drops glacial acetic acid) on SPE sub-fraction eluted with 4% EtOAc in hexane yielded compounds **5** (7 mg) and **6** (4 mg), whilst preparative TLC (mobile phase 4% EtOAc in hexane plus two drops glacial acetic acid) on SPE sub-fraction eluted with 4% EtOAc in hexane led to the isolation of compounds **6** (5 mg) and **7** (7 mg). Similarly, VLC fraction of chloroform extract was subjected to SPE and preparative TLC for the purification of compounds. SPE on VLC fraction eluted with 15% EtOAc in *n*-hexane followed PTLC (mobile phase 4% EtOAc in hexane) on SPE sub-fraction eluted with 4% EtOAc in hexane yielded **8** (4 mg), while compound **1** (6.5 mg) was isolated from the VLC fraction eluted with 25% EtOAc in *n*-hexane, followed by PTLC (mobile phase 15% EtOAc in hexane) on SPE sub-fraction eluted with 10% EtOAc in hexane.

3.4. Antibacterial Assay against Clinical Isolates of Multi-Drug Resistant and Methicillin Resistant Staphylococcus Aureus

The antibacterial activity of crude extracts and the isolated compounds were tested against clinical isolates of MRSA strains by microtitre assay using 96 well plates to determine the minimum inhibitory concentrations (MICs). Mueller–Hinton broth (MHB) used in this study was purchased from Oxoid, Hamshire, UK and prepared as instructed by the supplier; however, MHB was adjusted to contain 20 mg/L and 10 mg/L of Ca^{2+} and Mg^{2+}, respectively. The clinical isolates of *S. aureus* strains used in this study included ATCC25923, SA1199B, RN4220, XU212, EMRSA15, MRSA340702, and MRSA274829. A standard laboratory strain, ATCC25923, was also used in this study, which is sensitive to antibiotics, like tetracycline [31]. SA1199B over-expresses the NorA MDR efflux pump [32], RN4220 possesses the MsrA macrolide efflux protein [33], XU212 is a Kuwaiti hospital isolate that is an MRSA strain possessing the TetK tetracycline efflux pump [31], whilst the EMRSA15 strain [34] is epidemic in the UK. All *S. aureus* strains were subcultured on nutrient agar (Oxoid) and incubated for approximately 24 h at 37 °C prior to MIC determination. All of the bacterial strains were prepared in 9 g/L saline water with an inoculum density of 5×10^5 colony forming unit (cfu/mL) by comparison with the 0.5 MacFarland turbidity standard.

The stock solution of control positive (Norfloxacin) was prepared by dissolving the antibiotic (2.0 mg) in DMSO (244 µL) and diluting 16 fold with MHB to obtain the desired concentration of the antibiotic stock solution (512 µg/mL). Similarly, stock solutions of crude extract (2048 µg/mL) and isolated compounds (256–512 µg/mL) were prepared by dissolving in required amount of DMSO, followed by 16 fold dilution with MHB.

During the experiment, using a multi-channel pipette an aliquot of 100 µL of MHB was dispensed into each well of 96-well plate except those in the last column. Then 100 µL of stock solution of crude extract or isolated compounds and antibiotic was added in duplicate to the wells of the first column of 96-well plate (total content 200 µL), followed by mixing the content thoroughly and transferring 100 µL of this content to the wells of the second column of 96-well plate using a multi-channel pipette. This two-fold serial dilution process was continued to the 10th well, followed by the addition of the final 100 µL solution to the empty wells of 12th column of 96 well-plate. The inoculum (100 µL) of each bacterium at a density of 5×10^5 cfu/mL was added to all wells, except those in the final (12th) column. The contents of the wells in the 11th and 12th columns represented growth control (bacteria, but no antibiotic, extract, or compounds) and sterility control (antibiotic, extract, or compounds but no bacteria), respectively. Every assay was performed in duplicate. The plates were incubated for 18 h at 37 °C. For the measurement of MIC, 20 µL of a 5 mg/mL methanolic solution of 3-[4,5-dimethylthiazol-2-yl]-2,5-iphenyltetrazolium bromide (MTT; Sigma) was added to each of the wells, followed by incubation for 20–30 min at 37 °C. Bacterial growth was indicated by a colour change from yellow (colour of MTT) to dark blue. The MIC was recorded as the lowest concentration at which no growth (yellow color) was observed [13]. If no growth was observed at any of the concentrations tested, the assay was repeated starting from a stock solution of lower concentration. If growth was observed at all of the concentrations tested, the assay was repeated, starting with a stock solution of higher concentration.

4. Conclusions

In this study, the crude extracts of the rhizomes of *Z. montanum* and compounds that were isolated from active extracts were assessed against a panel of clinical isolates of multi-drug resistant (MDR) and methicillin resistant *Staphylococcus aureus* (MRSA), including SA1199B, XU212, RM4221, EMRSA15, MRSA27819, and MRSA340702. Bioassay directed isolation using a range of chromatographic techniques, including vacuum liquid chromatography (VLC), solid phase extraction (SPE), column chromatography over Sephadex LH20, and preparative thin layer chromatography (PTLC) led to the identification of two monoterpenes, borneol (**7**) and camphor (**8**), and five sesquiterpnes, zerumbol (**2**), zerumbone (**3**), buddledone A (**4**), furanodienone (**5**), germacrone (**6**), and a diterpene, (*E*)-8(17),12-labdadiene-15,16-dial (**1**). Among these terpenes, compounds **1** and **2** displayed significant activity with MICs of 32–128 µg/mL (0.145–0.291 mM) against the clinical isolates of MRSA strains tested. Such activity encourages the authors to carry out bioassay guided phytochemical investigation on related members of Zingiberacae family for the identification of lead anti-Staphylococcal compounds.

Supplementary Materials: NMR and mass spectra of active compounds together with results of antibacterial activity using 96 well plates are available in supplementary materials.

Author Contributions: H.S. as part of her PhD project collected and extracted the plant, isolated and identified the compounds, carried out antibacterial assay and prepared the initial draft manuscript. M.M.R. and B.P. supervised the various aspects of the project, helped with the interpretation of data and contributed towards the preparation of the manuscript. M.M.R. as the Director of Study and Corresponding Author also prepared and submitted the final version of the manuscript.

Funding: This research received no external funding.

Acknowledgments: Holly Siddique is grateful to the School of Health, Sport and Bioscience at the University of East London for the award of PhD studentship. The authors sincerely thank to Simon Gibbons at the UCL School of Pharmacy for providing the MRSA strains for antibacterial assay and Satyajit D Sarker at Liverpool John Moores University for providing mass spectrometry data of compounds.

Conflicts of Interest: The authors declare no conflict of interest.

References

1. O'Neill, J. Tackling Drug-Resistant Infections Globally: Final Report and Recommendations the Review on Antimicrobial Resistance. 2016. Available online: https://amr-review.org/sites/default/files/160518_Final%20paper_with%20cover.pdf (accessed on 15 February 2018).

2. World Health Organization (WHO). Antibiotic Resistance. 2018. Available online: https://www.who.int/news-room/fact-sheets/detail/antibiotic-resistance (accessed on 18 November 2018).

3. Faridi, A.; Kareshk, A.T.; Fatahi-Bafghi, M.; Ziasistani, M.; Ghahraman, M.R.K.; Seyyed-Yousefi, S.Z.; Shakeri, N.; Kalantar-Neyestanaki, D. Detection of methicillin-resistant Staphylococcus aureus (MRSA) in clinical samples of patients with external ocular infection. *Iran. J. Microbiol.* **2018**, *10*, 215–219. [PubMed]

4. Hollman, A. Digoxin comes from Digitalis lanata. *Br. Med. J.* **1996**, *312*, 912. [CrossRef]

5. Alam, M.M.; Naeem, M.; Khan, M.M.A.; Uddin, M. Vincristine and Vinblastine Anticancer *Catharanthus* Alkaloids: Pharmacological Applications and Strategies for Yield Improvement. In *Catharanthus Roseus*; Springer: Cham, Switzerland, 2017; pp. 277–307.

6. Patrick, G. The Opioid Analgesic. In *An Introduction to Medicinal Chemistry*, 5th ed.; Oxford University Press: Oxford, UK, 2013; p. 632.

7. Butler, A.; Hensman, T. Drugs for the fever. *Educ. Chem.* **2000**, *37*, 151.

8. Mahdi, J.G. Medicinal potential of willow: A chemical perspective of aspirin discovery. *J. Saudi Chem. Soc.* **2010**, *14*, 317–322. [CrossRef]

9. Gautam, R.; Saklani, A.; Jachak, S.M. Indian medicinal plants as a source of antimycobacterial agents. *J. Ethnopharmacol.* **2007**, *110*, 200–234. [CrossRef] [PubMed]

10. Wang, Z.; Yu, P.; Zhang, G.; Xu, L.; Wang, D.; Wang, L.; Zeng, X.; Wang, Y. Design, synthesis and antibacterial activity of novel andrographolide derivatives. *Bioorg. Med. Chem.* **2010**, *18*, 4269–4274. [CrossRef] [PubMed]

11. Oluwatuyi, M.; Kaatz, G.W.; Gibbons, S. Antibacterial and resistance modifying activity of Rosmarinus officinalis. *Phytochemistry* **2004**, *65*, 3249–3254. [CrossRef] [PubMed]

12. Schempp, C.M.; Pelz, K.; Wittmer, A.; Schöpf, E.; Simon, J.C. Antibacterial activity of hyperforin from St John's wort against multi-resistant Staphylococcus aureus and Gram-positive bacteria. *Lancet* **1999**, *353*, 2129. [CrossRef]

13. Shiu, W.K.P.; Rahman, M.M.; Curry, J.; Stapleton, P.D.; Zloh, M.; Malkinson, J.P.; Gibbons, S. Antibacterial acylphloroglucinols from *Hypericum olympicum*. *J. Nat. Prod.* **2012**, *75*, 336–343. [CrossRef] [PubMed]

14. Khare, C.P. *Indian Medicinal Plants: An Illustrated Dictionary*; Springer: Berlin/Heidelberg, Germany, 2007; p. 733.

15. Farnsworth, N.R.; Bunyapraphatsara, N. *Thai Medicinal Plants: Recommended for Primary Health Care System*; Medicinal Plants Information Center: Bangkok, Thailand, 1992.

16. Singh, C.B.; Manglembi, N.; Swapana, N.; Chanu, S.B. Ethnobotany, phytochemistry and pharmacology of *Zingiber cassumunar* Roxb. (Zingiberaceae). *J. Pharmacogn. Phytochem.* **2015**, *4*, 1–6.

17. Sharma, G.J.; Thokchom, D.S. Antioxidant and radioprotective properties of *Zingiber montanum* (J. König) A. Dietr. *Planta Med.* **2011**, *77*, 127. [CrossRef]

18. Thokchom, D.S.; Sharma, T.D.; Sharma, G.J. Radioprotective effect of rhizome extract of *Zingiber montanum* in Rattus norvegicus. *Radiat. Environ. Biophys.* **2012**, *51*, 311–318. [CrossRef] [PubMed]

19. Al-Amin, M.; Sultana, G.N.; Hossain, C.F. Antiulcer principle from *Zingiber montanum*. *J. Ethnopharmacol.* **2012**, *141*, 57–60. [CrossRef] [PubMed]

20. Masuda, T.; Jitoe, A.; Mabry, M.J. Isolation and structure determination of cassumunarins A, B and C: New anti-inflammatory antioxidants from a tropical ginger, *Zingiber cassumunar*. *J. Am. Oil Chem. Soc.* **1995**, *72*, 1053–1057. [CrossRef]

21. Verma, R.S.; Joshi, N.; Padalia, R.C.; Singh, V.R.; Goswami, P.; Verma, S.K.; Iqbal, H.; Chanda, D.; Verma, R.K.; Darokar, M.P.; et al. Chemical composition and antibacterial, antifungal, allelopathic and acetylcholinesterase inhibitory activities of cassumunar-ginger. *J. Sci. Food Agric.* **2018**, *98*, 321–327. [CrossRef] [PubMed]

22. Sy, K.L.; Brown, D.G. Labdane diterpenoids from *Alpinia chinensis*. *J. Nat. Prod.* **1997**, *60*, 904–908. [CrossRef]

23. Firman, K.; Kinoshita, T.; Itai, A.; Sankawa, U. Terpenoids from Curcuma Heyneana. *Phytochemistry* **1988**, *27*, 3887–3891. [CrossRef]

24. Takashi, K.; Nagao, R.; Masuda, T.; Hill, R.K.; Morita, M.; Takatani, M.; Sawada, S.; Okamoto, T. The chemistry of Zerumbone IV: Asymmetric synthesis of Zerumbol. *J. Mol. Catal. B Enzym.* **2002**, *17*, 75–79.

25. Nathaniel, C.; Elaine-Lee, Y.L.; Yee, B.C.; How, C.W.; Yim, H.S.; Rasadee, A.; Ng, H.S. Zerumbone-loaded nanostructured lipid carrier induces apoptosis in human colorectal adenocarcinoma (Caco-2) cell line. *Nanosci. Nanotechnol. Lett.* **2016**, *8*, 294–302. [CrossRef]

26. Cai, Z.; Yongpruksa, N.; Harmata, M. Total synthesis of the terpenoid buddledone A: 11-membered ring-closing metathesis. *Org. Lett.* **2012**, *14*, 1661–1663. [CrossRef] [PubMed]

27. Simova, S.D.; Bozhkova, N.V.; Orahovats, A.S. ^{1}H and ^{13}C NMR studies of some germacrones and isogermacrones. *Org. Magn. Reson.* **1984**, *22*, 431–433. [CrossRef]

28. Brieskorn, C.H.; Noble, P. Furanosesquiterpenes from the essential oil of myrrh. *Phytochemistry* **1983**, *22*, 1207–1211. [CrossRef]

29. Uchio, Y. Constituents of the essential oil of *Chrysanthemum japonense*. Nojigiku alcohol and its acetate. *Bull. Chem. Soc. Jpn.* **1978**, *51*, 2342–2346. [CrossRef]

30. Da Silva, T.M.; Pinheiro, C.D.; Orlandi, P.P.; Pinheiro, C.C.; Sontes, G.S. Zerumbone from *Zingiber zerumbet* (L.) smith: A potential prophylactic and therapeutic agent against the cariogenic bacterium *Streptococcus mutans*. *BMC Complement. Altern. Med.* **2018**, *18*, 301.

31. Gibbons, S.; Udo, E.E. The effect of reserpine, a modulator of multidrug efflux pumps, on the in vitro activity of tetracycline against clinical isolates of methicillin resistant *Staphylococcus aureus* (MRSA) possessing the *tet*(K) determinant. *Phytother. Res.* **2000**, *14*, 139–140. [CrossRef]

32. Kaatz, G.W.; Seo, S.M.; Ruble, C.A. Efflux-mediated fluoroquinolone resistance in *Staphylococcus aureus*. *Antimicrob. Agents Chemother.* **1993**, *37*, 1086–1094. [CrossRef]

33. Ross, J.L.; Farrell, A.M.; Eady, E.A.; Cove, J.H.; Cunliffe, W.J.J. Characterisation and molecular cloning of the novel macrolide-streptogramin B resistance determinant from *Staphylococcus epidermidis*. *Antimicrob. Agents Chemother.* **1989**, *24*, 851–862. [CrossRef]

34. Richardson, J.F.; Reith, S. Characterization of a strain of methicillin-resistant *Staphylococcus aureus* (EMRSA-15) by conventional and molecular methods. *J. Hosp. Infect.* **1993**, *25*, 45–52. [CrossRef]

Sample Availability: Samples of the compounds **1–8** are available from the authors.

molecules

MDPI

Article

Biotransformed Metabolites of the Hop Prenylflavanone Isoxanthohumol

Hyun Jung Kim [1], Soon-Ho Yim [2], Fubo Han [3], Bok Yun Kang [3], Hyun Jin Choi [4], Da-Woon Jung [5], Darren R. Williams [5], Kirk R. Gustafson [6], Edward J. Kennelly [7] and Ik-Soo Lee [3,*]

[1]   College of Pharmacy and Natural Medicine Research Institute, Mokpo National University, Muan, Jeonnam 58554, Korea; hyunkim@mokpo.ac.kr
[2]   Department of Pharmaceutical Engineering, Dongshin University, Naju, Jeonnam 58245, Korea; virshyim@dsu.ac.kr
[3]   College of Pharmacy, Chonnam National University, Gwangju 61186, Korea; hanfubo0306@gmail.com (F.H.); bykang@chonnam.ac.kr (B.Y.K.)
[4]   College of Pharmacy and Institute of Pharmaceutical Sciences, CHA University, Seongnam, Gyeonggi-do 13488, Korea; hjchoi3@cha.ac.kr
[5]   New Drug Targets Laboratory, School of Life Sciences, Gwangju Institute of Science and Technology, Gwangju 61005, Korea; jung@gist.ac.kr (D.-W.J.); darren@gist.ac.kr (D.R.W.)
[6]   Molecular Targets Program, Center for Cancer Research, National Cancer Institute, Frederick, MD 21702-1201, USA; gustafki@mail.nih.gov
[7]   Department of Biological Sciences, Lehman College, City University of New York, Bronx, NY 10468, USA; edward.kennelly@lehman.cuny.edu
*   Correspondence: islee@chonnam.ac.kr; Tel.: +82-62-530-2932

Academic Editors: Muhammad Ilias and Charles L. Cantrell
Received: 30 December 2018; Accepted: 20 January 2019; Published: 22 January 2019

Abstract: A metabolic conversion study on microbes is known as one of the most useful tools to predict the xenobiotic metabolism of organic compounds in mammalian systems. The microbial biotransformation of isoxanthohumol (**1**), a major hop prenylflavanone in beer, has resulted in the production of three diastereomeric pairs of oxygenated metabolites (**2–7**). The microbial metabolites of **1** were formed by epoxidation or hydroxylation of the prenyl group, and HPLC, NMR, and CD analyses revealed that all of the products were diastereomeric pairs composed of ($2S$)- and ($2R$)- isomers. The structures of these metabolic compounds were elucidated to be ($2S,2''S$)- and ($2R,2''S$)-4'-hydroxy-5-methoxy-7,8-(2,2-dimethyl-3-hydroxy-2,3-dihydro-4H-pyrano)-flavanones (**2** and **3**), ($2S$)- and ($2R$)-7,4'-dihydroxy-5-methoxy-8-(2,3-dihydroxy-3-methylbutyl)-flavanones (**4** and **5**) which were new oxygenated derivatives, along with ($2R$)- and ($2S$)-4'-hydroxy-5-methoxy-2''-(1-hydroxy-1-methylethyl)dihydrofuro[2,3-h]flavanones (**6** and **7**) on the basis of spectroscopic data. These results could contribute to understanding the metabolic fates of the major beer prenylflavanone isoxanthohumol that occur in mammalian system.

Keywords: microbial transformation; hop prenylflavanone; isoxanthohumol

1. Introduction

Isoxanthohumol (**1**) (5-methoxy-8-prenylnaringenin, $C_{21}H_{22}O_5$) is a well-known prenylated flavanone which occurs together with the prenylated chalcone xanthohumol in the female inflorescences (cones) of *Humulus lupulus* L. (hops) (Cannabaceae), which are added during the beer brewing process [1,2]. This flavanone has been specifically regarded as a beer prenylflavanone, since it is the main isomeric product of xanthohumol cyclization formed during hop processing and brewing. Hops naturally contain only minor quantities of isoxanthohumol compared with those of the most abundant hop chalcone xanthohumol, whereas beer contains much higher levels of isoxanthohumol than xanthohumol [3,4]. Biological and pharmacological properties of isoxanthohumol (**1**) have been less characterized than those of xanthohumol, but **1** has shown moderate estrogenic activity [5], antiproliferative and anticancer activities [6–8], cancer chemoprevention properties [9], and modifying effects in ontogenetic steroidogenesis [10].

Despite its importance in beer, only a few metabolism studies have been carried out with isoxanthohumol (**1**) to identify its metabolic fate in humans. Nikolic and colleagues investigated the oxidative metabolism of **1** using human liver microsomes in vitro and described several metabolites on the basis of liquid chromatography-tandem mass spectrometry [11]. Modification on one of the two terminal methyl groups of the prenyl moiety into the *cis*- and *trans*-hydroxymethyl analogues, respectively, followed by further oxidation to the *cis*- and *trans*-aldehydes, and double bond migration and subsequent hydroxylation to give an *exo*-methylene with an allylic alcohol was observed. Derivatives formed by hydroxylation or oxidation on the A- or B-ring, and *O*-demethylation were also reported [11]. A recent metabolism study of **1** showed that it was transformed by microorganisms into derivatives with a dihydrofuran ring or a methoxyglucosyl group [12]. In addition, some metabolic studies have focused on the conversion or activation of isoxanthohumol into 8-prenylnaringenin, a potent phytoestrogen [13–15].

Microbial biotransformation studies are regarded as one of the most useful tools to mimic and predict the xenobiotic metabolism of compounds in mammalian systems. Clark and Hufford have systematically summarized and reviewed the potential for the microorganisms as tools in the study of drug metabolisms with a number of specific examples that demonstrated the similarity in microbial and mammalian metabolism of xenobiotics [16,17]. They noted that microbial systems could offer a reliable, reproducible alternative to small laboratory animals for preliminary drug metabolism studies to identify the structural modifications by enzymatic reactions. General techniques and methods utilized in microbial metabolism studies clearly offer the practical advantages of convenient and inexpensive maintenance, production of metabolites in high yields and considerable amounts, and curtail the sacrifice of animals in biomedical research [16–18]. However, despite all the strengths and interesting parallels enumerated, microbial biotransformation could not ever completely replace the validity of xenobiotic metabolism studies with animals as well as liver microsomes or perfused livers. This model is a recently accounted practical tool with high potential for the creation of molecular diversity far beyond the metabolic changes observed in mammals [18].

We previously reported that microbial biotransformation of hop prenylflavonoids, including xanthohumol and 8-prenylnaringenin, produced several glucosylated, acyl-glucosylated, and cyclized metabolites [19,20], while biotransformation of **1** provided a diastereomeric pair of metabolites that resulted from 7-*O*-glucosylation on the A-ring via microbial Phase II conjugation reaction [21]. In our ongoing metabolism study of hop prenylflavonoids, a preparative-scale microbial transformation of isoxanthohumol (**1**) by the fungi, *Rhizopus oryzae* KCTC 6399 and *Fusarium oxysporum* f.sp. *lini* KCTC 16325, afforded three pairs of oxygenated metabolites (**2–7**) (Figure 1). The production of these microbially biotransformed metabolites of **1** and their structure elucidation are reported herein.

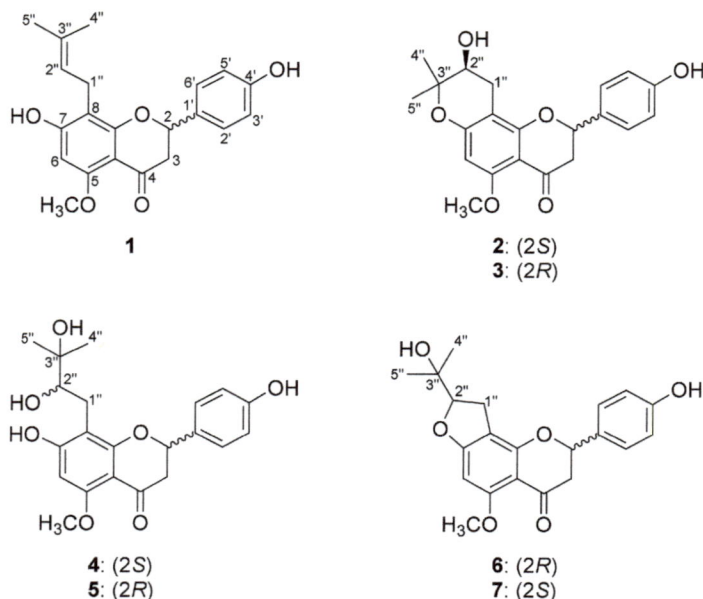

Figure 1. The structures of isoxanthohumol (**1**) and its metabolites (**2**–**7**).

2. Results and Discussion

2.1. Preparation and Microbial Biotransformation of Isoxanthohumol

Preparation of the substrate isoxanthohumol (**1**) was achieved by both chemical cyclization of xanthohumol in aqueous alkali solution [4] and enzymatic cyclization using a microbial transformation method [19]. The isoxanthohumol produced by these methods was a racemic mixture of (2S)- and (2R)-flavanones, which was confirmed by spectroscopic data analysis showing no absorption in the CD spectrum and no optical rotation.

A total of forty-one microbial cultures were screened for their ability to metabolize **1** and two fungi, *Rhizopus oryzae* KCTC 6399 and *Fusarium oxysporum* f.sp. *lini* KCTC 16325, were selected for further scale-up fermentation studies. For each microbe, separate substrate and culture control studies were carried out under the same fermentation conditions, which showed that the metabolites were produced as a result of enzymatic activity by the fungi, and not as a consequence of chemical or non-metabolic conversion. The ability to biotransform **1** was confirmed on the basis of reversed-phase (C_{18}) TLC analyses. The R_f values of the three new diasteromeric mixtures (**2,3**: R_f 0.32, **4,5**: 0.42, and **6,7**: 0.28) were significantly larger than that of **1** (R_f 0.15), which indicated that the fungi produced metabolites with higher polarity.

2.2. Structure Elucidation of Isoxanthohumol Metabolites

Metabolites **2** and **3** of isoxanthohumol were produced by the fungus *R. oryzae* KCTC 6399 and obtained as a pale yellow amorphous powder. HPLC analyses of the mixture of **2** and **3** revealed that each metabolite had the same ultraviolet (UV) spectral data at 190–400 nm and a minor difference of retention time (t_R) (**2**: 21.56 min and **3**: 22.45 min). Their ^{1}H and ^{13}C-NMR spectra also exhibited nearly identical chemical shift and coupling constant values, which suggested that they consisted of a pair of diastereomers. HRESIMS of metabolites **2** and **3** exhibited $[M + H]^+$ peaks at m/z 371.1504 and m/z 371.1482 (calcd for $C_{21}H_{23}O_6$, 371.1495) respectively, which established their molecular formula as $C_{21}H_{22}O_6$ and indicated that they were mono-oxygenated metabolites of **1**. The UV spectrum of both compounds displayed typical absorptions for a flavanone, with a maximum absorption peak at ~287 nm and an inflection at 318 nm, which were similar to those observed for **1**. However, their ^{1}H-NMR spectra showed major differences in the isoprenyl group, which showed two methyl signals shifted upfield (H-4″ and H-5″) at δ_H 1.31, 1.28 (3H, s) and 1.33, 1.34 (3H, s) as well as a new oxymethine signal (H-2″) at δ_H 3.76 (brt, J = 6.0 Hz) and 3.73 (1H, dd, J = 7.3, 5.8 Hz) in **2** and **3**, respectively. These observations suggested that the two sp^2 carbons of the isoprenyl group in **1** had been metabolized to oxygenated sp^3 carbons. The ^{13}C-NMR data of both compounds showed resonances for an sp^3 oxymethine at δ_C 69.6 and 69.9 (C-2″) and an oxygenated sp^3 quaternary carbon at δ_C 79.9 (C-3″), and the loss of the olefin signals at δ_C 124.0 and 131.9 that were observed in **1**. An oxygenated aromatic carbon at δ_C 161.9 (C-7) in metabolites **2** and **3** was shifted upfield relative to the corresponding carbon in **1** (δ_C 164.0). These results indicated that a dihydropyran ring had been formed by oxidation and subsequent cyclization of the isoprenyl moiety, possibly going through an epoxide intermediate. Two-dimensional NMR data, including HSQC and HMBC experiments, supported the presence of a dihydropyran ring. The *gem*-dimethyl protons H$_3$-4″ and H$_3$-5″ correlated with C-2″ and C-3″, and the H$_2$-1″ methylene protons correlated with C-8 and C-3″. Based on these spectroscopic analyses, the planar structure of diastereomeric metabolites **2** and **3** was assigned as 4′-hydroxy-5-methoxy-7,8-(2,2-dimethyl-3-hydroxy-2,3-dihydro-4*H*-pyrano)-flavanone. The absolute configurations of the two asymmetric carbons, C-2 and C-2″, were established by circular dichroism [22,23], and the Mosher's ester method [24,25], respectively. Metabolite **2** displayed positive and negative Cotton effects at 331 (n → π* transition) and 288 nm (π → π* transition), respectively, which corresponded to a 2*S* configuration. In contrast, metabolite **3** showed a negative Cotton effect at 334 nm and a positive one at 288 nm which established a 2*R* configuration (See Supplementary Materials). The absolute configuration at C-2″ was determined by the modified Mosher's method. Metabolite **3** was converted into the (*S*)- and (*R*)-methoxytrifluoromethylphenylacetic acid (MTPA) esters, **3a** and **3b**, by treatment with (*R*)- and (*S*)-MTPA chloride, respectively. The Δδ values (δ_H = δ_S − δ_R, ppm) calculated for the two esters (Figure 2) indicated that C-2″ had an *S* configuration. From these results, unambiguous structures of **2** and **3** were assigned as (2*S*,2″*S*)- and (2*R*,2″*S*)-4′-hydroxy-5-methoxy-7,8-(2,2-dimethyl-3-hydroxy-2,3-dihydro-4*H*-pyrano)-flavanones, respectively.

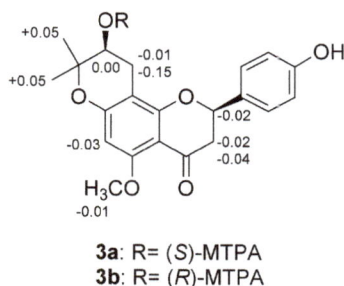

3a: R= (*S*)-MTPA
3b: R= (*R*)-MTPA

Figure 2. The Δδ$_{S-R}$ values (ppm) from Mosher ester derivatives of **3**.

Metabolites **4** and **5** of isoxanthohumol were also obtained by the enzymatic activity of *R. oryzae* KCTC 6399 and isolated as a pale yellow amorphous powder. As observed with metabolites **2** and **3**, HPLC analyses (t_R, **4**: 11.65 min and **5**: 12.56 min) and spectroscopic data including UV, infrared (IR), ^{1}H- and ^{13}C-NMR spectra indicated that two diastereomeric isomers were produced. Metabolites **4** and **5** exhibited HRESIMS [M + H]$^+$ peaks at m/z 389.1612 and m/z 389.1600 (calcd for $C_{21}H_{25}O_7$, 389.1600) respectively, which suggested a molecular formula of $C_{21}H_{24}O_7$ corresponding to dihydroxylated derivatives of **1**. The UV spectra of **4** and **5** displayed characteristic absorptions of a flavanone moiety at 225, 285 and ~310 nm, which were similar to their substrate **1**. However, the ^{1}H-NMR spectra of **4** and **5** exhibited major differences in the isoprenyl group resonances. They showed two pairs of methyl signals shifted upfield (H-4″ and H-5″) at δ_H 1.12, 1.15 (3H, s) and 1.14, 1.16 (3H, s), as well as oxymethine signals (C-2″) at δ_H 3.52 (dd, J = 10.0, 2.3 Hz) and 3.55 (dd, J = 10.0, 2.0 Hz), suggesting formation of two hydroxylated sp^3 carbons from the olefinic group of **1**. The ^{13}C-NMR spectra of both compounds showed the presence of a hydroxymethine signal (C-2″) at δ_C 80.2 in **4** and δ_C 80.1 in **5**, and an oxygenated quaternary carbon at δ_C 74.1 (C-3″) in both. In addition, ^{13}C-NMR signals of a *gem*-dimethyl group at δ_C 25.0 and 26.0, were consistent with their substitution on an oxygenated carbon. HMBC correlations from H$_3$-4″ and H$_3$-5″ to C-2″ and C-3″, as well as from H$_2$-1″ to C-8 and C-2″, and H-2″ to C-8 helped to establish the structure of diastereomers **4** and **5** as 7,4′-dihydroxy-5-methoxy-8-(2,3-dihydroxy-3-methylbutyl)-flavanone. The absolute configuration of the asymmetric center at C-2 was assigned by CD studies [22,23]. Metabolite **4** displayed positive and negative Cotton effects at 331 (n → π* transition) and 289 nm (π → π* transition), respectively, corresponding to a 2S configuration, while C-2 in metabolite **5** was established to have a 2R configuration from the opposite Cotton effects at 331 (negative) and 288 nm (positive). Therefore, the structures of metabolites **4** and **5** were assigned as (2S)- and (2R)-7,4′-dihydroxy-5-methoxy-8-(2,3-dihydroxy-3-methylbutyl)-flavanones, respectively. Regarding the determination of absolute configuration at C-2″, an asymmetric center of oxygenated C-prenyl side chain in aryl prenyl derivatives, X-ray crystallography has been considered as the most powerful approach as shown in a case of a prenylated coumarin, meranzin hydrate [26]. Adequate single crystals of **4** and **5** suitable for X-ray diffraction studies, however, were not available due to the limitations of their physicochemical properties and limited quantities. Chiroptical tools including electronic circular dichroism (ECD) were not effectively applicable on account of the strong Cotton effect curves obtained from (2S)- and (2R)-flavanones [23,27] (See Supplementary Materials). In an effort to determine the absolute configuration at C-2″, Mosher's ester derivatives were prepared, however the results were ambiguous. The *bis*-MTPA esters of **4** and **5** provided Δδ values ($\delta_H = \delta_S - \delta_R$) that had a non-uniform distribution of positive and negative signs, which meant that Mosher's analysis was not valid in this case [24,25]. Further, chemical synthesis may be necessary to determine absolute configuration of the chiral carbon C-2″ at the prenyl moiety.

Metabolites **6** and **7** were also obtained as a pair of diastereomers by microbial transformation of **1** using *F. oxysporum* f.sp. *lini*, which was confirmed by HPLC (t_R, **6**: 18.63 min and **3**: 19.42 min) and NMR experiments. HRESIMS of **6** and **7** showed [M + Na]$^+$ peaks at m/z 393.1307 and m/z 393.1316 (calcd for $C_{21}H_{23}O_6Na$, 393.1314), respectively, which established their molecular formula as $C_{21}H_{22}O_6$ which was isomeric with the mono-oxygenated metabolites **2** and **3**. However, several significant differences were observed in the ^{1}H- and ^{13}C-NMR spectra of **6** and **7**. Highly downfield shifted oxymethine proton (δ_H 4.73, brt, J = 8.8 Hz and 4.74, dd, J = 9.3, 8.0 Hz) and carbon (δ_C 92.5) resonances of C-2″ indicated that a dihydrofuran ring was formed from the isoprenyl group substituted on the A-ring. The NMR data of **6** and **7** were in good agreement with those of (2S)-4′-hydroxy-5-methoxy-7,8-[2-(1-hydroxy-1-methylethyl)-2,3-dihydrofurano]flavanone, a metabolite of xanthohumol by *Pichia membranifaciens* (ATCC 2254) [28], and (2R)-4′-hydroxy-5-methoxy-7,8-[2-(1-hydroxy-1-methylethyl)-2,3-dihydrofurano]flavanone, a metabolite of isoxanthohumol by *F. equiseti* (AM15) [12]. The structures of compounds **6** and **7** were assigned as 4′-hydroxy-5-methoxy-2″-(1-hydroxy-1-methylethyl)dihydrofuro[2,3-h]flavanones, and the absolute configurations of the asymmetric center at C-2 were identified by CD [22,23]. Metabolite **6** showed negative and positive Cotton effects at 325 and 289 nm, respectively, corresponding to 2R, while C-2 in metabolite **7** was established to be 2S from the opposite sign Cotton effects at 324 (positive) and 290 nm (negative). The configuration at C-2″ in compounds **6** and **7** could not be assigned from the NMR or CD measurements.

3. Materials and Methods

3.1. General Experimental Procedures

Optical rotations were recorded with a JASCO DIP 1000 digital polarimeter (JASCO, Tokyo, Japan). UV spectra were recorded on a JASCO V-530 spectrophotometer (JASCO, Tokyo, Japan), and CD spectra were recorded on a JASCO J-810 spectrometer (JASCO, Tokyo, Japan). IR spectra were obtained on a JASCO FT/IR 300-E spectrometer (JASCO, Tokyo, Japan). ^{1}H-, ^{13}C-, HSQC, and HMBC NMR experiments were recorded using a Varian Unity INOVA 500 spectrometer (Agilent Technologies, Inc., Santa Clara, CA, USA). HRESIMS were determined on Waters Synapt HDMS LC/MS mass spectrometer (Waters Corp., Milford, MA, USA). TLC was carried out on Merck silica gel F_{254}-precoated glass and RP-18 F_{254S} plates (Merck, Darmstadt, Germany). Medium pressure liquid chromatography (MPLC) was performed using Lobar^{TM} C_{18} column (10 × 240 mm, 40–63 µm, Merck, Darmstadt, Germany) and silica gel (40–63 µm, Merck). HPLC was performed on a Hewlett-Packard Agilent 1100 Series (Agilent Technologies, Inc., Santa Clara, CA, USA) HPLC System composed of a degasser, a binary mixing pump, a column oven and a DAD detector using Waters SunFire^{TM} (Waters Corp., Milford, MA, USA) (4.6 × 150 mm, 5 µm) and SunFire^{TM} Prep C_{18} column (10 × 150 mm, 5 µm) with acetonitrile (solvent A) and water containing 0.1% formic acid (solvent B).

3.2. Chemicals and Ingredients

Isoxanthohumol was prepared by chemical cyclization in aqueous NaOH solution at 0 °C as described by Stevens et al. [4], and also by a microbial transformation method using the fungus *R. oryzae* KCTC 6946 as previously reported by Kim and Lee [19]. Isoxanthohumol prepared by both methods was extracted with EtOAc and then purified by chromatographic methods including silica gel and reversed-phase C_{18} MPLC. The spectroscopic data of isoxanthohumol (**1**) were in good agreement with data in the literature [1] and its structure was also confirmed by 2D NMR experiments. Optical rotation and CD measurements revealed that the substrate isoxanthohumol was a racemic mixture of (2*S*)- and (2*R*)-isoxanthohumol. Ingredients for media including D-glucose, peptone, malt extract, yeast extract, and potato dextrose medium were purchased from Becton, Dickinson and Co. (Sparks, MD, USA), and sucrose was purchased from Sigma-Aldrich Co. (St Louis, MO, USA).

3.3. Microorganisms and Fermentation

Forty-one microbial strains were obtained from the Korean Collection for Type Cultures (KCTC) and cultured for preliminary screening. Microorganisms and culture broth composition were described in the previous literature in detail [20,21].

3.4. Biotransformation Screening Procedure

All of the microbial cultures were grown according to the two-stage procedure [16,17]. In the screening studies, the actively-growing microbial cultures were inoculated in 100 mL flasks containing 20 mL of media, and incubated with gentle agitation (200 rpm) at 25 °C in a temperature-controlled shaking incubator. Isoxanthohumol (**1**) (2 mg/0.1 mL in EtOH) was added to each flask 24 h after inoculation, and further fermented under the same condition for 3 d. Sampling and TLC monitoring were generally carried out on RP-18 TLC$_{254S}$ with 60% MeOH at 24 h intervals. Two control studies were performed for identification of metabolites produced by enzymatic transformation. Substrate controls consisted of **1** and each sterile medium incubated without microorganisms. Culture controls consisted of fermentation cultures in which the microorganisms were grown without addition of **1**.

3.5. Biotransformation of Isoxanthohumol (**1**) by R. oryzae KCTC 6399

Preparative-scale fermentations were carried out under the same condition with two 1 L flasks each containing 250 mL of medium and 20 mg of isoxanthohumol (**1**) for 10 d. The cultures were extracted with EtOAc two times and the organic layers were combined and concentrated at reduced pressure. The EtOAc extract (730 mg) was subjected to silica gel (70–230 mesh, Merck) column chromatography with a CHCl$_3$–MeOH (9:1) to give three fractions. Fraction 1 (108 mg) was chromatographed by RP-MPLC (Lobar^{TM}, 10 × 240 mm) using 55% MeOH isocratic solvent system to give a mixture of isoxanthohumol metabolites **2** and **3** (13.2 mg, 15.8% yield). An aliquot of compounds (7.8 mg) was further chromatographed by HPLC with a gradient solvent system of 20% solvent A to 33% solvent A for 25 min to afford two isomers **2** (1.6 mg, t_R 21.56 min) and **3** (2.2 mg, t_R 22.45 min). Fraction 2 (203 mg) was also chromatographed by RP-MPLC (Lobar^{TM}, 10 × 240 mm) using aqueous MeOH solvent (50 → 55%) to give a mixture of metabolites **4** and **5** (7.7 mg, 8.9% yield). A portion of mixture (4.3 mg) was further purified by HPLC with a gradient solvent system of 20% A to 35% A for 25 min to afford two isomers **4** (1.4 mg, t_R 11.65 min) and **5** (1.6 mg, t_R 12.56 min).

(2S,2″S)-4′-Hydroxy-5-methoxy-7,8-(2,2-dimethyl-3-hydroxy-2,3-dihydro-4H-pyrano)-flavanone (**2**): pale yellow amorphous powder, [α]$_D$ +37.0° (*c* 0.2, MeOH); UV λ$_{max}$ (MeOH) (log ε) 224 (4.35) 285 (4.18), 319 (3.65) nm; CD (MeOH) λ$_{ext}$ (Δε): 288 (−13.5), 311 (0.0), 331 (+4.7); IR (KBr) ν$_{max}$: 3421, 1658, 1608, 1579, 1519, 1485, 1338, 1206, 1133, 1106, 835 cm^{-1}; ^{1}H-NMR (CD$_3$OD, 500 MHz) δ 7.33 (2H, d, *J* = 8.3 Hz, H-2′, 6′), 6.82 (2H, d, *J* = 8.3 Hz, H-3′, 5′), 6.09 (1H, s, H-6), 5.37 (1H, brd, *J* = 12.5 Hz, H-2), 3.80 (3H, s, 5-OCH$_3$), 3.76 (1H, brt, *J* = 6.0 Hz, H-2″), 3.02 (1H, dd, *J* = 16.5, 13.0 Hz, H-3a), 2.82 (1H, dd, *J* = 17.0, 5.0 Hz, H-1″a), 2.70 (1H, dd, *J* = 16.5, 3.0 Hz, H-3b), 2.55 (1H, dd, *J* = 17.0, 6.5 Hz, H-1″b), 1.33 (3H, s, H-5″), 1.31 (3H, s, H-4″); ^{13}C-NMR (CD$_3$OD, 125 MHz) δ 192.7 (C-4), 164.1 (C-8a), 162.0 (C-5), 161.9 (C-7), 159.1 (C-4′), 131.4 (C-1′), 129.0 (C-2′,6′), 116.5 (C-3′,5′), 106.4 (C-4a), 101.9 (C-8), 94.9 (C-6), 80.4 (C-2), 79.9 (C-3″), 69.6 (C-2″), 56.3 (5-OCH$_3$), 46.2 (C-3), 26.7 (C-1″), 25.7 (C-5″), 22.0 (C-4″); ESIMS *m/z* 371 [M + H]$^+$; HRESIMS *m/z* 371.1504 [M + H]$^+$ (calcd for C$_{21}$H$_{23}$O$_6$, 371.1495).

(2R,2″S)-4′-Hydroxy-5-methoxy-7,8-(2,2-dimethyl-3-hydroxy-2,3-dihydro-4H-pyrano)-flavanone (**3**): pale yellow amorphous powder, [α]$_D$ +66.4° (*c* 0.2, MeOH); UV λ$_{max}$ (MeOH) (log ε) 224 (4.47), 288 (4.30), 318 (3.77) nm; CD (MeOH) λ$_{ext}$ (Δε): 288 (+16.4), 313 (0.0), 334 (−5.4); IR (KBr) ν$_{max}$: 3421, 1658, 1608, 1578, 1519, 1485, 1337, 1206, 1133, 1106, 835 cm^{-1}; ^{1}H-NMR (CD$_3$OD, 500 MHz) δ 7.32 (2H, d, *J* = 8.5 Hz, H-2′, 6′), 6.82 (2H, d, *J* = 8.5 Hz, H-3′, 5′), 6.08 (1H, s, H-6), 5.36 (1H, dd, *J* = 13.0, 2.5 Hz, H-2), 3.79 (3H, s, 5-OCH$_3$), 3.73 (1H, dd, *J* = 7.3, 5.8 Hz, H-2″), 3.00 (1H, dd, *J* = 16.5, 13.0 Hz, H-3a), 2.83 (1H, dd, *J* = 17.0, 5.5 Hz, H-1″a), 2.69 (1H, dd, *J* = 16.5, 3.0 Hz, H-3b), 2.50 (1H, dd, *J* = 17.0, 7.0 Hz, H-1″b), 1.34 (3H, s, H-5″), 1.28 (3H, s, H-4″); ^{13}C-NMR (CD$_3$OD, 125 MHz) δ 192.7 (C-4), 163.9 (C-8a), 162.0 (C-5), 161.9 (C-7), 159.1 (C-4′), 131.4 (C-1′), 129.0 (C-2″,6′), 116.5 (C-3′,5′), 106.4 (C-4a), 102.2 (C-8), 94.9 (C-6), 80.3 (C-2), 79.9 (C-3″), 69.9 (C-2″), 56.3 (5-OCH$_3$), 46.2 (C-3), 26.7 (C-1″), 25.9 (C-5″), 21.3 (C-4″); ESIMS *m/z* 371 [M + H]$^+$; HRESIMS *m/z* 371.1482 [M + H]$^+$ (calcd for C$_{21}$H$_{23}$O$_6$, 371.1495).

(2S)-7,4′-Dihydroxy-5-methoxy-8-(2,3-dihydroxy-3-methylbutyl)-flavanone (**4**): pale yellow amorphous powder, [α]$_D$ −31.7° (*c* 0.2, MeOH); UV λ$_{max}$ (MeOH) (log ε): 225 (4.42), 285 (4.17), 310 (3.73) nm; CD (MeOH) λ$_{ext}$ (Δε): 289 (−9.1), 313 (0.0), 331 (+4.0); IR (KBr) ν$_{max}$: 3424, 1600, 1515, 1463, 1351, 1280, 1150, 1099, 828 cm^{-1}; ^{1}H-NMR (CD$_3$OD, 500 MHz) δ 7.34 (2H, d, *J* = 8.3 Hz, H-2′, 6′), 6.81 (2H, d, *J* = 8.3 Hz, H-3′, 5′), 6.16 (1H, s, H-6), 5.33 (1H, brd, *J* = 13.5 Hz, H-2), 3.81 (3H, s, 5-OCH$_3$), 3.52 (1H, dd, *J* = 10.0, 2.3 Hz, H-2″), 3.00 (1H, dd, *J* = 16.8, 13.5 Hz, H-3a), 2.93 (1H, brd, *J* = 14.0 Hz, H-1″a), 2.69 (1H, brd, *J* = 16.8 Hz, H-3b), 2.62 (1H, dd, *J* = 14.0, 10.0 Hz, H-1″b), 1.14 (3H, s, H-5″), 1.12 (3H, s, H-4″); ^{13}C-NMR (CD$_3$OD, 125 MHz) δ 192.9 (C-4), 165.5 (C-7), 164.2 (C-8a), 162.5 (C-5), 159.0 (C-4′), 131.5 (C-1′), 129.0 (C-2′,6′), 116.4 (C-3′,5′), 108.5 (C-8), 105.9 (C-4a), 94.6 (C-6), 80.4 (C-2), 80.2 (C-2″), 74.1 (C-3″), 56.1 (5-OCH$_3$), 46.3 (C-3), 26.6 (C-1″), 26.0 (C-4″), 25.0 (C-5″); ESIMS *m/z* 389 [M + H]$^+$; HRESIMS *m/z* 389.1612 [M + H]$^+$ (calcd for C$_{21}$H$_{25}$O$_7$, 389.1600).

(2R)-7,4′-Dihydroxy-5-methoxy-8-(2,3-dihydroxy-3-methylbutyl)-flavanones (**5**): pale yellow amorphous powder, [α]$_D$ −53.0° (*c* 0.2, MeOH); UV λ$_{max}$ (MeOH) (log ε): 225 (4.30), 285 (4.08), 318 (3.65) nm; CD (MeOH) λ$_{ext}$ (Δε): 288 (+10.8), 311 (0.0), 331 (−3.8); IR (KBr) ν$_{max}$: 3422, 2975, 1599, 1514, 1463, 1349,

1281, 1210, 1150, 1101, 829 cm^{-1}; ^{1}H-NMR (CD$_3$OD, 500 MHz) δ 7.34 (2H, d, *J* = 8.5 Hz, H-2', 6'), 6.81 (2H, d, *J* = 8.5 Hz, H-3', 5'), 6.17 (1H, s, H-6), 5.34 (1H, dd, *J* = 13.0, 2.5 Hz, H-2), 3.82 (3H, s, 5-OCH$_3$), 3.55 (1H, dd, *J* = 10.0, 2.0 Hz, H-2''), 2.98 (1H, dd, *J* = 16.5, 13.0 Hz, H-3a), 2.95 (1H, dd, *J* = 14.0, 2.5 Hz, H-1''a), 2.70 (1H, dd, *J* = 16.5, 3.0 Hz, H-3b), 2.61(1H, dd, *J* = 14.0, 10.0 Hz, H-1''b), 1.16 (3H, s, H-5''), 1.15 (3H, s, H-4''); ^{13}C-NMR (CD$_3$OD, 125 MHz) δ 192.9 (C-4), 165.5 (C-7), 164.0 (C-8a), 162.6 (C-5), 159.0 (C-4'), 131.6 (C-1'), 128.9 (C-2',6'), 116.4 (C-3',5'), 108.6 (C-8), 105.8 (C-4a), 94.7 (C-6), 80.5 (C-2), 80.1 (C-2''), 74.1 (C-3''), 56.1 (5-OCH$_3$), 46.4 (C-3), 26.7 (C-1''), 26.0 (C-4''), 25.0 (C-5''); ESIMS *m*/*z* 389 [M + H]$^+$; HRESIMS *m*/*z* 389.1600 [M + H]$^+$ (calcd for C$_{21}$H$_{25}$O$_7$, 389.1600).

3.6. Biotransformation of **1** by F. oxysporum f.sp. lini KCTC 16325

Scale-up fermentations were carried out under the same condition with two 1 L Erlenmeyer flasks each containing 250 mL medium and 25 mg isoxanthohumol (**1**) for 5 d. Production of a metabolite was monitored by reversed-phase C$_{18}$ TLC (MeOH 70%). The red-colored cultures were extracted with EtOAc two times and the organic layers were combined and concentrated in vacuo. The EtOAc extract (280 mg) was subjected to silica gel (70–230 mesh, Merck) column chromatography with *n*-hexane-EtOAc (2:1) mixture to give three fractions. Fraction 2 (89 mg) containing a metabolite was chromatographed with MPLC (Lobar^{TM}, 10 × 240 mm) using MeOH 45% isocratic solvent system to afford isoxanthohumol as a mixture of metabolites **6** and **7** (18 mg, 34.3% yield). An aliquot of compound mixture (3.4 mg) was further chromatographed by HPLC with a gradient solvent system of 20% A to 35% A for 25 min to afford two isomers **6** (1.1 mg, *t*$_R$ 18.63 min) and **7** (1.1 mg, *t*$_R$ 19.42 min).

3.7. Determination of Absolute Configuration by Modified Mosher's Method

Compound **3** (1.0 mg in 0.2 mL pyridine) was treated with 15 μL (20.3 mg) of (R)-(−)-α-methoxy-α-(trifluoromethyl)phenylacetyl chloride (MTPA-chloride), and stirred overnight at room temperature under nitrogen gas. The reaction mixture was evaporated in vacuo, and chromatographed by HPLC using a SunFire^{TM} Prep C$_{18}$ (10 × 150 mm, Waters) with a gradient solvent MeCN–H$_2$O (70:30 → 90:10) at 3 mL/min for 20 min, afforded the (S)-Mosher ester derivative **3a** (*t*$_R$ 17.30, 0.9 mg). The derivative of (R)-Mosher ester **3b** (*t*$_R$ 17.49, 1.0 mg) was prepared by reaction of **3** (1.0 mg) with 15 μL of (S)-(−)-MTPA-chloride reagent under the same condition as described above.

3a: ^{1}H-NMR (CDCl$_3$, 500 MHz) δ 6.03 (1H, s, H-6), 5.44 (1H, dd, *J* = 12.6, 3.0 Hz, H-2), 5.16 (1H, dd, *J* = 6.0, 5.4 Hz, H-2''), 3.86 (3H, s, 5-OCH$_3$), 2.99 (1H, dd, *J* = 17.4, 5.4 Hz, H-1''), 2.93 (1H, dd, *J* = 16.2, 12.6 Hz, H-3), 2.83 (1H, dd, *J* = 16.2, 3.0 Hz, H-3), 2.65 (1H, dd, *J* = 17.4, 6.0 Hz, H-1''), 1.36 (3H, s, H-5''), 1.32 (3H, s, H-4''). **3b**: ^{1}H-NMR (CDCl$_3$, 500 MHz) δ 6.06 (1H, s, H-6), 5.46 (1H, dd, *J* = 12.6, 3.0 Hz, H-2), 5.16 (1H, dd, *J* = 6.0, 5.4 Hz, H-2''), 3.87 (3H, s, 5-OCH$_3$), 3.00 (1H, dd, *J* = 17.4, 5.4 Hz, H-1''), 2.95 (1H, dd, *J* = 16.2, 12.6 Hz, H-3), 2.87 (1H, *J* = 16.2, 3.0 Hz, H-3), 2.80 (1H, dd, *J* = 17.4, 5.4 Hz, H-1''), 1.31 (3H, s, H-5''), 1.27 (3H, s, H-4'').

4. Conclusions

Microbial transformation studies of isoxanthohumol (**1**) resulted in the production of three oxygenated pairs of metabolites (**2**–**7**). These include two novel metabolite pairs possessing a dihydropyran (**2**,**3**) and a dihydroxymethylbutane (**4**,**5**) substituted flavanone core structure. It was widely reported that cyclization at the prenyl side chain resulted in forming a five-membered furan or a six-membered pyran heterocycle attached to the A-ring in prenylated flavanones [29,30]. The formation of dihydroxymethylbutyl group has frequently occurred by dihydroxylation on the double bond of the prenyl group substituted to the A ring in biotransformation of prenylated flavanones [29]. These transformation and derivatization have been commonly encountered in natural products [31], and the presence of cyclized and dihydroxylated derivatives of the prenyl substituent has been revealed in the bark of the Amazonian tree *Brosimum acutifolium*, a rich source of 8-prenylated flavonoids [32].

Unlike the microbial metabolites **2–7**, the regioselectively mono-hydroxylated prenyl side chain metabolites of isoxanthohumol, namely *cis*- and *trans*-prenyl alcohols, were previously identified as the most abundant oxidation metabolites of isoxanthohumol during in vitro metabolism studies using human liver microsomes, which was catalyzed by hepatic cytochrome P450 enzymes [11,33]. However, knowledge of the in vitro microbial metabolic conversions of isoxanthohumol may contribute to the detection and identification of the metabolic products of isoxanthohumol that occur in mammalian systems. Conjugation reactions including sulfation and glucuronidation are involved in the major metabolic pathways of polyphenols in mammals, which metabolize polyphenols into very hydrophilic conjugates [34,35]. It was previously proved that microbial aryl sulfotransferase was capable of producing Phase II metabolites of flavonoids, rather close to the mammalian enzyme [35]. Considering that isoxanthohumol is a significant component of beer and that beer is consumed by a large number of people world-wide, these findings could have potential health and nutritional implications.

Supplementary Materials: The following are available online, Figures S1–S18: 1D, 2D-NMR spectra of isoxanthohumol (**1**) and metabolites **2–7**, Scheme S1: HPLC Profiles of oxygenated metabolites (**2–7**) of isoxanthohumol (**1**), Figure S19: High resolution ESIMS data of metabolites **2–7**, Figure S20: Circular dichroism (CD) profiles of metabolites **2–7**.

Author Contributions: Conceptualization, I.-S.L.; Data curation, H.J.K., S.-H.Y., F.H. and B.Y.K.; Funding acquisition, I.-S.L.; Investigation, H.J.K., F.H. and I.-S.L.; Methodology, H.J.K. and I.-S.L.; Project administration, I.-S.L.; Resources, H.J.K. and I.-S.L.; Supervision, I.-S.L.; Validation, H.J.K.; Writing-original draft, H.J.K.; Writing-review&editing, H.J.C., D.-W.J., D.R.W., K.R.G., E.J.K. and I.-S.L.

Funding: This work was supported by a Korea Research Foundation Grant (KRF-2008-220-E00042) and a National Research Foundation of Korea Grant (NRF-2012R1A1A4A01009908) funded by the Korean government.

Acknowledgments: We gratefully thank the Gwangju branch of Korea Basic Science Institute (KBSI) for running NMR, ESIMS and HRESIMS experiments.

Conflicts of Interest: The authors declare no conflict of interest.

References

1. Stevens, J.F.; Ivancic, M.; Hsu, V.L.; Deinzer, M.L. Prenylflavonoids from *Humulus lupulus*. *Phytochemistry* **1997**, *44*, 1575–1585. [CrossRef]
2. Stevens, J.F.; Page, J.E. Xanthohumol and related prenylflavonoids from hops and beer: To your good health! *Phytochemistry* **2004**, *65*, 1317–1330. [CrossRef] [PubMed]
3. Stevens, J.F.; Taylor, A.W.; Clawson, J.E.; Deinzer, M.L. Fate of xanthohumol and related prenylflavonoids from hops to beer. *J. Agric. Food Chem.* **1999**, *47*, 2421–2428. [CrossRef] [PubMed]
4. Stevens, J.F.; Taylor, A.W.; Deinzer, M.L. Quantitative analysis of xanthohumol and related prenylflavonoids in hops and beer by liquid chromatography-tandem mass spectrometry. *J. Chromatogr. A* **1999**, *832*, 97–107. [CrossRef]
5. Milligan, S.R.; Kalita, J.C.; Heyerick, A.; Rong, H.; De Cooman, L.; De Keukeleire, D. Identification of a potent phytoestrogen in hops (*Humulus lupulus* L.) and beer. *J. Clin. Endocrinol. Metab.* **1999**, *84*, 2249–2252. [CrossRef]
6. Delmulle, L.; Bellahcène, A.; Dhooge, W.; Comhaire, F.; Roelens, F.; Huvaere, K.; Heyerick, A.; Castronovo, V.; DeKeukeleire, D. Anti-proliferative properties of prenylated flavonoids from hops (*Humulus lupulus* L.) in human prostate cancer cell lines. *Phytomedicine* **2006**, *13*, 732–734. [CrossRef] [PubMed]
7. Miranda, C.L.; Stevens, J.F.; Helmrich, A.; Henderson, M.C.; Rodriguez, R.J.; Yang, Y.H.; Deinzer, M.L.; Barnes, D.W.; Buhler, D.R. Antiproliferative and cytotoxic effects of prenylated flavonoids from hops (*Humulus lupulus*) in human cancer cell lines. *Food Chem. Toxicol.* **1999**, *37*, 271–285. [CrossRef]
8. Żołnierczyk, A.K.; Mączka, W.K.; Grabarczyk, M.; Wińska, K.; Woźniak, E.; Anioł, M. Isoxanthohumol-Biologically active hop flavonoid. *Fitoterapia* **2015**, *103*, 71–82. [CrossRef]
9. Gerhäuser, C.; Alt, A.P.; Klimo, K.; Knauft, J.; Frank, N.; Becker, H. Isolation and potential cancer chemopreventive activities of phenolic compounds of beer. *Phytochem. Rev.* **2002**, *1*, 369–377. [CrossRef]
10. Izzo, G.; Söder, O.; Svechnikov, K. The prenylflavonoid phytoestrogens 8-prenylnaringenin and isoxanthohumol differentially suppress steroidogenesis in rat Leydig cells in ontogenesis. *J. Appl. Toxicol.* **2011**, *31*, 589–594. [CrossRef]

11. Nikolic, D.; Li, Y.; Chadwick, L.R.; Pauli, G.F.; van Breeman, R.B. Metabolism of xanthohumol and isoxanthohumol, prenylated flavonoids from hops (*Humulus lupulus* L.), by human liver microsomes. *J. Mass Spectrom.* **2005**, *40*, 289–299. [CrossRef] [PubMed]

12. Bartmańska, A.; Huszcza, E.; Tronina, T. Transformation of isoxanthohumol by fungi. *J. Mol. Catal. B Enzym.* **2009**, *61*, 221–224. [CrossRef]

13. Possemiers, S.; Bolca, S.; Grootaert, C.; Heyerick, A.; Decroos, K.; Dhooge, W.; De Keukeleire, D.; Rabot, S.; Verstraete, W.; Van de Wiele, T. The prenylflavonoid isoxanthohumol from hops (*Humulus lupulus* L.) is activated into the potent phytoestrogen 8-prenylnaringenin in vitro and in the human intestine. *J. Nutr.* **2006**, *136*, 1862–1867. [CrossRef] [PubMed]

14. Possemiers, S.; Heyerick, A.; Robbens, V.; De Keukeleire, D.; Verstraete, W. Activation of proestrogens from hops (*Humulus lupulus* L.) by intestinal microbiota; conversion of isoxanthohumol into 8-prenylnaringenin. *J. Agric. Food Chem.* **2005**, *53*, 6281–6288. [CrossRef] [PubMed]

15. Fu, M.L.; Wang, W.; Chen, F.; Dong, Y.C.; Liu, X.J.; Ni, H.; Chen, Q.H. Production of 8-prenylnaringenin from isoxanthohumol through biotransformation by fungi cells. *J. Agric. Food Chem.* **2011**, *59*, 7419–7426. [CrossRef] [PubMed]

16. Clark, A.M.; McChesney, J.D.; Hufford, C.D. The use of microorganisms for the study of drug metabolism. *Med. Res. Rev.* **1985**, *5*, 231–253. [CrossRef] [PubMed]

17. Clark, A.M.; Hufford, C.D. Use of microorganisms for the study of drug metabolism: An update. *Med. Res. Rev.* **1991**, *11*, 473–501. [CrossRef]

18. Venisetty, R.K.; Cidii, V. Application of microbial biotransformation for the new drug discovery using natural drugs as substrates. *Curr. Pharm. Biotechnol.* **2003**, *4*, 153–167. [CrossRef]

19. Kim, H.J.; Lee, I.-S. Microbial metabolism of the prenylated chalcone xanthohumol. *J. Nat. Prod.* **2006**, *69*, 1522–1524. [CrossRef]

20. Kim, H.J.; Kim, S.-H.; Kang, B.Y.; Lee, I.-S. Microbial metabolites of 8-prenylnaringenin, an estrogenic prenylflavanone. *Arch. Pharm. Res.* **2008**, *31*, 1241–1246. [CrossRef]

21. Kim, H.J.; Kang, M.-A.; Lee, I.-S. Microbial transformation of isoxanthohumol, a hop prenylflavonoid. *Nat. Prod. Sci.* **2008**, *14*, 269–273.

22. Gaffield, W. Circular dichroism, optical rotatory dispersion and absolute configuration of flavanones, 3-hydroxyflavanones and their glycosides: Determination of aglycone chirality in flavanone glycosides. *Tetrahedron* **1970**, *26*, 4093–4108. [CrossRef]

23. Slade, D.; Ferreira, D.; Marais, J.P.J. Circular dichroism, a powerful tool for the assessment of absolute configuration of flavonoids. *Phytochemistry* **2005**, *66*, 2177–2215. [CrossRef] [PubMed]

24. Ohtani, I.; Kusumi, T.; Kashman, Y.; Kakisawa, H. High-field FT NMR application of Mosher's method. The absolute configuration of marine terpenoids. *J. Am. Chem. Soc.* **1991**, *113*, 4092–4096. [CrossRef]

25. Shi, G.; Gu, Z.; He, K.; Wood, K.V.; Zeng, L.; Ye, Q.; MacDougal, J.M.; McLaughlin, J.L. Applying Mosher's method to acetogenins bearing vicinal diols. The absolute configurations of muricatetrocin C and rollidecins A and B, new bioactive acetogenins from *Rollinia mucosa*. *Bioorg. Med. Chem.* **1996**, *4*, 1281–1286. [CrossRef]

26. Julaeha, E.; Supratman, U.; Mukhtar, M.R.; Awang, K.; Ng, S.W. Meranzin hydrate from *Muraya paniculata*. *Acta Crystallogr. Sect. E Struct. Rep. Online* **2010**, *66*, o620. [CrossRef] [PubMed]

27. Vázquez, J.T. Features of electronic circular dichroism and tips for its use in determining absolute configuration. *Tetrahedron Asymmetry* **2017**, *28*, 1199–1211. [CrossRef]

28. Herath, W.H.M.W.; Ferreira, D.; Khan, I.A. Microbial transformation of xanthohumol. *Phytochemistry* **2003**, *62*, 673–677. [CrossRef]

29. Cao, H.; Chen, X.; Jassbi, A.R.; Xiao, J. Microbial biotransformation of bioactive flavonoids. *Biotechnol. Adv.* **2015**, *33*, 214–223. [CrossRef]

30. Bartmańska, A.; Tronina, T.; Popłoński, J.; Huszcza, E. Biotransformations of prenylated hop flavonoids for drug discovery and production. *Curr. Drug Metab.* **2013**, *14*, 1083–1097. [CrossRef]

31. Dewick, P.M. *Medicinal Natural Products, A Biosynthetic Approach*, 3rd ed.; John Wiley & Sons Ltd.: Chichester, UK, 2009; pp. 161–178. ISBN 978-0-470-74168-9.

32. Takashima, J.; Ohsaki, A. Brosimacutins A-I, nine new flavonoids from *Brosimum acutifolium*. *J. Nat. Prod.* **2002**, *65*, 1843–1847. [CrossRef] [PubMed]

33. Guo, J.; Nikolic, D.; Chadwick, L.R.; Pauli, G.F.; van Breemen, R.B. Identification of human hepatic cytochrome P450 enzymes involved in the metabolism of 8-prenylnaringenin and isoxanthohumol from hops (*Humulus lupulus* L.). *Drug Metab. Dispos.* **2006**, *34*, 1152–1159. [CrossRef] [PubMed]

34. Jeong, E.J.; Liu, X.; Jia, X.; Chen, J.; Hu, M. Coupling of conjugating enzymes and efflux transporters: Impact on bioavailability and drug interactions. *Curr. Drug Metab.* **2005**, *6*, 455–468. [CrossRef] [PubMed]

35. Purchartová, K.; Valentová, K.; Pelantová, H.; Marhol, P.; Cvačka, J.; Havlíček, L.; Křenková, A.; Vavříková, A.; Biedermann, D.; Chambers, C.S.; et al. Prokaryotic and eukaryotic aryl sulfotransferases: Sulfation of quercetin and its derivatives. *Chem. Cat. Chem.* **2015**, *7*, 3152–3162. [CrossRef]

Sample Availability: Samples of the compounds are currently unavailable from the authors.

![molecules logo] *molecules*

MDPI

Article

4-*O*-Methylhonokiol Influences Normal Cardiovascular Development in Medaka Embryo

Santu K. Singha [1], Ilias Muhammad [2], Mohamed Ali Ibrahim [2,3], Mei Wang [2], Nicole M. Ashpole [1,2] and Zia Shariat-Madar [1,2,4,*]

[1] Department of Biomolecular Sciences, Division of Pharmacology, University of Mississippi, University, MS 38677, USA; sksingha@go.olemiss.edu (S.K.S.); nmashpol@olemiss.edu (N.M.A.)

[2] The National Center for Natural Products Research, Research Institute of Pharmaceutical Sciences, University of Mississippi, University, MS 38677, USA; milias@olemiss.edu (I.M.); mmibrahi@olemiss.edu (M.A.I.); meiwang@olemiss.edu (M.W.)

[3] Chemistry of Natural Compounds Department, National Research Centre, Dokki-Giza 12622, Egypt

[4] Light Microscopy Core, University of Mississippi, University, MS 38677, USA

[*] Correspondence: madar@olemiss.edu; Tel.: +662-915-5150

Received: 14 December 2018; Accepted: 27 January 2019; Published: 29 January 2019

Abstract: Although 4-*O*-Methylhonokiol (MH) effects on neuronal and immune cells have been established, it is still unclear whether MH can cause a change in the structure and function of the cardiovascular system. The overarching goal of this study was to evaluate the effects of MH, isolated from *Magnolia grandiflora*, on the development of the heart and vasculature in a Japanese medaka model in vivo to predict human health risks. We analyzed the toxicity of MH in different life-stages of medaka embryos. MH uptake into medaka embryos was quantified. The LC_{50} of two different exposure windows (stages 9–36 (0–6 days post fertilization (dpf)) and 25–36 (2–6 dpf)) were 5.3 ± 0.1 µM and 9.9 ± 0.2 µM. Survival, deformities, days to hatch, and larval locomotor response were quantified. Wnt 1 was overexpressed in MH-treated embryos indicating deregulation of the Wnt signaling pathway, which was associated with spinal and cardiac ventricle deformities. Overexpression of major proinflammatory mediators and biomarkers of the heart were detected. Our results indicated that the differential sensitivity of MH in the embryos was developmental stage-specific. Furthermore, this study demonstrated that certain molecules can serve as promising markers at the transcriptional and phenotypical levels, responding to absorption of MH in the developing embryo.

Keywords: cardiomyogenesis; factor VII; factor X; inflammation; thrombosis; vasculogenesis; herbal medicine

1. Introduction

Magnolia bark extract has been used as a component of dietary supplements and cosmetic products [1]. One specific compound found in *Magnolia* species, 4-*O*-Methylhonokiol (MH), is recognized to have multifunctional activities both in vitro and in vivo, similarly to other honokiol analogs [2,3]. Magnolias appear to naturally produce a significant amount of these biphenyl-type neolignan compounds, many of which show tissue specific distribution [4]. MH is expressed throughout the plant, with high amounts found in the leaves and seeds [5], whereas honokiol with its isomer magnolol are largely limited to the bark [6]. MH and 2-*O*-Methylhonokiol are isomers [7]. The optimal ratio of these two isomers and their mechanism of synthesis have not been fully characterized. Moreover, the effects of MH on the cardiovascular system remain poorly understood. Due to its low hydrophilicity, MH exhibits poor pharmacokinetics [8], which may lead to increased accumulation in the organs of the body. Some evidence has suggested beneficial effects of

MH, as it has been associated with anti-inflammatory [9], anti-osteoclastogenic, anti-oxidative [10,11], and neuroprotective [12] effects. Considering the apparent non-specificity of these effects, it is likely that MH has low targeting efficacy and exerts its cellular protective properties through a wide range of mechanisms. In contrast to the observed beneficial effects, the co-treatment of compounds with MH counterparts, both magnolol and honokiol, can exert synergistic cytotoxicity [13]. The incongruity between the protective and detrimental effects of MH thus far in the literature highlights the importance of understanding how MH or other compounds extracted from magnolia affect the development and function of tissues and organs, like the cardiovascular system.

Both angiogenesis and vasculogenesis (de novo blood vessel formation from embryonic precursors) have many features in common, and impairment of these processes can in turn cause damage to organs and influence blood circulation. We aim to identify exposure windows, which can provide insight into the potentially toxic effects of MH on the development of the heart, angiogenesis, and vasculogenesis, beyond its other potential action on multiple sites through different toxicity pathways. Identifying the critical stage of MH-induced cardiovascular toxicity lays down a basis for further elucidation of an adverse outcome pathway for MH and provides a starting point for future studies on the mechanisms of MH toxicity.

Due to its inherent low concentration in plant extracts containing honokiol and magnolol, MH has not been fully characterized. MH, like propofol, has a phenol ring, which produces side effects causing hypertension and altering both heartbeat and heart rate. Since magnolia bark extract is gaining widespread popularity as a preventive and alternative to medical treatments [14,15], it is vital to understand the molecular mechanisms of MH and characterize whether it causes embryotoxic and teratogenic effects in the cardiovascular system in a variety of vertebrates. By accounting for heterogeneities typical of Japanese medaka (*Oryzias latipes*), which shares 58% homology with its human counterpart, our first vertebrate model for the embryonic lethality of MH enables us to investigate the effects of MH on the embryo's developing heart and deformities, as well as alterations in inflammatory and parameters of coagulation.

In this study, we assessed the toxicity of MH in different life-stages of the medaka cardiovascular system. We hypothesized that the differential susceptibility of the stage-specific embryo might identify critical exposure windows to MH, which could in turn produce lethal and sublethal thresholds for toxicity because of differences in uptake of MH and subsequent internal concentrations.

2. Results

2.1. Toxicity of MH in Medaka Embryo

The toxicity of MH exposure on medaka was assessed at various stages of development (see MH-treatment in Materials and Methods section). While 10 μM MH was not toxic to embryos after 48 h, it caused 70% mortality within 96 h. A full concentration-response analysis of toxicity following six days of MH exposure time point revealed significant mortality of MH-treated medaka embryo/larvae (LC_{50} = 5.3 ± 0.1 μM) (Figure 1A). Larval toxicity was also evident, with the most common effects being spine malformations and edema. Observation of overall mortality with increasing concentrations over time revealed a reduction in larval survivability following exposure to concentrations of 2 μM MH and higher (Figure 1B). Some spinal deformities were observed in the larvae (Appendix A, Figure S1 under Supplementary Materials) which were associated with 5 μM and 10 μM MH. Embryos exposed to 10 μM MH showed delayed growth and high mortality rates during late larval- and juvenile-life stages.

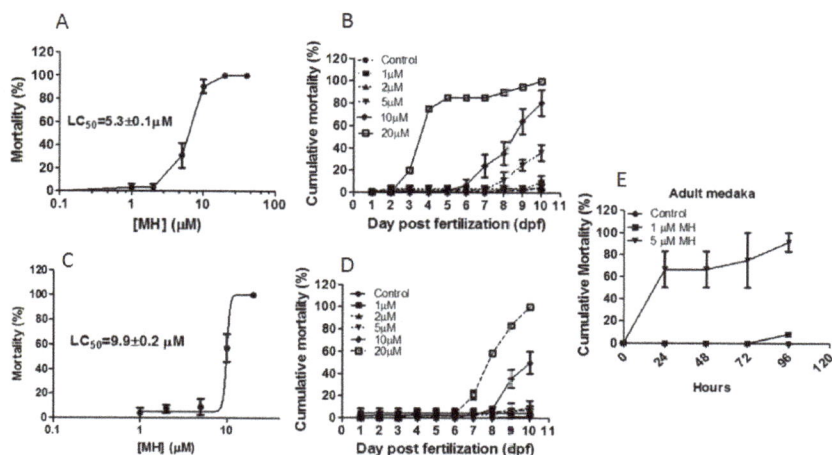

Figure 1. Cumulative mortality of medaka embryos. Embryos were exposed to 1, 2, 5, 10, and 20 μM 4-*O*-Methylhonokiol (MH) for 0–6 days post fertilization (dpf) (**A,B**) or 2–6 dpf (**C,D**). (**A**) Embryos were treated with MH for six days from hour 5, and the LC_{50} was calculated based on mortality observed at 10 dpf. (**B**) Cumulative mortality of all embryos until 10 dpf treated with various concentrations of MH for 0–6 dpf. (**C**) Embryos were treated with MH for four days from 2 dpf and the LC_{50} was calculated based on mortality observed at 10 dpf. (**D**) Cumulative mortality until 10 dpf of all embryos treated with various concentrations of MH for 2–6 dpf. The LC_{50} was calculated by log transformed data using nonlinear regression (curve-fit) (GraphPad Prism). Each value represents mean ± SEM ($n = 12$, replicated five times). (**E**) The relationship between MH concentration and mean adult medaka mortality. Eighteen male medakas were randomly divided into three groups ($n = 6$). They were exposed to 1 or 5 μM MH, and monitored for 96 h. Controls were exposed to 0.02% DMSO.

A second batch of embryos exposed to MH later in development (2–6 dpf) also exhibited concentration-dependent toxicity (10 dpf, LC_{50} = 9.9 ± 0.2 μM) (Figure 1C). Like the first batch of embryos, survivability was significantly reduced in later-staged larvae exposed to 10 and 20 μM (Figure 1D). There was a direct relationship between the changes in MH concentration and changes in hatching efficiency. Behavioral alterations such as difficulty in swimming and equilibrium loss were seen with both 5 μM MH and 10 μM MH. However, while termination from MH treatment allowed 5 μM MH-treated embryos to recover normal behavior, stopping MH treatment at 6 dpf from hour 5 was ineffective in the 10 μM MH group. The control group treated with DMSO, ranging from 0.02 to 0.04%, exhibited normal embryo development and normal hatching. Embryo mortality for this group was always below 10%, contrasting clearly with the experimental groups. The sensitivity of adult medaka to MH toxicity was then assessed. Exposure of 5 μM MH in adults caused 65% mortality within 24 h, whereas both the control group and fish exposed to 1 μM MH survived (Figure 1E). The survival time of 5 μM MH was significantly less than that of 1 μM MH-immersed adult medaka. This result demonstrates that 5 μM MH reduces survivability in the adult medaka model and further underscores its importance for additional health-related research.

The membrane penetration properties of MH on the medaka chorion are unknown. Although the chorion of the egg acts as a barrier within the embryo, the 80% mortality of embryos indicated an incorporation of MH into the egg. Mass spectrometry analysis of MH levels in conditioned growth media from the medaka revealed that the concentrations of MH decreased to non-detectable levels from time 0 h to 24 h (Figure 2A,B). However, a second molecule with a retention time of 2.628 min was present following 24 h incubation when MS was operated in scan mode (Figure 2C). Since the molecular and biochemical basis of formation of this molecule from MH is unknown, the identity of this unknown molecule was not characterized. Nevertheless, incubation of embryos with MH for 24 h led to 100% MH disappearance from the media (Figure 2D), suggesting either uptake of MH into the fish and/or degradation of the molecule.

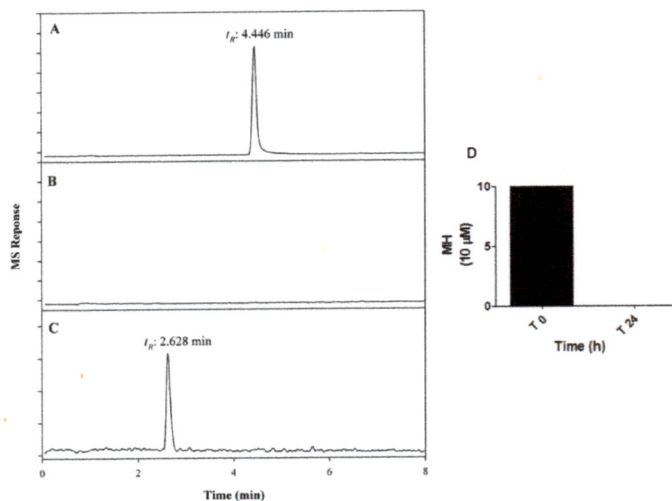

Figure 2. Quantification of MH uptake by medaka embryos by liquid chromatography-mass spectrometry (MS). This figure is a typical MS profile for the analysis of 10 µM MH uptake following 24 h-incubation with embryos, and maintained at 26 ± 1 °C. The amount of MH disappearance from the conditioned media was measured at time 0 h and time 24 h. Selected ion monitoring (SIM) at m/z 281 shows signals for MH at time 0 h (**A**) and time 24 h (**B**). (**C**) The unknown molecule in MS scan mode. (**D**) The amount of disappearance of MH following 24 h incubation with embryo.

2.2. Exposure to MH during Early Development Affects Cardiovascular Structure and Function

Cardiovascular changes in response to MH were then assessed. We demonstrated here that MH treatment resulted in reduced blood flow which peaked at around an 80% decrease with 10 µM MH at the end of the 0–6 dpf treatment session (Figure 3A). The reductions in blood flow were associated with corresponding changes in blood vessel occlusion (Figure 3B). Reduced blood flow (Figure 3C) and blood vessel occlusion (Figure 3D) were noticeable in the 2–6 dpf embryos also, at the same concentrations as compared to those of control embryos from day six (stages 36–38).

As shown in Figure 4A, the resting heart rate for both the control and DMSO-treated (0.02%) embryos increased with development. MH-treated embryos had lower heart rates than the control group at three-days and six-days post-treatment. MH treatment was effective in reducing the heart-beat of late embryos (6 dpf) ($p \leq 0.05$, $n = 27$) by 9%, suggesting MH influences the normal functioning of the heart. High-speed time-lapse analysis of the heartbeat showed that bradycardia occurred in the 0–6 day embryos treated with 10 µM MH when treatments were initiated early (from hour 5) (Figure 4B,C; see videos for better visualization of the differences in heart malfunction between the control (video 1) and MH-treated embryo (video 2) in Appendix B under Supplementary Materials). The average heart rate and blood flow for the 10 µM MH-treated embryos were lower than for the DMSO-treated controls.

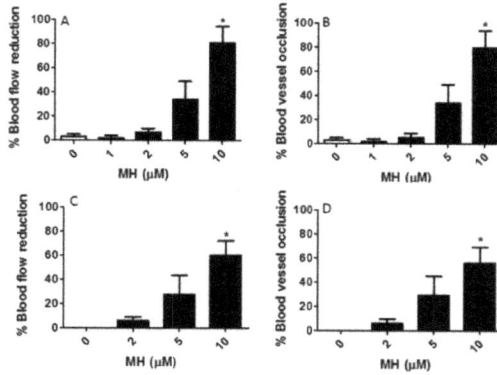

Figure 3. Reduction of blood flow and vascular occlusion by MH. Embryos were exposed to 1, 2, 5, and 10 μM MH from 5 h post fertilization (hpf) to 6 dpf (0–6 dpf, (**A,B**)) or from 2 dpf to 6 dpf (2–6 dpf, (**C,D**)). This figure showed that the reduction in blood flow was proportionally correlated with blood vessel occlusion. (**A**) Blood flow was observed on days 2, 3, 4, 5, and 6 to verify the duration of the reduction of blood flow. This panel exemplifies the percent reduction in blood flow until 10 dpf in response to various concentrations of MH for 0-6 dpf. (**B**) This panel shows percent occlusion following MH treatment. During the treatment, blood flow was significantly reduced compared to the control immediately after occlusion. (**C**) Thirty-two embryos (2 dpf) were randomly divided into four groups (*n* = 8). Embryos were treated with the indicated concentrations of MH and 0.02% DMSO (control) in embryo medium and were maintained at 26 ± 1 °C for 4 days. This figure shows reduced blood flow in MH-treated embryos (2–6 dpf). (**D**) This figure demonstrates a blockage that prevents normal flow of blood for the MH-treated embryos exposed to various concentrations of MH from 2 dpf to 6 dpf. Each bar represents data pooled from 4–5 independent experiments. Statistical analysis was performed by one-way analysis of variance (ANOVA) followed by Tukey's post-hoc multiple comparison test. *p* < 0.05 was considered as significant. The asterisk (*) indicates values significantly different from the control.

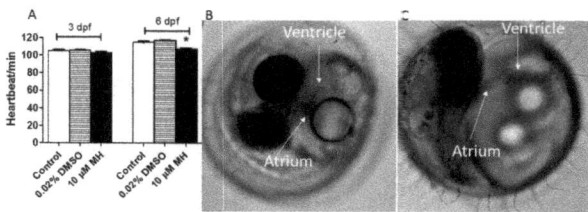

Figure 4. MH is responsible for the defect in cardiac function in vivo. Embryos were exposed to 10 μM MH or 0.02% DMSO (which was used as solvent in the treated condition) as the control from 5 hpf to 6 dpf. Heartbeat, blood flow, and heart structure were evaluated. (**A**) Heart beat was recorded and expressed as beats per min. Bar graphs of results obtained by counting heart beats. Results are given as the mean percentage of heart beat ± SEM (*n* > 8 embryos for each condition). Statistical analysis was performed by one-way ANOVA followed by Tukey's post-hoc multiple comparison test (*p* < 0.05). (**B**) Representative image of heart function in the control. (**C**) Representative bright field images of MH-induced heart ventricle malfunction in embryos treated with 10 μM MH for six days from hour 5. See videos for better visualization of the differences in heart malfunction between the control (video 1) and MH-treated embryo (video 2). Images were acquired on a Nikon TI2-E inverted microscope with a white light LED illuminator and a sCMOS Cooled Monochrome Camera. Videos were captured using a Nikon Elements automated acquisition device.

2.3. MH Is Implicated in Cardiovascular Dysfunction

To examine the mechanism of these vascular changes, expression levels of three key coagulation factors, factor XI (FXI, a protease of intrinsic pathway), factor VII (FVII, a protease of extrinsic pathway), and factor X (FX, a protease of common pathway) were quantified (Figure 5A). These factors are essential for hemostasis and indispensable for thrombosis. FXI, FX, and FVII circulate in the blood in zymogen forms. Activation of them leads to the formation of blood clots. As shown in Figure 5, these three protease transcripts were significantly ($p \leq 0.05$, $n = 100$, replicated four times) elevated in embryos treated with 10 μM MH for six days from hour 5.

Since MH reduced blood flow (Figure 3A,C), we hypothesized that the reduced blood flow activated the endothelium to synthesize and release tissue plasminogen activator (tPA, a fibrinolytic peptide) and plasminogen activator inhibitor 1 (PAI-1, a prothrombotic peptide) and thus decrease the ratio of tPA/PAI-1, ultimately promoting thrombosis. PAI-1 cDNA product was significantly elevated at the mRNA level with 10 μM MH, while tPA and endothelin B mRNA levels were not altered (Figure 5A). However, the ratio of the steady-state tPA/PAI-1 mRNA was significantly decreased in the embryos, suggesting increased levels of fibrin fragments, as previously described [16]. We measured the quantitative expression pattern of urokinase plasminogen activator (uPA) in 0–6 day embryos treated with 10 μM MH from hour 5. uPA mRNA levels were significantly increased (Figure 5A). To confirm that vascular endothelial cells are activated in response to MH, we examined the expression profiles of endothelin B receptors that are located primarily in vascular endothelial cells and angiotensin II type 1 receptor-associated protein (ATRAP), which is highly expressed in the kidney [17] and large vessel [18]. Vascular expression of ATRAP was significantly enhanced in response to MH (Figure 5A). Taken together, the data of Figure 5 demonstrated a previously unrecognized effect of MH on the control of the cardiovascular system and suggested that FXI, FX, FVII, PAI-1, uPA, and ATRAP are targets of MH.

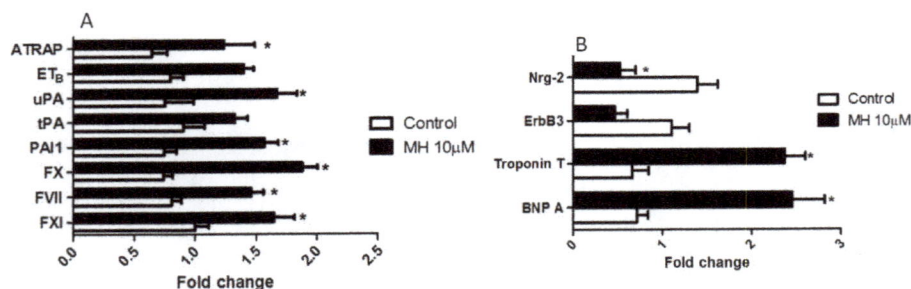

Figure 5. Embryos treated with MH lead to altered cardiac biomarker gene expression and altered expression of genes involved in thrombosis, fibrinolysis, and vascular tone. Real-time RT-qPCR was performed on total RNA isolated from each group of embryos using described primers in Table 1 leading to the amplification of the target gene. (**A**) RT-qPCR analysis of factor XI (FXI), factor VII (FVII), factor X (FX), plasminogen activator inhibitor 1 (PAI-1), tissue plasminogen activator (tPA), urokinase plasminogen activator (uPA), endothelin B (ETB), and angiotensin type 1 receptor associated protein (ATRAP) isolated from control and MH-treated embryos. (**B**) RT-qPCR analysis of natriuretic peptide A, Troponin T, ErbB3, and Nrg2. Data were normalized to the eukaryotic elongation factor 1-alpha (eEf1α) polymerase chain reaction signal. Each value represents mean ± SEM ($n = 100$, replicated four times). * indicates a value is significant versus the respective control group ($p < 0.05$).

To further confirm that MH influences the normal functioning of the heart, we measured the expression patterns of heart development-related genes. Controls were exposed to 0.01% DMSO for the RT-qPCR study. After treatment with 10 μM MH, the embryos showed highly up-regulated expression of brain natriuretic peptide A and troponin T (Figure 5B). Further investigations were performed to explore the effect of MH on Nrg-2 and ErbB3, the molecules involved in the synthesis of acetylcholine at the neuromuscular junction [19]. Nrg-2, a cardiac chamber maturation marker, and ErbB3, which is involved in proper heart morphogenesis and function, were attenuated (Figure 5B). These findings conferred that the cardioprotective property of the Nrg molecule was compromised in the presence of MH.

2.4. MH Possesses the Proinflammatory and Pro-Oxidative Properties

We looked into the effect of MH on the expression levels of catalase, glutathione peroxidase (GPX), glutathione-S-transferase (GST), and superoxide dismutase (SOD) (Figure 6). Forkhead boxO1 (FoxO1) was overexpressed with 0–6 dpf MH exposure beginning from hour 5 (Figure 6). Catalase, GPX, and GST mRNAs were significantly overexpressed, whereas SOD mRNA was not statistically significant (Figure 6). This finding suggested that MH-dependent increased expression of these anti-oxidant enzymes in the embryo was a mechanism by which they could eliminate excess reactive oxygen species (ROS). Since the Wnt/β-catenin signaling pathway is capable of regulating inflammatory cell migration and macrophage phenotypes in zebrafish [20], we then assessed the expression profiles of tissue necrosis factor-alpha (TNF-α) and interleukin 1 beta (IL-1 β) (Figure 6). Expression analysis showed significant upregulation of IL-1β and TNF-α in MH-treated embryos at the transcription level (Figure 6). This observation suggests that the overexpression of FoxO1 might cause a decrease in endothelial cell sprouting and migration, as seen in a previous report [21]. On the basis of these results, MH appears to possess proinflammatory properties at the embryo level.

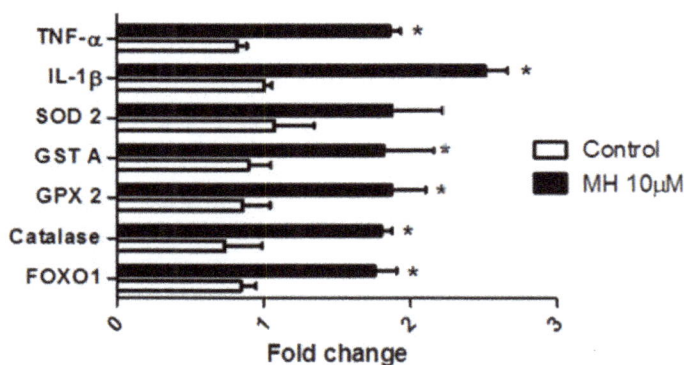

Figure 6. Embryos treated with MH lead to altered expression of genes involved in cell regulation, oxidative stress, and inflammation. Real-time RT-qPCR was performed on total RNA isolated from each group of embryos using described primers in Table 1 leading to the amplification of the target gene. RT-qPCR analysis of FoxO1, catalase, glutathione peroxidase (GPX2), glutathione-s-transferase (GSTA), superoxide dismutase 2 (SOD-2), Interleukin 1 beta (IL-1β), and tissue necrosis factor-alpha (TNF-α) isolated from control and MH-treated embryos. Data were normalized to the eEf1α polymerase chain reaction signal. Each value represents mean ± SEM ($n = 100$, replicated four times). Statistical analysis was performed by two-way ANOVA followed by post-hoc Bonferroni test. * indicates values which are significant versus the respective control group ($p < 0.05$).

2.5. MH Reduces the Normal Hatching Process of Medaka Embryos

We hypothesized that decreased heart rate might reduce the embryos' activity, leading to a reduction in the distribution of nutrients and both coagulation and hatching enzymes, as has been previously suggested [22]. To test this hypothesis, we explored the effects of MH on hatching. MH caused a reduction in hatching in embryos exposed to 10 μM MH starting from the hours immediately after fertilization (Figure 7A) or starting two days later (Figure 7B). Our findings demonstrate that MH can influence both early and late developmental stages of medaka, leading to reduced hatchability of eggs at higher concentrations.

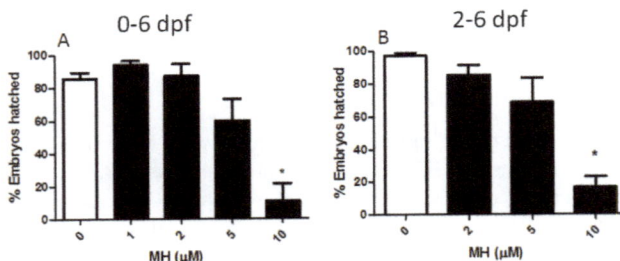

Figure 7. The effectiveness of MH at different concentrations at altering hatching of medaka embryos. Under pure culture conditions, embryo hatching efficacy was altered by increasing concentrations of MH. (**A**) Embryos were treated with increasing concentrations of MH for six days from hour 5 (0–6 dpf). MH concentrations of 5 µM and higher were required to cause a delay in hatching. Hatchability of embryos from six days MH (10 µM)-treatment was significantly decreased in comparison to untreated controls until 10 dpf. The external differences in appearance of the newly hatched control frys on day ten and the 10 µM MH-treated embryos were significantly obvious. (**B**) Embryos were treated with increasing concentrations of MH for four days from stage 25 (2–6 dpf). MH concentrations of 10 µM and higher were required to significantly cause a delay in hatching. Results are given as the mean percentage of hatching efficiency ± SEM (*n* > 8 embryos for each condition). Statistical analysis was performed by one-way ANOVA followed by Tukey's post-hoc multiple comparison test. *p* < 0.05 was considered as significant. The asterisk (*) indicates values significantly different from the control.

2.6. MH Regulates Wnt/β-Catenin Pathway during Cardiomyogenesis

Transcript levels of several Wnt/β-catenin signaling pathway proteins were altered in embryos treated with 10 µM MH for six days (Figure 8). Wnt 1 mRNA expression was significantly increased by MH. As shown in Figure 8, control and MH-treated embryos had similar levels of TGF-β2 mRNA expression, suggesting that MH could potentially promote cardiac fibrogenesis through the Wnt signaling pathway.

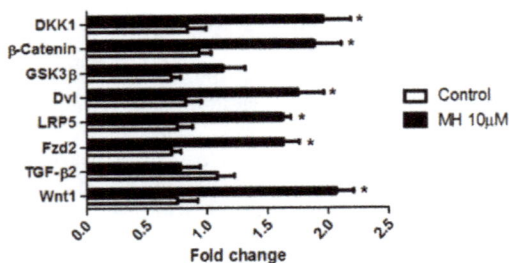

Figure 8. MH controls the Wnt signal transduction pathway, a main regulator of development. Real-time RT-qPCR was performed on total RNA isolated from each group of embryos using described primers in Table 1, leading to the amplification of the target gene. RT-qPCR analysis of Wnt 1, transforming growth factor beta 2 (TGF-β2), frizzled 2 (Fzd2), low-density lipoprotein receptor-related protein 5 (LRP5), dishevelled (Dvl), glycogen synthase kinase 3 beta (GSK-3β), β-catenin, and dickkopf 1 (DKK1) isolated from the control and MH-treated embryos. Each value represents mean ± SEM (*n* = 100, replicated four times). Statistical analysis was performed by two-way ANOVA followed by post-hoc Bonferroni test. * indicates values which are significant versus the respective control group (*p* < 0.05).

High expression levels of Fzd 2 (Wnt 1 receptor), LRP5 (Wnt 1 co-receptor), and Dvl were observed in MH-treated embryos, but there was no difference in expression levels of glycogen synthase kinase-3 beta (GSK-3β), which is involved in the suppression of the Wnt/β-catenin and subsequent degradation of beta-catenin (Figure 8). A significant increase was apparent in levels of β-catenin transcript of the MH group. Increased β-catenin is implicated in ventricular myocyte proliferation control, while its decrease leads to differentiation [23]. Thus, MH might inhibit cardiac differentiation via increased endogenous β-catenin-mediated signaling during normal cardiac development.

Expression levels of dickkopf 1 (DKK1, a secreted Wnt/β-catenin pathway inhibitor) in the MH-treated embryos were significantly higher compared to those of the control group (Figure 8). MH-induced Wnt 1 overexpression enhanced the expression of several downstream molecules involved in heart development. These molecules included Fzd 2, Dvl, LRP5, β-catenin (a downstream target of LRP5), and DKK1. Collectively, MH clearly influenced the Wnt/β-catenin signaling pathway, which had roles during various stages of cardiac development.

2.7. MH Prolongs Swimming Duration

The locomotion of embryos exposed to a sublethal concentration of MH was assessed two days post hatching (Figure 9). Control fish showed clear response to the light-dark cycle while MH treatment blunted this locomotor response (Figure 9).

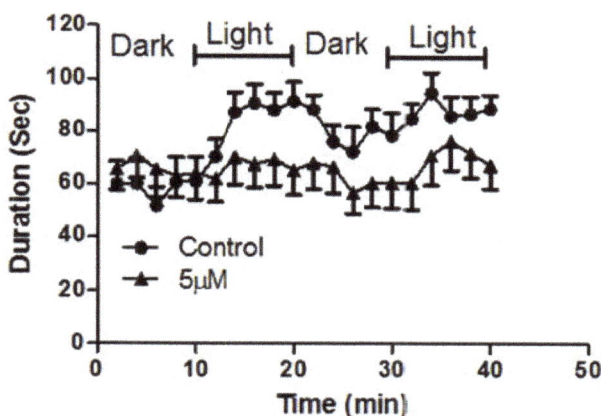

Figure 9. MH altered larvae' locomotor responses to light and dark stimuli. Medaka larvae (2 days post hatching (dph)) were monitored for 40 min (0–10 min dark, 10–20 min light, 20–30 min dark, 30–40 min light) using a ViewPoint Zebrabox. Duration of movements was measured at 2 min intervals at a velocity of ≥2 mm/s. Larvae were treated with 5 μM, a sub-lethal concentration; MH and their activity were measured in 10 min windows at the indicated time points. No significant (two-tailed *t*-test, $p < 0.05$) increase in activity in MH-treated larvae (solid closed triangles, $n = 14$) was noted for the time window during the light-cycle compared to baseline levels and that of the control group (solid closed circles, $n = 16$). Each datum represents mean ± SEM of 16 observations.

3. Discussion

Much of our knowledge of the function of MH has been extrapolated from its analogs' action and its usage in the form of plant extracts in health and disease, which could provide a bias leading to untoward side-effects. Compounds/drugs often have diverse and/or mixed effects from one organ to another as well as across different disease states. The present study was aimed at demonstrating the effects of MH and to establish its possible therapeutic utility. Many plant extracts have MH-like molecules, but the effects of MH on cardiogenesis and neurogenesis have not been comprehensively studied in contrast to cancer and inflammation. With increasing interest in honokiol related compound therapy, and their usage as supplements in diets and cosmetics, investigations were performed to answer the following questions: (1) What was the effect of MH on known genes involved in cardiac development and function in medaka embryos? (2) What were the morphological and physiological effects of daily MH treatment in medaka embryos? (3) What was the effect on embryonic survival and hatching of subjecting medaka embryos at different stages of cardiovascular development to various concentrations of MH for different lengths of time? (4) Was MH a prothrombotic agent? (5) What was the impact of sublethal concentration of MH on locomotion?

Fertilized eggs of medaka at two developmental stage windows (9–36 or 25–36) were exposed to increasing concentrations of MH in an embryo medium. Our data showed that MH-induced embryonic fatality was developmental-stage-specific. The embryos were more sensitive to MH at early stages of development (9–25) than in late stages of development. MH-treated embryos exhibited cardiovascular complications with a spectrum ranging from reduced blood flow to blood vessel occlusion, thrombus formation, and slow heart rate.

In recent years, it has become apparent that the Wnt/β-catenin pathway (Wnt signaling pathway) is essential for the regulation of numerous genes in embryogenesis [24], adult cell biology [25], tissue homeostasis [26], and disease [27]. The canonical (known as the morphogen pathway) signaling of this pathway is capable of upregulating the expression of a wide range of genes, which have roles in giving rise to the majority of cardiomyocytes [28]. Targeted pathways related to Wnt include cardiac differentiation [29] and cardiac remodeling [30]. Increased activated Wnt signaling has been recognized as a major pathomechanism in heart and blood vessels [31]. Uncontrolled activation of the Wnt signaling pathway has been implicated in the pathogenesis of cardiovascular disease [32] and inflammation [33]. The Wnt/β-catenin pathway not only has a major role in cardiovascular development, but it has been also proven that a specific isoform of Wnt, Wnt-5A, functions in the process of neurogenesis and establishment of functional connectivity [34], suggesting its tissue specificity. To understand why chronic MH treatment may have undesirable effects in embryos, we investigated the effect of MH on the components of the Wnt signaling pathway and its downstream targeted molecules, which may offer novel mechanistic insights that could pave the way to enhancing its clinical utility.

Due to difficulties in the pharmacological approach and the absence of antibodies, quantification of gene expression profiling of medaka cardiovascular tissues was the alternative approach to clarify the role(s) MH plays in embryogenesis, and to assess the clinical utility of MH. MH-treated embryos showed increased expression of the Wnt 1 gene (Figure 8). This deregulation of the Wnt signaling pathway prompted us to target key members of this pathway, which had exhibited distinct temporal and spatial profiles of expression during normal embryogenesis, post-surgery [35], and disease states [36]. It has been established that when both the Fzd receptor and LRP5/6 form a complex with Wnt ligands, the Wnt/β-catenin signaling pathway is activated [37]. Thus, we assessed the effect of MH on the expression profile of Fzd receptors and co-receptor LRP5. MH upregulated the expression levels of both Fzd and LRP5, suggesting Wnt/β-catenin signaling activation is sensitive to MH concentration. Most importantly, the transcript level of Fzd was positively correlated with the Wnt transcript expression level. Wnt-mediated Fzd/LRP5 stimulation leads to cytosolic β-catenin accumulation. However, its accumulation and signaling are tightly regulated via Wnt-dependent and Wnt-independent mechanisms [37]. Evidence shows that β-catenin signaling is essential for proper vascular formation and the development and functioning of the heart [38]. Next, investigations were performed to measure the expression pattern of β-catenin mRNA in MH-treated embryos. There was an increase in β-catenin mRNA. Since β-catenin has a rapid turnover [37], we explored the expression of GSK3β transcript. Evidence indicates that GSK3β activity is reduced/inactivated in the presence of Wnt signal [37]. As shown in Figure 8, GSK3β expression was unaltered, which is critical to the activation of β-catenin-mediated signaling. Moreover, our data suggested that MH had no direct effect on GSK3β gene expression and the unaltered level of GSK3β might be due to overexpression of Wnt levels, its binding to the complex of Fzd-LRP5, and stabilization of β-catenin [37]. Inactivation of

GSK3β can result in the translocation of β-catenin to the nucleus and, subsequently, the induction of β-catenin-dependent downstream target genes.

Urokinase plasminogen activator receptor (uPAR) [39] is one of these target genes downstream of β-catenin that has roles in both thrombosis and complement system. tPA and uPA are two plasminogen activators which are capable of catalyzing the activation of plasminogen. While uPA is required for the generation of plasmin activity in tissue undergoing pathological remodeling, tPA is associated with plasmin-induced activation of latent TGF-β in the vessel wall [40]. TGFβ2, a cytokine, is involved in vascular function, and mutations in TGF β2 are found to be implicated in cardiovascular diseases such as vascular complications and aortic disease [41]. The activity of uPA is regulated by its specific receptor, uPAR [42]. The mRNA of PAI-1 and uPA were enhanced by MH treatment in medaka embryos, whereas tPA mRNA steady-state levels were unaffected (Figure 5A). Similarly, TGFβ2 mRNA was unaffected by MH (Figure 8). This coincides well with the finding in the vascular wall [40] that there is a potential direct correlation between tPA expression and TGFβ2 gene expression and/or activation. The tPA/PAI-1 mRNA ratio was significantly decreased in the 10 μM MH group compared to the control. This suggests that MH appears to influence the fibrinolytic system during embryogenesis. In addition, Dvl mRNA was significantly elevated in MH-treated embryos. Overexpression of Dvl can inhibit the phosphorylation of β-catenin by GSK3β leading to β-catenin stabilization, and can consequently promote expression of the downstream targets. Overexpression of Dvl-1 in an atorvastatin-treated rat model of balloon-injured carotid artery has been shown to reverse the treatment effects of atorvastatin on vascular smooth muscle cells and collagen expression [43], suggesting the anti-restenosis action of Dvl. Collectively, these results also confirm that there is a positive correlation between β-catenin increases and the level of downstream target gene expression.

While DKK-1 is a transcriptional target of the p53 tumor suppressor [44] and β-catenin [45], it plays a functionally redundant but protective role [46]. It is capable of suppressing the expression of Wnt target genes [47] during postnatal angiogenesis [48]. MH causes an overexpression of the DKK-1 gene in embryos. Our data has demonstrated that both the Wnt/β-catenin signaling and its modulator, DKK-1, are increased in MH-treated embryos, highlighting their imbalance pattern of expression. Evidence indicates that the overexpression of DKK-1 is markedly associated with reduced cell proliferation [45], endothelial dysfunction, and concomitant platelet activation [49]. There is a positive correlation between DKK-1 expression and the recovery period following acute myocardial infarction (MI) [50]. FVII, FX, and FXI were elevated in MH-treated embryos, suggesting their consumption due to the activation of platelets. Activation of platelets lead to the formation of thrombus and reduced blood flow, which was apparent in MH-treated embryos. It is well-established that reduced blood supply to the myocardial tissues can result in ischemia and subsequent MI. On the basis of plaque rupture, platelets are found to be the cellular source of DKK-1 in patients with acute ST segment-elevated MI [51]. In addition, our study also demonstrates that MH is capable of upregulating the gene expression of brain type natriuretic peptide (a secretory cardiac neurohormone, BNP) and Troponin T, which has cardiac specificity and is essential for cardiac contractility [52]. There is a positive correlation between elevated troponin levels and ST-segment elevation MI (STEMI) [52]. Elevated BNP levels are a strong predictive marker of heart failure [53]. It is plausible to suggest that overexpression of the components of the Wnt1/β-catenin dependent pathway by MH is due to heart failure and ischemic areas in the embryo heart during the wound healing process following acute MI, whereas MH-induced increased DKK-1 gene expression is mediated via platelet-induced endothelial activation.

Although this is the first work in the literature to have explored the effects of MH on inflammatory cytokines in medaka, MH surprisingly caused robust increased levels of inflammatory markers such as TNF-α and IL-1β (Figure 6) in contrast to those of honokiol and magnolol. Our study did not address the differences we observed in the MH-mediated increased inflammatory mediators. We do not have a ready explanation for the discrepancy regarding the cited anti-inflammatory action of MH and that of our study. However, it is tempting to suggest that the elevation of these cytokines could be the result of at least two potential explanation: (1) elevated inflammatory mediators in medaka embryos were the result of genetic differences passed down from parents in our medaka colony or (2) the additive effect of MH on constitutive expression of the aforementioned cytokines in medaka as previously seen in mouse embryos [54]. The underlying mechanism(s) for the increase in these cytokines in MH-treated embryos remains uncertain. The increases in TNF-α and IL-1β are positively correlated with the presence and extent of cardiac biomarkers level (Figure 5). There is a wealth of knowledge about the association between heart failure and circulating inflammatory cytokines [55]. It is tempting to suggest that MH-induced inflammation might be the result of local tissue injury due to lack of oxygen/nutrients or systemic inflammasome activation. However, further investigation is needed to determine the potential causative role these inflammatory cytokines play in the progression of MH-induced cardiac injury.

4. Materials and Methods

4.1. Medaka Maintenance and Breeding

All fish work was performed in compliance with animal ethics guidelines as given by the Institutional Animal Care and Use Committee (IACUC) at the University of Mississippi according to the Association for Assessment and Accreditation of Laboratory Animal Care International (AAALAC). Embryos were collected in the morning and maintained on a 14 h:10 h light:dark cycle. The medium or test solutions as well as plates were autoclaved or sterilized. All experiments with medaka embryos were conducted in an embryo medium (17 mM NaCl, 0.4 mM KCl, 0.36 mM CaCl$_2$, 0.6 mM MgSO$_4$, pH 7.4, and 0.0002% methylene blue) at 26 ± 1 °C.

Natural breeding or in vitro fertilization (known as squeezing) was used as a preferred method of choice for generating embryos for pharmacological treatment experiments. The latter techniques were used to generate embryos for most of our experiments, unless otherwise stated.

4.2. Extraction and Isolation of MH

The seeds of *Magnolia grandiflora* were collected at the University of Mississippi campus (MS 38677). Air-dried powdered seeds (107 g) were soaked in EtOH (200 mL × 2 × 24 h each). The combined extract afforded oily material, where 2 g was subjected to centrifugal preparative thin layer chromatography (CPTLC, Chromatotron®, Analtech Inc., Newark, DE, USA), using a 6 mm silica gel rotor. The sample was dissolved in dichloromethane (DCM) and applied to the rotor under a rotation of 700 rpm, and subsequently eluted with *n*-hexane, then DCM, to end up with MeOH (200 mL each). Eighteen fractions were monitored and collected via TLC analysis (silica gel; solvents: *n*-hexane-EtOAc; 75:25). The fractions were visualized by spraying the TLC plates, with 1% vanillin-H$_2$SO$_4$ used as a spray reagent, where fractions 6 (361 mg) and 7 (36.0 mg) contained MH with a purity of 95% and 85%, respectively, via LC and NMR analyses. Further purification of fraction 6 was completed using a silica gel solid phase extraction cartridge (SPE) and was eluted with gradient of *n*-hexane/EtOAc (100:0→99:1) with 0.1% increments to afford 10 fractions. The fractions were monitored and pooled by TLC analysis (silica gel; solvents: n-Hex-EtOAc; 75:25), where fractions 7–9 afforded MH with a purity of 95% via LC and NMR analyses. LC analysis was conducted using an Agilent 1100 high performance liquid chromatography (HPLC) system equipped with a degasser (G1379A), quaternary pump (G13311A), auto sampler (G1313A), column oven (G1316A), and UV-Diode detector (G1315B) controlled by Chemstation software. Analysis of the fractions was carried out on an RP-C18 column (150 × 4.6 mm; particle size 5 μm; Luna) with column oven temperature set to 25 °C and a gradient system of eluent water (A) and acetonitrile (B) used. The gradient condition was as follows: 0–2 min (10% B), 2–30 min (10% B→90% B), 30–35 min (100% B). The flow rate of the solvent was 1.0 mL/min and the injection volume was 20 μL. All the analysis was carried out at wavelengths of 254, 280, and 325 nm with a run time of 35 min. HPLC-grade acetonitrile and water solvents were used. Acetic acid was added as a modifier to achieve a final concentration of 0.1% in each solvent. NMR spectra were acquired on a Varian Mercury 400 MHz spectrometer at 400 (^{1}H) and 100 (^{13}C) MHz in CDCl$_3$, using the residual solvent as an internal standard. Multiplicity determinations (DEPT) and 2D-NMR spectra (HMQC, HMBC, and NOESY) were obtained using standard Bruker pulse programs.

4.3. MH Treatment

The up-and-down procedure (UDP) testing approach was used to determine the toxicity of MH on medaka, beginning at stage 9 (hour 5) of embryological development (as delineated by [56,57]) (Figure 10). For the analysis of the effects of pharmacological treatments on cardiac rate, thrombus generation, and blood vessel occlusion, fertilized eggs were collected within 5 h of mating and immersed in embryo medium containing various concentrations of MH (1, 2, 5, 10, and 20 μM) and 0.02–0.04% DMSO (control group) in 48-well culture plates. The first batch of embryos (0–6 dpf, $n = 8$ to 12 embryos/group) was exposed to five different concentrations (1, 2, 5, 10, and 20 μM) of MH and vehicle sample, 0.02–0.04% DMSO, which is known to be safe [58] and increase the permeability [59] of the embryo's chorion. The second batch of embryos (2–6 dpf) was exposed to similar dilutions of MH (1, 2, 5, 10, and 20 μM) and 0.02–0.04% DMSO. All embryos were placed in a 48-well plate with 1 mL of their respective dilutions. Both batches were monitored for changes in morphology, delayed growth, behavior alteration, and mortality throughout the embryonic and larval-, juvenile-, and adult-life stages.

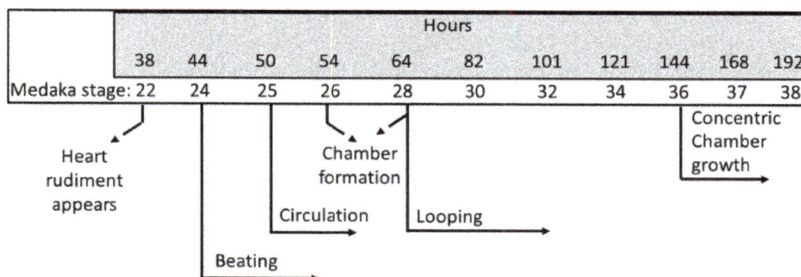

Figure 10. Stages with regard to hours post fertilization of the medaka heart. This scheme outlines the embryological development of the heart during the first 144 h and subsequent formation of concentric chamber growth. The medaka heart begins beating and pumping around 44 (stage 24) to 50 (stage 25) hours post fertilization. Medaka embryos were treated with MH either from 0–6 days post fertilization or 2–6 dpf at 26 ± 1 °C.

Embryos and larvae were raised in 48-well plates and maintained at a density of 1/1.0 mL in embryo medium with daily medium change. Groups of medaka embryos, often numbering 8 to 12, were exposed for an exposure period from 0 to 6 days post fertilization (organogenesis period at stages 9 to 36–38), and 2 to 6 dpf (stages 25 to 36–38), with an interval of at least 24 h. Embryos/hatched fry were reared in normal embryo medium at 6–10 dpf without MH (washed groups). Survivability, hatching efficiency, blood vessel occlusion, and blood flow frequency were observed until 10 dpf. Heart beats were counted on 3 and 6 dpf, and RNA isolation was immediately performed after 0–6 dpf exposure. Behavioral experiments were conducted two days post hatching (dph) after 0–6 dpf MH exposure. Larvae with no obvious malformation were used for locomotion experiments. Both embryos and larvae were monitored daily by imaging.

4.4. Determination of MH Purity and MH Absorption in Medaka Embryos

The absorption of MH from embryo medium to embryos were ascertained after a single concentration of MH (10 μM) for a period of 24 h using an Agilent 1290 Infinity series UHPLC with a diode array detector and an Agilent 6120 quadrupole mass spectrometer (Agilent Technologies,

Santa Clara, CA, USA). The highly purified MH reference standard showed a base peak of $[M + H]^+$ 281 with the APCI positive ionization mode; this ion was thus used in a selected ion monitoring mode (SIM) to detect MH in the samples. Embryos were exposed to embryo medium containing 10 μM MH from 5 h post fertilization (hpf) to 24 hpf. The disappearance of MH in the embryo's bathing (conditioned) medium was regarded as the absorption characterization of MH in the embryo toxicity test and the cardiovascular toxicity test. The identity and concentration of MH in conditioned medium was measured by UHPLC/MS.

The toxicity testing involved multi-stage exposure with repeated MH concentrations for six consecutive days with an interval of at least 24 h. Six groups of embryos ($n = 12$/group) were treated concurrently with 1, 2, 5, 10, and 20 μM MH and 0.02–0.04% DMSO (control group) at 5 hpf. There was a washout period between each treatment to clear any remaining free MH and extruded molecules into the conditioned embryo medium. The stock solution of MH was prepared in 100% DMSO (Sigma-Aldrich, St. Louis, MO, USA). Treatment was done according to the design depicted in Figure 10. In another set of experiments, hatched fry were reared in normal embryo medium without MH until 10 dpf (washed groups).

4.5. Microscopy Study

Control and treated fish were sampled at various stages of development according to the treatment schedule. Cardiovascular structure, blood flow, and heartbeat were analyzed under a microscope. Images were acquired on a Nikon TI2-E inverted microscope with a white light LED illuminator and a sCMOS Cooled Monochrome Camera. Videos were captured using Nikon Elements automated acquisition at a rate of 20 frames per second for one minute.

4.6. Locomotion Study

The free-swimming behavior of MH-treated larvae was compared to that of the control group. MH was prepared at a 5 μM concentration in embryo medium. This concentration was selected on the basis of LC_{50} studies. The effects of acute MH exposure on locomotor activity in larvae were examined at 2 dph. We assessed activity in 24-well plates. Larvae were placed individually in wells containing 2 mL of embryo medium. The larvae acclimatized to the darkness of the Zebrabox over 20 min (Viewpoint, Montreal, Canada) before the start of the locomotion experiment. The duration of movements was measured at a velocity of ≥ 2 mm/s.

4.7. RNA Extraction

The isolated heart of the medaka embryo from the 6 dpf control and MH-treated embryos was too small to obtain from it enough total RNA for the quantification of target genes. To overcome this problem, 100 embryos from the 6 dpf control and MH-treated embryos were pooled into one sample and RNA was extracted using an RNAeasy micro kit (QIAGEN GmbH) following the manufacturer's protocol. However, samples were replicated four times.

4.8. Reverse Transcription—Quantitative Polymerase Chain Reaction (RT-qPCR)

One microgram of total RNA extracted from each sample were reverse-transcribed into cDNA using a Quantitect RT kit (QIAGEN GmbH). Quantitative expression profiles of the genes of interest were analyzed using SYBR Green (Invitrogen, Carlsbad, CA, USA) according to the instructions of the manufacturer. Primers used for the amplification of each gene have been tabulated in Table 1.

Table 1. Oligonucleotide sequences used in this study.

Gene	Sense (5'-3')	Antisense (3'-5')	Product Size (bp)	Reference
eEf1α	GGAGGCCAGCGACAAGAT	GCGAGAAGGTGGCAGGAT	115	NM_001104662.1
FX	TGTCAAAGCCCTGTGTGAAT	AGAAATGTTCAGCCACCA	147	XM_020710825.1
FXI	GAAGGATAATGCAGACCAGTGTC	GATGACACCCTTCAAGTAGCATC	127	XM_004074394.4
FVII	GTTCTGTCGGATAGGTGGATTT	CCTCCAGGTCATGTTACCTAC	97	XM_004066449.3
PAI-1	ATGCCGAGGTTTCTCTGAAC	GTTGAACATGTCTCCCAGTCC	78	XM_020711407.1
uPA	ACTGTGTTTCTGGGAAAGAGTG	GGATGATCATTTTCTCCACGGT	82	XM_004077409.4
tPA	CAGCCCGATCCAAGC	CCCTTCCATCGCAGCC	185	XM_011478844.3
ATRAP	CATGTGGGGAACTTCAGC	GCCCACCAGAAACATGAGG	91	XM_011475678.2
ETB	CTGATCTTTGTGGTGGCAT	CCCATTCCTCATGCACTTGT	78	NM_001104844.1
IL-1β	CTGTTTCTGAGGAGGTGG	AGAAGAGGAAGGCGCACATT	79	XM_011478737.2
TNF-α	AACCGAAGAGTCTGAGAGGG	AGCTGAAGAAGAGTACCGCT	105	XM_004074335.3
Catalase	TGCTAGCAGTTGATTGTCTGT	CACAGATCCACTGAAACAGGA	100	XM_004069460.2
GPX2	TCAACGGAGTAAACACGCAT	GATCCTGCATGAGAGAGCTG	90	XM_004082594.3
GST A	CTGAAGGAGAGCGGCAC	CAGGAACGAGCCAGAGC	107	XM_020710769.1
SOD2	AAATGTGCGTCCTGACTATGT	TTTTGGCTATCTGAAGACGCT	83	XM_004083471.3
Wnt1	CCAGAAAACCCAGCTCACAA	TTGTGGGAGCAGAAGTTTGG	80	XM_020704658.1
FZd2	CACATGACCCCAGACTTCAC	AGAAACCAGAAGTGATGCCG	76	XM_020705151.1
LRP5	GAAGGCCCGAGCAGTTCA	AAGACATGGCTCCGTCGT	101	XM_011472833.2
Dvl	TGCTGAAACAAAGCCCAAAGT	ACCTCAAGGATCTGAGTGAGC	87	XM_011490628.3
β-catenin	CACAGAACTCCTACACAGCC	AGGCGCTTCTTGTAGTCTTG	102	XM_004077778.3
DKK1	GTGACACATGCCTGAGATCG	CACACCCTTACACATGCCAG	83	XM_020709512.1
GSK3β	AGCTGCAGATTATGAGGAAGTTG	TAGACGGTCTCTGGAACATAGTC	130	XM_023950884.1
ErbB3	GAGGTTGAGAAGGATGGCGT	CTACCTGGACTTCCTGTGCC	86	XM_011474665.3
NRG-2	CTCGTCACTGTGGGGGA	CTCGTCAGTGGGGTCCA	93	XM_020708867.2
BNP A	GAGCTCTGTTGATGAGGAGG	CAGTCCTGGCTCATCTTCTC	88	NM_001104685.2
Troponin T	GAGCTCTGTTGATGAGGAGG	GCTGATCCGGTTTCTGAGT	362	XM_004068713.4
FoxO1	GCCCATGCCAGTTCTGAGTA	ATCCTCCGTGTTGGTGGATG	102	XM_011485361.2
TGF-β2	GTTACTCCGACCTGAGGAAGATAG	TGACACCCAATCTTTAACGTTTC	127	XM_004073149.3

Quantification of target genes were done using an ABI 7000 real-time PCR machine (Applied Biosystems, Inc., Foster City, CA, USA). To measure the relative quantity of target genes per 1 µg of the total RNA from each group, a $2^{-\Delta\Delta Ct}$ method was used. On the basis of McCurley and Callard's 2008 study [60] as well as our own observations, eEf1α was included as an internal control. Samples were replicated four times. Quantitative expression data were used as the basis for making the major interpretations of this study.

4.9. Statistical Analyses

Statistical analysis was performed using Graph Pad Prism V6.0. Data are presented as mean ± SEM. Morphological data were analyzed by one-way ANOVA followed by Tukey's post-hoc multiple comparison test where more than two groups were compared. The LC_{50} was calculated by log transformed data using nonlinear regression (curve-fit) (GraphPad Prism). Statistical analysis for all RT-qPCR data was performed by two-way ANOVA followed by post-hoc Bonferroni test. Data for locomotion study were analyzed by two-tailed *t*-test. A difference between two means was considered to be significant when $p < 0.05$ ($* p < 0.05$).

5. Conclusions

Natural product extracts play complex roles in cardiovascular homeostasis. They can have protective and exacerbating effects on diseases due to the presence and complexity of plant extract composition. There has been very little information produced about the pharmacodynamics and pharmacokinetics [61] of MH, particularly in regard to both its teratogenic and cardiovascular effects. The typical recommended levels of magnolia bark extracts range from 200–800 mg/day/person [62]. The potential effects of MH on embryonic development are prominently apparent in our current study (Figure 11). Since medaka can be used to model the human cardiovascular system [57], here we report that MH alone is harmful to embryos because of its proinflammatory and prothrombotic properties as well as its effects on the Wnt signaling pathway. Unfortunately, little is known about the contribution(s) of MH once the trajectory has been set following its ingestion in humans. Our evidence is not sufficiently robust to support and extrapolate its deleterious effects in humans. However, we suggest that its inclusion in plant extracts could potentially retard the beneficial effects of other components of magnolia bark extracts or others. Herbal medicine optimization research must take MH levels into consideration in order to prevent the stimulation of both stress-induced pathways and the Wnt signaling pathway, which plays major roles in the control of all facets of embryonic development. Complementary approaches are needed to have a better understanding of the effects of maternal use of plant extracts containing MH on the offspring's health during pregnancy.

The potential causal effect of MH on medaka embryos

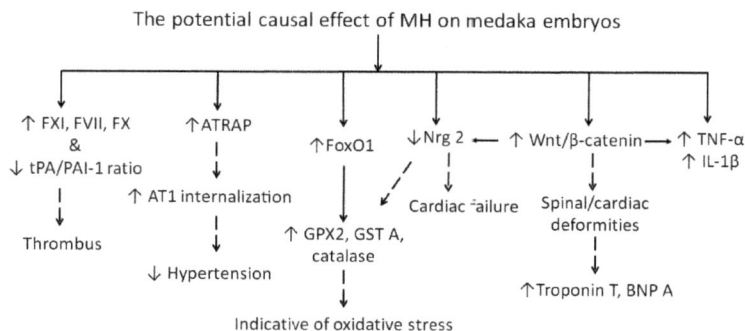

Figure 11. Flowchart depicting MH treatment induces cellular and morphological changes in medaka embryos that lead to inflammation, thrombosis, spinal, and cardiac deformities. Solid arrows indicate the direction of the effects of MH on indicators of increased inflammation/cardiovascular risk and the anatomic locations as well as a link with the prior object/event. The flow of physiological changes that require further investigations are shown in dashed arrows. Legend: ATRAP, angiotensin receptor-associated protein; AT1, angiotensin type I receptor; β-catenin, beta catenin; BNPA, brain natriuretic peptide type A; FVII, Factor VII; FX, Factor X; FXI, Factor XI; FoxO1, Forkhead box O1; GPX2, glutathione peroxidase 2; GST A, glutathione S-transferase A; IL-1β, interleukin-1 beta; Nrg 2, neuregulin 2; tPA. urokinase plasminogen activator; PAI-1, plasminogen activator inhibitor 1; Wnt, wingless/integrated oncogenes.

Supplementary Materials: The following are available online.

Author Contributions: Conceptualization, Z.S.-M.; Data curation, S.K.S.; Funding acquisition, Z.S.-M.; Investigation, S.K.S. and M.W.; Methodology, M.A.I.; Resources, I.M. and Z.S.-M.; Validation, I.M., M.A.I., M.W., and Z.S.-M.; Visualization, N.M.A.; Writing—original draft, Z.S.-M.; Writing—review & editing, S.K.S., I.M., M.A.I., M.W., N.M.A., and Z.S.-M.

Funding: This work was supported by the University of Mississippi overhead account to ZSM and partially supported by the United State Department of Agriculture, Agricultural Research Service (USDA ARS) Specific Cooperative Agreement No. 58-6408-2-00.

Acknowledgments: The authors thank Yan-Hong Wang (University of Mississippi, USA) for the assistance with HPLC. The authors also thank A.K. Dasmahapatra (Jackson State University, Mississippi, USA) for his challenging advice and comments. Part of the data from this article were presented at the Society of Toxicology Annual Meeting, San Antonio Meeting, 11–15 March 2018.

Conflicts of Interest: The authors declare no competing interests. The contents of this article are the sole responsibility of the authors and do not necessarily represent the official University of Mississippi or the grant agency officials.

Abbreviations

ATRAP	AT II type 1 receptor-associated protein
BNP A	Brain natriuretic peptide A
CAT	Catalase
DKK1	Dickkopf 1
dpf	Days post fertilization
dph	Days post hatching
Dvl	Dishevelled
eEf1α	Eukaryotic elongation factor 1 alpha
ErbB	Erythroblastic leukemia viral oncogene homolog
ET$_B$	Endothelin receptor B
FoxO1	Forkhead box O1
Fzd	Frizzled
GPX 2	Glutathione peroxidase 2
GSK-3β	Glycogen synthase kinase 3 beta

GST A	Glutathione S-transferase A
hpf	Hours post fertilization
LRP5	Low-density lipoprotein receptor-related protein 5
MH	4-O-methylhonokiol
MI	Myocardial infarction
Nrg	Neuregulin
PAI-1	Plasminogen activator inhibitor 1
SOD2	Superoxide dismutase 2
STEMI	ST-segment elevation MI
tPA	Tissue plasminogen activator
UHPLC/MS	Ultra-high performance liquid chromatography/mass spectrometer
u-PA	Urokinase plasminogen activator
uPAR	Urokinase plasminogen activator receptor

Appendix

Figure A1 shows the images representing the medaka phenotype following MH treatment.

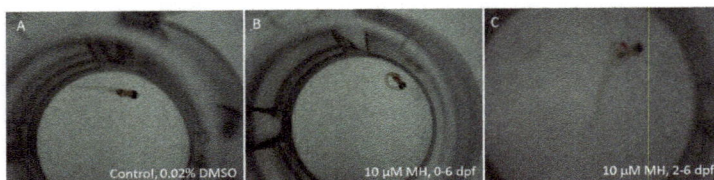

Figure A1. Examples of images representing the medaka phenotype following MH treatment. Medaka embryos were treated with 10 μM MH or an untreated control batch (0.02% DMSO) for 0–6 dpf or 2–6 dpf following the same protocol as described in the method. (**A**) Control. No phenotypic abnormality was detected with the control for the two groups. This first image displays an example of a typical control obtained immediately for both treatment plans. (**B**) This is an example of a typical medaka treated with 10 μM MH for 0–6 dpf. Curved tail phenotypes were observed in many. (**C**) Edema phenotypes were observed in medaka treated with 10 μM MH for 2–6 dpf. Bent tail phenotypes were observed in few.

Appendix

The effect of MH on heart function. These videos show how the medaka heart functions normally (Video 1) and when it is exposed to 10 μM MH for six days of treatment from 5 h.

References

1. Li, N.; Song, Y.; Zhang, W.; Wang, W.; Chen, J.; Wong, A.W.; Roberts, A. Evaluation of the in vitro and in vivo genotoxicity of magnolia bark extract. *Regul. Toxicol. Pharmacol.* **2007**, *49*, 154–159. [CrossRef]
2. Shen, J.L.; Man, K.M.; Huang, P.H.; Chen, W.C.; Chen, D.C.; Cheng, Y.W.; Liu, P.L.; Chou, M.C.; Chen, Y.H. Honokiol and magnolol as multifunctional antioxidative molecules for dermatologic disorders. *Molecules* **2010**, *15*, 6452–6465. [CrossRef] [PubMed]
3. Chicca, A.; Gachet, M.S.; Petrucci, V.; Schuehly, W.; Charles, R.P.; Gertsch, J. 4'-O-methylhonokiol increases levels of 2-arachidonoyl glycerol in mouse brain via selective inhibition of its COX-2-mediated oxygenation. *J. Neuroinflamm.* **2015**, *12*, 89. [CrossRef] [PubMed]
4. Dominguez, F.; Chavez, M.; Garduno-Ramirez, M.L.; Chavez-Avila, V.M.; Mata, M.; Cruz-Sosa, F. Honokiol and magnolol production by in vitro micropropagated plants of Magnolia dealbata, an endangered endemic Mexican species. *Nat. Prod. Commun.* **2010**, *5*, 235–240. [PubMed]
5. Schuhly, W.; Khan, S.I.; Fischer, N.H. Neolignans from North American Magnolia species with cyclooxygenase 2 inhibitory activity. *Inflammopharmacology* **2009**, *17*, 106–110. [CrossRef] [PubMed]

6. Bernaskova, M.; Schoeffmann, A.; Schuehly, W.; Hufner, A.; Baburin, I.; Hering, S. Nitrogenated honokiol derivatives allosterically modulate GABAA receptors and act as strong partial agonists. *Bioorg. Med. Chem.* **2015**, *23*, 6757–6762. [CrossRef]

7. Lin, C.F.; Hwang, T.L.; Al-Suwayeh, S.A.; Huang, Y.L.; Hung, Y.Y.; Fang, J.Y. Maximizing dermal targeting and minimizing transdermal penetration by magnolol/honokiol methoxylation. *Int. J. Pharm.* **2013**, *445*, 153–162. [CrossRef]

8. Fang, J.Y.; Huang, T.H.; Hung, C.F.; Huang, Y.L.; Aljuffali, I.A.; Liao, W.C.; Lin, C.F. Derivatization of honokiol by integrated acetylation and methylation for improved cutaneous delivery and anti-inflammatory potency. *Eur. J. Pharm. Sci.* **2018**, *114*, 189–198. [CrossRef]

9. Kim, H.S.; Ryu, H.S.; Kim, J.S.; Kim, Y.G.; Lee, H.K.; Jung, J.K.; Kwak, Y.S.; Lee, K.; Seo, S.Y.; Yun, J.; et al. Validation of cyclooxygenase-2 as a direct anti-inflammatory target of 4-*O*-methylhonokiol in zymosan-induced animal models. *Arch. Pharm. Res.* **2015**, *38*, 813–825. [CrossRef]

10. Chiu, J.H.; Ho, C.T.; Wei, Y.H.; Lui, W.Y.; Hong, C.Y. In vitro and in vivo protective effect of honokiol on rat liver from peroxidative injury. *Life Sci.* **1997**, *61*, 1961–1971. [CrossRef]

11. Huang, L.; Zhang, K.; Guo, Y.; Huang, F.; Yang, K.; Chen, L.; Huang, K.; Zhang, F.; Long, Q.; Yang, Q. Honokiol protects against doxorubicin cardiotoxicity via improving mitochondrial function in mouse hearts. *Sci. Rep.* **2017**, *7*, 11989. [CrossRef] [PubMed]

12. Lee, Y.J.; Choi, D.Y.; Lee, Y.K.; Lee, Y.M.; Han, S.B.; Kim, Y.H.; Kim, K.H.; Nam, S.Y.; Lee, B.J.; Kang, J.K.; et al. 4-*O*-methylhonokiol prevents memory impairment in the Tg2576 transgenic mice model of Alzheimer's disease via regulation of beta-secretase activity. *J. Alzheimer's Dis.* **2012**, *29*, 677–690. [CrossRef] [PubMed]

13. Bunel, V.; Antoine, M.H.; Stevigny, C.; Nortier, J.; Duez, P. New in vitro insights on a cell death pathway induced by magnolol and honokiol in aristolochic acid tubulotoxicity. *Food Chem. Toxicol.* **2016**, *87*, 77–87. [CrossRef] [PubMed]

14. Greenberg, M.; Urnezis, P.; Tian, M. Compressed mints and chewing gum containing magnolia bark extract are effective against bacteria responsible for oral malodor. *J. Agric. Food Chem.* **2007**, *55*, 9465–9469. [CrossRef] [PubMed]

15. Hsu, H.T.; Chi, C.W. Emerging role of the peroxisome proliferator-activated receptor-gamma in hepatocellular carcinoma. *J. Hepatocell. Carcinoma* **2014**, *1*, 127–135. [PubMed]

16. Rijken, D.C.; Hoylaerts, M.; Collen, D. Fibrinolytic properties of one-chain and two-chain human extrinsic (tissue-type) plasminogen activator. *J. Biol. Chem.* **1982**, *257*, 2920–2925. [PubMed]

17. Kobayashi, R.; Wakui, H.; Azushima, K.; Uneda, K.; Haku, S.; Ohki, K.; Haruhara, K.; Kinguchi, S.; Matsuda, M.; Ohsawa, M.; et al. An angiotensin II type 1 receptor binding molecule has a critical role in hypertension in a chronic kidney disease model. *Kidney Int.* **2017**, *91*, 1115–1125. [CrossRef]

18. Wakui, H.; Dejima, T.; Tamura, K.; Uneda, K.; Azuma, K.; Maeda, A.; Ohsawa, M.; Kanaoka, T.; Azushima, K.; Kobayashi, R.; et al. Activation of angiotensin II type 1 receptor-associated protein exerts an inhibitory effect on vascular hypertrophy and oxidative stress in angiotensin II-mediated hypertension. *Cardiovasc. Res.* **2013**, *100*, 511–519. [CrossRef]

19. Ford, B.D.; Liu, Y.; Mann, M.A.; Krauss, R.; Phillips, K.; Gan, L.; Fischbach, G.D. Neuregulin-1 suppresses muscarinic receptor expression and acetylcholine-activated muscarinic K+ channels in cardiac myocytes. *Biochem. Biophys. Res. Commun.* **2003**, *308*, 23–28. [CrossRef]

20. Petrie, T.A.; Strand, N.S.; Yang, C.T.; Rabinowitz, J.S.; Moon, R.T. Macrophages modulate adult zebrafish tail fin regeneration. *Development* **2014**, *141*, 2581–2591. [CrossRef]

21. Potente, M.; Urbich, C.; Sasaki, K.; Hofmann, W.K.; Heeschen, C.; Aicher, A.; Kollipara, R.; DePinho, R.A.; Zeiher, A.M.; Dimmeler, S. Involvement of Foxo transcription factors in angiogenesis and postnatal neovascularization. *J. Clin. Investig.* **2005**, *115*, 2382–2392. [CrossRef] [PubMed]

22. Hagenmaier, H.E. The hatching process in fish embryos. IV. The enzymological properties of a highly purified enzyme (chorionase) from the hatching fluid of the rainbow trout, *Salmo gairdneri* Rich. *Comp. Biochem. Physiol. Part B Comp. Biochem.* **1974**, *49*, 313–324. [CrossRef]

23. Buikema, J.W.; Mady, A.S.; Mittal, N.V.; Atmanli, A.; Caron, L.; Doevendans, P.A.; Sluijter, J.P.; Domian, I.J. Wnt/beta-catenin signaling directs the regional expansion of first and second heart field-derived ventricular cardiomyocytes. *Development* **2013**, *140*, 4165–4176. [CrossRef]

24. Bajoghli, B.; Aghaallaei, N.; Jung, G.; Czerny, T. Induction of otic structures by canonical Wnt signalling in medaka. *Dev. Genes Evol.* **2009**, *219*, 391–398. [CrossRef] [PubMed]

25. Loh, N.Y.; Neville, M.J.; Marinou, K.; Hardcastle, S.A.; Fielding, B.A.; Duncan, E.L.; McCarthy, M.I.; Tobias, J.H.; Gregson, C.L.; Karpe, F.; et al. LRP5 regulates human body fat distribution by modulating adipose progenitor biology in a dose- and depot-specific fashion. *Cell Metab.* **2015**, *21*, 262–273. [CrossRef] [PubMed]

26. Ackers, I.; Malgor, R. Interrelationship of canonical and non-canonical Wnt signalling pathways in chronic metabolic diseases. *Diabetes Vasc. Dis. Res.* **2018**, *15*, 3–13. [CrossRef]

27. Nguyen, G. Renin, (pro)renin and receptor: An update. *Clin. Sci.* **2011**, *120*, 169–178. [CrossRef]

28. Bisson, J.A.; Mills, B.; Paul Helt, J.C.; Zwaka, T.P.; Cohen, E.D. Wnt5a and Wnt11 inhibit the canonical Wnt pathway and promote cardiac progenitor development via the Caspase-dependent degradation of AKT. *Dev. Biol.* **2015**, *398*, 80–96. [CrossRef] [PubMed]

29. Ueno, S.; Weidinger, G.; Osugi, T.; Kohn, A.D.; Golob, J.L.; Pabon, L.; Reinecke, H.; Moon, R.T.; Murry, C.E. Biphasic role for Wnt/beta-catenin signaling in cardiac specification in zebrafish and embryonic stem cells. *Proc. Natl. Acad. Sci. USA* **2007**, *104*, 9685–9690. [CrossRef]

30. Sklepkiewicz, P.; Shiomi, T.; Kaur, R.; Sun, J.; Kwon, S.; Mercer, B.; Bodine, P.; Schermuly, R.T.; George, I.; Schulze, P.C.; et al. Loss of secreted frizzled-related protein-1 leads to deterioration of cardiac function in mice and plays a role in human cardiomyopathy. *Circ. Heart Fail.* **2015**, *8*, 362–372. [CrossRef]

31. Foulquier, S.; Daskalopoulos, E.P.; Lluri, G.; Hermans, K.C.M.; Deb, A.; Blankesteijn, W.M. WNT Signaling in Cardiac and Vascular Disease. *Pharmacol. Rev.* **2018**, *70*, 68–141. [CrossRef] [PubMed]

32. Askevold, E.T.; Aukrust, P.; Nymo, S.H.; Lunde, I.G.; Kaasboll, O.J.; Aakhus, S.; Florholmen, G.; Ohm, I.K.; Strand, M.E.; Attramadal, H.; et al. The cardiokine secreted Frizzled-related protein 3, a modulator of Wnt signalling, in clinical and experimental heart failure. *J. Intern. Med.* **2014**, *275*, 621–630. [CrossRef]

33. Barandon, L.; Casassus, F.; Leroux, L.; Moreau, C.; Allieres, C.; Lamaziere, J.M.; Dufourcq, P.; Couffinhal, T.; Duplaa, C. Secreted frizzled-related protein-1 improves postinfarction scar formation through a modulation of inflammatory response. *Arterioscler. Thromb. Vasc. Biol.* **2011**, *31*, e80–e87. [CrossRef] [PubMed]

34. Halleskog, C.; Dijksterhuis, J.P.; Kilander, M.B.; Becerril-Ortega, J.; Villaescusa, J.C.; Lindgren, E.; Arenas, E.; Schulte, G. Heterotrimeric G protein-dependent WNT-5A signaling to ERK1/2 mediates distinct aspects of microglia proinflammatory transformation. *J. Neuroinflamm.* **2012**, *9*, 111. [CrossRef]

35. Kasaai, B.; Moffatt, P.; Al-Salmi, L.; Lauzier, D.; Lessard, L.; Hamdy, R.C. Spatial and temporal localization of WNT signaling proteins in a mouse model of distraction osteogenesis. *J. Histochem. Cytochem.* **2012**, *60*, 219–228. [CrossRef]

36. Tortelote, G.G.; Reis, R.R.; de Almeida Mendes, F.; Abreu, J.G. Complexity of the Wnt/betacatenin pathway: Searching for an activation model. *Cell Signal.* **2017**, *40*, 30–43. [CrossRef] [PubMed]

37. Luu, H.H.; Zhang, R.; Haydon, R.C.; Rayburn, E.; Kang, Q.; Si, W.; Park, J.K.; Wang, H.; Peng, Y.; Jiang, W.; et al. Wnt/beta-catenin signaling pathway as a novel cancer drug target. *Curr. Cancer Drug Targets* **2004**, *4*, 653–671. [CrossRef]

38. Dawson, K.; Aflaki, M.; Nattel, S. Role of the Wnt-Frizzled system in cardiac pathophysiology: A rapidly developing, poorly understood area with enormous potential. *J. Physiol.* **2013**, *591*, 1409–1432. [CrossRef]

39. Koch, A.; Waha, A.; Hartmann, W.; Hrychyk, A.; Schuller, U.; Wharton, K.A., Jr.; Fuchs, S.Y.; von Schweinitz, D.; Pietsch, T. Elevated expression of Wnt antagonists is a common event in hepatoblastomas. *Clin. Cancer Res.* **2005**, *11*, 4295–4304. [CrossRef]

40. Grainger, D.J.; Kemp, P.R.; Liu, A.C.; Lawn, R.M.; Metcalfe, J.C. Activation of transforming growth factor-beta is inhibited in transgenic apolipoprotein(a) mice. *Nature* **1994**, *370*, 460–462. [CrossRef]

41. Drera, B.; Ritelli, M.; Zoppi, N.; Wischmeijer, A.; Gnoli, M.; Fattori, R.; Calzavara-Pinton, P.G.; Barlati, S.; Colombi, M. Loeys-Dietz syndrome type I and type II: Clinical findings and novel mutations in two Italian patients. *Orphanet J. Rare Dis.* **2009**, *4*, 24. [CrossRef] [PubMed]

42. Senthilkumar, K.; Arunkumar, R.; Elumalai, P.; Sharmila, G.; Gunadharini, D.N.; Banudevi, S.; Krishnamoorthy, G.; Benson, C.S.; Arunakaran, J. Quercetin inhibits invasion, migration and signalling molecules involved in cell survival and proliferation of prostate cancer cell line (PC-3). *Cell Biochem. Funct.* **2011**, *29*, 87–95. [CrossRef] [PubMed]

43. Hua, J.; Xu, Y.; He, Y.; Jiang, X.; Ye, W.; Pan, Z. Wnt4/beta-catenin signaling pathway modulates balloon-injured carotid artery restenosis via disheveled-1. *Int. J. Clin. Exp. Pathol.* **2014**, *7*, 8421–8431.

44. Shou, J.; Ali-Osman, F.; Multani, A.S.; Pathak, S.; Fedi, P.; Srivenugopal, K.S. Human Dkk-1, a gene encoding a Wnt antagonist, responds to DNA damage and its overexpression sensitizes brain tumor cells to apoptosis following alkylation damage of DNA. *Oncogene* **2002**, *21*, 878–889. [CrossRef] [PubMed]

45. Li, J.; Gong, W.; Li, X.; Wan, R.; Mo, F.; Zhang, Z.; Huang, P.; Hu, Z.; Lai, Z.; Lu, X.; et al. Recent Progress of Wnt Pathway Inhibitor Dickkopf-1 in Liver Cancer. *J. Nanosci. Nanotechnol.* **2018**, *18*, 5192–5206. [CrossRef] [PubMed]

46. Combes, A.N.; Bowles, J.; Feng, C.W.; Chiu, H.S.; Khoo, P.L.; Jackson, A.; Little, M.H.; Tam, P.P.; Koopman, P. Expression and functional analysis of Dkk1 during early gonadal development. *Sex. Dev.* **2011**, *5*, 124–130. [CrossRef]

47. Foley, A.C.; Mercola, M. Heart induction by Wnt antagonists depends on the homeodomain transcription factor Hex. *Genes Dev.* **2005**, *19*, 387–396. [CrossRef]

48. Peghaire, C.; Bats, M.L.; Sewduth, R.; Jeanningros, S.; Jaspard, B.; Couffinhal, T.; Duplaa, C.; Dufourcq, P. Fzd7 (Frizzled-7) Expressed by Endothelial Cells Controls Blood Vessel Formation Through Wnt/beta-Catenin Canonical Signaling. *Arterioscler. Thromb. Vasc. Biol.* **2016**, *36*, 2369–2380. [CrossRef]

49. Lattanzio, S.; Santilli, F.; Liani, R.; Vazzana, N.; Ueland, T.; Di Fulvio, P.; Formoso, G.; Consoli, A.; Aukrust, P.; Davi, G. Circulating dickkopf-1 in diabetes mellitus: Association with platelet activation and effects of improved metabolic control and low-dose aspirin. *J. Am. Heart Assoc.* **2014**, *3*, e001000. [CrossRef]

50. Chen, L.; Wu, Q.; Guo, F.; Xia, B.; Zuo, J. Expression of Dishevelled-1 in wound healing after acute myocardial infarction: Possible involvement in myofibroblast proliferation and migration. *J. Cell. Mol. Med.* **2004**, *8*, 257–264. [CrossRef]

51. Ueland, T.; Otterdal, K.; Lekva, T.; Halvorsen, B.; Gabrielsen, A.; Sandberg, W.J.; Paulsson-Berne, G.; Pedersen, T.M.; Folkersen, L.; Gullestad, L.; et al. Dickkopf-1 enhances inflammatory interaction between platelets and endothelial cells and shows increased expression in atherosclerosis. *Arterioscler. Thromb. Vasc. Biol.* **2009**, *29*, 1228–1234. [CrossRef] [PubMed]

52. Babuin, L.; Jaffe, A.S. Troponin: The biomarker of choice for the detection of cardiac injury. *Can. Med. Assoc. J.* **2005**, *173*, 1191–1202. [CrossRef] [PubMed]

53. Gaggin, H.K.; Januzzi, J.L., Jr. Natriuretic peptides in heart failure and acute coronary syndrome. *Clin. Lab. Med.* **2014**, *34*, 43–58. [CrossRef]

54. Kohchi, C.; Noguchi, K.; Tanabe, Y.; Mizuno, D.; Soma, G. Constitutive expression of TNF-alpha and -beta genes in mouse embryo: Roles of cytokines as regulator and effector on development. *Int. J. Biochem.* **1994**, *26*, 111–119. [PubMed]

55. Dick, S.A.; Epelman, S. Chronic Heart Failure and Inflammation: What Do We Really Know? *Circ. Res.* **2016**, *119*, 159–176. [CrossRef] [PubMed]

56. Iwamatsu, T. Stages of normal development in the medaka *Oryzias latipes. Mech. Dev.* **2004**, *121*, 605–618. [CrossRef] [PubMed]

57. Fujita, M.; Isogai, S.; Kudo, A. Vascular anatomy of the developing medaka, *Oryzias latipes*: A complementary fish model for cardiovascular research on vertebrates. *Dev. Dyn.* **2006**, *235*, 734–746. [CrossRef] [PubMed]

58. Gonzalez-Doncel, M.; Okihiro, M.S.; Torija, C.F.; Tarazona, J.V.; Hinton, D.E. An artificial fertilization method with the Japanese medaka: Implications in early life stage bioassays and solvent toxicity. *Ecotoxicol. Environ. Saf.* **2008**, *69*, 95–103. [CrossRef]

59. Kais, B.; Schneider, K.E.; Keiter, S.; Henn, K.; Ackermann, C.; Braunbeck, T. DMSO modifies the permeability of the zebrafish (Danio rerio) chorion-implications for the fish embryo test (FET). *Aquat. Toxicol.* **2013**, *140–141*, 229–238. [CrossRef]

60. McCurley, A.T.; Callard, G.V. Characterization of housekeeping genes in zebrafish: Male-female differences and effects of tissue type, developmental stage and chemical treatment. *BMC Mol. Biol.* **2008**, *9*, 102. [CrossRef]

61. Li, M.Y.; Tang, Y.H.; Liu, X.; Lu, H.Y.; Shi, X.Y. Sensitive determination of 4-*O*-methylhonokiol in rabbit plasma by high performance liquid chromatography and application to its pharmacokinetic investigation. *J. Pharm. Anal.* **2011**, *1*, 108–112. [CrossRef]
62. Poivre, M.; Duez, P. Biological activity and toxicity of the Chinese herb Magnolia officinalis Rehder & E. Wilson (Houpo) and its constituents. *J. Zhejiang Univ. Sci. B* **2017**, *18*, 194–214. [PubMed]

Sample Availability: Samples of the compounds are not available from the authors.

molecules

Article

Antiplasmodial and Cytotoxic Cytochalasins from an Endophytic Fungus, *Nemania* sp. UM10M, Isolated from a Diseased *Torreya taxifolia* Leaf

Mallika Kumarihamy [1,2], Daneel Ferreira [2], Edward M. Croom Jr. [2], Rajnish Sahu [1], Babu L. Tekwani [1], Stephen O. Duke [3], Shabana Khan [1,2], Natascha Techen [1] and N. P. Dhammika Nanayakkara [1,*]

[1] National Center for Natural Products Research, Research Institute of Pharmaceutical Sciences, University of Mississippi, University, MS 38677, USA; mkumarih@olemiss.edu (M.K.); rsahu@alasu.edu (R.S.); btekwani@southenresearch.org (B.L.T.); skhan@olemiss.edu (S.K.); ntechen@olemiss.edu (N.T.)
[2] Department of BioMolecular Sciences, Division of Pharmacognosy, School of Pharmacy, The University of Mississippi, University, MS 38677, USA; dferreir@olemiss.edu (D.F.); emcroom@olemiss.edu (E.M.C.J.)
[3] Natural Products Utilization Research Unit, USDA-ARS, University, MS 38677, USA; stephen.duke@ars.usda.gov
*   Correspondence: dhammika@olemiss.edu; Tel.: +1-662-915-1019

Academic Editors: Muhammad Ilias and Charles L. Cantrell
Received: 31 December 2018; Accepted: 16 February 2019; Published: 21 February 2019

Abstract: Bioassay-guided fractionation of an EtOAc extract of the broth of the endophytic fungus *Nemania* sp. UM10M (Xylariaceae) isolated from a diseased *Torreya taxifolia* leaf afforded three known cytochalasins, 19,20-epoxycytochalasins C (**1**) and D (**2**), and 18-deoxy-19,20-epoxy-cytochalasin C (**3**). All three compounds showed potent in vitro antiplasmodial activity and phytotoxicity with no cytotoxicity to Vero cells. These compounds exhibited moderate to weak cytotoxicity to some of the cell lines of a panel of solid tumor (SK-MEL, KB, BT-549, and SK-OV-3) and kidney epithelial cells (LLC-PK$_{11}$). Evaluation of in vivo antimalarial activity of 19,20-epoxycytochalasin C (**1**) in a mouse model at 100 mg/kg dose showed that this compound had weak suppressive antiplasmodial activity and was toxic to animals.

Keywords: *Nemania*; Xylariaceae; *Torreya taxifolia*; plant pathogenic and endophytic fungi; cytochalasins; malaria; cytotoxicity; phytotoxicity

1. Introduction

Malaria remains a serious threat to human health in many parts of the tropics and subtropics, and an estimated 216 million cases and 445,000 deaths occurred worldwide in 2016 [1]. This disease is caused by apicomplexan parasites of the genus *Plasmodium*. *P. falciparum*, which is responsible for the majority of malaria deaths, has developed resistance to all currently available drugs, and mosquito vectors have become resistant to one or more insecticides in all WHO regions [1]. There is an urgent need to develop novel drugs with new modes of action for effective control of this disease.

Apicomplexan parasites such as *Plasmodia* contain the apicoplast organelle that is similar to plastids found in the cells of photosynthetic organisms with the same genetic complements [2]. Apicoplasts have some essential plant-like metabolic pathways which are absent in mammalian hosts. Over the last decade, a number of validated targets in plant-like metabolic pathways in the apicoplast of malaria parasites have been reported [3–7]. Apicoplasts perform vital metabolic functions and are essential for both erythrocytic and hepatic development of the parasites in mammalian hosts [3–7].

Therefore, plant-like metabolic pathways in apicoplasts serve as promising targets for antimalarial drug discovery [3–7].

Most phytotoxins disrupt metabolic pathways in plants, including those in plastids [8]. Several studies have been carried out to evaluate the in vitro antiplasmodial activity of herbicides and phytotoxins that act on known apicoplast metabolic pathways [9–11], and some of them have shown activity against *Plasmodium*.

We have screened a number of plant pathogenic fungal extracts with phytotoxic activity for their antiplasmodial activity. Some of these extracts yielded compounds with good antiplasmodial activity [12–14]. As part of this program, we investigated endophytic fungi isolated from a diseased leaf of cultivated *Torreya taxifolia* Arnott (Figure 1).

Figure 1. Symptoms of diseased needles of cultivated *T. taxifolia*.

One of the fungal extracts showed potent phytotoxic activity and selective antiplasmodial activity. This fungus (UM10M) (Figure 2) was identified as a species of the genus *Nemania* (Xylariaceae). Some members of this family have been identified as endophytes and plant pathogens [15].

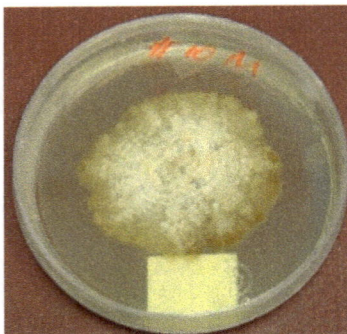

Figure 2. Potato dextrose agar plate of the fungus UM10M.

Bioassay guided fractionation of this extract led to the isolation of three cytochalasins, 19,20-epoxycytochalasins C (**1**) and D (**2**), and 18-deoxy-19,20-epoxycytochalasin C (**3**) (Figure 3), as the compounds responsible for both the phytotoxic and antimalarial activity.

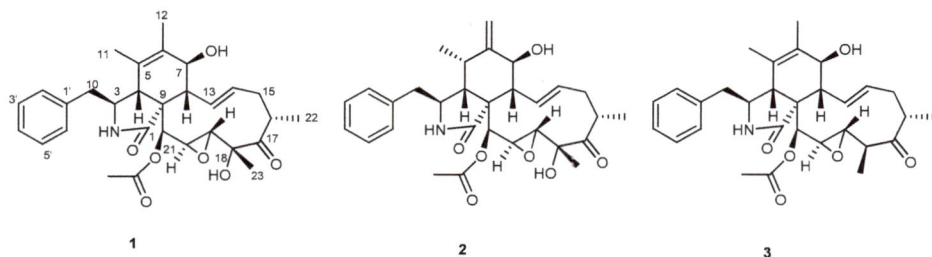

Figure 3. Structures of compounds **1–3** isolated from the fungus UM10M.

A number of cytochalasins have been reported from Xylariaceae and several other plant pathogenic and endophytic ascomycete and basidiomycete genera (16). Cytochalasins have shown diverse biological activities including antiplasmodial [16,17], antitoxoplasma [18], and phytotoxic [19–21] activities. Cytochalasin B, which is known to inhibit actin filament formation, is capable of interfering with several cellular processes including cytokinesis, intracellular motility, and exo and endocytosis [22,23]. This compound is extensively used to study cytoskeletal mechanisms [22,23].

Compounds **1** and **2** showed potent antimalarial activity against both chloroquine-sensitive and -resistant clones and were not cytotoxic to Vero cells at the highest concentration tested (4760 ng/mL) whereas compound **3** showed moderate activity. Compound **1** which was isolated as the major compound was evaluated for in vivo antimalarial activity in a mouse model.

2. Results and Discussion

An EtOAc extract of the fermentation broth of the endophytic fungus isolated from a diseased *T. taxifolia* leaf showed good phytotoxic activity against both a dicot (lettuce, *Lactuca sativa* L.) and a monocot (bentgrass, *Agrostis stolonifera* L.) and potent antiplasmodial activity against chloroquine-sensitive (D6) and -resistant (W2) strains of *P. falciparum* (IC_{50} = 40 and 44 ng/mL respectively) with low cytotoxicity (6800 ng/mL) to mammalian kidney fibroblasts (Vero cells).

Analysis of the ITS genomic region (partial 18S ribosomal RNA gene, internal transcribed spacer 1, 5.8S ribosomal RNA gene, and internal transcribed spacer 2, and partial 28S ribosomal RNA gene) of this fungus (Figure 4 and Supplementary, Table S1) indicated that it is related to the genus *Nemania* of the family Xylariaceae with 545 of 557 nucleotides (98% sequence identity) to the *Nemania* sp. genotype 547 isolate NC0453 (GenBank accession: JQ761479.1) which represents an endolichenic fungus of the host *Pseudevernia consocians*. Since it does not show a 100% identity to other already published *Nemania* species in GenBank, this fungus was designated as a new isolate, UM10M (GenBank accession: MK321315).

The EtOAc extract of the culture broth of *Nemania* sp. UM10M was fractionated over silica gel and subsequently separated over Sephadex LH-20 and PTLC to afford three compounds. Compound **1** was identified as 19,20-epoxycytochalasin C by spectroscopic methods. This compound has previously been reported from several *Xylaria* species [24–26] and an unidentified endophytic fungus [27]. Our ^{1}H and ^{13}C NMR assignments were comparable to reported data, but the chemical shift for H-11 needs to be revised, and the ^{13}C assignments for C-11 and C-12, and C-13 and C-14 need to be interchanged [25] (Table 1). Compound **2** has been isolated from *Xylaria* [25,28] *Engleromyces* [29,30], and an unidentified endophytic fungus [27], and its NMR data were comparable to the reported data [30].

Compound **3** (18-deoxy-19,20-epoxycytochalasin C) has been reported from an unidentified endophytic fungus, KL-1.1, and isolated from *Psidium guajava* leaves [27]. However, in the structure drawn for this compound, the absolute configurations of C-4 (*R*) and C-19 (*S*) were not correctly represented [30]. The NMR data for compound **3** were comparable to the reported data but the ^{13}C assignments for C-12 and C-23 need to be interchanged (Table 1).

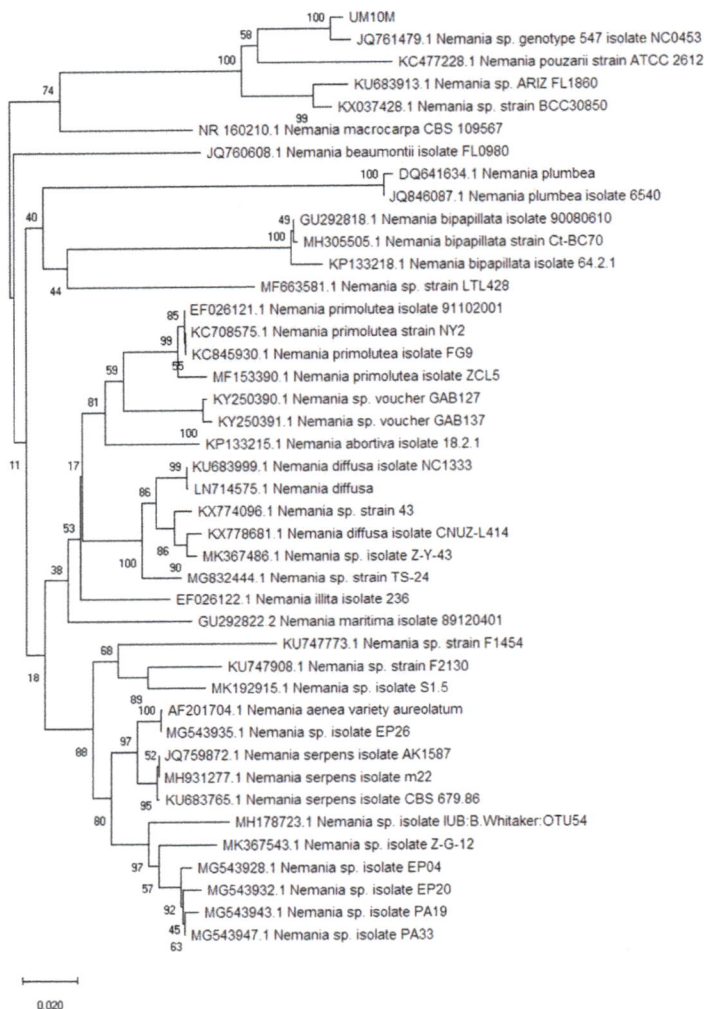

0.020

Figure 4. Constructed tree using Neighbor-Joining method employing MEGA X software to match UM10M to already published sequences to help identifying close relatives. Numbers displayed on branches are the percentage of replicate trees in which the associated taxa clustered together in the bootstrap test obtained through 500 replications. Sequences used are shown with GenBank accession numbers (Supplementary Table S1). UM10M is closely related to *Nemania* sp. genotype 547 isolate NC0453.

The configuration depicted in the reported structures for these compounds was not consistent [25–30]. In some cases, proper guidelines for using configurational descriptors have not been followed when drawing and defining stereogenic centers, and in others the absolute configuration of some stereogenic centers have been drawn incorrectly. The correct structure and the absolute configuration of 19,20-epoxycytochalasin D (**2**) have previously been determined by spectroscopic and X-ray analysis [30]. Our spectroscopic and physical data of compound **2** matched well with those reported for 19,20-epoxycytochalasin D [30], and we deduced that the structure in this reference represented the correct absolute configuration of compound **2**. Since compounds **1–3** presumably share the same biosynthetic precursor, they all possess the same absolute configuration. Therefore the structures and absolute configurations of these compounds were determined as (3*S*,4*R*,7*S*,8*S*,9*R*,16*S*,18*R*,19*R*,20*S*,21*S*)-19,20-epoxycytochalasin C (**1**), (3*S*,4*R*,5*S*,7*S*,8*S*,9*R*,16*S*,18*R*, 19*R*,20*S*,21*S*)-19,20-epoxycytochalasin D (**2**), and (3*S*,4*R*,7*S*,8*S*,9*R*,16(*S*),18*S*,19*S*,20*R*,21*S*)-18-deoxy-19,20-epoxycytochalasin C (**3**) and are correctly represented in the structures shown in Figure 3.

Table 1. ^{1}H and ^{13}C NMR data of **1** and **3** in CDCl$_3$-methanol-d_4.

Carbon	1		3	
	δ_C [b]	δ_H [a] (*J* in Hz)	δ_C [b]	δ_H [a] (*J* in Hz)
1	175.0		175.1	
3	61.0	3.35, t (7.3)	61.1	3.32, m
4	49.9	2.50, brs	49.9	2.43, brs
5	126.4		128.5	
6	131.8		131.7	
7	68.1	3.76, brd (9.9)	68.1	3.77, d (9.7)
8	48.7	2.25, dd (10.1,10.0)	49.0	2.27, dd, (10.1, 10.1)
9	52.0		51.8	
10	44.1	3.03, d (7.4)	44.4	3.04, bd, (7.5)
11	16.7	1.19, s	14.1	1.26, s
12	13.8	1.62, s	13.9	1.65, s
13	131.4	5.99, dd (15.5, 10.2)	131.0	6.24, dd (15.5, 10.4)
14	132.9	5.68, ddd (15.6, 9.7, 5.8)	133.7	5.61, m
15	37.5	2.63, dd (22.7, 11.9), 2.14, m [c]	37.6	2.53, dd (25.3, 12.2), 2.14 m [c]
16	41.7	3.27, m [c]	42.9	2.98, d (1.8)
17	215.3		216.9	
18	76.3		52.1	2.18, m [c]
19	59.9	3.25, brs [c]	58.6	3.01, m [c]
20	53.2	3.38, brs	57.6	3.38, d (1.7)
21	72.0	5.75, s	72.3	5.69, s
22	18.9	1.21, d (7.6)	18.8	1.12, d (6.7)
23	21.5	1.54, s	17.1	1.32, d (6.9)
24	170.4		170.5	
25	20.5	2.18, s	20.7	2.15, s
1'	137.3		136.9	
2',6'	129.2	7.28, m	129.4	7.26, m
3',5'	128.6	7.33, m	128.7	7.33, m
4'	126.8	7.25, m	126.9	7.25, m

[a] ^{1}H NMR spectra recorded at 400 MHz, [b] ^{13}C NMR spectra recorded at 100 MHz, [c] overlapped signal.

Compounds **1–3** showed potent selective (calculated as a ratio of IC$_{50}$ for cytotoxicity to Vero cells and IC$_{50}$ for antimalarial activity) in vitro antiplasmodial activity against chloroquine-sensitive (D6) and -resistant (W2) strains of *P. falciparum* (Table 2) and nonspecific moderate phytotoxic activity against both a monocot (bentgrass, *Agrostis stolonifera*) and a dicot (lettuce, *Lactuca sativa cv.* L., iceberg) in the presence of light (Table 3).

Table 2. Antiplasmodial activity of 1–3.

Compound	Chloroquine-Sensitive (D6)-Strain		Chloroquine-Resistant (W2)-Strain		Cytotoxicity to Vero Cells
	IC$_{50}$ μM (ng/mL)	S. I.	IC$_{50}$ μM (ng/mL)	S. I.	IC$_{50}$ μM
19,20-Epoxycytochalasin C (1)	0.07 (37)	>128.6	0.05 (28)	>170	NC
19,20-Epoxycytochalasin D (2)	0.04 (20)	>216.3	0.04 (20)	>238	NC
18-Deoxy-19,20-epoxy-cytochalasin C (3)	0.56 (280)	>17	0.19 (100)	>47.6	NC
Chloroquine [a]	0.03 (16)	>297.5	0.31 (160)	>29.8	NC
Artemisinin [a]	0.02 (5.6)	>850	0.01 (3.0)	>1586.6	NC

[a] Positive controls; NC: not cytotoxic at the highest concentration tested (4760 ng/mL); S. I. (selectivity index) = IC$_{50}$ for cytotoxicity/IC$_{50}$ for antiplasmodial activity.

Table 3. Phytotoxic activity of 1–3 [a].

Compound	Lettuce	Bentgrass
19,20-Epoxycytochalasin C (1)	3	2
19,20-Epoxycytochalasin D (2)	3	3
18-Deoxy-19,20-epoxycytochalasin C (3)	3	4

[a] Concentration (mM) = 1 mg/mL. Ranking based on scale of 0 to 5; 0 = no effect; 5 = no growth; Solvent used, 10% acetone; Concentration used, 1 mg/mL.

The in vitro cytotoxic potential of 1–3 was further evaluated against a panel of solid tumor cell lines (SK-MEL, KB, BT-549, SK-OV-3) and kidney epithelial cells (LLC-PK$_{11}$) (Table 4). Compounds 1 and 3 showed moderate toxicity to cell lines SK-MEL and BT-549, respectively, whereas compound 2 exhibited moderate toxicity to BT-549 and LLC-PK$_{11}$ cell lines.

Table 4. Cytotoxic activity [IC$_{50}$ (μM)] of 1–3.

Compound	SK-MEL	KB	BT-549	SK-OV-3	LLC-PK$_{11}$
19,20-Epoxycytochalasin C (1)	8.02	NC	NC	NC	NC
19,20-Epoxycytochalasin D (2)	NC	NC	7.84	NC	8.4
18-Deoxy-19,20-epoxycytochalasin C (3)	NC	NC	6.89	NC	NC
Doxorubicin [a]	1.29	2.12	1.83	1.47	1.28

[a] Positive control. NC: not cytotoxic at 10 μM. IC$_{50}$ = concentration causing 50% growth inhibition. SK-MEL = human malignant melanoma; KB = human epidermal carcinoma; BT-549 = human breast carcinoma (ductal); SK-OV-3 = human ovary carcinoma; LLC-PK$_{11}$ = pig kidney epithelial.

In vitro antiplasmodial [16,17] and antitoxoplasma [18] activities of cytochalasin derivatives have been reported. These compounds have been shown [17,18] to inhibit the actin-based gliding motility in extracellular parasites and impair erythrocyte invasion of apicomplexan parasites of *Toxoplasma gondii* [18,31] and *P. falciparum* [32,33]. Both host cells and the parasites contain actin-based cytoskeletons but due to their divergence, parasite actin-1 (PfACT-1) has been identified as a viable drug target [33]. The information on the activity of this class of compound indicates that the observed potent in vitro antimalarial activity with high selectivity indices for compounds 1 and 2 is most probably due to their selective disruption of parasite actin-1 and not due to disruption of plant-like metabolic pathways in apicoplasts. Even though in vitro antimalarial activity of this class of compounds has previously been reported [16,17], their in vivo activity has not yet been reported. Cytochalasins B, D, and E have been extensively studied for their anticancer effects. In animal models, no toxicity has been observed at the therapeutic doses when administered in intraperitoneal, subcutaneous, or intravenous injections up to 100 mg/kg/day [34–36].

Compound 1, which was isolated as the major compound, was evaluated for antimalarial activity against *P. berghei* in mice at 100 mg/kg/day for 3 days through the oral route. Chloroquine was used as the positive control. A control group of mice were treated with vehicle only. Three out of 5 animals treated with compound **1** (Table 5) died due to drug induced toxicity. In animals that survived, 33.9% and 71.4% suppression of parasitemia was observed on days 5 and 7, respectively (Table 5). Even though compound **1** appears to have some suppressive activity at this dose, its high toxicity would preclude it from being a viable antimalarial lead.

Table 5. In vivo antimalarial activity of **1**.

Treatment (PO)	Dose (mg/kg × # days Post Infection)	% Parasitemia Suppression [b] Day 5	Day 7	Survival [c]	Day of Death	MST [d]	Cure [f]
Vehicle	×3	-	-	0/5	14/14/13/13/5	11.8	0/5
Chloroquine [a]	100 × 3	100	100	5/5	28/28/28/28/28	28	2/5
1	100 × 3	33.9	71.4	0/5	17/5/3/1/0	5.2	0/2

[a] Positive control; [b] % suppression in parasitemia is calculated by considering the mean parasitemia in the vehicle control as 100%, Parasitemia suppression < 80% is considered as non-significant; [c] Number of animals that survived day 28/total animals in group (the day of the death-post infection); [d] % MST—mean survival time (days); [f] Number of mice without parasitemia (cured) till day 28 post-infection.

3. Materials and Methods

3.1. General Experimental Procedures

Optical rotations were measured using a Autopol IV automatic polarimeter model 589-546 (Rudolph Research Analytical, Flanders, NJ, USA). NMR spectra were recorded on a Bruker 400 MHz. spectrometer (Rheinstetten, Germany) using $CDCl_3$/methanol-d_4 as the solvent unless otherwise stated. MS analyses were performed on an Agilent Series 1100 SL equipped with an ESI source (Agilent Technologies, Palo Alto, CA, USA). Column chromatography was carried out on silica gel 60 (230-400 mesh) (Sigma-Aldrich, St. Louis, MO, USA) and Sephadex LH-20 (GE Healthcare Bio-Sciences, Marlborough, MA, USA). Preparative TLC was carried out using silica gel GF plates (20 × 20 cm, thickness 0.25 mm). Spots were detected under UV light and by heating after spraying with anisaldehyde reagent.

3.2. Fungal Material

Diseased leaves were collected from a cultivated *T. taxifolia* plant from Oxford, MS. A voucher of the *T. taxifolia* Arn. plant was identified by E. M. Croom, Jr. and deposited in the University of Mississippi Pullen Herbarium. The voucher accession number is MISS 83490. Leaves were surface sterilized with 5% Chlorox for 5 min and rinsed with sterile water (×3). Transverse sections from the dried leaf were cut aseptically into small portions and immersed in potato dextrose plates. The plates were incubated for two weeks and fungal colonies were subcultured on PDA plates to isolate pure fungal strains.

3.3. Fermentation, Extraction, and Purification

Fungus UM10M was cultured in four conical flasks (2 L) containing 500 mL of PDB and incubated at 27 °C for 30 days on an orbital shaker at 100 rpm. Mycelia and broth were separated by filtration and extracted with EtOAc (×3). The organic layer from the broth was evaporated to give a brown/black residue (625 mg).

This extract (600 mg) was chromatographed on Sephadex LH-20 and eluted with 80% MeOH in $CHCl_3$ to give 10 fractions. Fractions 2, 3, and 4 showed antimalarial activity. Fraction 2 was subjected to silica gel gravity column chromatography using hexanes, CH_2Cl_2, and MeOH gradient as the mobile phase to yield five fractions. Subfraction 3 which showed antimalarial activity was further separated by preparative thin layer chromatography (PTLC) using 0.4% MeOH in $CHCl_3$ (×3) as the developing solvent to obtain 19,20-epoxycytochalasin C (**1**, 85.0 mg). Fractions 3 and 4 were also subjected to

PTLC using 0.4% MeOH in CHCl$_3$ (×3) as the developing solvent to afford 19,20-epoxycytochalasin D (**2**, 4 mg), and 18-deoxy-19,20-epoxycytochalasin C (**3**, 2 mg).

19,20-Epoxycytochalasin C (**1**); white amorphous powder; ^{1}H, ^{13}C NMR data (Table 1) $[\alpha]_D^{26}$ −13 (*c* 0.5, CHCl$_3$) lit [24] −6.8; HRESIMS [M + H]$^+$ *m/z* 524.2642.

19,20-Epoxycytochalasin D (**2**); white amorphous powder; ^{1}H, and ^{13}C NMR, data were consistent with literature values [30]. $[\alpha]_D^{26}$ −113 (*c* 0.1, CHCl$_3$) lit [34] −190; HRESIMS [M + H]$^+$ *m/z* 524.2837.

18-Deoxy-19,20-epoxycytochalasin C (**3**); white amorphous powder; ^{1}H and ^{13}C NMR (see Table 1) $[\alpha]_D^{26}$ −2.4 (*c* 0.1, CHCl$_3$); HRESIMS [M + H]$^+$ *m/z* 508.2703 (calcd for [C$_{24}$H$_{33}$NO$_4$ + H]$^+$ 508.26201.

3.4. Biological Assays

3.4.1. In Vitro Antiplasmodial Assay

The antiplasmodial assay was performed against D6 (chloroquine sensitive) and W2 (chloroquine resistant) strains of *P. falciparum* using the in vitro assay as reported [37]. Artemisinin and chloroquine were included as the drug controls, and IC$_{50}$ values were computed from the dose-response curves.

3.4.2. In Vitro Phytotoxicity Assay

Herbicidal or phytotoxic activity of the extract and compounds was performed according to the published procedure [38] using bentgrass (*Agrostis stolonifera*) and lettuce (*Lactuca sativa* cv. L., Iceberg), in 24-well plates. Phytotoxicity was ranked visually. The ranking of phytotoxic activity was based on a scale of 0 to 5 with 0 showing no effect and 5 no growth.

3.4.3. In Vitro Cytotoxicity Assay

In vitro cytotoxicity was determined against a panel of mammalian cells that included kidney fibroblast (Vero), kidney epithelial (LLC-PK$_{11}$), malignant melanoma (SK-MEL), oral epidermal carcinoma (KB), breast ductal carcinoma (BT-549), and ovary carcinoma (SK-OV-3) cells [39] Doxorubicin was used as a positive control.

3.4.4. In Vivo Antimalarial Assay

The protocol for in vivo antimalarial evaluation was approved by the University of Mississippi Institutional Animal Care and Use Committee (IACUC). The in vivo antimalarial activity was determined in mice infected with *P. berghei* (NK-65 strain). Male mice (Swiss Webster strain) weighing 18-20 g were intraperitoneally inoculated with 2 × 10^7 parasitized red blood cells obtained from a highly infected donor mouse. Mice were divided into different groups with five mice in each group. Compound stocks were prepared in DMSO and administered orally to the mice through gavage two hours after the infection (day 0). The animals were treated once daily for consecutive days (day 0 to 2) and were closely observed for at least 2 h after every dose for any apparent signs of toxicity. Blood smears were prepared on different days starting from 5 days post infection (through 28 days) by clipping the tail end, stained with Giemsa, and the slides were observed under microscope for determination of parasitemia. Mice without parasitemia through day 28 postinfection were considered as cured. Also, suppression in development of parasitemia was computed by comparing the parasitemia in the control vehicle treated group and groups treated with compound. The mean survival time was also computed for control and treated groups. The results are presented as parasitemia suppression of day 5 post treatment, mean survival time of mice in each group, cure and survival graphs computed by Prism 6.0. (San Diego, CA, USA).

3.5. DNA Analysis

A 0.5 by 1.5 cm piece of fungus grown on solid media was transferred into a 2 mL microcentrifuge tube. Two 3 mm in diameter stainless steel balls were added and the sample ground for 1 min in a MM2000 mixer mill (Retsch, Haan, Germany). DNA was extracted with the DNeasy Plant mini kit (Qiagen, Valencia, CA, USA) according to the manufacturer's recommendation. DNA was eluted with 50 μL buffer AE. DNA quality and quantity was determined with the NanoDrop 1000 Spectrophotometer (Thermo Fisher, Wilmington, DE, USA). DNA dilutions were prepared to achieve a 10 ng/μL solution to be used as template for PCR.

The internal transcribed spacer region (ITS) was amplified in a 25 μL reaction containing 10 ng DNA, 1X PCR reaction buffer, 0.2 mM dNTP mixture, 0.2 μM of each forward and reverse primers ITS1 and ITS4 [40], 1.5 mM $MgSO_4$ and 1 U of Taq Polymerase High Fidelity (Invitrogen, Carlsbad, CA, USA). The PCR program consisted of 40 cycles with denaturation at 94 °C for 3 min, 50 °C annealing temperature for 30 s, 68 °C for 90 s, and a final extension at 68 °C for 3 min. After amplification, an aliquot was analyzed by electrophoresis on a 0.7% TAE agarose gel. Successfully amplified PCR products were cleaned up with MinElute PCR Purification Kit (Qiagen) according to the manufacturer's instructions. PCR products were cloned into pJet.1.2 blunt using the CloneJet PCR cloning kit (Fermentas, Glen Burnie, MD, USA). Eight individual colonies were transferred into 4 mL liquid LB carbenicillin media and grown under constant shaking overnight at 37 °C. Plasmid DNA from the overnight cultures was isolated with the Qiagen plasmid purification kit (Qiagen) according to the manufacturer's instructions. Sequencing was performed at the Genomics and Bioinformatics Research Facility in Stoneville, MS. Resulting sequences were analyzed with DNASTAR (DNASTAR, Madison, WI, USA). Homology searches were performed with the Basic Local Alignment Search Tool (BLAST) [41]. The UM10M sequence was submitted to GenBank (Accession: MK321315).

To visualize sequence alignment the software Genedoc [42] was used. Shading is according to alignment consensus as given by GeneDoc (black, 100%; dark grey, 80%; light grey, 60%) (Supplementary, Figure S19). To identify close relatives of UM10M a phylogenetic analysis was performed (Supplementary, Figure S19) using three best hits from the BLAST analysis and already published sequences in Genebank of various taxa of the genus *Nemania*. The configuration for the alignment analysis was maximum likelihood with the neighbor joining statistical method [43]. The percentage of replicate trees in which the associated taxa clustered together in the bootstrap test (500 replicates) is shown next to the branches [44]. The evolutionary distances were computed using the Kimura 2-parameter method [45] with nucleotide transitions and transversions included and are in the units of the number of base substitutions per site. The analysis involved 42 nucleotide sequences (Supplementary Table S1). All positions containing gaps and missing data were eliminated by the software for the final dataset analysis. There were a total of 692 positions in the final dataset. Evolutionary analyses were conducted in MEGA X [46].

4. Conclusions

Three cytochalasins were isolated from an endophytic fungus (UM10M) of the genus *Nemania* identified from a diseased leaf of *T. taxifolia* Arnott. All compounds showed potent in vitro activity against *P. falciparum* D6 and W2 strains with no cytotoxicity to Vero cells but displayed moderate to weak cytotoxicity to some of the human solid tumor cell lines and kidney epithelial cells. Compound **1** was evaluated in a mouse model for activity against *P. berghei* at 100 mg/kg/day. It showed weak suppressive activity but its high toxicity to animals would preclude it as a potential malaria drug lead.

Supplementary Materials: The following are available online. Figures S1–S18 are NMR spectra of compounds 1–3, Figure S19 is UM10M sequence alignment, Table S1 is Sequences used for alignment analysis and to identify close relatives.

Author Contributions: N.P.D.N., B.L.T., and S.O.D. conceived and designed experiments and reviewed the manuscript. M.K. carried out fungal culture, extraction, isolation, and identification of compounds and wrote the original draft. R.S. carried out in vivo antimalarial assay under B.L.T.'s supervision. S.K. supervised in vitro antimalarial and cytotoxic assays, analyzed the data and reviewed the manuscript. D.F. assisted with structure elucidation and critically reviewed the manuscript. E.M.C.J. identified and provided the infected plant material and reviewed the manuscript. N.T. identified the fungus using ITS DNA analysis and reviewed the manuscript.

Funding: This work was supported in part, by the United States Department of Agriculture, ARS, Specific Cooperative Agreement No. 58-6408-2-009.

Acknowledgments: We thank Bharathi Avula NCNPR, University of Mississippi, for acquiring the MS data and Marsha Wright, John Trott, and Robert Johnson for biological testing.

Conflicts of Interest: The authors declare no conflict of interest.

References

1. World Health Organization (WHO). *World Malaria Report*; WHO: Geneva, Switzerland, 2017.
2. Lim, L.; McFadden, G.I. The evolution, metabolism, and functions of the apicoplast. *Philos. Trans. R. Soc. B* **2010**, *365*, 749–763. [CrossRef] [PubMed]
3. Ralph, S.A.; D'Ombrain, M.C.; McFadden, G.I. The apicoplast as an antimalarial drug target. *Drug Resist. Updates* **2001**, *4*, 145–151. [CrossRef] [PubMed]
4. Ralph, S.A.; van Dooren, G.G.; Waller, R.F.; Crawford, M.J.; Fraunholz, M.J.; Foth, B.J.; Tonkin, C.J.; Roos, D.S.; McFadden, G.I. Tropical infectious diseases: Metabolic maps and functions of the *Plasmodium falciparum* apicoplast. *Nat. Rev. Microbiol.* **2004**, *2*, 203–216. [CrossRef] [PubMed]
5. Waller, R.F.; McFadden, G.I. The apicoplast: A review of the derived plastid of apicomplexan parasites. *Curr. Issues Mol. Biol.* **2005**, *7*, 57–80. [PubMed]
6. Dhal, E.L.; Rosenthal, P.J. Multiple antibiotics extert delayed effects against the *Plasmodium falciparum* apicoplast. *Antimicrob. Agents Chemother.* **2007**, *51*, 3485–3490. [CrossRef] [PubMed]
7. Botté, C.Y.; Dubar, F.; McFadden, G.I.; Maréchal, E.; Biot, C. *Plasmodium falciparum* Apicoplast Drugs: Targets or Off-Targets? *Chem. Rev.* **2012**, *112*, 1269–1283. [CrossRef] [PubMed]
8. Franck, E.D.; Duke, S. Natural compounds as next-generation herbicides. *Plant Physiol.* **2014**, *166*, 1090–1105. [CrossRef]
9. Roberts, F.; Roberts, C.W.; Johnson, J.J.; Kyle, D.E.; Krell, T.; Coggins, J.R.; Coombs, G.H.; Milhous, W.K.; Tzipori, S.; Ferguson, D.J.; et al. Evidence for the shikimate pathway in apicomplexan parasites. *Nature* **1998**, *393*, 801–805. [CrossRef]
10. Lichtenthaler, H.K. Non-mevalonate isoprenoid biosynthesis enzymes, genes, and inhibitors. *Biochem. Soc. Trans.* **2000**, *28*, 785–789. [CrossRef]
11. Bajsa, J.; Singh, K.; Nanayakkara, D.; Duke, S.O.; Rimando, A.M.; Evidente, A.; Tekwani, B.L. A Survey of synthetic and natural phytotoxic compounds and phytoalexins as potential antimalarial compounds. *Biol. Pharm. Bull.* **2007**, *30*, 1740–1744. [CrossRef]
12. Herath, H.M.T.B.; Herath, W.H.M.W.; Carvalho, P.; Khan, S.I.; Tekwani, B.L.; Duke, S.O.; Tomaso-Peterson, M.; Nanayakkara, N.P.D. Biologically active tetranorditerpenoids from the fungus *Sclerotinia homoeocarpa* causal agent of dollar spot in turfgrass. *J. Nat. Prod.* **2009**, *72*, 2091–2097. [CrossRef] [PubMed]
13. Kumarihamy, M.; Fronczek, F.R.; Ferreira, D.; Jacob, M.; Khan, S.I.; Nanayakkara, N.P.D. Bioactive 1,4-dihydroxy-5-phenyl-2-pyridinone alkaloids from *Septoria pistaciarum*. *J. Nat. Prod.* **2010**, *73*, 1250–1253. [CrossRef] [PubMed]
14. Kumarihamy, M.; Khan, S.I.; Jacob, M.; Tekwani, B.L.; Duke, S.O.; Ferreira, D.; Nanayakkara, N.P.D. Antiprotozoal and antimicrobial compounds from *Septoria pistaciarum*. *J. Nat. Prod.* **2012**, *75*, 883–889. [CrossRef] [PubMed]
15. Edwards, R.L.; Jonglaekha, N.; Anandini, K.; Maitland, D.J.; Mekkamol, S.; Nugent, L.K.; Phosri, C.; Rodtong, S.; Ruchichachorn, N.; Sangvichien, E.; et al. The Xylariaceae as phytopathogens. *Recent Res. Dev. Plant Sci.* **2003**, *1*, 1–19.
16. Isaka, M.; Jaturapat, A.; Kladwang, W.; Punya, J.; Lertwerawat, Y.; Tanticharoen, M.; Thebtaranonth, Y. Antiplasmodial compounds from the wood-decayed fungus *Xylaria* sp. BCC 1067. *Planta Med.* **2000**, *66*, 473–475. [CrossRef] [PubMed]
17. Dieckmann-Schuppert, A.; Franklin, R.M. Compounds binding to cytoskeletal proteins are active against *Plasmodium falciparum* in vitro. *Cell Biol. Int. Rep.* **1989**, *13*, 411–418. [CrossRef]

18. Dobrowolski, J.M.; Sibley, L.D. Toxoplasma invasion of mammalian cells is powered by the actin cytoskeleton of the parasite. *Cell* **1996**, *84*, 933–939. [CrossRef]
19. Zhang, Q.; Xiao, J.; Sun, Q.-Q.; Qin, J.-C.; Pescitelli, G.; Gao, J.-M. Characterization of cytochalasins from the endophytic sp. and their biological functions. *J. Agric. Food Chem.* **2014**, *62*, 10962–10969. [CrossRef]
20. Hussain, H.; Kliche-Spory, C.; Al-Harrasi, A.; Al-Rawahi, A.; Abbas, G.; Green, I.R.; Schulz, B.; Krohn, K.; Shah, A. Antimicrobial constituents from three endophytic fungi. *Asian Pac. J. Trop. Med.* **2014**, *7*, S224–S227. [CrossRef]
21. Berestetskiy, A.; Dmitriev, A.; Mitina, G.; Lisker, I.; Andolfi, A.; Evidente, A. Nonenolides and cytochalasins with phytotoxic activity against *Cirsium arvense* and *Sonchus arvensis*: A structure-activity relationships study. *Phytochemistry* **2008**, *69*, 953–960. [CrossRef]
22. Scherlach, K.; Boettger, D.; Remme, N.; Hertweck, C. The chemistry and biology of cytochalasans. *Nat. Prod. Rep.* **2010**, *27*, 869–886. [CrossRef] [PubMed]
23. Trendowski, M. Using cytochalasins to improve current chemotherapeutic approaches. *Anti-Cancer Agents Med. Chem. (Former. Curr. Med. Chem.-Anti-Cancer Agents)* **2015**, *15*, 327–335. [CrossRef]
24. Abate, D.; Abraham, W.-R.; Meyer, H. Cytochalasins and phytotoxins from the fungus *Xylaria obovata*. *Phytochemistry* **1997**, *44*, 1443–1448. [CrossRef]
25. Espada, A.; Rivera-Sagredo, A.; de la Fuente, J.M.; Hueso-Rodríguez, J.A.; Elson, S.W. New cytochalasins from the fungus *Xylaria hypoxylon*. *Tetrahedron* **1997**, *53*, 6485–6492. [CrossRef]
26. Song, Y.X.; Wang, J.; Li, S.W.; Cheng, B.; Li, L.; Chen, B.; Liu, L.; Lin, Y.C.; Gu, Y.C. Metabolites of the mangrove fungus *Xylaria* sp. BL321 from the South China Sea. *Planta Med.* **2012**, *78*, 172–176. [CrossRef]
27. Okoyea, F.B.C.; Nworuc, C.S.; Debbaba, A.; Esimone, C.O.; Proksch, P. Two new cytochalasins from an endophytic fungus, KL-1.1 isolated from *Psidium guajava* leaves. *Phytochem. Lett.* **2015**, *14*, 51–55. [CrossRef]
28. Chen, Z.; Chen, Y.; Huang, H.; Yang, H.; Zhang, W.; Sun, Y.; Wen, J. Cytochalasin P1, a new cytochalasin from the marine-derived fungus *Xylaria* sp. SOF11. *Z. Naturforsch. C* **2017**, *72*, 129–132. [CrossRef]
29. Jikai, L.; Zejun, D.; Zhihui, D.; Xianghua, W.; Peiqui, L. Neoengleromycin, a novel compound from *Engleromyces goetzii*. *Helv. Chim. Acta* **2002**, *85*, 1439–1442. [CrossRef]
30. Shi, L.-M.; Zhan, Z.-J. Structural revision of 19, 20-epoxycytochalasin D and its cytotoxic activity. *J. Chem. Res.* **2007**, *3*, 144–145. [CrossRef]
31. Drewry, L.D.; Sibley, L.D. *Toxoplasma* actin is required for efficient host cell invasion. *mBio* **2015**, *6*, e00557-15. [CrossRef]
32. Baum, J.; Papenfuss, A.T.; Baum, B.; Speed, T.P.; Cowman, A.F. Regulation of apicomplexan actin-based motility. *Nat. Rev. Microbiol.* **2006**, *4*, 621–628. [CrossRef] [PubMed]
33. Das, S.; Lemgruber, L.; Tay, C.L.; Baum, J.; Meissner, M. Multiple essential functions of *Plasmodium falciparum* actin-1 during malaria blood-stage development. *BMC Biol.* **2017**, *15*, 70. [CrossRef] [PubMed]
34. Bousquet, P.F.; Paulsen, L.A.; Fondy, C.; Lipski, K.M.; Loucy, K.J.; Fondy, T.P. Effects of cytochalasin B in culture and in vivo on murine Madison 109 lung carcinoma and on B16 melanoma. *Cancer Res.* **1990**, *50*, 1431–1439. [PubMed]
35. Trendowski, M.; Mitchell, J.M.; Corsette, C.M.; Acquafondata, C.; Fondy, T.P. Chemotherapy in vivo against M109 murine lung carcinoma with cytochalasin B by localized, systemic, and liposomal administration. *Investig. New Drugs* **2015**, *33*, 280–289. [CrossRef] [PubMed]
36. Trendowski, M.; Mitchell, J.M.; Corsette, C.M.; Acquafondata, C.; Fondy, T.P. Chemotherapy with cytochalasin congeners in vitro and in vivo against murine models. *Investig. New Drugs* **2015**, *33*, 290–299. [CrossRef] [PubMed]
37. Bharate, S.B.; Khan, S.I.; Yunus, N.A.M.; Chauthe, S.K.; Jacob, M.R.; Tekwani, B.L.; Khan, I.A.; Singh, I.P. Antiprotozoal and antimicrobial activities of *O*-alkylated and formylated acylphloroglucinols. *Bioorg. Med. Chem.* **2007**, *15*, 87–96. [CrossRef]
38. Dayan, F.E.; Romagni, J.G.; Duke, S.O. Investigating the mode of action of natural phytotoxins. *J. Chem. Ecol.* **2000**, *26*, 2079–2094. [CrossRef]
39. Mustafa, J.; Khan, S.I.; Ma, G.; Walker, L.; Khan, I.A. Synthesis and anticancer activities of fatty acid analogs of podophyllotoxin. *Lipids* **2004**, *39*, 167–172. [CrossRef]
40. White, T.J.; Bruns, T.; Lee, S.; Taylor, J.W. Amplification and direct sequencing of fungal ribosomal RNA genes for phylogenetics. In *PCR Protocols: A Guide to Methods and Applications*; Innis, M.A., Gelfand, D.H., Sninsky, J.J., Whit, T.J., Eds.; Academic Press, Inc.: New York, NY, USA, 1990; pp. 315–322.

41. Altschul, S.F.; Gish, W.; Myers, E.W.; Lipman, D.J. Basic local alignment search tool. *J. Mol. Biol.* **1990**, *215*, 403–410. [CrossRef]
42. Nicholas, K.B.; Nicholas, H.B.; Deerfield, D.W., II. A Tool for Editing and Annotating Multiple Sequence Alignments, Version 2.7. Distributed by the Authors. 1997. Available online: http://www.psc.edu/biomed/genedoc (accessed on 15 October 2006).
43. Saitou, N.; Nei, M. The neighbor-joining method: A new method for reconstructing phylogenetic trees. *Mol. Biol. Evol.* **1987**, *4*, 406–425. [CrossRef]
44. Felsenstein, J. Confidence limits on phylogenies: An approach using the bootstrap. *Evolution* **1985**, *39*, 783–791. [CrossRef] [PubMed]
45. Kimura, M. A simple method for estimating evolutionary rate of base substitutions through comparative studies of nucleotide sequences. *J. Mol. Evol.* **1980**, *16*, 111–120. [CrossRef] [PubMed]
46. Kumar, S.; Stecher, G.; Li, M.; Knyaz, C.; Tamura, K. MEGA X: Molecular evolutionary genetics analysis across computing platforms. *Mol. Biol. Evol.* **2018**, *35*, 1547–1549. [CrossRef] [PubMed]

Sample Availability: Samples of the compounds are not available from the authors.

molecules

MDPI

Article

Selective Inhibition of Human Monoamine Oxidase B by Acacetin 7-Methyl Ether Isolated from *Turnera diffusa* (Damiana)

Narayan D. Chaurasiya [1,†], Jianping Zhao [1], Pankaj Pandey [2], Robert J. Doerksen [1,2], Ilias Muhammad [1,*] and Babu L. Tekwani [1,2,*,†]

[1] National Center for Natural Products Research, Research Institute of Pharmaceutical Sciences, School of Pharmacy, The University of Mississippi, Oxford, MS 38677, USA; nchaurasiya@southernresearch.org (N.D.C.); jianping@olemiss.edu (J.Z.); rjd@olemiss.edu (R.J.D.)
[2] Department of BioMolecular Sciences, Division of Medicinal Chemistry and Research Institute of Pharmaceutical Sciences, School of Pharmacy, The University of Mississippi, Oxford, MS 38677, USA; ppandey@olemiss.edu
*   Correspondence: milias@olemiss.edu (I.M.); btekwani@southernresearch.org (B.L.T.)
†   Present address: Department of Infectious Diseases, Division of Drug Discovery, Southern Research, Birmingham, AL 35205, USA.

Received: 29 December 2018; Accepted: 19 February 2019; Published: 23 February 2019

Abstract: The investigation of the constituents that were isolated from *Turnera diffusa* (damiana) for their inhibitory activities against recombinant human monoamine oxidases (MAO-A and MAO-B) in vitro identified acacetin 7-methyl ether as a potent selective inhibitor of MAO-B (IC_{50} = 198 nM). Acacetin 7-methyl ether (also known as 5-hydroxy-4′, 7-dimethoxyflavone) is a naturally occurring flavone that is present in many plants and vegetables. Acacetin 7-methyl ether was four-fold less potent as an inhibitor of MAO-B when compared to acacetin (IC_{50} = 50 nM). However, acacetin 7-methyl ether was >500-fold selective against MAO-B over MAO-A as compared to only two-fold selectivity shown by acacetin. Even though the IC_{50} for inhibition of MAO-B by acacetin 7-methyl ether was ~four-fold higher than that of the standard drug deprenyl (i.e., Selegiline[TM] or Zelapar[TM], a selective MAO-B inhibitor), acacetin 7-methyl ether's selectivity for MAO-B over MAO-A inhibition was greater than that of deprenyl (>500- vs. 450-fold). The binding of acacetin 7-methyl ether to MAO-B was reversible and time-independent, as revealed by enzyme-inhibitor complex equilibrium dialysis assays. The investigation on the enzyme inhibition-kinetics analysis with varying concentrations of acacetin 7-methyl ether and the substrate (kynuramine) suggested a competitive mechanism of inhibition of MAO-B by acacetin 7-methyl ether with Ki value of 45 nM. The docking scores and binding-free energies of acacetin 7-methyl ether to the X-ray crystal structures of MAO-A and MAO-B confirmed the selectivity of binding of this molecule to MAO-B over MAO-A. In addition, molecular dynamics results also revealed that acacetin 7-methyl ether formed a stable and strong complex with MAO-B. The selective inhibition of MAO-B suggests further investigations on acacetin 7-methyl as a potential new drug lead for the treatment of neurodegenerative disorders, including Parkinson's disease.

Keywords: acacetin 7-methyl ether; acacetin; monoamine oxidase-A; monoamine oxidase-B; molecular docking; molecular dynamics; neurological disorder; *Turnera diffusa*

1. Introduction

The monoamine oxidases (EC.1.4.3.4; MAO-A and MAO-B) are FAD (flavin adenine dinucleotide)-dependent enzymes that are responsible for the metabolism of neurotransmitters, such as dopamine, serotonin, adrenaline, and noradrenaline, and for the oxidative deamination of exogenous arylalkyl amines [1,2]. Due to their central role in neurotransmitter metabolism, these enzymes represent attractive drug targets in the pharmacological therapy of neurodegenerative diseases and neurological disorders [3,4]. Recent efforts toward the development of MAO inhibitors have focused on selective MAO-A or MAO-B inhibitors. Selective MAO-A inhibitors are effective in the treatment of depression and anxiety [5], whereas the MAO-B inhibitors are useful for treatment Parkinson's disease and in combination for treatment of Alzheimer's Disease [4,6–9]. In recent studies, acacetin was reported as a potent inhibitor of MAO-A and MAO-B isolated from *Calea urticifolia* [10]. Similarly, acacetin and its derivative, acacetin 7-O-(6-O-malonylglucoside), from *Agatache rugosa* was also reported as a selective potent MAO-B reversible inhibitor [11].

Turnera diffusa Willd. ex Schult (known as damiana), which is a native plant to America and Africa, is traditionally used for the treatment of various diseases, including sexual impotence, neurasthenia, diabetes mellitus, urine retention, malaria, diarrhea, and peptic ulcer [12]. The plant is particularly used as a stimulant, an aphrodisiac, and generally as a tonic in neurasthenia and impotency with long tradition in Central America [13]. Phytochemical studies have revealed flavonoids, cyanogenic glycosides, terpenoids, and other secondary metabolites as prominent constituents in *Turnera diffusa* [14,15]. Several flavonoids from plant sources have been identified as inhibitors of MAO-A and MAO-B [16]. Based on the fact that damiana contains many flavonoids and that it is traditionally used as a tonic herb, we postulated that some components of the plant might be associated with the inhibition of MAOs. In the present study, the constituents that were isolated from damiana were evaluated for their inhibitory activities against recombinant human MAO-A and MAO-B. Acacetin 7-methyl ether showed the most potent selective inhibition of the MAO-B enzyme. The studies were further extended to investigate the selective binding and mode of interaction of acacetin 7-methyl ether with human MAO-B.

2. Results

2.1. Determination of MAO-A and -B Inhibition Activity

A series of flavonoids and flavonoid glycosides, namely acacetin, acacetin 7-methyl ether, vetulin (Figure 1), apigenin-7-O-β-D-(6-O-pcoumaroyl) glucoside, echinaticin, tetraphyllin B, tricin-7-glucoside, diffusavone, turneradiffusin, rhamnosylorientin, rhamnosylvitexin, and turneradin were isolated from *T. diffusa* [12].

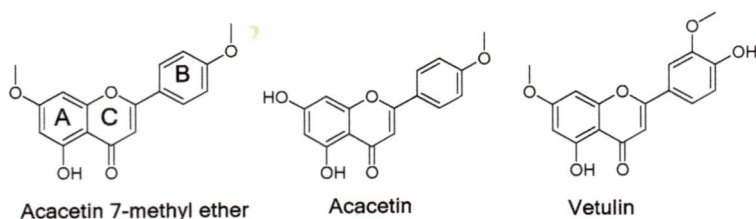

Figure 1. Chemical structure of acacetin, acacetin 7-methyl ether and vetulin.

Besides acacetin, acacetin 7-methyl ether and vetulin also showed selective concentration-dependent inhibition of MAO-B (Table 1, Figure 2). The MAO inhibitory properties of acacetin have been recently reported [10]. Acacetin 7-methyl ether was >500-fold selective for MAO-B (IC$_{50}$ = 0.198 µM) as compared to MAO-A (IC$_{50}$ = >100 µM) (Table 1). The concentration-dependent inhibition of MAO-B with acacetin 7-methyl ether showed a plateau at ~80% inhibition (~20% activity remaining). This may potentially be due to the low solubility of acacetin 7-methyl ether in assay buffer medium at higher concentrations. Even though the potency of inhibition of human MAO-B by acacetin 7-methyl ether was about four-fold lower when compared to the standard drug deprenyl (a selective MAO-B inhibitor), its selectivity for MAO-B was higher (>500 fold) as compared to deprenyl (450 fold). Deprenyl (available with the trade name selegiline) is a clinically used drug for the treatment of Parkinson's disease and major depressive disorder [17]. Other constituents that were isolated from *T. diffusa* only showed moderate inhibition of MAO-A and MAO-B (IC_{50} in the range of 13–61 µM), with no significant selectivity towards MAO-B or MAO-A (Table 1). The studies were extended to investigate the kinetics of inhibition of MAO-B by acacetin 7-methyl ether. Further, characteristics of the interaction and the putative binding mode of acacetin 7-methyl ether were also investigated using computational docking of the ligand to the X-ray crystal structures of MAO-A and MAO-B.

Table 1. Inhibition of recombinant human Monoamine Amine Oxidase-A and -B by constituents from *T. diffusa*.

Compound	MAO-A IC_{50} (µM) *	MAO-B IC_{50} (µM) *	SI MAO-A/-B
Acacetin 7-methyl ether	>100.00	0.198 ± 0.001	>505.051
Vetulin	18.799 ± 0.291	0.447 ± 0.010	42.056
Apigenin-7-O-β-D-(6-O-p-coumaroyl) glucoside	22.508 ± 4.440	22.001 ± 1.759	1.023
Echinaticin	21.830 ± 4.367	13.404 ± 0.148	1.703
Tetraphyllin B	-	-	-
Tricin-7-glucoside	50.901 ± 0.506	27.444 ± 0.819	1.854
Diffusavone	19.091 ± 1.450	14.518 ± 0.214	1.314
Turneradiffusin	61.427 ± 3.568	37.250 ± 2.933	1.649
Rhamnosylorientin	34.909 ± 3.887	25.541 ± 2.020	1.366
Rhamnosylvitexin	40.940 ± 6.810	26.286 ± 0.277	1.557
Turneradin	19.169 ± 0.802	13.027 ± 2.142	1.471
Acacetin	0.115 ± 0.004	0.050 ± 0.0025	2.30
Clorgyline	0.0052 ± 0.0001	2.30 ± 0.0570	-
Deprenyl	23.00 ± 1.00	0.051 ± 0.002	450.980
Safinamide [#] (ref)	90.00 ± 2.470	0.060 ± 0.005	1500.00

* The IC_{50} values, computed from the dose response inhibition curves, are mean ± S.D. of at least triplicate observations. SI—Selectivity Index-IC_{50} MAO-A/IC_{50} MAO-B inhibition. [#] Safinamide data from ref. [18].

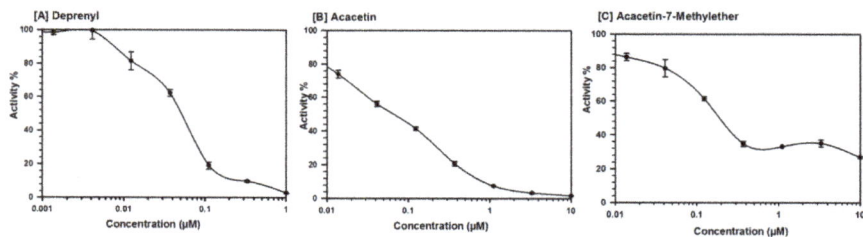

Figure 2. Concentration-dependent inhibition of recombinant human MAO-B by (**A**) deprenyl, (**B**) acacetin, and (**C**) acacetin 7-methyl ether. Each point shows mean ± SD of three observations.

2.2. Enzyme Kinetics and Mechanism Studies

Acacetin 7-methyl ether was tested against MAO-B at varying concentrations of kynuramine, which is a nonselective substrate, to investigate the nature of inhibition of the enzyme. Based on dose-response inhibition, five concentrations of acacetin 7-methyl ether were selected, two below the IC_{50} value (100 and 150 nM), one around the IC_{50} value (200 nM), and two above the IC_{50} value (300 nM and 500 nM) for the inhibition kinetics experiments. The assays were run at varying concentrations of the substrate and fixed concentrations of the inhibitor. The Ki (i.e., inhibition/binding affinity) values were computed with double reciprocal Lineweaver–Burk plots The binding of acacetin 7-methyl ether to human MAO-B affected the K_m value (i.e., the affinity of the substrate for the enzyme) without much effects on the V_{max} (maximum enzyme activity), indicating that the inhibition of MAO-B by acacetin 7-methyl ether was competitive (Figure 3). Acacetin 7-methyl ether inhibited the enzymatic activity of MAO-B with considerably high affinity (Ki = 45 nM) (Table 2).

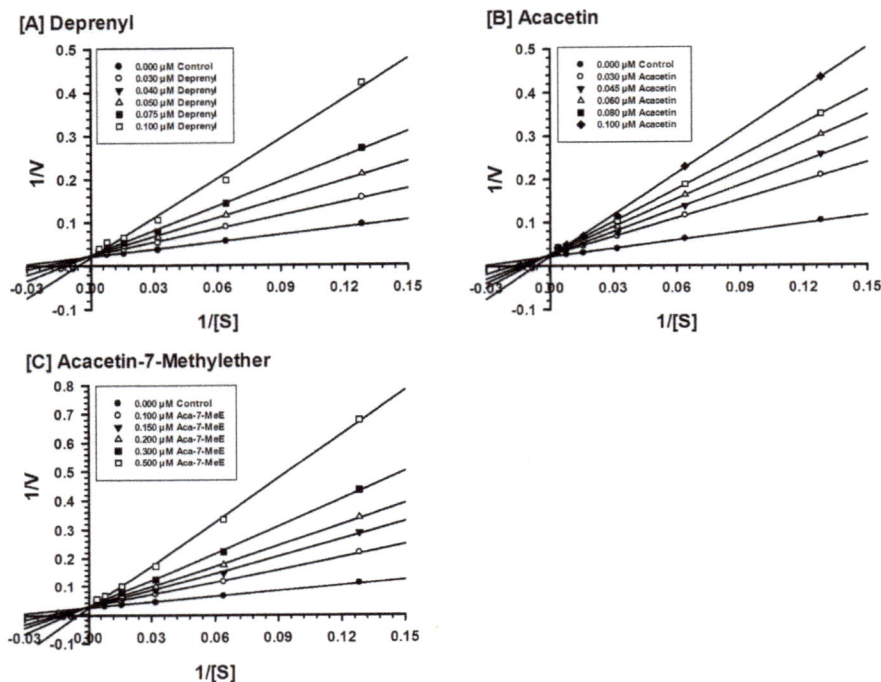

Figure 3. Kinetics characteristics of inhibition of recombinant human MAO-B by (**A**) deprenyl, (**B**) acacetin and (**C**) acacetin 7-methyl ether. Each point shows the mean value of three observations.

Table 2. Inhibition/binding affinity constants (Ki) for inhibition of recombinant human MAO-A and MAO-B by acacetin, acacetin 7-methyl ether, and deprenyl.

Compounds	Monoamine Oxidase-B	
	Ki (nM) *	Type of Inhibition
Acacetin 7-methyl ether	45.0 ± 3.0	Competitive/Partially Reversible
Acacetin	36 ± 4.0	Competitive/Reversible
Deprenyl	29 ± 6.3	Mixed/Irreversible

* Values are mean ± S.D. of triplicate experiments.

2.3. Analysis of Reversibility of Binding of Inhibitor

The characteristics of binding of acacetin 7-methyl ether to MAO-B were also investigated using an equilibrium-dialysis assay. The inhibitor acacetin 7-methyl ether was incubated at1.0 and 2.0 µM) with the MAO-B enzyme for 20 min and the resulting enzyme-inhibitor complex preparation was dialyzed overnight in phosphate buffer. The activity of the enzyme was analyzed before and after the dialysis (Figure 4). The incubation of MAO-B with 1.0 and 2.0 µM of acacetin 7-methyl ether caused >60% inhibition of activity and only ~80% activity of the enzyme was recovered after equilibrium dialysis. Thus, the binding of acacetin 7-methyl ether to MAO-B was partially reversible. The binding of selective MAO-B inhibitor deprenyl was confirmed to be irreversible (Figure 4).

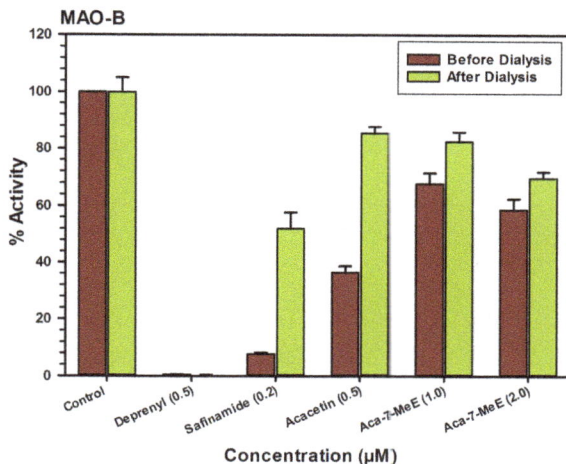

Figure 4. Reversibility assay of recombinant human MAO-B with acacetin (0.5 μM), acacetin 7-methyl ether (Aca 7-MeE) (1.0 and 2.0 μM), deprenyl (0.5 μM), and safinamide (0.2 μM). The remaining activity was expressed as % of activity. Each point shows the mean ± S.D. value of three observations.

2.4. Time-Dependent Assay for Enzyme Inhibition

The time dependence of binding of acacetin 7-methyl ether to MAO-B was analyzed. The recombinant enzyme was incubated for different times with the test compounds, namely, deprenyl, (0.100 μM), acacetin 7-methyl ether (0.500 μM), and acacetin (C.100 μM) (Figure 5). The control without inhibitors was also run simultaneously. The activity of the enzyme was determined, as described above, and the percentage of enzyme activity remaining was plotted against the pre-incubation time to determine the time-dependence of inhibition. The binding/inhibition of MAO-B by acacetin 7-methyl ether was not dependent on the pre-incubation time (Figure 5).

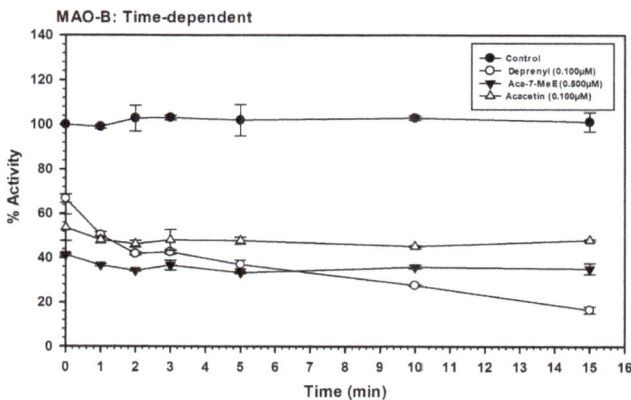

Figure 5. Time-dependent inhibition of recombinant human MAO-B by deprenyl (0.100 μM), acacetin 7-methyl ether (Aca-7-MeE) (0.500 μM), and acacetin (0.100 μM. Each point represents mean ± S.D. of triplicate values.

2.5. Computational Docking Study of Acacetin 7-Methyl Ether

Selective interactions of acacetin 7-methyl ether were investigated by molecular docking to understand its binding pose to *h*MAO-A and *h*MAO-B. Three-dimensional (3D) protein-ligand interactions of acacetin (Figure 6A,C, magenta, stick model) and acacetin 7-methyl ether (Figure 6B,D, cyan, stick model), respectively, with the X-ray crystal structures of MAO-A and MAO-B are presented in Figure 6. The docking scores and binding free-energies of acacetin 7-methyl-ether and acacetin at the active sites of the *h*MAO-A and *h*MAO-B X-ray crystal structures are shown in Table 3 and their corresponding putative binding poses are shown in Figure 6. The docking protocol was validated by self-docking, in which the native ligands, harmine, and pioglitazone, from the X-ray structures were docked into their corresponding protein's structures of MAO-A and MAO-B, respectively. The calculated RMSD between the docked and experimental poses were <0.6 Å, which verified that the docking protocol was appropriate for this use.

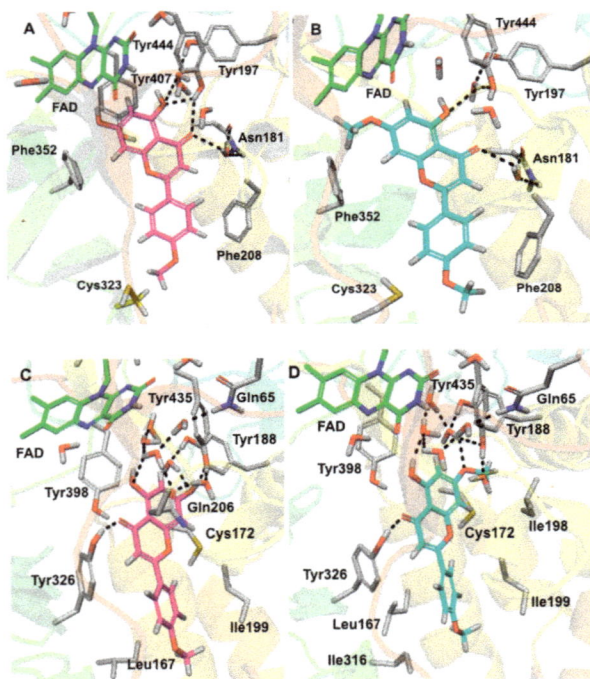

Figure 6. Three-dimensional (3D) protein-ligand interactions of acacetin (C magenta, stick model) and acacetin 7-methyl ether (C cyan, stick model) with the X-ray crystal structures of MAO-A and MAO-B. (**A**) Acacetin with MAO-A, (**B**) acacetin 7-methyl ether with MAO-A, (**C**) acacetin with MAO-B, and (**D**) acacetin 7-methyl ether with MAO-B. FAD (C green, stick model), some crystallographic waters (O red, H gray, stick model), and the important residues of MAO-A and MAO-B (C gray) are also shown. The black dashed lines represent H-bonding.

Table 3. Glide docking scores and binding free energies of acacetin and acacetin 7-methyl ether to MAO-A and MAO-B.

Compound	IC$_{50}$ (μM) *		Glide Docking Score (kcal/mol)		Binding Free-Energies (kcal/mol)	
	MAO-A	MAO-B	MAO-A	MAO-B	MAO-A	MAO-B
Acacetin	0.115	0.049	−10.685	−11.890	−56.303	−67.205
Acacetin 7-methyl ether	>100	0.198	−9.085	−10.708	−31.791	−67.494

* Values are mean ± S.D. of triplicate experiments.

According to the docking and binding free-energy results, acacetin 7-methyl ether showed better binding affinity (Docking score = −10.708 kcal/mol, ΔG = −67.494 kcal/mol) to *h*MAO-B than *h*MAO-A (Docking score = −9.085 kcal/mol, ΔG = −31.791 kcal/mol). The experimental data also supported this (Table 3). The docking results revealed that the binding orientation of acacetin 7-methyl ether and acacetin were similar to the orientations of the native ligands of the X-ray crystal structures of MAO-A (PDB ID: 2Z5X) and MAO-B (PDB ID: 4A79). The computational docking studies provided further insights into selective interactions of acacetin-7-methyl ether with the human MAO-B over MAO-A. Acacetin 7-methyl ether tightly binds to MAO-B by forming π-π stacking and H-bonding (at its C4 carbonyl) with Tyr326 and strong hydrophobic interactions with nearby amino acid residues. Ile199 and Ile316, which are known as critical residues for MAO-B selectivity [19], showed strong hydrophobic interaction with the acacetin 7-methyl ether. The additional methoxy moiety at the C7 position of acacetin 7-methyl ether compared to acacetin participated in the hydrophobic interactions with the Ile198. However, for MAO-A, this additional methoxy moiety exhibited bad contact with Phe352 and Lys305 and this may be the reason for the poorer binding with MAO-A and the selectivity of acacetin-7-methyl ether for MAO-B over MAO-A. Additionally, the conserved water molecules within the enzyme active site were important for enhancing the interaction of acacetin-7-methyl ether with MAO-B. Water-mediated hydrogen bonding was found between acacetin-7-methyl ether and Tyr188, Tyr435, and Gln65, which was similar to what was found for acacetin binding with MAO-B (Figure 6C). We also observed water-mediated H-bonding with the carbonyl of Cys172 and the oxygen at the C7 position of acacetin and acacetin-7-methyl ether (the interaction is not shown in Figure 6). A previous study on the interaction of acacetin with Cys172 reported direct (not water-mediated) interaction with Cys172 [11]. Interestingly, in the docked pose of acacetin with MAO-B that was reported in that study, the chromene moiety was flipped, with its carbonyl pointed towards Cys172. By contrast, in the docked pose that was determined in the present study, the carbonyl was pointed away from Cys172. To help determine which docked pose is more likely correct, we studied the available structures in the Protein Data Bank and found the X-ray crystal structure of MAO-B with co-crystalized ligand dimethylphenyl-chromone-carboxamide (PDB ID: 6FVZ) to be helpful. In 6FVZ, two alternate rotamers of Cys172 (side chain dihedral angle chi1 = −158° or −86°), each with an average occupancy of 50%, were included in the reported structure. The different values for the side chain dihedral angle of Cys172 open up the possibility of a direct H-bond interaction between Cys172 and acacetin or acacetin-7-methyl ether. The docked pose of acacetin that was reported in this study adopted a similar orientation for its chromene moiety (with the chromene oxygen being located close to Cys172) to that of the chromene moiety of the co-crystallized ligand of 6FVZ. These properties support the reliability of the docked pose that is presented here.

2.6. Molecular Dynamics Study of Acacetin 7-Methyl Ether and MAO-B

Docking alone cannot provide full insight into the binding mode and dynamics of acacetin 7-methyl ether within *h*MAO-B. Therefore, we carried out a 10 ns MD simulation of the complex of acacetin 7-methyl ether with MAO-B using Desmond software [20]. The MD simulation suggests that acacetin-7-methyl ether fits tightly into the binding pocket of MAO-B, since there was very little deviation in the ligand Root Mean Square Deviation (RMSD) (Figure 7) during the simulation.

No significant flexibility (as measured by Root Mean Square (RMS) fluctuation, RMSF) was observed in the secondary structure elements (SSE) (α-helices and β-strand: Helix = 28.63%, Strand = 15.78%, and Total SSE = 44.40%) of the protein model, with major fluctuations only in residues 490–520 and the loop regions (Figure 8). The %SSE remained close to 45% throughout the simulation. The amino acids that interacted with acacetin-7-methyl ether did not show any major fluctuations in their RMSF values (Figure 8). The interaction histograms (Figure 9A) and protein-ligand contact graphs (Figure 9B) were analyzed to study the time-dependent changes in the interaction of the ligands with key residues of the protein. The interaction histogram shows that crucial amino acids for interaction with acacetin-7-methyl ether are Leu171 (hydrophobic and water bridges), Cys172 (hydrogen bond, hydrophobic, and water bridges), Tyr188 (hydrophobic and water bridges), Ile199 (hydrophobic), Gln206 (hydrogen bond and water bridges), Tyr326 (hydrophobic and water bridges), Tyr398 (water bridges), and Tyr435 (hydrogen bond, hydrophobic, and water bridges) (Figure 9A). The stabilization of the complex can be mainly attributed to several H-bond contacts and π-π interactions. H-bond contacts were observed between oxygen at the C7 position of acacetin-7-methyl ether and Cys172 and between the C5 OH and Tyr435.

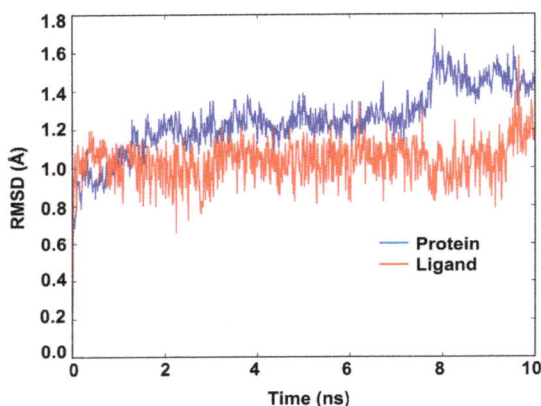

Figure 7. Root Mean Square Deviation (RMSD) plot of atom locations vs. simulation time of MAO-B (protein) and acacetin 7-methyl ether (ligand) for the molecular dynamics (MD) simulation of their interaction complex.

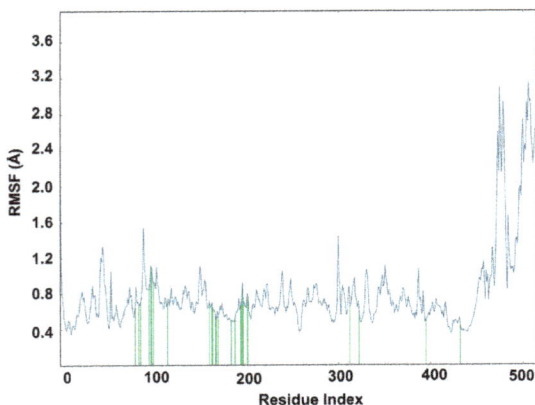

Figure 8. The Root Mean Square Fluctuation (RMSF) plot based on Cα atoms of MAO-B. Protein residues that interact with the acacetin 7-methyl ether is marked with green vertical bars.

Strong π-π interactions were observed between Tyr326 and Ring B and Ring C of acacetin 7-methyl ether. In addition, the protein-ligand interaction diagram (Figure 9B) indicates that acacetin 7-methyl ether forms H-bond interactions with Cys172 (70% contribution), Gln206 (24% contribution), and Tyr435 (30% contribution), and water-mediated hydrogen bond interactions with Leu171 (32% contribution) throughout the 10 ns MD simulations. The contribution of π-π interactions with Tyr326 was found to be 72% (with Ring C) and 24% (with Ring B). In conclusion, the interaction histograms and contact graphs show that acacetin 7-methyl ether forms a significantly stable complex with MAO-B.

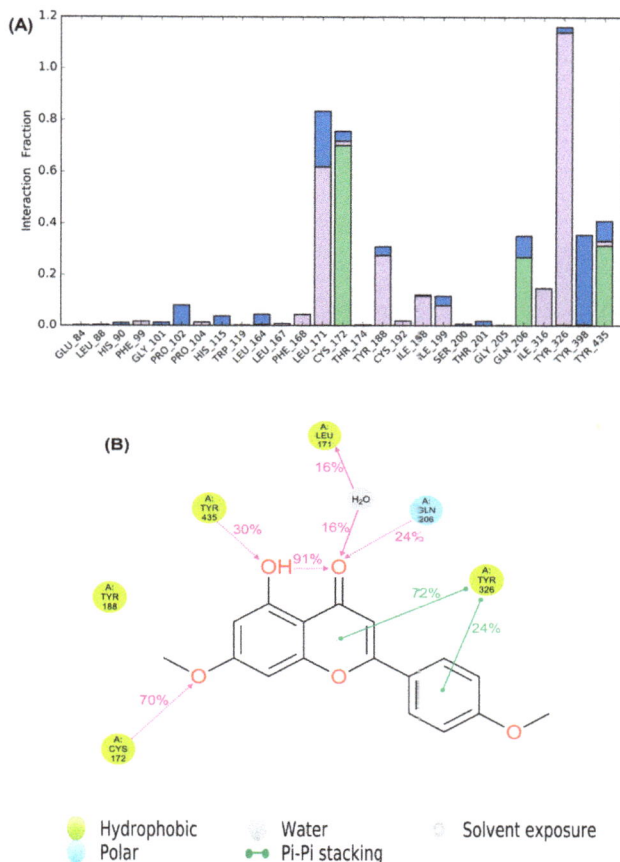

Figure 9. SID (Simulation Interactions Diagram) plots showing the protein-ligand interactions between the amino acid residues of the *h*MAO-B binding site and acacetin 7-methyl ether during the MD simulation. (**A**) The stacked bar charts are categorized as follows: hydrogen bonding (green), hydrophobic interactions (violet), and water bridges (blue) formed. (**B**) A schematic of detailed ligand atom interactions with the protein residues.

3. Discussion

Acacetin 7-methyl ether (Figure 1) was identified as a potent and highly selective MAO-B inhibitory constituent of *Turnera diffusa*. Acacetin 7-methyl ether has also been isolated from a few other plants and showed antidiabetic, antiproliferative, and antioxidative activities [21–23]. MAO inhibitors, as well as oxidative stress reducers, have significantly contributed in the treatment of neurodegenerative disorders, including Parkinson's disease [24]. However, MAO inhibitory action of acacetin 7-methyl ether has not been reported earlier.

The selective MAO-B inhibitors are used alone as well as in combination with carbidopa-levodopa for the treatment of motor symptoms of Parkinson's disease [25]. Reduction in dopamine levels is responsible for the motor symptoms of Parkinson's disease. The neurons that produce dopamine are damaged during the progression of Parkinson's disease. Treatment with levodopa provides the precursor for the biosynthesis of dopamine. The addition of a MAO-B inhibitor to levodopa therapy results in a slowing of the breakdown of levodopa and dopamine in the brain, and it may boost the effect of levodopa. The current battery of clinically used MAO-B inhibitors includes selegiline (L-deprenyl), rasagiline, and safinamide. Selegiline and rasagiline are irreversible inhibitors of MAO and they may be associated with significant side effects [26,27]. Both acacetin, reported earlier, [10] and acacetin 7-methyl ether reported here show the potent inhibition of MAO-B. However, acacetin 7-methyl ether showed >500-fold selectivity, compared to only two-fold selectivity of acacetin for MAO-B over MAO-A. The 7-methyl substitution marginally reduces the potency of MAO-B inhibition, relative to acacetin (IC_{50} 198 nM vs. 50 nM, respectively), but it dramatically improves the selectivity (>500) for MAO-B. This suggests a significant scope for further optimization of the 5, 7-dihydroxy-4′-methoxyflavone pharmacophore to achieve high potency and selectivity against the human MAO-B. 5, 7-Dihydroxy-4′-methoxyflavones represent a new class of selective reversible MAO-B inhibitors with potential therapeutic application for treatment of neurodegenerative disorders, including Parkinson's disease in combination with standard drugs [28]. A recent study has also reported MAO-A and MAO-B inhibition activity by acacetin 7-O-(6-O-malonylglucoside), a derivative of acacetin that was isolated from *Agastache rugosa* plant leaves [11]. The reversibility of MAO-B inhibition with acacetin 7-methyl ether adds significant value to its potential therapeutic use. A recent pharmacokinetics study of acacetin metabolism in vitro and in vivo that analysed its metabolites showed that acacetin is metabolized by the sulfotransferase 1A1 enzyme into acacetin-7-sulfate, which was detected in vivo in plasma samples and also in vitro from incubation of the Liver S9 fraction [29].

Further, computational analysis of the binding of acacetin 7-methyl ether to human MAO-A and MAO-B confirmed its selective and stronger interaction with MAO-B when compared to MAO-A. Both acacetin and acacetin 7-methyl ether interact with Gln65, Cys172, Tyr188, and Tyr435 in human MAO-B. However, less prominent interactions of acacetin 7-methyl ether with Tyr197 and Tyr407 in MAO-A as compared to acacetin afforded the higher inhibition of MAO-A with acacetin (IC_{50} = 0.115 μM) than with acacetin 7-methyl ether (IC_{50} = >100 μM). To shed light on the interaction profile, binding mechanism, and dynamic properties of acacetin 7-methyl ether with MAO-B, we carried out a 10 ns MD simulation of their protein-ligand complex. The structural and dynamical properties of acacetin 7-methyl ether were analyzed using RMSD plots of the ligand and of the Cα atoms of the protein (Figure 7), and by RMSF (Figure 8) plots of protein that was bound to the ligand. The RMSD plot of atom locations vs. simulation time indicated that the protein and ligand maintained a significantly stable state throughout the simulation (Figure 7). The most significant deviation in RMSD was a result of the flexible motion of the modeled C-terminal residues (residues 498–520) (not included in Figure 7). The MD simulations revealed that Leu171, Cys172, Tyr188, Ile199, Gln206, Tyr326, Tyr398, and Tyr435 are crucial amino acids for the interaction of acacetin 7-methyl ether and MAO-B. These results show us that acacetin 7-methyl ether forms a stable and strong complex with MAO-B.

These studies support the further in vivo evaluation of acacetin, when considering the MAO inhibitory activities that were reported earlier [10] and those of acacetin 7-methyl ether reported herein, in experimental animal models for neurological and neurodegenerative diseases.

4. Materials and Methods

4.1. Enzymes and Chemicals

Recombinant human monoamine oxidase-A and monoamine oxidase-B enzymes were obtained from the BD Biosciences (Bedford, MA, USA). Kynuramine, clorgyline, deprenyl, acacetin, and DMSO were purchased from Millipore Sigma (St. Louis, MO, USA). Safinamide was purchased from TCI Chemicals, Portland, OR, USA.

Acacetin 7-methyl ether (purity >98%), vetulin (>97%), apigenin-7-*O*-β-D-(6-*O*-pcoumaroyl) glucoside (>97%), echinaticin (>96%), tetraphyllin B (>98%), tricin-7-glucoside (>98%), diffusavone (>97%), turneradiffusin (>96%), rhamnosylorientin (>97%), rhamnosylvitexin (>95%), turneradin (>95%), and acacetin (>96%) were isolated from *T. diffusa* at the National Center for Natural Products Research (NCNPR), University of Mississippi, University, MS, USA. Their identities and purities were confirmed by chromatographic (TLC, HPLC) and spectroscopic (IR, one-dimensional (1D)- and two-dimensional (2D)-NMR, HR-ESI-MS) methods, as well as by comparison with the published spectroscopic data [14,15].

4.2. MAOs Inhibition Assay

To investigate the effect of the isolated constituents from *T. diffusa* on recombinant human MAO-A and MAO-B, the kynuramine oxidative deamination assay was performed in 384-well plates, as previously reported, with minor modifications [30,31]. A single fixed concentration of kynuramine substrate and varying concentrations of inhibitor were used to test enzyme inhibition and determine the IC$_{50}$ values. Kynuramine concentrations for MAO-A and MAO-B were 80 µM and 50 µM, respectively. These concentrations of kynuramine were twice the apparent K$_M$ value for substrate binding [10,31]. The concentrations that were tested for pure constituents were from 0.01 to 100 µM, for MAO-A and MAO-B inhibition assays. Reactions were performed in a clear 384-well microplate (50 µL) with 0.1 M potassium phosphate buffer, pH 7.4. The inhibitors and compounds were dissolved in DMSO, diluted in the buffer solution, and pre-incubated at 37 °C for 10 min (no more than 1.0% DMSO final concentration). The reactions were initiated by the addition of 12.50 µL MAO-A (to 5 µg/mL) or MAO-B (to 12.5 µg/mL), incubated for 20 min at 37 °C, and then terminated by the addition of 18.8 µL of 2N NaOH. The enzyme product 4-hydroxyquinoline formation was recorded fluorometrically using a SpectraMax M5 fluorescence plate reader (Molecular Devices, Sunnyvale, CA, USA) with an excitation (320 nm) and emission (380 nm) wavelength, using the SoftMaxPro 6.0 program. The inhibition effects of enzyme activity were calculated as a percent of product formation when compared to the corresponding control (enzyme-substrate reaction) without inhibitors. Controls, including samples in which the enzyme or the substrate was added after stopping the reactions, were simultaneously checked to determine the interference of inherent fluorescence of the test compounds with the measurements. IC$_{50}$ values for MAO-A and MAO-B inhibition were calculated from the concentration dependent inhibition curves using XLfit, a Microsoft Excel-based plug-in which performs Regression, curve fitting and statistical analysis (IDBS, Bridgewater, NJ, USA).

4.3. Determination of IC$_{50}$ Values

To determine the IC$_{50}$ values for the inhibition of MAO-A and MAO-B by acacetin 7-methyl ether, the enzyme assay was performed at a fixed concentration of the substrate kynuramine for MAO-A (80 µM) and for MAO-B (50 mM) and with varying concentrations of inhibitor/test compounds for MAO-A and MAO-B (0.01 µM to 100 µM). The dose-response curves were generated using Microsoft® Excel and the IC$_{50}$ values were analyzed using XLfit software.

4.4. Enzyme Kinetics and Mechanism Studies

For the determination of the binding affinity of the inhibitor (Ki) with MAO-A and MAO-B, the enzyme assays were carried out at different concentrations of kynuramine substrate (1.90 µM to 500 μM) and varying concentrations of the inhibitors/compound. Compounds acacetin and acacetin 7-methyl ether were tested at 0.030–0.100 μM and 0.100–0.500 μM, respectively, for MAO-B to determine the Km and Vmax values of the enzymes in the presence of the inhibitor. The controls without inhibitor were also simultaneously run. The results were analyzed by standard double reciprocal Line–Weaver Burk plots for computing Km and Vmax values, which were further analyzed to determine Ki values [10].

4.5. Analysis of Reversibility of Binding of Inhibitor

The inhibitors bind with the target enzyme through the formation of an enzyme-inhibitor complex. The formation of the enzyme-inhibitor complex may be accelerated in the presence of the high concentration of the inhibitor. The reversibility of binding of MAO inhibitory compound acacetin 7-methyl ether was determined by the formation of the complex by incubating the enzyme with a high concentration of the inhibitor, followed by extensive dialysis of the enzyme-inhibitor complex and the recovery of catalytic activity of the enzymes. The MAO-B (0.2 mg/mL protein) enzyme was incubated with each inhibitor: acacetin (0.5 μM), acacetin 7-methyl ether (1.0, 2.0 μM), deprenyl (0.5 μM), and safinamide (0.2 μM) in a total volume of 1 mL, 100 mM potassium phosphate buffer (pH 7.4). After 20 min incubation at 37 °C, the reaction was stopped by chilling in an ice bath. All of the samples were dialyzed against potassium phosphate buffer (25 mM; pH 7.4) at 4 °C for 14–18 h (three buffer changes). Control enzyme (without inhibitor) was also run through the same procedure and the activity of the enzyme was determined before and after the dialysis [18].

4.6. Time-Dependent Assay for Enzyme Inhibition

To analyze whether the binding of the inhibitor with MAO-B was time-dependent, the enzyme was pre-incubated for different time periods (0–15 min) with the inhibitors at a concentration that produces approximately 70–80% inhibition. The inhibitor concentrations used to test time-dependent inhibition were acacetin (0.1 μM), acacetin 7-methyl ether (0.5 μM), and deprenyl (0.1 μM) with MAO-B (12.5 μg/mL). The controls without inhibitors were also simultaneously run. The activities of the enzymes were determined, as described above, and the percentage of enzyme activity remaining was plotted against the pre-incubation time to determine the time-dependence of inhibition.

4.7. Computational Methods

The X-ray crystal structures of MAO-A (PDB accession number: 2Z5X [32]) and MAO-B (PDB accession number: 4A79 [33]) were downloaded from the Protein Data Bank website. These proteins were prepared by adding hydrogens, adjusting bond orders and proper ionization states, and refining overlapping atoms. The water molecules beyond 5 Å from the co-crystalized ligands were removed and the ligand states were generated using Epik at pH 7.4. During the refinement process, the water molecules with less than two H-bonds to non-waters were also removed. Acacetin and acacetin-7-methyl ether were sketched in Maestro [34], prepared, and energy-minimized at a physiological pH of 7.4 using the LigPrep [35] module of the Schrödinger software (Cambridge, MA, USA) [36]. Acacetin was used as a positive control for the docking studies. For protein and ligand preparation, we used the Optimized Potentials for Liquid Simulations 3 (OPLS3) force field. The active sites of the MAO-A and MAO-B proteins were each defined by the centroid of the co-crystallized ligands that were present in 2Z5X and 4A79, respectively. We did not remove cofactor FAD during protein preparation and docking. Acacetin and acacetin 7-methyl ether were docked using the Induced Fit docking [37] protocol, applying the standard-precision (SP) docking method. The top 10 poses were kept for analysis. The best docking poses were subjected to binding free-energy calculations using the Prime MM-GBSA module of Schrödinger software.

4.8. Molecular Dynamics (MD) Simulations

Monoamine oxidase B consists of a two-domain molecular architecture. Each identical monomer consists of 520 amino acids. In this study, we used one monomer (chain A) for the docking and MD simulations. It is reported [38] that the C-terminal helix (residues 498–520) of MAO-B is located in the outer membrane of mitochondria. The MAO-B X-ray crystal structure that we used (PDB ID: 4A79) contains only 501 residues; therefore, we modeled its missing residues that were known to be embedded in lipid bilayer using the Maestro molecular modeling suite [39]. The Protein Preparation Wizard [40] of Maestro was used to prepare the protein structure. We used the Desmond [20] Molecular Dynamics System, ver. 3.6 (Schrödinger) with the OPLS-3 force field and RESPA [41] integrator to perform a 10 ns MD simulation. The best scoring pose of acacetin-7-methyl ether in complex with MAO-B (PDB ID: 4A79) was selected and the C-terminal residues of the protein-ligand complex between amino acids 498 and 520 were embedded into a pre-equilibrated 1-palmitoyl-2-oleoyl-sn-glycero-3-phosphocholine (POPC) membrane and the rest of the system was solvated with TIP3P [42] explicit waters. The whole system was neutralized using sodium chloride (NaCl) and it was set to an ionic strength of 0.15 M. The buffer dimensions of the orthorhombic simulation box were set to $30 \times 30 \times 70$ Å^3. The solvated system was energy-minimized with the DESMOND minimization algorithm for 2000 iterations when considering a convergence threshold of 1.0 kcal/mol/Å. The constructed system was simulated with the relaxation protocol in Desmond [20]. The protocol involved an initial minimization of the solvent, while keeping restraints on the solute, followed by short MD simulations, including the following steps: (1) Simulation (100 ps) using Brownian dynamics, in the NVT ensemble at 10 K with solute heavy atoms restrained; (2) Simulation (12 ps) in the NVT ensemble using a Berendsen thermostat (10 K) with solute heavy atoms restrained; (3) Simulation (12 ps) in the NPT ensemble using a Berendsen thermostat (10 K) and a Berendsen barostat (1 atm) with non-hydrogen solute atoms restrained; (4) Simulation (12 ps) in the NPT ensemble using a Berendsen thermostat (300 K) and a Berendsen barostat (1 atm) with non-hydrogen solute atoms restrained; and, (5) Simulation (24 ps) in the NPT ensemble using a Berendsen thermostat (300 K) and a Berendsen barostat (1 atm) with no restraints. The production run was carried out using an NPT ensemble at 300 K with Nosé–Hoover temperature coupling [43] and at a constant pressure of 1.01 bar via Martyna–Tobias–Klein pressure coupling [44]. The simulation trajectories (frames) were sampled at intervals of 4.8 ps. We used a timestep of 2.0 fs. The resulting trajectory was analysed using the Simulation Interactions Diagram (SID) utility of Desmond [20]. Finally, the PyMOL 1.4.1 and Maestro 11.5.011 molecular graphics systems were used to visualize the protein-ligand complex and to generate all of the figures.

5. Conclusions

Biological screening of constituents of *Turnera diffusa* (damiana) identified three O-methyl flavones, namely, acacetin, acacetin 7-methyl ether, and vetulin as selective inhibitors of human MAO-B. Acacetin 7-methyl ether was a highly potent and selective MAO-B inhibitor with >500-fold selectivity towards MAO-B when compared to MAO-A. Further studies suggested that acacetin 7-methyl ether is a reversible competitive inhibitor of MAO-B. Computational docking analysis confirmed the selective binding of acacetin 7-methyl ether to human MAO-B as compared to MAO-A. Acacetin 7-methyl ether may have a potential therapeutic application for the treatment of neurodegenerative disorders, including Parkinson's disease.

Author Contributions: I.M. and B.L.T. conceived these studies. J.Z. performed extraction, isolation, purification and identification of constituents. N.D.C. and B.L.T. planned the experiments. N.D.C. performed MAO-inhibition assays. P.P. and R.J.D. planned and performed the computational studies. N.D.C., B.L.T. and I.M. performed the MAO-inhibition data analysis. All authors contributed to writing the draft and review of the final draft of the manuscript.

Funding: This work was supported by the USDA-ARS Specific Cooperative Agreement No. 58-6408-1-603. This research was funded by grant numbers P20GM104932 from the National Institute of General Medical Sciences (NIGMS), a component of the National Institutes of Health (NIH). Its contents are solely the responsibility of the authors and do not necessarily represent the official view of NIGMS or NIH. This investigation was conducted

Molecules **2019**, *24*, 810

in part in a facility constructed with support from the Research Facilities Improvements Program (C06RR14503) from the National Institutes of Health (NIH) National Center for Research Resources.

Acknowledgments: The authors are thankful to the National Center for Natural Products Research for the facilities to support this work.

Conflicts of Interest: The authors declare no conflict of interest.

References

1. Shih, J.C.; Chen, K.; Ridd, M.J. Monoamine oxidase: From genes to behavior. *Annu. Rev. Neurosci.* **1999**, *22*, 197–217. [CrossRef] [PubMed]
2. Abell, C.W.; Kwan, S.W. Molecular characterization of monoamine oxidase A and B. *Prog. Nucleic Acids Res. Mol. Biol.* **2001**, *65*, 129–156.
3. Cesura, A.M.; Pletscher, A. The New Generation of Monoamine Oxidase Inhibitors. *Prog. Drug Res.* **1992**, *38*, 171–297. [PubMed]
4. Yamada, M.; Yasuhara, H. Clinical pharmacology of MAO inhibitors: Safety and future. *Neurotoxicology* **2004**, *25*, 215–221. [CrossRef]
5. Yudim, M.B.; Fridkin, M.; Zheng, H. Novel bifunctional drugs targeting monoamine oxidase inhibition and iron chelation as an approach to neuroprotection in Parkinson's disease and other neurodegenerative diseases. *J. Neural Transm.* **2004**, *111*, 1455–1471. [CrossRef] [PubMed]
6. Youdim, M.B.; Edmondson, D.; Tipton, K.F. The therapeutic potential of monoamine oxidase inhibitors. *Nat. Rev. Neurosci.* **2006**, *7*, 295–309. [CrossRef] [PubMed]
7. Youdim, M.B.; Bakhle, Y.S. Monoamine oxidase: Isoforms and inhibitors in Parkinson's disease and depressive illness. *Br. J. Pharmacol.* **2006**, *147*, 287–296. [CrossRef] [PubMed]
8. Herraiz, T.; Chaparro, C. Human monoamine oxidase enzyme inhibition by coffee and beta-carbolines norharman and harman isolated from coffee. *Life Sci.* **2006**, *78*, 795–802. [CrossRef] [PubMed]
9. Borroni, E.; Bohrmann, B.; Grueninger, F.; Prinssen, E.; Nave, S.; Loetscher, H.; Chinta, S.J.; Rajagopalan, S.; Rane, A.; Siddiqui, A.; et al. Sembragiline: A Novel, Selective Monoamine Oxidase Type B Inhibitor for the Treatment of Alzheimer's Disease. *J. Pharmacol. Exp. Ther.* **2017**, *362*, 413–423. [CrossRef] [PubMed]
10. Chaurasiya, N.D.; Gogineni, V.; Elokely, K.M.; Leon, F.; Nunez, M.J.; Klein, M.L. Isolation of Acacetin from *Calea urticifolia* with Inhibitory Properties against Human Monoamine Oxidase-A and -B. *J. Nat. Prod.* **2016**, *79*, 2538–2544. [CrossRef] [PubMed]
11. Lee, H.W.; Ryu, H.W.; Baek, S.C.; Kang, M.G.; Park, O.D.; Han, H.Y.; An, J.H.; Oh, S.R.; Kim, H. Potent inhibitions of monoamine oxidase A and B by acacetin and its 7-*O*-(6-*O*-malonylglucoside) derivative from *Agastache rugosa*. *Int. Biol. Macromol.* **2017**, *104*, 547–553. [CrossRef] [PubMed]
12. Szewczyk, K.; Zidorn, C. Ethnobotany, phytochemistry, and bioactivity of the genus Turnera (Passifloraceae) with a focus on damiana—*Turnera diffusa*. *J. Ethnopharm.* **2014**, *152*, 424–443. [CrossRef] [PubMed]
13. Otsuka, R.D.; Ghilardi Lago, J.H.; Rossi, L.; Fernandez Galduróz, J.C.; Rodrigues, E. Psychoactive plants described in a Brazilian literary work and their chemical compounds. *Cent. Nerv. Syst. Agents Med. Chem.* **2010**, *10*, 218–237. [CrossRef] [PubMed]
14. Zhao, J.; Pawar, R.S.; Ali, Z.; Khan, I.A. Phytochemical investigation of *Turnera diffusa*. *J. Nat. Prod.* **2007**, *70*, 289–292. [CrossRef] [PubMed]
15. Zhao, J.; Dasmahapatra, A.K.; Khan, S.I.; Khan, I.A. Anti-aromatase activity of the constituents from damiana (*Turnera diffusa*). *J. Ethnopharm.* **2008**, *120*, 387–393. [CrossRef] [PubMed]
16. Jäger, A.K.; Saaby, L. Flavonoids and the CNS. *Molecules* **2011**, *16*, 1471–1485.
17. Moore, J.J.; Saadabadi, A. *Selegiline*; Pearls Publishing LLC: Treasure Island, FL, USA, 2018.
18. Pandey, P.; Chaurasiya, N.D.; Tekwani, B.L.; Doerksen, R.J. Interactions of endocannabinoid virodhamine and related analogs with human monoamine oxidase-A and-B. *Biochem. Pharmacol.* **2018**, *155*, 82–91. [CrossRef] [PubMed]
19. Hubálek, F.; Binda, C.; Khalil, A.; Li, M.; Mattevi, A.; Castagnoli, N.; Edmondson, D.E. Demonstration of isoleucine 199 as a structural determinant for the selective inhibition of human monoamine oxidase B by specific reversible inhibitors. *J. Biol. Chem.* **2005**, *280*, 15761–15766. [CrossRef] [PubMed]
20. Desmond Molecular Dynamics System. *D.E. Shaw Research*; Schrödinger, LLC: New York, NY, USA, 2018.

21. Krishna, M.S.; Joy, B.; Sundaresan, A. Effect on oxidative stress, glucose uptake level and lipid droplet content by Apigenin 7, 4′-dimethyl ether isolated from *Piper longum* L. *J. Food. Sci. Technol.* **2015**, *52*, 3561–3570. [CrossRef] [PubMed]

22. Sghaier, M.B.; ISkandrania, I.; Nasr, N.; Franca, M.C.D.; Chekir-Ghedira, L.; Ghedira, K. Flavonoids and sesquiterpenes from *Tecurium ramosissimum* promote antiproliferation of human cancer cells and enhance antioxidant activity: A structure–activity relationship study. *Environ. Toxicol. Pharmacol.* **2011**, *32*, 336–348. [CrossRef] [PubMed]

23. Thao, N.P.; Luyen, B.T.T.; Lee, S.H.; Jang, H.D.; Kim, Y.H. Anti-osteoporotic and Antioxidant Activities by Rhizomes of Kaempferia parviflora Wall. ex Baker. *Nat. Prod. Sci.* **2016**, *22*, 13–19. [CrossRef]

24. Dezsi, L.; Vecsei, L. Monoamine Oxidase B Inhibitors in Parkinson's Disease. *CNS Neurol. Disord. Targets* **2017**, *16*, 425–439. [CrossRef] [PubMed]

25. Riederer, P.; Müller, T. Monoamine oxidase-B inhibitors in the treatment of Parkinson's disease: Clinical-pharmacological aspects. *J Neural. Transm.* **2018**, *125*, 1751–1757. [CrossRef] [PubMed]

26. Hwang, O. Role of oxidative stress in Parkinson's disease. *Exp. Neurobiol.* **2013**, *22*, 11–17. [CrossRef] [PubMed]

27. Finberg, J.P.M.; Rabey, J.M. Inhibitors of MAO-A and MAO-B in Psychiatry and Neurology. *Front. Pharmacol.* **2016**, *7*, 340. [CrossRef] [PubMed]

28. Riederer, P.; Laux, G. MAO-inhibitors in Parkinson's Disease. *Exp. Neurobiol.* **2011**, *20*, 1–17. [CrossRef] [PubMed]

29. Zhang, Q.; Zhu, L.; Gong, X.; Ruan, Y.; Yu, J.; Jiang, H.; Wang, Y.; Qi, X.X.; Lu, L.; Liu, Z. Sulfonation Disposition of Acacetin: In Vitro and in Vivo. *J. Agric. Food Chem.* **2017**, *65*, 4921–4931. [CrossRef] [PubMed]

30. Chaurasiya, N.D.; Leon, F.; Ding, Y.; Gomez-Betancur, I.; Benjumea, D.; Walker, L.A.; Cutler, S.J.; Tekwani, B.L. Interactions of Desmethoxyyangonin, a Secondary Metabolite from Renealmia alpinia, with Human Monoamine Oxidase-A and Oxidase-B. *Evid. Based Complement. Altern. Med.* **2017**, *2017*, 4018724. [CrossRef] [PubMed]

31. Parikh, S.; Hanscom, S.; Gagne, P.; Crespi, C.; Patten, C. *A Fluorescent-Based, High-Throughput Assay for Detecting Inhibitors of Human Monoamine Oxidase A and B*; BD Biosciences Discovery Labware, Inc.: Woburn, MA, USA, 2002.

32. Son, S.Y.; Ma, J.; Kondou, Y.; Yoshimura, M.; Yamashita, E.; Tsukihara, T. Structure of human monoamine oxidase A at 2.2-Å resolution: The control of opening the entry for substrates/inhibitors. *Proc. Natl. Acad. Sci. USA* **2008**, *105*, 5739–5744. [CrossRef] [PubMed]

33. Binda, C.; Aldeco, M.; Geldenhuys, W.J.; Tortorici, M.; Mattevi, A.; Edmondson, D.E. Molecular insights into human monoamine oxidase B inhibition by the glitazone antidiabetes drugs. *ACS Med. Chem. Lett.* **2011**, *3*, 39–42. [CrossRef] [PubMed]

34. *Maestro*, version 10.6; Schrödinger, LLC: New York, NY, USA, 2016.

35. *LigPrep*, version 3.8; Schrödinger, LLC: New York, NY, USA, 2016.

36. *Schrödinger*, version 2016-2; Schrödinger, LLC: New York, NY, USA, 2016.

37. Sherman, W.; Day, T.; Jacobson, M.P.; Friesner, R.A.; Farid, R. Novel procedure for modeling ligand/ receptor induced fit effects. *J. Med. Chem.* **2006**, *49*, 534–553. [CrossRef] [PubMed]

38. Binda, C.; Hubalek, F.; Li, M.; Edmondson, D.E.; Mattevi, A. Crystal Structure of Human Monoamine Oxidase B, a Drug Target Enzyme Monotopically Inserted into the Mitochondrial Outer Membrane. *FEBS Lett.* **2004**, *564*, 225–228. [CrossRef]

39. *Maestro*, version 11.5.011; Schrodinger, LLC: New York, NY, USA, 2018.

40. Sastry, G.M.; Adzhigirey, M.; Day, T.; Annabhimoju, R.; Sherman, W. Protein and ligand preparation: Parameters, protocols, and influence on virtual screening enrichments. *J. Comput. Aid. Mol. Des.* **2013**, *27*, 221–234. [CrossRef] [PubMed]

41. Zhou, R.H. Exploring the protein folding free energy landscape: Coupling replica exchange method with P3ME/RESPA algorithm. *J. Mol. Graph. Model.* **2004**, *22*, 451–463. [CrossRef] [PubMed]

42. Li, G.S.; Martins-Costa, M.T.C.; Millot, C.; Ruiz-Lopez, M.F. AM1/TIP3P molecular dynamics simulation of imidazole proton-relay processes in aqueous solution. *Chem. Phys. Lett.* **1998**, *297*, 38–44. [CrossRef]

43. Evans, D.J.; Holian, B.L. The Nose-Hoover Thermostat. *J. Chem. Phys.* **1985**, *83*, 4069–4074. [CrossRef]

44. Martyna, G.J.; Tobias, D.J.; Klein, M.L. Constant- Pressure Molecular-Dynamics Algorithms. *J. Chem. Phys.* **1994**, *101*, 4177–4189. [CrossRef]

Sample Availability: Samples of the compounds are available from the authors after execution of Intellectual Property and Material Transfer Agreements with the University of Mississippi.

MDPI

St. Alban-Anlage 66

4052 Basel

Switzerland

Tel. +41 61 683 77 34

Fax +41 61 302 89 13

www.mdpi.com

Molecules Editorial Office

E-mail: molecules@mdpi.com

www.mdpi.com/journal/molecules

www.ingramcontent.com/pod-product-compliance
Lightning Source LLC
Chambersburg PA
CBHW051722210326
41597CB00032B/5573